Developments in Soil Science 5B

SOIL CHEMISTRY
B. Physico-Chemical Models

Further Titles in this Series

Developments in Soil Science 5B

SOIL CHEMISTRY
B. Physico-Chemical Models

EDITED BY

G.H. BOLT

CONTRIBUTING AUTHORS:

J. BEEK
G.H. BOLT
M.G.M. BRUGGENWERT
F.A.M. DE HAAN
K. HARMSEN
A. KAMPHORST
W.H. VAN RIEMSDIJK

Department of Soil Science and Plant Nutrition,
Agricultural University of Wageningen, The Netherlands

R. BRINKMAN

Department of Soil Science and Geology,
Agricultural University of Wageningen, The Netherlands

R.W. CLEARY
M.Th. VAN GENUCHTEN

Water Resources Program, Department of Civil Engineering,
Princeton University, Princeton, NJ 08540, U.S.A.

A. CREMERS
A. MAES

Faculty of Agronomy,
Catholic University of Louvain, Belgium

ELSEVIER SCIENTIFIC PUBLISHING COMPANY
Amsterdam—Oxford—New York 1979

ELSEVIER SCIENTIFIC PUBLISHING COMPANY
335 Jan van Galenstraat
P.O. Box 211, 1000 AE Amsterdam, The Netherlands

Distributors for the United States and Canada:

ELSEVIER/NORTH-HOLLAND INC.
52, Vanderbilt Avenue
New York, N.Y. 10017

ISBN 0-444-41668-4 (Vol. 5B)
ISBN 0-444-40882-7 (Series)

Printed in The Netherlands

PREFACE

The present description of physico-chemical models covering a number of transport and accumulation phenomena occurring in soils was originally envisaged as a set of addenda to the chapters of part A of this text on soil chemistry. It soon became clear, however, that the need to include details of derivations, if all assumptions and approximations delimiting the validity of the models were to be exposed, would require a separate volume B. Once this was decided the material was reorganized in an effort to create a continuous and self-supporting treatise that could also be used in a graduate teaching program.

Nevertheless the ties with part A remain obvious. The present chapter (B)1 elaborates the theory of the diffuse double layer, introduced summarily in chapter (A)3 of part A. The previous chapter (A)4 on cation adsorption finds its extension here in chapters (B)2—(B)6. Of these, (B)2 covers some details about the choice of standard states and the definition of activity coefficients needed when formulating the thermodynamics of exchange equilibria. Chapter (B)3 and (B)4 treat the models and (B)5 gives experimental information available from the substantial amount of literature covering cation exchange involving soils and soil constituents. Some newer developments including the adsorption of heavy metal complexes are described in (B)6. Chapters (B)7 and (B)8 tie in with (A)5 and partly with (A)6, the first one elaborating co-ion exclusion, the second one covering the retention of orthophosphate ions in soil. The schematic presentation of adsorption chromatography in the previous (A)7 has been given a generalized theoretical treatment in (B)9, followed in (B)10 by an exposure of computer calculated case histories stressing the significance of transport and accumulation coefficients and including some experimental verifications involving the movement of pesticides. The electro-chemical aspects of transport phenomena in soil covered in the present (B)11 were mentioned only superficially in part A. Finally (B)12 may be seen as an extension of A(8), covering the interaction between solid and liquid phase of soil as reflected in changes of certain constituents of the solid phase.

While ample reference has been given to sources of experimental information, the discussion of such information has been limited largely to presumed trends of some general validity. In the particular case of cation exchange equilibria, where generalization appeared difficult, a considerable amount of data was condensed into standardized tables (cf. chapter 5). As indicated by its title, however, this text is primarily meant to be an exposure of theoretical developments in soil chemistry. The need for such developments seems undisputable, as theory should provide the basis for interpretation and generalization of practical experience.

The difficulty encountered when attempting to create a theoretical framework in an applied science like soil chemistry is obvious. The natural system "soil" is a complex and often ill-defined system, while physico-chemical theory has been based largely on the study of fairly simple and well-defined ones. In line with this, the "pioneers" in Soil Science were seldom (if ever) discoverers of new laws of science, as was pointed out already in the preface of part A. Instead, the "theory" in, for example, soil chemistry usually consists of extending theoretical knowledge, derived elsewhere for simpler systems, to soils.

The challenging task of the theoretical soil chemist then consists of analyzing in detail existing theory in basic chemistry, exposing assumptions and approximations and weighing their validity for application to soil constituents. Obviously, a judgement "invalid", though often correct, is then usually of little help. Instead it must be attempted to specify the limits of uncertainty arising because of expected deviations from the underlying assumptions.

Many of the detailed efforts to derive and extend certain theories in the present text may be viewed against the above considerations. Though not denying some hobbyism of the authors involved, it was felt that exposure of many details, including the consequences if a reasoning is followed till a sometimes far-fetched endpoint, would allow one to recognize in what respect certain models differ from other ones and to form a judgement as to when and where one model is to be preferred over another.

Following these derivations it was attempted in a number of cases to recombine and simplify the theory in terms of generalized graphs, tables or equations, together with a guide line for quick guessing procedures. Often such guesses are amply satisfactory and perhaps all one will ever be able to do with the given models in view of the complexity of the system involved. In this context, the derivations mentioned constitute the defence of the guessing procedures proposed. If deemed satisfactory in this manner, the derivations could be forgotten thereafter. If not, they might be of help to develop other and better simplifications.

Returning to the first paragraph, it is pointed out that although this part B is meant to stand on its own, prior knowledge of soil chemistry on the level of that covered in the first eight chapters of part A has been taken for granted. In order to avoid confusion the many symbols used here check with the few used in part A. Practically all symbols used have been tabulated in front of the book. Accepting the need for uniformity of symbols throughout a text if cross-reference between different chapters be possible, some unusual choices of symbols could not be avoided.

Units conform to SI, with the notable exception of the "indispensable" *equivalent* concentration of ions, specified in kilo-equivalent (keq) per m^3.

Again it is a pleasure to acknowledge all those who contributed to this book: the group of authors for their willingness to participate and then comply with editor's wishes as to division of subject material and uniformity of presentation; Dr. H.C. Thomas for reading chapter 2; Ir. J. Blom, Ir. T. Breimer and Mr. Th.A. Vens who assisted in different stages with the compilation of cation exchange data given in chapter 5; Dr. P.A.C. Raats and Mr. G. Menelik, M. Sc for reading very carefully chapter 9; Dr. D. Zaslavsky, whose comments on chapter 11 influenced considerably its final form; Ir. P. Koorevaar, who performed the calculations for the many Tables and graphs in chapters 1, 3, 9 and 11; Dr. L. Stroosnijder, who made the computer simulation for Figs. 9.12—9.14; Dr. M.G.M. Bruggenwert for his help in compiling the subject index; Mr. B.W. Matser, who made the drawings of almost all Figures; Miss G.G. Gerding, who typed the Editor's chapters from original, abominable handwriting to final draft; Mrs. D.J. Eleveld-Hoftijser, who took over for other chapters and late last-minute repairs; Miss K. de Kat and Miss M. Bubberman who finished off with indexes. Finally the involvement of the Editor's family, ranging from friendly tolerance of his prolonged occupation with "the book" to active support in preparing indexes and proof reading, is mentioned with pride.

Wageningen, 1978 G.H. Bolt

CONTENTS

LIST OF SYMBOLS

Symbol	Description	Equation(s) or page(s) where introduced	Unit
a	anion species	(7.1)	
a_k	activity of ion species 'k' (in solution), relative to that in the chosen standard state	(2.2)	
$\tilde{a}_k$	same, in a solution phase subject to force fields	(2.4b)	
$\bar{a}_k$	activity of ion species 'k' in an adsorbed phase, relative to that in a chosen standard state	(2.4c)	
a_s	activity of the solvent (water) in a solution	(2.28)	
b	friction coefficient	p. 393	N sec/m^4
$^m c_k$	molar concentration of species 'k'	(1.2)(11.4)	kmol/m^3
c_k	(equivalent) concentration of ion species 'k' in solution		keq/m^3 = eq/l
c_o	concentration in an equilibrium solution	(1.2)	keq/m^3 = eq/l
c_t	concentration at the plane of truncation of a DDL	(1.39)	keq/m^3 = eq/l
c_s	concentration at the charges surface bordering a DDL	(1.39)	keq/m^3 = eq/l
c_i	concentration initially present in a soil column	(9.12)	keq/m^3 = eq/l
c_f	concentration in a feed solution entering a soil column	(9.12)	keq/m^3 = eq/l
$\bar{c}$	$= (c - c_i)/(c_f - c_i)$, scaled concentration in a soil column	p. 294	
d	distance		m, or Å
d_l	thickness of a liquid layer on charged surface	p. 20	m, or Å
$d_{ex,a}$	distance of exclusion of anion species 'a' in a freely extended DDL	(7.4)	m, or Å
$d^t_{ex,a}$	same, in a truncated DDL	(7.2)	m, or Å
e_{kl}	energy of interaction between ions 'k' and 'l' occupying neighboring sites on an exchanger surface	(4.2)	J
e_m	energy of mixing for ions 'k' and 'l' as above	(4.2)	J
f_k	activity coefficient of ion species 'k' in solution, here chosen on the molarity scale such that $a_k \equiv f_k {}^m c_k$ (where $^m c_k$ then indicates the dimensionless ratio of the molar concentration to the standard unit molar concentration at the same temperature)	(2.3)	

$\tilde{f}_k$	(total) activity coefficient in a solution phase subject to force fields	(2.4b)	
$\bar{f}_k$	activity coefficient of ion species 'k' in an adsorbed phase, here chosen on the equivalent fraction scale such that $\bar{a}_k \equiv \bar{f}_k N_k$	(2.9),(2.17)	
$g(m_k, m_l)$	nr. of distinguishable configurations of species 'k' and 'l' in an adsorbed phase	(4.96)	
j_k	flux of ion species 'k', per unit cross section of a column	(9.3)	$keq/m^2 sec$
j_k^D	diffusion/dispersion flux, as above	(9.3)	$keq/m^2 sec$
${}^m j_k$	molar flux of species 'k', as above	(11.1)	$kmol/m^2 sec$
k	proportionality constants in adsorption equations	p. 118	
kT	kinetic energy per molecule	(4.5)	J
m_k	nr. of molecules of species 'k'	p. 80	
n_k	nr. of moles of species 'k', adsorbed per equivalent of exchanger	(4.32)	kmol/keq
n_s	same, of solvent	(2.28)	kmol/keq
p	a fraction, as locally specified		
p_k	constant governing the formation of site-ion pairs in a Stern layer	(3.65)	m^3/keq
∇p_{e+}	$= z_+ {}^m c_+ F \nabla E_+$, a scaled electric force	(11.16)	N/m^3
q_k	amount adsorbed of ion species 'k', per unit bulk volume in a soil column	(9.2)	keq/m^3
q_i	same, initially present in a soil column	(9.12)	keq/m^3
q_f	same, finally present after passage of a feed solution with concentration c_f	(9.12)	keq/m^3
$\bar{q}$	$= (q - q_i)/(q_f - q_i)$, scaled value of q	p. 294	
$q'(c)$	$= dq/dc$, differential capacity of an exchanger (for a particular ion) at a given value of c	(9.7)	
${}^w q_k$	amount adsorbed, per unit mass of solid phase	(10.1)	keq/kg
r	$= {}^m c_{o,+}/\sqrt{{}^m c_{o,2+}}$, reduced concentration ratio of mono- and divalent cations in solution	p. 70	$\sqrt{kmol/m^3}$ $= \sqrt{mol/l}$
$\bar{r}$	$= r/\sqrt{C_o}$ $(= f_+\sqrt{2/f_{2+}})$, scaled value of r	(3.59)	
s_{kl}	distance between ions 'k' and 'l', adsorbed on a regular array of sites	(4.73)	m, Å
t	time		sec
$\hat{t}$	scaled time, used in column processes	(9.83)	
u	= exp (y), Boltzmann accumulation factor for monovalent counterions in a DDL	(1.8)	
$\bar{u}$	same, in a phase with constant electric potential	(3.2),(11.17)	

v	$= J^V/\theta$, carrier velocity in a column	(10.2a)	m/sec
v_ι	$= v(1 + \iota)$, adjusted value of v, allowing for a decay process	(10.7)	m/sec
v^*	$= J^V/(\theta + \Delta q/\Delta c)$, mean rate of propagation of a solute front in a column	(9.51)	m/sec
v_k	$= {}^m j_k/{}^m c_k$, equivalent "filter" velocity of component 'k' in a liquid mixture passing through a porous medium	(11.12)	m/sec
v_k'	$= {}^m j_k/{}^m c_k'$, actual "filter" velocity in case the concentration ${}^m c_k'$ deviates from the local equilibrium concentration ${}^m c_k$	(11.18)	m/sec
x	position coordinate		m, Å
x_t	point of truncation of a DDL	(1.14)	m, Å
x_c	position in a column at which a particular concentration resides	(9.10)	m, Å
x_c^o	first approximation to x_c, neglecting diffusion/dispersion	(9.71)	m, Å
$\langle x_c \rangle$	average value of x_c in a concentration range from 0 to c	(9.63)	m, Å
x_q	position in a column at which a particular value of q resides	p. 316	
$\hat{x}$	scaled position coordinate in a column	(9.83)	
x_p	$= J^V t\Delta c/(\Delta q + \theta\Delta c)$, equivalent depth of penetration of a solute front in a column	(9.22)	m
y	$= - F\psi/RT$, Boltzmann exponent for monovalent counterions in a DDL	(1.4)	
z_k	valence of ion species 'k'		
a	proportionality constant, also used in Langmuir equation as the adsorption maximum	(5.11)	
b	function of composition and energy of mixing used in regular mixture theory	(4.9)	
d	charge distribution parameter used in mixture theory	(4.67)	
e	electronic charge	(4.73)	Coul
f_k	equivalent fraction of ion species 'k' in a mixed electrolyte solution	(1.9)	
f	(without subscript), refers to f_{2+} in mixture of mono- and divalent cations	(1.11)	
g	geometry factor	(9.39),(9.42)	
h_k	adjustment factor determining the number of nearest neighbors to an ion of valence z_k adsorbed on a regular array of sites	(4.1)	
k	first-order rate constant (decay constant)	(9.103)	sec^{-1}
k_a	same, covering diffusion equilibration of a stagnant phase	(9.78)	sec^{-1}

k^*	$= k/(R_D + 1)$	(9.104)	sec^{-1}
$2l$	nr. of nearest neighbors of one site in a regular array of sites	(4.1)	
m	membrane passage factor in pressure filtration	(7.35 m)	
$\bar{m}$	fraction mobile of countercharge as evident from electrokinetic phenomena	(1.42)	
$\tilde{m}$	same, as evident from AC-conductance	p. 421	
n	exponent in adsorption equation, particularly of the Freundlich equation	(10.3)	
p	$= Q_m/Q$, fraction of total adsorption capacity that equilibrates instantaneously with a mobile liquid phase in a column	(10.29)	
r	$= k_r x/v$, scaled position parameter for columns	(10.20)	
s	$= vx/D$, scaled position parameter for columns	(10.21)	
t	$= vt/x$, scaled time parameter for columns	(10.21)	
x	position coordinate relative to the plane of shear in a moving liquid layer	(11.29)	m, Å
z	$= \theta_m R_{f,m}/\theta R_f$, adjusted ratio of retardation factors in a column comprising mobile and stagnant liquid phases	(10.40)	
A_k	amount of species 'k' present per unit volume in a soil column	(9.1)	keq/m^3
A, B	cation species		
$\bar{A}$, $\bar{B}$	same, in an adsorbed phase	(2.1)	
C	total electrolyte concentration		$keq/m^3 = eq/l$
C_o	same, in equilibrium solution	(1.9)	$keq/m^3 = eq/l$
C_p	same, in a pressure filtrate	(7.31)	$keq/m^3 = eq/l$
D_o	diffusion coefficient in a free solution	(9.21)	m^2/sec
D	$= D_o/\lambda$, same in a solution phase inside a porous medium subject to a tortuosity factor λ	(9.21)	m^2/sec
D^*	$= D/(R_D + 1) = D/R_f$, effective diffusion/dispersion coefficient of a solute subject to adsorption in a column	(9.51)	m^2/sec
D_{dis}	dispersion coefficient in a porous medium	(9.28),(9.45)	m^2/sec
E	energy		J
E_{kl}	energy of interaction between cation species 'k' and 'l' when adsorbed on one equivalent of exchanger	(4.99)	J/eq
E_{mix}	energy of mixing of cations adsorbed on one equivalent of exchanger	(4.3)	J/eq

F_k	friction force on component 'k' moving through a porous medium	(11.18)	N/m^3
F_e	electric field strength	(1.1)	V/m
G	$= E + PV - TS$, free enthalpy (Gibbs function)		J,kcal
$\bar{G}^0_{\bar{A}}$	partial molar free enthalpy of a homoionic A-exchanger in its standard state	(2.46)	kcal/mol
$G_{\bar{A}}$	free enthalpy of *one equivalent* of a homoionic A-exchanger	(2.51a)	kcal/eq
$G_{\overline{AB}}$	same, for a mixed AB-exchanger	(2.38)	kcal/eq
$G^X_{\overline{AB}}$	excess free enthalpy of the same	(2.40)	kcal/eq
ΔG^0_{ex}	standard free enthalpy of an exchange reaction, *per equivalent* of exchanger	(2.10)	kcal/eq
ΔG_{el}	electrical free enthalpy associated with the separation of charges in an adsorbed phase, c.q. DDL	(3.17),(3.18)	J/m^2, J/eq
I	ionic strength		
I_o	same, in the equilibrium solution of a DDL	(1.16)	
I_t	same, at the plane of truncation of a DDL	(1.16)	
J_k	$= {}^m j_k / {}^m c_k$, "filter" velocity of species 'k' in a column		m/sec
J^D_k	$= J_k - {}^m j_w / {}^m c_w$, same in excess of that of the solvent water	(11.4b)	m/sec
$J_\pm$	$= {}^m j_+ / {}^m c_+ - {}^m j_- / {}^m c_-$, same of cations in excess of anions ($\equiv I / z_+ {}^m c_+ F$)	(11.15)	m/sec
J^V	$= \sum_k {}^m j_k \bar{V}_k$, "filter" velocity of the liquid phase in a column	(9.3),(11.4a)	m/sec
K	selectivity coefficient governing exchange equilibria, specified for *one equivalent* of exchanger and using solution phase activities		
K'	same, but using solution phase concentrations		
K_N	selectivity coefficient based on the use of equivalent fractions adsorbed	(2.14)	
K_M	same, using mole fractions adsorbed	(4.42)	
$K_{\mathcal{M}}$	same, using adjusted mole fractions adsorbed	(4.43)	
K_c	same, using concentrations in the adsorbed phase	(2.7)	
K^0_{ex}	thermodynamic exchange constant, specified for *one equivalent* of exchanger	(2.16)	
K^0_L	same, involving a metal-ion complex L_nM	(6.5)	
L	ligand species	(6.3)	
$\bar{L}$	same, in an adsorbed phase	(6.3)	
L_nM	metal-ion complex	(6.3)	

$\overline{L_nM}$	same, in adsorbed phase	(6.3)	
L_D	$= D/v$, diffusion/dispersion length parameter in columns	(9.48)	m, cm
L_{dif}	component of L_D caused by longitudial diffusion	(9.24)	m, cm
L_{dis}	same, caused by velocity dispersion	(9.38)	m, cm
L_r	same, caused by a finite rate of equilibration	(9.46)	m, cm
L_d	$= v/k\theta$, decay length parameter	(9.105)	m, cm
M	metal ion species	(6.4)	
$\bar{M}$	same in adsorbed phase	(6.4)	
M_k	mole fraction of ion species 'k' in an adsorbed phase	(4.11)	
N_k	equivalent fraction of ion species 'k' in an adsorbed phase	(2.9), (2.36)	
N'_k	$= dN_k/df_k$, slope of normalized exchange isotherm	p. 295	
P	pressure		N
P_k	rate of production of component 'k' in a soil column	(9.1)	$kmol/m^3 sec$
Q	adsorption (exchange) capacity per unit bulk volume of soil	p.294	keq/m^3
R	gas constant		
R_D^{Δ}	$= \Delta q/\theta \Delta c$, distribution ratio of ions between adsorbed and liquid phase in a column, for a given concentration increment Δc	(9.72)	
R_D	$= q/\theta c$, same, if a linear adsorption isotherm applies		
$\bar{R}_D$	$= Q/\theta C$, overall value of R_D^{Δ} covering complete exchange	(9.18)	
R_D^{am}	$= (\Delta q_a + \theta_a \Delta c)/\theta_m \Delta c$, distribution ratio between a stagnant phase 'a' and a mobile phase 'm'	(9.80)	
R_f	$= \{q'(c)/\theta + 1\}$, retardation factor	(10.5)	
^vS	entropy, per unit volume of a porous system	(11.1)	J/m^3 K
S	specific surface area	(2.34)	m^2/kg
$S_{ex,a}$	same, exhibiting anion exclusion	(7.29)	m^2/kg
T	temperature		K
T	$= J^V t/L\theta$, nr. of pore volumes that has entered a column	(10.12)	
T_k	total amount of component 'k' present per unit mass of solid phase in soil	(2.34)	keq/kg
U	$= I/I_0$, ion accumulation factor in a DDL, expressed as relative ionic strength	p. 7	
U_t	same, at plane of truncation of a DDL	(1.16)	
U_t^L	Langmuir approximation to U_t	p. 10	

V	volume		m^3
V^α	volume of adsorbed phase per unit amount of adsorbed ions	p. 30	m^3/kg
$V_{ex,a}$	volume of exclusion of anion species 'a', per unit mass of soil	(7.21)	m^3/kg
$\bar{V}_k$	partial molar volume of component 'k'	(2.4)	$m^3/kmol$
V	$= \int J^V dt$, feed volume that has entered a unit cross-section of a column	(9.6)	m^3/m^2
$V_0(c)$	inverse feed function	(9.11)	m^3/m^2
W	liquid volume present, per unit mass of solid phase in soil	(2.34)	m^3/kg
A	$= (R_D^a + 1)/(R_D + 1)$	(9.102)	
B	$= R_D/(R_D + 1)$	(9.60)	
C	relative volume conductance of a solid phase	(1.62)	
E_+	electric potential registered by a cation-reversible electrode	(1.6)	V
F	Faraday constant	p. 2	Coul/eq, Coul/keq
H	$= 2F/\beta\eta$, Helmholtz-Smoluchowski constant	(11.49)	$m^2/Vsec$
I	electric current per unit cross-section of a porous medium	(11.4c)	$Coul/m^2$ sec
L	mixing length, column length	p.303 (10.12)	m
L_{sub}	transport coefficient in generalized flux-force scheme, the subscript indicating particular processes	(11.2)	$m^4/Nsec$
L'_{sub}	same, using practical units	(11.49)	
M_k	adjusted mole fraction of ion species 'k' in an adsorbed phase	(4.8)	
N_{Av}	Avogadro's number		
Q	coefficient in anion exclusion equations, pertaining to freely extended DDL	(7.4)	
Q^t	same, for a truncated DDL	(7.10)	
R	radius of pore or capillary	(9.39)	m
R_a	radius of aggregate containing a stagnant liquid phase	(9.42)	m, cm
S	$= (\Delta q/\Delta c + \theta)/\{q'(c) + \theta\}$, shape factor relating x_c to x_p for a solute front covering a concentration increment Δc, in the absence of diffusion/dispersion	(9.72)	
T_k	Hittorf transport number of ion species 'k'	(11.73)	eq/96500 Coul
T_k^*	$= {}^m j_k F/I$, reduced transport number of species 'k'	(11.74)	mol/96500 Coul
V	$= (u + 1)/(u - 1)$	(1.38)	

X_{kl}	nr. of pairs of sites with mixed occupation, per equivalent of exchanger	(4.5)	
α^2	$= (v^*t)^2/D^*t = x_p v/D = x_p/L_D$, reduced depth of penetration of a solute front in a column, Peclet number as pertains to depth of penetration	(9.54)	
α_r^2	$= x_p/L_r$, same, as governed by the rate of equilibration with a stagnant phase	(9.92)	
α_ι^2	$= \alpha^2(1 + 4L_D/L_d)$, adjusted value of α^2, allowing for a decay process	(9.105)	
β	$= 8\pi F^2/\epsilon' RT$, constant in equation of a DDL	(1.5)	m/kmol
$\bar{\gamma}_a$	anion exclusion per unit mass of soil	(7.21)	keq/kg
δ	$\simeq 4/z_+\beta\Gamma$, cut-off distance on the x-axis used in equations of a DDL	(1.35),(1.37)	m, Å
δ_m	same, as pertaining to a given mobile countercharge, Γ_m	(11.46)	m, Å
$\epsilon'/4\pi$	$= \epsilon_0\epsilon_r$, permittivity of a dielectric	(1.1)	Coul/Vm
ϵ_0	permittivity of empty space		Coul/Vm
ϵ_r	dielectric constant		
ζ	zeta potential as used in electrokinetics	p. 409	V
η	viscosity	(1.29)	$Nsec/m^2$
θ	volumetric liquid (water) content in a soil column	(9.2)	
θ_a	same, present as a stagnant liquid phase, presumably inside an aggregate	(9.38)	
θ_m	same, present as a mobile phase	(9.38)	
ι	$= \sqrt{1 + 4L_D/L_d} - 1$, decay-diffusion/dispersion interaction parameter used in column transport	(9.105)	
κ	$= \sqrt{\beta I}$, Debye-Huckel reciprocal length		m^{-1}, $Å^{-1}$
κ_0	same, for an equilibrium solution, e.g. of a DDL	(1.15)	m^{-1}, $Å^{-1}$
κ_t	same, as applies to the plane of truncation of a DDL	p. 6	m^{-1}, $Å^{-1}$
λ	tortuosity factor in a porous medium	(9.21)	
λ_k	charge attenuation factor for ion species k used in regular mixture theory	(4.74)	
μ_k	chemical potential of ion species k in a solution phase	(2.2)	J/mol
μ_k^0	same, in a chosen standard state	(2.2)	J/mol
$\bar{\mu}_k$	chemical potential in an adsorbed phase	(2.4c)	J/mol
$\tilde{\mu}_k$	thermodynamic potential in a solution phase subject to force fields	(2.4),(11.1)	J/mol
μ_k^c	concentration dependent part of $\tilde{\mu}_k$	(11.5)	J/mol
ν	stochiometric reaction number	(11.9)	

ξ	$= (x - v^* t)/2\sqrt{D^* t}$, scaled position coordinate in a diffusion/dispersion front	(9.52)	
ρ	density		
ρ_s	density of solid phase		kg/m^3
ρ_b	bulk density of soil	(10.1)	kg/m^3
$\rho_\pm$	charge density	(1.1)	$Coul/m^3$
σ	surface charge density		$Coul/m^2$
σ	reflection coefficient	(11.25)	
σ'	same, at zero electric current	(11.35)	
υ	$= w/2\sqrt{D^* t}$, scaled front width	(9.76)	
ϕ	volume fraction		
ϕ_k	volume fraction of component k in a liquid mixture	(11.7)	
ϕ_m	$= \theta_m/\theta$, volume fraction of liquid phase in a column considered to be mobile	(10.41)	
ψ	electric potential	(1.1)	V
$\mathscr{B}_n$	cumulative or gross stability constant of a metal ion complex, L_nM, in solution	(6.8)	
$\overline{\mathscr{B}}_n$	same, in an adsorbed phase	(6.9)	
Γ	countercharge per unit surface area	(1.35)	keq/m^2
Γ_m	mobile part of the same	(11.42)	keq/m^2
Γ_k^+	surface excess of cation species 'k'	(2.34),(3.31)	keq/m^2
Γ_a^-	surface deficit of anion species 'a'	(7.1)	keq/m^2
$\overline{\Gamma}$	$= \Gamma\sqrt{\beta}/\sqrt{C_0}$, scaled value of Γ	(3.35)	
${}^v\Gamma$	countercharge per unit volume in a homogeneous Donnan phase	(3.6)	keq/m^3
K	specific conductance (conductivity)		mho/m
K_0	conductivity of the equilibrium solution permeating a porous medium	p. 418	mho/m
$\overline{\overline{K}}$	DC-conductivity of a porous medium	p. 417	mho/m
$\widetilde{K}$	AC-conductivity of the same	p. 417	mho/m
$\mathring{K}$	$= \overline{\overline{K}} - \widetilde{K}$, drag component of the conductivity, due to electro-osmosis	p. 417	mho/m
K_a'	contribution of adsorbed ions to $\widetilde{K}$	(11.58)	mho/m
K_s'	contribution of dissolved salt to $\widetilde{K}$	(11.58)	mho/m
K_{iso}	isoconductivity value characteristic for a charged solid phase	p. 420	mho/m
K_S	surface conductance	(11.65)	mho
K_{wt}	weight conductance	(11.66)	$mho\ m^2/kg$
Λ_k'	mobility of ion species k	p. 417	$m^2/V\ sec$
Λ_k	same, expressed as equivalent conductance	p. 400	$mho\ m^2/(k)eq$
Λ_a	same, of adsorbed ions	(11.57)	$mho\ m^2/(k)eq$
Ξ	$= \kappa_t x_t$, truncation distance expressed in *local* D-H units, which approaches the constant $\pi/\sqrt{2}$ at small values of x_t	p. 6	
Π	osmotic pressure	(11.9)	N/m^2, bar

Φ	formation factor governing electric conductance	p. 418
Ω_k	partition function of ion species k adsorbed on z_k sites	(4.15)

THE IONIC DISTRIBUTION IN THE DIFFUSE DOUBLE LAYER

G.H. Bolt

1.1. HISTORICAL NOTE

Already in the early nineteen hundreds Gouy (1910) and Chapman (1913) derived an equation describing the ionic distribution in the diffuse layer formed on a charged planar surface. Applied at that time to a charged mercury surface, the theory turned out to meet with little success as the calculated electric capacity of the double layer exceeded grossly the values found in reality. About ten years later Stern (1924) suggested appropriate corrections accounting for the special nature and limitations of the first layers of counterions against the charged surface. Eventually Grahame (1947) refined the combined Gouy-Stern model of the diffuse double layer to such a degree that a realistic description of the double layer (DL) on the Hg-surface became possible.

When quantitative colloid chemistry came into existence, particularly through the works in the "Kruyt" school (cf. Verwey and Overbeek, 1948; Kruyt, 1952) the Gouy-Chapman theory and the Stern modification were routinely applied to the description of the diffuse layer of counterions formed on colloids possessing a surface reversible to certain so-called potential determining ions. Treated most extensively was the AgI-surface, where Ag^+- and I^--ions serve as potential determining ions.

Aside from the above development, Schofield (1946) re-introduced the Gouy-Chapman theory for the calculation of the thickness of water films on mica surfaces. One year later the same theory was used to calculate the negative adsorption of anions in a bentonite suspension (Schofield, 1947). The work of Schofield (and collaborators) at the Rothamsted Experimental Station, UK, and also at Cornell University, USA, has contributed significantly towards the introduction of the Gouy-Chapman theory as a basic tool in Soil Chemistry. Particularly, Schofield's recognition of the fact that often soil colloids possess a fairly constant surface charge density (much in contrast to most other colloids studied till then) enabled him to use a rather simple mathematical boundary condition to solve the relevant equations. The following treatment is a generalized form of Schofield's approach of constant charge surfaces.

1.2. BASIC THEORY

Referring to chapter 3 of part A of this text the ion swarm in the neighborhood of a surface with a given charge density may be regarded as a self-limiting "atmospheric distribution" of counterions. Thus the accumulated counterions serve to annihilate the field strength of the electric field emanating from the surface charge, and the latter is reduced to zero beyond the point where the accumulated countercharge just balances the surface charge. Using Poisson's equation this is expressed for a linear electric field as:

$$\frac{dF_e}{dx} = \frac{4\pi\rho_{\pm}(x)}{\epsilon'} \tag{1.1}$$

where F_e is the electric field strength, $\rho_{\pm}(x)$ indicates the local density of the countercharge and $\epsilon'/4\pi \equiv \epsilon_r\epsilon_0$ is the permittivity of the medium with ϵ_r indicating the dielectric constant. The local density of countercharge is determined by the local concentration of ions, $\Sigma_k z_k\, {}^{m}c_k$, where ${}^{m}c_k$ is the molar concentration of species k, while z_k indicates its valence (negative for anions!). It is recalled now that the local ionic concentration may be related to the (reference) equilibrium concentration of the ion in the system (i.e. outside the range of influence of the charged surface) via the Boltzmann equation, cf. eqn. A3.3). The difference in potential energy of the ion when comparing the equilibrium solution with a particular point inside the diffuse layer may then be taken as $z_k F\psi(x)$ for the point under consideration, with $z_k F$ constituting the charge in Coulomb per kmol and $\psi(x)$ indicating the local electric potential (in Volts). The local molar concentration of species k thus equals:

$${}^{m}c_k = {}^{m}c_{0,k} \exp(-z_k F\psi(x)/RT) \tag{1.2}$$

Equating F_e with $-(d\psi/dx)$ and combining eqns. (1.1) and (1.2) gives:

$$\frac{d^2\psi}{dx^2} = -\frac{4\pi F}{\epsilon'} \Sigma_k z_k\, {}^{m}c_{0,k} \exp(-z_k F\psi(x)/RT) \tag{1.3}$$

as the constituting Poisson-Boltzmann equation of the planar double layer. Introducing now a scaled electric potential, $y \equiv -F\psi/RT$ (i.e. the potential in units of about -25 mV at room temperature, giving positive numbers for y in the double layer formed on negatively charged surfaces, as prevalent in soil systems), (1.3) may be written as:

$$\frac{d^2y}{dx^2} = \frac{\beta}{2} \Sigma_k z_k\, {}^{m}c_{0,k} \exp(z_k y) \tag{1.4}$$

The constant β is then given by:

$$\beta = \frac{8\pi F^2}{\epsilon' RT} = 1.07 \cdot 10^{19} \text{ m/kmol at } 15^\circ\text{C} \tag{1.5}$$

and the concentration must thus be specified in $kmol/m^3$ (i.e. as the molar concentration) to yield d^2y/dx^2 in m^{-2}. Integrating both sides of (1.4) over y now yields the central equation of double layer theory:

$$\left(\frac{dy}{dx}\right)^2 = \beta \Sigma_k {}^m c_{0,k} \exp(z_k y) - \text{constant} \tag{1.6}$$

It is interesting to note that the RHS of (1.6) specifies the local *molar* concentration in the double layer, multiplied by β.

In the case of ideal behavior (as was presumed by using concentrations in the Boltzmann equation) the latter may be expressed by the local osmotic pressure, $\Pi = RT\,\Sigma_k {}^m c_k$. If now the double layer is bisected at a particular point one may visualize the local (linear) electric field as acting between the charges present on both sides of the plane of sectioning. Thus, one finds on one side the surface charge, σ_s, minus part of the counter charge, $\alpha\sigma_c$, and on the other side the remainder of the counter charge, $(1-\alpha)\sigma_c$, the two being equal and opposite in sign. The electric attraction between the two causes a contractive force proportional to $(dy/dx)^2$. At equilibrium this contractive force is kept in balance by the local excess osmotic pressure (compared to the osmotic pressure in the equilibrium solution where (dy/dx) equals zero), i.e. by the local liquid pressure build-up corresponding to this excess osmotic pressure.

Keeping in mind that in the presence of a limited liquid content of the system, the value of y at the point where (dy/dx) — i.e. the electric field strength — vanishes is not necessarily zero, the constant in (1.6) is expressed in a terminal value of y. Putting by definition $y \equiv y_t$ for $(dy/dx) = 0$, eqn. (1.6) becomes:

$$\left(\frac{dy}{dx}\right)^2 = \beta \Sigma_k {}^m c_{0,k} \{\exp(z_k y) - \exp(z_k y_t)\} \tag{1.7}$$

Introducing now the Boltzmann accumulation factor for a monovalent cation, $u \equiv \exp(y)$, the last equation may be converted into:

$$\sqrt{\beta}dx = -\frac{du}{u\sqrt{\Sigma_k {}^m c_{0,k} \{u^{z_k} - u_t^{z_k}\}}} \tag{1.8}$$

where the minus sign is chosen as u decreases with increasing x and in which u_t indicates the Boltzmann accumulation factor at the plane of truncation of the double layer. It is finally convenient to specify the equilibrium concentrations of the different ion species in terms of their equivalent fraction, $f_k \equiv |z_k|^m c_{0,k}/C_0$. This then gives the fully scaled version of (1.8) as:

$$\sqrt{\beta C_0}\,dx = -\frac{du}{u\sqrt{\Sigma_k \{f_k u^{z_k}/|z_k| - f_k u_t^{z_k}/|z_k|\}}} \tag{1.9}$$

The LHS of this equation specifies the differential of the distance in units of $1/\sqrt{\beta C_0}$ (related to the Debye-Hückel (D-H) unit of $1/\kappa_0 \equiv 1/\sqrt{\beta I_0}$, with I_0 = ionic strength of the equilibrium solution), the RHS is the differential of a composite function of u, the Boltzmann accumulation factor for monovalent cations in the double layer formed on negatively charged surfaces. In order to obtain an explicit relation between u (or $y = \ln u$) and x, (1.9) must be integrated with suitable boundary conditions. As will be shown in the following section a convenient general solution is obtained by putting $u \to \infty$ for $x \to 0$. In fact this condition results in x(u) for the infinitely highly charged surface.

Although this may look like a rather strained condition as all assumptions made are likely to break down in the region where u approaches infinity, this is later easily repaired by introducing a cut-off value of x, below which the actual course of u(x) becomes immaterial (cf. section 1.5).

The infinitely highly charged surface then results in the explicit relationship:

$$\sqrt{\beta C_0}\,x = -\int_\infty^u \frac{du}{u\sqrt{\Sigma_k\{f_k u^{z_k}/|z_k| - f_k u_t^{z_k}/|z_k|\}}} \tag{1.10}$$

where u in the integral expression should be treated as a "dummy-variable". Using u_t as the upper limit yields $\sqrt{\beta C_0}\,x_t$, with x_t constituting the distance at which the terminal plane of the liquid layer is located. Eqn. (1.10) may easily be programmed for numerical solution, for any chosen composition of the equilibrium solution and a given value of u_t. Such a program was executed by Bresler (1972), resulting in a large set of composite graphs. As a superficial inspection of these graphs indicates a considerable uniformity of the lines representing individual cases, it appears worth while to attempt to summarize the solution of (1.10) by means of an analytical solution.

1.3. GENERALIZED ANALYTICAL SOLUTION OF THE DL EQUATION

Inspection of (1.10) shows that for small values of $\sqrt{\beta C_0}\,x$ the terms in the denominator representing the co-ion contribution are entirely negligible in comparison to those representing the counterion contribution. This follows from the consideration that the parameter u carries an exponent equal to the valence, which differs in sign for counter- and co-ions. Thus for a negatively charged surface, u is very large close to the surface, and $u^{z+} \gg u^{z-}$. The shape of u(x) close to the charged surface is thus derived safely by neglecting the presence of co-ions (cf. Langmuir, 1938).

Moving away from the surface this approach must obviously fail, i.e. once u^{z+} approaches u^{z-}, the co-ionic terms must be included. It still remains true, however, that the co-ionic *valence* influences the denominator very little.

This may be demonstrated by comparing the magnitude of the denominator in a system containing a mono-monovalent salt with one containing a mono-divalent salt. Taking $u_t = 1$, corresponding to a freely developed double layer, one finds at given u for the mono-monovalent system, the co-ionic contribution to the denominator as $(1/u - 1)$.

In the mono-divalent system the latter equals $(1/2u^2 - 1/2)$. Taking the difference between the two, relative to the total ionic concentration, one finds the error resulting from taking the wrong anionic valence for the purpose of the calculation as $1/4u$. Thus the error in the denominator is only about 10% at $u = 2$ for $u_t = 1$. For the calculated distance corresponding to $u = 2$ the overall error is very much smaller, as (1.10) is integrated over the entire range of u down to $u = 2$. Obviously the error is further decreased if $u_t > 1$ and if the anionic composition is approximated by taking a standardized mixture of species with different valencies.

With the above in mind, while admitting that the numerous assumptions involved in deriving the Gouy theory (cf. below) preclude great precision as a goal, it appears absolutely warranted to manipulate the co-ionic composition in such a manner that (1.10) may easily be solved analytically. A fairly safe manipulation in this respect is to use a symmetric composition of the equilibrium solution. For the mono-divalent system this implies that if the equivalent fraction of divalent cations equals f, the same must hold for the divalent anions, while the fraction of monovalent cations and anions equals $(1 - f)$. Introducing this condition, (1.10) reverts to:

$$\sqrt{\beta C_0}\,x = -\int_{\infty}^{u} \frac{du}{u\sqrt{A(f,u) - A(f,u_t)}} \tag{1.11}$$

in which $A(f, u) \equiv fu^2/2 + (1-f)u + (1-f)u^{-1} + fu^{-2}/2$. Bringing u inside the square root sign results in a fourth-degree polynomial in u, which leads to an elliptic integral of the first kind. Substituting $v \equiv u + 1/u$ allows easy factorizing of the denominator, according to:

$$\sqrt{\beta C_0}\,x = -\int_{\infty}^{v} \frac{dv}{\sqrt{f/2}\sqrt{(v-v_t)(v-2)(v+2)\{v+v_t+2(1-f)/f\}}} \tag{1.12}$$

The solution is found as (cf. Bateman, 1953):

$$\sqrt{\beta C_0}\,x = \mu[F(\alpha, \phi_\infty) - F(\alpha, \phi)] \tag{1.13}^*$$

with

$$\mu = 2/\sqrt{(v_t + 2)\{fv_t/2 + 1\}}$$

$$\alpha = \arcsin\sqrt{\frac{4\{fv_t + (1-f)\}}{(fv_t/2 + 1)(v_t + 2)}}$$

$$\phi = \arcsin\sqrt{\frac{(fv_t/2 + 1)}{\{fv_t + (1-f)\}}\,\frac{(v - v_t)}{(v-2)}}$$

---

* If only a single symmetric salt of valence $(z, -z)$ is present (c.q. $f = 0$ or $f = 1$), (1.11) may also be written as an easily factorizable function of the variable $(u_t/u)^z$. This leads to the slightly more convenient expression $\sqrt{\beta C_0}\,x = (2/\sqrt{zu^z})\,F(\alpha, \phi)$ with $\alpha = \arcsin(u_t^{-z})$, $\phi = \arcsin\sqrt{(u_t/u)^z}$. The corresponding expression for x_t in terms of the complete elliptic integral has been used in chapter 7, eqns. (7.12) and (7.32).

$$\phi_\infty = \arcsin \sqrt{\frac{(fv_t/2 + 1)}{\{fv_t + (1-f)\}}} \qquad \text{(i.e. } \phi = \phi_\infty \text{ for v} \to \infty)$$

$$= \pi/2 \text{ for } f = 0$$

$$= \pi/4 \text{ for } f = 1$$

$$v_t \equiv u_t + 1/u_t$$

$F(\alpha, \phi)$ = (incomplete) elliptic integral of the first kind.

While eqn. (1.13) — expressing a scaled distance as a function of the Boltzmann factor for monovalent counterions, u, in terms of the tabulated elliptic integral of a composite function of u for any chosen set of values of f and u_t — contains all the necessary information to describe the double layer configuration for most systems of practical interest, it is still not easy to use. The main reason for this is lack of a priori information for the value of u_t. In fact, this parameter is often the one primarily sought for given conditions with respect to C_0, f and the thickness of the liquid layer terminated at $x = x_t$. Using (1.13) for $x = x_t$ now gives:

$$\sqrt{\beta C_0}\, x_t = \mu F(\alpha, \phi_\infty) \tag{1.14}$$

Replacing the scaled distance parameter by $\kappa_0 x_t$, in which $\kappa_0 = \sqrt{\beta I_0}$ — the D-H parameter of the equilibrium solution — now equals $\sqrt{\beta C_0 (1+f)}$, one finds:

$$\kappa_0 x_t = \mu \sqrt{1+f}\, F(\alpha, \phi_\infty) \equiv G(u_t, f) \tag{1.15}$$

Introducing finally the relative ionic strength in the terminal plane of the truncated double layer, according to:

$$U_t \equiv I_t/I_0 = \frac{f(u_t^2 + 1/u_t^2) + (1-f)(u_t + 1/u_t)/2}{(1+f)} \tag{1.16}$$

one may plot U_t as a function of $\kappa_0 x_t$. As is shown in Fig. 1.1 this relationship has the fortunate property that it is almost (i.e. within limits of the accuracy usually required) independent of f, thus yielding *one* curve with a certain "band width", valid for all mono-divalent systems. Moreover, as is elaborated below, this curve corresponds to a reasonably constant value of the product of both scaled parameters at large values of U_t, i.e. $\kappa_t x_t \equiv \kappa_0 x_t \sqrt{U_t} \simeq$ constant for $U_t > 5$. This product, to be indicated by the symbol Ξ hereafter, expresses the thickness of the liquid layer on a planar surface (which, for the time being, is considered to possess an infinitely high charge density, cf. section 1.5) in units of $1/\kappa_t$, i.e. the D-H length parameter corresponding to the ionic strength at the terminal plane of the double layer. As is shown in Fig. 1.1, the curves of Ξ pertaining to *all* mixed mono-divalent systems cover a band of comparatively narrow width, forming a gentle transition between $\Xi \approx 2.2$ for $\kappa_0 x_t < 2.2$ and $\Xi = \kappa_0 x_t$ for $\kappa_0 x_t > 2.2$. In

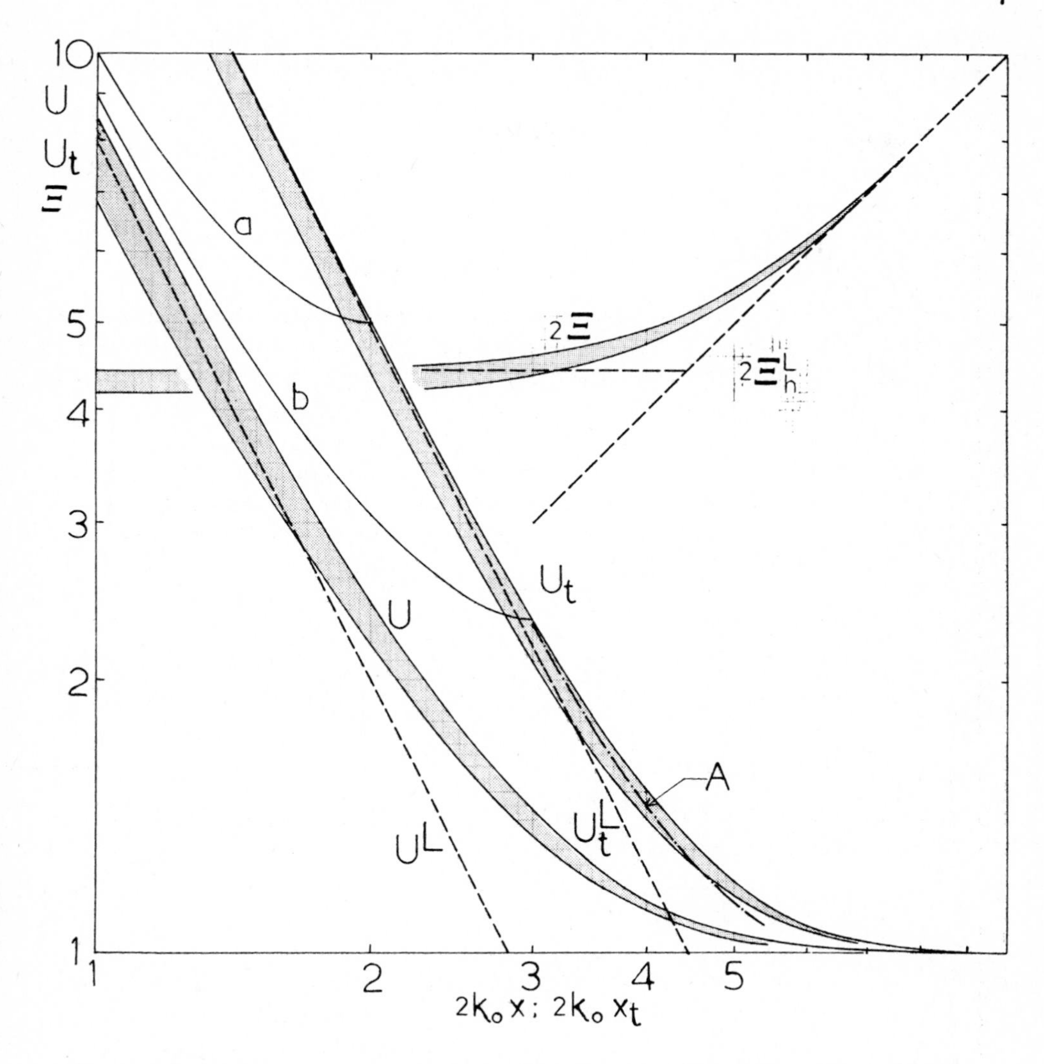

Fig. 1.1. The relative ionic strength, $U \equiv I/I_0$, in a freely extended double layer and its value at the terminal plane of a truncated double layer, U_t, as a function of $2\kappa_0 x$ and $2\kappa_0 x_t$, respectively, for systems containing a symmetric mixture of mono- and divalent cations and anions. The shaded "band" covers different values of the composition parameter f, the homovalent systems falling on the upper edge (cf. also Fig. 1.2). Also given are: the product $2\kappa_0 x_t\sqrt{U_t} \equiv 2\Xi$; two examples of the course of U within a truncated double layer with $2\kappa_0 x_t = 2$ (curve a) and $2\kappa_0 x_t = 3$ (curve b), respectively; the limiting lines for high values of U (line U^L) and U_t (line U_t^L) and the corresponding limiting value of $2\Xi = \pi\sqrt{2}$ for homovalent systems. The broken line A represents the approximation to U_t according to eqn. (1.23).

practice this mean course of Ξ is easily reproduced by taking $\Xi = 2.5$ at $\kappa_0 x_t = 2.2$, $\Xi = 2.2$ at $\kappa_0 x_t < 1.1$ and $\Xi = \kappa_0 x_t$ at $\kappa_0 x_t > 3.3$, and drawing a symmetric transition curve on double log paper. The ensuing curve then provides an estimate of Ξ within about 5% accuracy. This then gives U_t within about 10% as $(\Xi/\kappa_0 x_t)^2$. As an example one may thus estimate that for $x_t = 30$ Å, $C_0 = 0.02$ normal (N) and $f = 0.25$, U_t equals about 2.3 as in this case $I_0 = 0.025$ and thus $1/\kappa_0 \rightarrow 1/\sqrt{\beta I_0} = 2.10^{-9}$ m. This then gives $\kappa_0 x_t = 1.5$ and $\Xi \approx 2.3$, such that $(\Xi/\kappa_0 x_t)^2 \approx 2.3$.

In fact the use of the above scaled parameters U_t and $\kappa_0 x_t$ implies complete match for the limiting curves corresponding to $f = 0$ (i.e. pure mono-monovalent) and $f = 1$ (di-divalent), respectively. Identifying the value of $\kappa_t x_t$ for these homovalent endpoints as Ξ_h, the actual value of Ξ_h is found to be $\pi/\sqrt{2} = 2.22$ within 1% for $U_t > 5$ or $\kappa_0 x_t < 1$. It then changes gradually to 2.5 at $\kappa_0 x_t = 2$, $U_t = 1.5$, whereafter it approaches the value of $\kappa_0 x_t$ (cf. Fig. 1.1). Here again a certain "band-width" is indicated, covering the situation where $0 < f < 1$. In Fig. 1.2 the value of Ξ is plotted on a larger scale, showing the error involved when using Ξ_h for all values of f. The maximum error is about 8% for all values of $f < 0.025$. For higher values of f the error is always smaller (cf. figure), its maximum to be found around $\kappa_0 x_t = 1$. For values of $f < 0.025$ the maximum (at 8%) becomes shifted towards lower values of $\kappa_0 x_t = 5.3\sqrt{f}$. In Fig. 1.2 (insert) the percentage deviation of Ξ from Ξ_h (= 2.22) in the range $\kappa_0 x_t$ from zero to unity is plotted on a linear scale for easy reading.

In summary, Fig. 1.1 lends itself to a quick estimate of U_t as a function of $\kappa_0 x_t$ within about 3% accuracy for all systems of concern. If $\kappa_0 x_t$ is smaller than about unity, U_t may be found directly as $\pi^2/2(\kappa_0 x_t)^2$ — if necessary corrected with the help of Fig. 1.2 — as will be commented on in the following section. Once U_t has been found the corresponding value of $v_t \equiv u_t + 1/u_t$, the parameter in eqn. (1.13), may be back-calculated from:

$$v_t = \frac{1-f}{4f}\left[\sqrt{1 + 16f\{2f + (1+f)U_t\}/(1-f)^2} - 1\right] \tag{1.17}$$

Using the above value of v_t, (1.13) gives the values of $\kappa_0 xt$ corresponding to chosen values of v. Back translation of v and v_t into u and u_t allows one to plot u or $y = \ln u$ as a function of $\kappa_0 x$ for a given value of $\kappa_0 x_t$.

Although the above calculation scheme based on the use of Fig. 1.1 is fairly rapid, it still involves looking up $F(\alpha, \phi)$ for a series of values of ϕ. As usually moderate precision is satisfactory, the similarity in the curves relating U to $\kappa_0 x$ — regardless of the value of f — may be used in obtaining a quick estimate. As is shown in Fig. 1.1, a plot of log U as a function of $\log \kappa_0 x$ may be obtained by simple translation of the curves indicated by a and b, starting at the appropriate value of U_t (provided $U_t \geqslant 2$). Using a transparent overlay with curve b as shown, one may thus read $U(\kappa_0 x)$. The relative ionic strength, U, may then be expressed in terms of $v \equiv u + 1/u$ (with the help of (1.17)), in u, or in $y \equiv \ln u$.

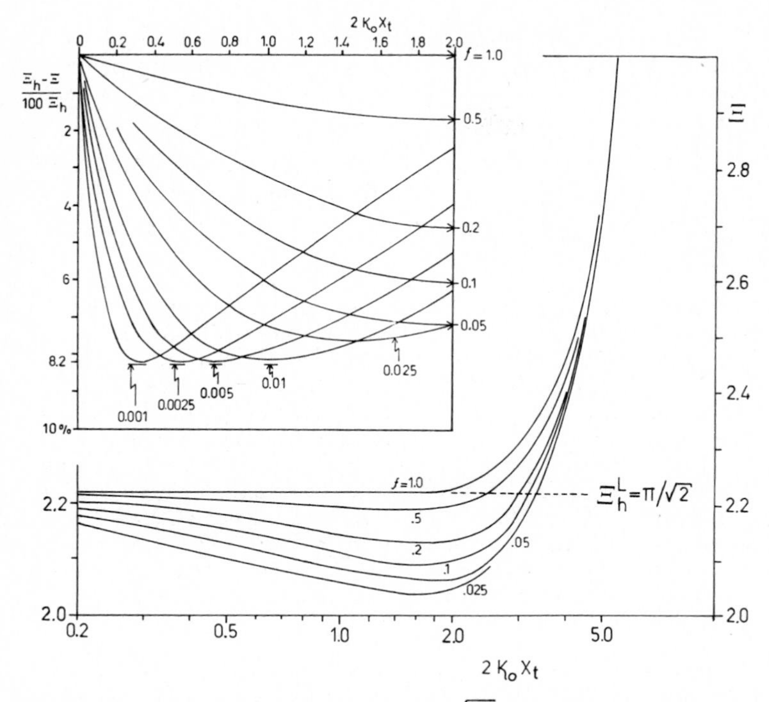

Fig. 1.2. Large scale plot of $\Xi \equiv \kappa_0 x_t \sqrt{U_t}$ against $2\kappa_0 x_t$, for different values of the fraction divalent ions, f, in symmetric salt systems. Insert: percentage deviation of Ξ from Ξ_h (i.e. Ξ for homovalent systems).

1.4. SIMPLIFIED SOLUTIONS FOR LIMITING CASES

If U_t, as read from Fig. 1.1, falls within the linear region of the curve, the use of elliptic integrals may be avoided. This linear region corresponds to the case where u_t is sufficiently larger than $1/u_t$ to warrant neglect of the presence of co-ions altogether. Eqn. (1.11) may then be written out as:

$$\sqrt{\beta C_0}\, x \;=\; -\sqrt{\frac{du}{u\sqrt{f(u^2 - u_t^2)/2 + (1-f)(u-u_t)}}} \tag{1.18}$$

This equation is a degenerated elliptic integral, with $\alpha = 0$, and may be expressed in terms of an inverse sine function, according to

$$\sqrt{\beta C_0}\, x \sqrt{\{fu_t/2 + (1-f)\}u_t} \;=\; 2(\phi_\infty^L - \phi^L) \tag{1.19}$$

with

$$\phi^L \;=\; \arcsin\sqrt{\frac{fu_t/2 + (1-f)}{fu_t + (1-f)}\;\frac{u - u_t}{u}}$$

$$\phi_\infty^L = \arcsin \sqrt{\frac{fu_t/2 + (1-f)}{fu_t + (1-f)}} \qquad \text{(i.e. } \phi_\infty^L = \phi^L \text{ for } u \rightarrow \infty)$$

$$= \pi/2 \text{ for } f = 0$$

$$= \pi/4 \text{ for } f = 1$$

Reconstructing the expression for $\kappa_0 x_t$ one finds:

$$\kappa_0 x_t^L = \sqrt{\frac{(1+f)}{\{fu_t/2 + (1-f)\}u_t}} \; 2\phi_\infty^L \tag{1.20}$$

where the superscript L has been added because the above approximation was used first by Langmuir (1938). If plotted against $U_t^L \equiv \{fu_t^2 + (1-f)u_t/2\}/(1+f)$, this yields the linear portion of the curve in Fig. 1.1. Indeed, one finds the expression for $\kappa_t x_t$ now as:

$$\kappa_t x_t^L \equiv \kappa_0 \sqrt{U_t^L}\, x_t^L = \sqrt{\frac{fu_t + (1-f)/2}{fu_t/2 + (1-f)}} \; 2 \arcsin \sqrt{\frac{fu_t/2 + (1-f)}{fu_t + (1-f)}} \tag{1.21}$$

For the homovalent endpoints this gives:

$$\kappa_t x_t^L (f = 0) = 2\sqrt{1/2} \arcsin (1) = \pi/\sqrt{2}$$

$$\kappa_t x_t^L (f = 1) = 2\sqrt{2} \arcsin (\sqrt{1/2}) = \pi/\sqrt{2}$$

The ensuing value of $\kappa_t x_t^L$ will be indicated as Ξ^L, which thus equals 2.22 for the homovalent systems. It differs slightly (i.e. 8% at maximum) for intermediate values of f, which has been shown already in Fig. 1.2. If an overall accuracy of 4% is satisfactory one may thus use for all systems $\kappa_t x_t = 2.22 - 4\% = 2.1^5$ in the range of $\kappa_0 x_t$ between zero and unity and $\kappa_t x_t = 2.2^0$ for $1.0 < \kappa_0 x_t < 1.5$.

For the homovalent endpoints the constancy of $\Xi^L = \pi/\sqrt{2}$ may be used to obtain a convenient approximation for u_t valid up to $\kappa_0 x_t = 2.5$. For these systems it may be verified that:

$$U_t = U_t^L + 1/4U_t^L \tag{1.22}$$

Substituting this into eqn. (1.21) gives:

$$U_t = \pi^2/2(\kappa_0 x_t)^2 + (\kappa_0 x_t)^2/2\pi^2$$

$$= 4.93/(\kappa_0 x_t)^2 + 0.55(\kappa_0 x_t)^2 \tag{1.23}$$

This approximation to U_t is marked with A in Fig. 1.1, indicating a validity within 6% for $1.5 < \kappa_0 x_t < 2.5$, for all values of f.

In the near linear region one may also use a simplified method to construct the complete trajectory of U in the double layer. Eqn. (1.19) is then written as:

$$\frac{x_t - x}{x_t} = \frac{\phi^L}{\phi_\infty^L} \tag{1.24}$$

Introducing now the parameter p, according to:

$$p \equiv \frac{1}{1 + fu_t/(1-f)}$$

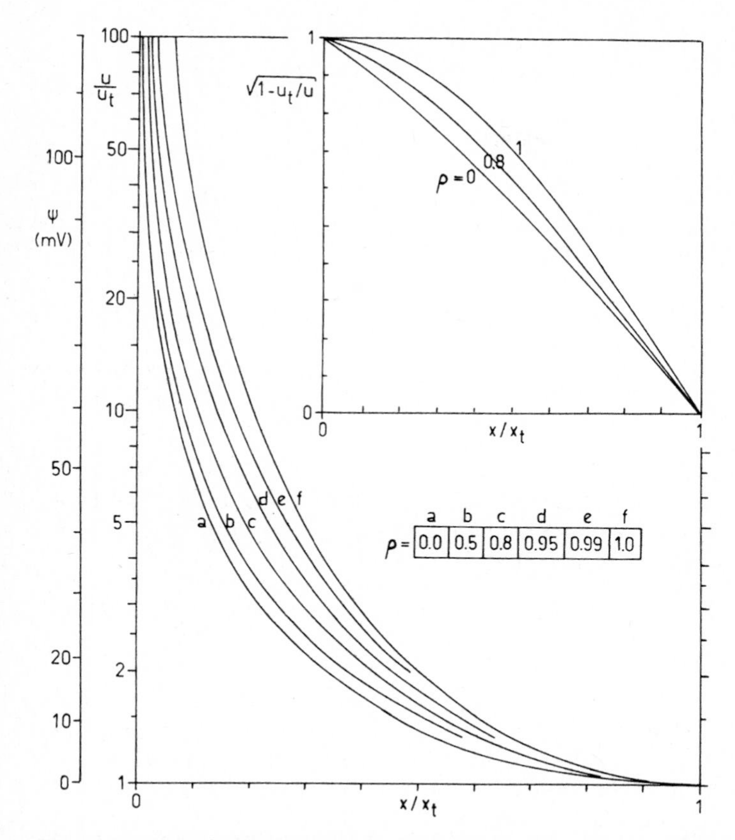

Fig. 1.3. The value of the Boltzmann factor u and the electric potential ψ inside the truncated double layer relative to its terminal value for different values of $p \equiv 1/\{1 + fu_t/(1-f)\}$. Insert: scaled plot of $\sqrt{1-u_t/u}$ against x/x_t.

one finds:

$$\frac{\phi^L}{\phi^L_\infty} = \frac{\arcsin\{\sqrt{(1+p)/2}\sqrt{1-u_t/u}\}}{\arcsin\sqrt{(1+p)/2}} \tag{1.25}$$

which allows one to plot $\phi^L/\phi^L_\infty = (x_t - x)/x_t$ as a function of u/u_t for chosen values of p (and thus of fu_t). In Fig. 1.3 the result is shown, where it may be noted that $\log u/u_t$ represents the electric potential in the double layer relative to its value at x_t, the terminal plane. It may also be pointed out that for the homovalent endpoints p equals 1 (for $f = 0$) and 0 (for $f = 1$). Thus one finds for these endpoints:

$$\frac{x_t - x}{x_t} = \frac{\arcsin\sqrt{1-u_t/u}}{\pi/2} \qquad (f = 0) \tag{1.26a}$$

and

$$\frac{x_t - x}{x_t} = \frac{\arcsin\sqrt{(1-u_t/u)/2}}{\pi/4} \qquad (f = 1) \tag{1.26b}$$

Inverting these equations then gives the convenient expressions:

$$\sqrt{u_t/u} = \sin(\pi x/2x_t) \qquad (f = 0) \tag{1.27a}$$

$$u_t/u = \sin(\pi x/2x_t) \qquad (f = 1) \tag{1.27b}$$

The above expressions may finally be used to explore the region very close to the charged surface, i.e. at $u \gg u_t$. For $f = 0$ this gives:

$$\frac{\pi}{2} \frac{\kappa_0 x \sqrt{u}}{\kappa_0 x_t \sqrt{u_t}} = 1 \qquad (f = 0) \tag{1.28}$$

Because in this case $U^L = u/2$, the denominator may be replaced by $\kappa_t x_t = \pi/\sqrt{2}$, such that:

$$\kappa_0 x \sqrt{U^L} (\equiv \kappa x) = \sqrt{2} \tag{1.29}$$

implying that close to the charged surface the local value of $1/\kappa$ (i.e. the D-H parameter corresponding to the *local* ionic strength) increases proportionally to the distance from the charged surface, giving the above constant value of κx. Using eqn. (1.27b) for $f = 1$ gives with $U^L = u^2/u$ the same result. The deviations from (1.29) for intermediate values of f are very small indeed if $U > 10U_t$, so this equation has again a general validity for the mixed mono-divalent systems. In Fig. 1.1 the extension of (1.29) into the region $U < 10$ is shown as the dashed line marked U^L, parallel to $U_t^L = \pi^2/2\kappa_0 x_t$. It constitutes the asymptote of all curves for U at different values of U_t.

Superposition of the information contained in Fig. 1.3 upon Fig. 1.1 allows one to plot such curves for U corresponding to a given value of U_t. The result is valid with sufficient accuracy at least up to $\kappa_0 x_t = 2$, provided one uses the *correct* value of U_t as found according to the procedure described above, i.e. *not* U_t^L. Two examples are shown in Fig. 1.1, corresponding to $U_t = 5$ and $U_t = 2.35$, respectively, and belonging to $f = 0$ or 1. In practice one is seldom interested in calculating the course of U itself; if the trajectory of u is sought it suffices to recalibrate the axes of Fig. 1.3 in terms of x and u. This then implies that first the appropriate value of U_t is read from Fig. 1.1, which may be translated for given f into v_t with (1.17), and then into the corresponding value of u_t according to:

$$2u_t = v_t + \sqrt{v_t^2 - 4} \tag{1.30}$$

The definition of p above (1.25) then allows the calculation of this parameter needed to select the appropriate curve of Fig. 1.3.

Another limiting case is the distribution of the ions in a freely extended double layer (i.e. the presence of excess liquid). Then $u_t = U_t = 1$; $v_t = 2$. Thus (1.11) becomes:

$$\sqrt{\beta C_0}\, x = -\int_\infty^u \frac{du}{(u-1)\sqrt{fu^2/2 + u + f/2}} \tag{1.31}$$

Integration yields:

$$\kappa_0 x = \ln \frac{(u+1)\sqrt{1+f} + \sqrt{2fu^2 + 4u + 2f}}{(u-1)(\sqrt{1+f} + \sqrt{2f})} \tag{1.32}$$

Introducing the auxiliary parameters

$V \equiv (u+1)/(u-1)$ and $g = \sqrt{(1+f)/(1-f)}$

a convenient expression is obtained:

$$\kappa_0 x = \ln[\{Vg + \sqrt{(Vg)^2 - 1}\}/\{g + \sqrt{g^2 - 1}\}] \tag{1.33}$$

$$= \text{arg cosh}\, Vg - \text{arg cosh}\, g \tag{1.34}$$

where the second equation becomes impractical for $g \to \infty$, i.e. for $f \to 1$. The corresponding curve in terms of the scaled parameters $\kappa_0 x$ and $U = I/I_0$ again becomes fairly independent of f, as is shown in Fig. 1.1. Obviously this curve must again approach the limiting straight line corresponding to $\kappa x = \kappa_0 x\sqrt{U} = \sqrt{2}$ for high values of U where the anions exert a negligible influence.

Because at low values of U the anionic valence gains in importance, the relationship $U(\kappa_0 x)$ remains more sensitive towards anionic valencies than the previous curve $U_t(\kappa_0 x_t)$: in the freely developed double layer these low values of U are always present in the "tail" end of the double layer, while in the truncated double layer U must remain larger than U_t. At the same time (1.33) is so much easier to handle than the equations for the truncated double layer, that direct calculation for a specific case is always possible.

It is finally noted that the deviations of U in mixed systems from the curve corresponding to the homovalent endpoints result from the sorting-out effect of the electric field on ions of different valence. Close to the surface the divalent counterions dominate while the tail end contains the majority of the monovalent ones. Where $(dy/dx)^2$ is connected with the excess molarity (cf. (1.7)) U, signifying the local ionic strength, increases rather slowly at the tail end because in that region U approaches the relative molarity. Close to the surface, where the divalent ions dominate, U approaches four times the relative molarity (for small values of f), and will thus steepen up more than in the case of the homovalent systems, where U remains equal to the relative molarity throughout the double layer.

In view of the expected increased influence of the anionic valence on the trajectory of $U(\kappa_0 x)$, some additional equations relating u to $\kappa_0 x$ in certain asymmetric systems are listed in Table (1.1) and plotted in Fig. 1.4. It is noted that $U(\kappa_0 x)$ coincides for all systems for $U > 4$, and remains useful up to $U \approx 2$. In the "tail" end differences are considerable, while the local ionic strength goes through a minimum value of about $0.8\, I_0$ at $\kappa_0 x = 2.5$ for the system containing monovalent cations with divalent anions (e.g. Na_2SO_4).

The above is easily verified by taking e.g. $u \approx 2$ in such a system, implying that $I = C_0(0 + 2 + 2/4 + 0)/2 = 1.25\, C_0$, while I_0 equals $1.5\, C_0$. For comparison also the values of u belonging to 1—1, 2—2, 2—1 and 1—2 systems, respectively, are plotted against $\kappa_0 x$. Here the symmetric systems vary widely: at the same position on the distance-axis the potential in the mono-monovalent system is twice that in the di-divalent system, corresponding to $(u_{2-2})^2 = u_{1-1}$. The asymmetric systems follow the symmetric ones having the same counterionic valence, though the distance coordinates at high values of u differ by a factor of $\sqrt{1.5}$ for the monovalent cations and $\sqrt{0.75}$ for the divalent ones. In the present single-salt systems one may indeed obtain a rather perfect match for all systems by plotting u^{z_+} against $\sqrt{z_+\beta C_0}\,x$, valid up to $u^{z_+} > 2$. Obviously this curve coincides with u against $\kappa_0 x$ for the mono-monovalent system. Unfortunately this type of plotting is useful only for single salt systems as the mixtures can only be brought together by plotting U against $\kappa_0 x$.

TABLE 1.1

Equations relating to the infinitely extended double layer (special cases)

Composition parameters (equivalent fractions in solution)

$f_{2+} = 0, f_+ = 1, f_- = 1, f_{2-} = 0$

$$\kappa_0 x = \text{arg cosh}\,\{(u+1)/(u-1)\} = 2\ \text{arg coth}\sqrt{u}$$

$$u = \coth^2(\kappa_0 x/2) = \{\cosh(\kappa_0 x) + 1\}/\{\cosh(\kappa_0 x) - 1\}$$

$$U = (u + 1/u)/2 = \{\cosh^2(\kappa_0 x) + 1\}/\{\cosh^2(\kappa_0 x) - 1\}$$

$$y = \ln u = 2 \ln \coth(\kappa_0 x/2);\ \kappa_0 x = -\ln\tanh(y/4) = \ln\coth(y/4)$$

$f_{2+} = 1, f_+ = 0, f_- = 0, f_{2-} = 1$

$$\kappa_0 x = \ln\{(u+1)/(u-1)\} = 2\ \text{arg coth}\ u = \text{arg cosh}\,\{(u^2+1)/(u^2-1)\}$$

$$u = \coth(\kappa_0 x/2)$$

$$U = (u^2 + 1/u^2)/2 = \{\cosh^2(\kappa_0 x) + 1\}/\{\cosh^2(\kappa_0 x) - 1\}$$

$$u = \ln\coth(\kappa_0 x/2);\ \kappa_0 x = -\ln\tanh(y/2) = \ln\coth(y/2)$$

$f_{2+} = 1, f_+ = 0, f_- = 1, f_{2-} = 0$

$$\kappa_0 x = \text{arg cosh}\,\{(2u+1)/(u-1)\} - \text{arg cosh}\ 2$$

$$u = \{\cosh(\kappa_0 x + \text{arg cosh}\ 2) + 1\}/\{\cosh(\kappa_0 x + \text{arg cosh}\ 2) - 2\}$$

$f_{2+} = 0, f_+ = 1, f_- = 0, f_{2-} = 1$

$$\kappa_0 x = \text{arg cosh}\,\{(u+2)/(u-1)\}$$

$$u = \{\cosh\kappa_0 x + 2\}/\{\cosh\kappa_0 x - 1\}$$

$f_{2+} = f, f_+ = 1 - f, f_- = 1, f_{2-} = 0$

$$\kappa_0 x = \text{arg cosh}\,[\{(1+f)u + 1\}/(u-1)] - \text{arg cosh}\,(1+f)$$

$$u = [\cosh\{\kappa_0 x + \text{arg cosh}\,(1+f)\} + 1]/[\cosh\{\kappa_0 x + \text{arg cosh}\,(1+f)\} - (1+f)]$$

1.5. THE SURFACE CHARGE DENSITY AS A BOUNDARY CONDITION

In contrast to surfaces that are reversible with respect to one or more potential determining ions, many clay minerals possess a fairly constant surface density of charge. Back calculations from the CEC and specific

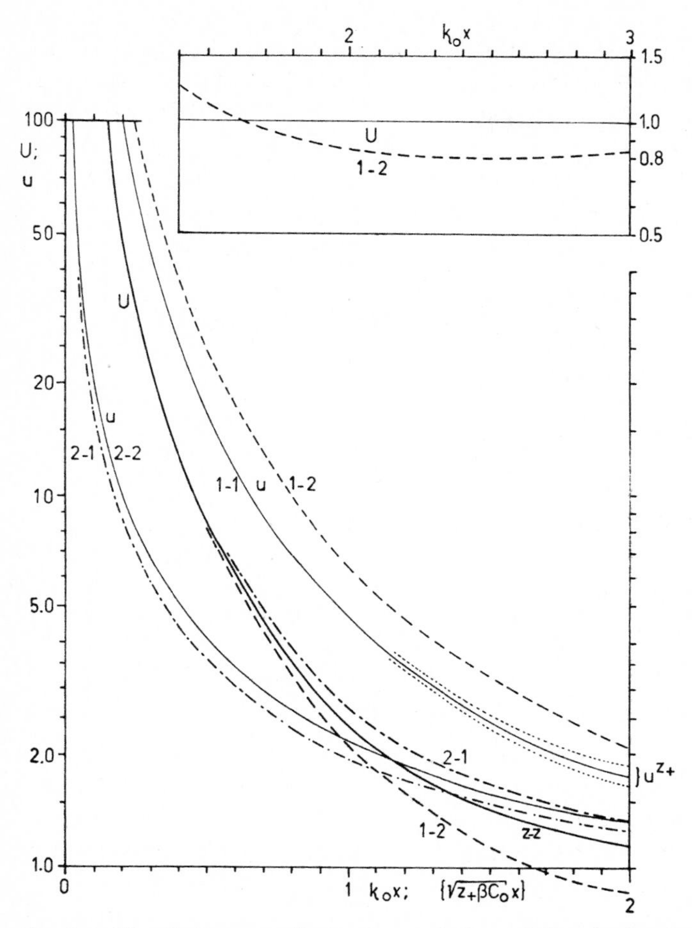

Fig. 1.4. The Boltzmann factor, u, and the relative ionic strength, U, as a function of $\kappa_0 x$, in single salt systems, as well as u^{z+} as a function of $\sqrt{z_+\beta C_0}\, x$. The hyphenated numbers indicate counter- and co-ionic valence. Insert shows minimum value of U in the 1—2 system.

surface area give values of one to several tenths of Coulomb per m^2, which translates to $1—3 \cdot 10^{-9}$ keq/m^2. The lowest figure given corresponds to one substitution charge per ca. 150 Å^2, i.e. about 2/3 substitution charge per unit cell containing 4 Al-atoms in the octahedral layer. This number corresponds also to a CEC* of about 0.8 meq/g for a surface area of 800 m^2/g as is often found for montmorillonite. Illites present in Dutch marine deposits possess a CEC of about 0.45 meq/g for a surface area of about 150 m^2/g (corresponding to a unit consisting of 5—6 individual platelets bonded together with e.g. K-ions). This gives a surface density of charge of 3.10^{-9} keq/m^2 (= 0.3 Coul/m^2) and amounts to two substitution charges per unit

* CEC refers to cation exchange capacity as used in chapter 4 of part A.

cell. If residing mainly in the tetrahedral layer this would amount to about 1/4 substitution of the Si-atoms on either side of the plate (by Al-atoms).

Returning to the generalized equations and curves representing the ionic distribution in diffuse layers as derived in the previous sections for the surface with infinitely high charge density, the given charge density may now be introduced as a boundary condition. Expressing the surface charge density, Γ, in keq/m^2, one finds for the freely developed double layer:

$$|F\Gamma| = \int_{\delta}^{\infty} \rho_{\pm}\, dx \tag{1.35}$$

in which $\rho_{\pm}$ must be specified in $Coul/m^3$ and where δ indicates the position of the charged surface on a distance axis whose origin is located at the point where $\rho_{\pm}$ (and u, y, etc.) reach infinity. In other words, δ indicates the lower cut-off value of the generalized distance coordinate — employed in the previous section — beyond which the total countercharge fits the chosen value of Γ. Using (1.1) and (1.3) gives:

$$\begin{aligned} |\Gamma| &= \frac{\epsilon'}{4\pi F} \int_{\delta}^{\infty} \left(\frac{d^2\psi}{dx^2}\right) dx \\ &= -\frac{2}{\beta} \int_{\delta}^{\infty} \frac{d^2 y}{dx^2}\, dx \\ &= -\frac{2}{\beta} \left(\frac{dy}{dx}\right)_{x=\delta} \end{aligned} \tag{1.36}$$

Accordingly the surface density of charge determines the value of $(dy/dx) \equiv d(\ln u)/dx$ at $x = \delta$. Making use of the expressions given for u (in terms of $\kappa_0 x$) in Table 1.1, yields for the homoionic mono- *and* divalent systems $u^{z_+} = \coth^2(\kappa_0 x/2)$, so for $x = \delta$ one finds:

$$|\Gamma| = \frac{2}{z_+\beta}\, \frac{2\kappa_0}{\sinh \kappa_0 \delta} \approx \frac{4}{z_+\beta\delta} \tag{1.37}$$

The validity of replacing $\sinh \kappa_0 \delta$ by its argument may be verified by using the value of Γ as indicated above, i.e. 10^{-9} keq/m^2. This gives δ as $4/z_+ \cdot 10^{-10}$ m = $4/z_+$ Å units, while $1/\kappa_0$ of a 0.1 *N* solution amounts to about 10 Å units. Thus for an equilibrium solution with a concentration less than 0.1 *N* the error introduced by the expansion of the hyperbolic sine is only about 2%. Certainly one has no reason to seek the value of δ with an accuracy better than 1 Å. Accordingly one may safely use a cut-off value on the distance axis as prescribed by (1.37), reading in reverse $\delta = 4/z_+\beta\Gamma$ (with z_+ = valence of the dominant counterion).

For mixed systems one may use (1.34), written in reverse as:

$$V = \frac{u+1}{u-1} = \frac{\cosh(\kappa_0 x + \arg\cosh g)}{g}$$

Differentiating u with respect to κ_0x then gives:

$$|\Gamma| \approx \frac{2}{\beta\delta} \frac{(1+f)\kappa_0\delta + \sqrt{2f(1-f)}}{(1+3f)\kappa_0\delta/2 + \sqrt{2f(1-f)}} \tag{1.38}$$

which neatly converges to (1.37) for $f = 0$ and $f = 1$. It is interesting to note that in dilute systems, where $\kappa_0\delta$ is very small (e.g. 0.02 in 10^{-3} normal electrolyte, for a mixture of mono- and divalent cations in equilibrium with a montmorillonite surface) the influence of the common second term of denominator and numerator is very dominant. Thus the fraction of divalent cations in the equilibrium solution must decrease to a value less than 10^{-4} before δ reaches 3 Å instead of 2 Å. This shows that in dilute solutions containing only a minute fraction of divalent cations in the presence of an excess of monovalent ones, the divalent cations remain dominant very close to the surface. Although (1.38) lends itself to an iterative calculation of δ in mixed systems, it seems hardly necessary to go through this effort. If the equilibrium solution is less than 0.1 *N* and if dealing with common clay minerals, (1.37) is satisfactorily accurate. This indicates that δ may be safely taken as e.g. 1 Å (illites containing at least some Ca), 2 Å (pure Na-illite, montmorillonite with some Ca), 4 Å (pure Na-montmorillonite)

One additional aspect of the use of δ as cut-off value should be mentioned. Whereas its value, as given within 1 Å unit, is certainly good enough for drawing potential- and other curves with the help of Figs. 1.1 and 1.4, the steepness of u(x) in this region prevents one from back calculating $u(\delta) \equiv u_s$. The latter may be found directly from (1.7), which couples (dy/dx) with the excess molarity at the colloid surface. Combination with (1.36) thus gives:

$$\Sigma_k ({}^m c_{s,k} - {}^m c_{t,k}) = C_0 \Sigma_k \{(u_s^{z_k} - u_t^{z_k}) f_k / |z_k|\} = \frac{\beta\Gamma^2}{4} \tag{1.39}$$

where ${}^m c_s$ refers to the molar concentration at the surface. For montmorillonite, with $\Gamma = 10^{-9}$ keq/m^2, this gives an excess molarity of about 2.6 molar. As long as the total molarity at x_t (or in the equilibrium solution, for freely expanded double layers) remains small in comparison to this figure, the *molar* surface concentration thus remains constant, *irrespective* of the valence of the cations. In turn this explains why the potential curve drops off much steeper in the divalent systems than in the monovalent ones: the ionic strength is much greater in the first case for the same total molarity. It also explains the fact that the ionic composition of the exchange complex is largely determined by the reduced ratio of the cationic concentrations in the equilibrium solution, as the reduced ratio is rather invariant with position in the double layer. So given its value, the cationic composition of the first layer (with constant molarity) is fixed, regardless of the total electrolyte level in the equilibrium solution (within certain limits).

Returning finally to the application of the theory exposed in sections 1.3 and 1.4, it is clear that the value of x_t needed for the application of the scaled graphs is found by adding δ (i.e. 1—4 Å) to the thickness of the liquid layer, d_l, as pertaining to the system. In practice d_l is often given with an accuracy hardly better than a few Å units.

1.6. DISCUSSION OF THE ASSUMPTIONS UNDERLYING THE GOUY-CHAPMAN THEORY

The Gouy-Chapman theory, as worked out for planar surfaces in the previous sections, involves many approximations and assumptions which deserve further attention. In this section these will be discussed in the light of conditions prevailing in soil systems.

First a remark must be made on the assumed "parallel array of the electric force lines" inside the double layer, as implied by the use of the one-dimensional form of the Poisson equation (1.1). Visualizing these force lines as originating from the counterions (which are assumedly cations in the present case), these will disappear at the negative substitution charge, which is located in a planar array a few Å-units below the surface of the clay plates. Although in principle the finite size of such plates must lead to a positive divergence of the field lines in the direction away from the surface, this effect becomes really significant only in the region outside a half sphere constructed on the charged surface. Taking another spherical surface located about half way between the above hemisphere and the charged surface one may safely treat the region inside this second spherical surface as a linear field. Taking the length and width of clay plates as about 1000 Å one finds that a double layer with an extent of around 100 Å (e.g. 10^{-3} N mono-monovalent electrolyte) is situated almost entirely within this linear region (cf. Fig. 1.5). Where the above considerations apply to a single plate, it should furthermore be recognized that even in dilute suspensions the fields of individual plates interact. As this interaction leads to more or less parallel arrangement of close neighbors, the diversion of field lines is further depressed.

Secondly, it should be pointed out that very close to the surface nonlinearity will arise because of the discrete nature of the surface charges. This effect will be noticeable only within distances of the same order as those between the individual charges. For clays (cf. section 5) this is about 10 Å spacing. As the surface charges are buried underneath the surface (e.g. about 5 Å for octahedral substitution) one must thus delineate the first few Å of the double layer as not following the linear Poisson equation. In total this leaves the entire zone between, say, 5 Å up to a few hundred Å from the surface as a safe domain for the application of the linear Poisson equation.

Leaving the surface region of the DL provisionally outside the discussion, the only approximation for the remainder is now centered on the use of concentrations in the Boltzmann equation. Put differently, it is the omission in the exponent of the Boltzmann equation of a term covering ionic interaction in the double layer which causes an error. As was shown by Loeb (1951) and Bolt (1955), this ionic interaction in a system containing an excess of ions of one type (i.e. the counterions) leads to an equation of the Davies type (cf. part A, eqn. (A2.12c)). However, in this case the denominator must be omitted because the regrouping of ions of like sign is not

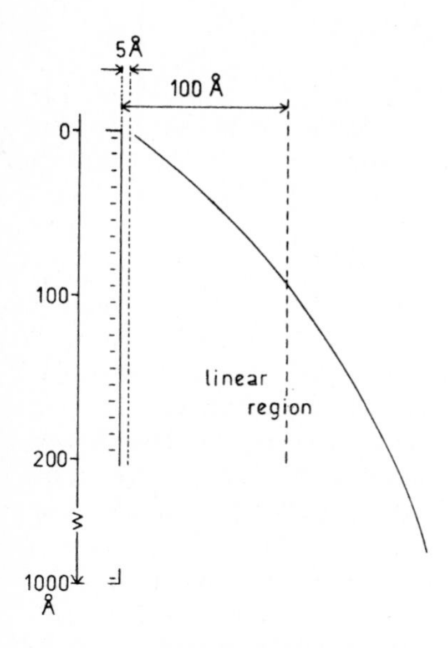

Fig. 1.5. Schematic presentation of the situation near a planar surface of 1000 × 1000 Å with point charges located about 10 Å apart. A diffuse layer of a hundred Å units or less is located almost entirely within the linear region of the electric field.

influenced by a distance of closest approach which delimits the clustering of ions of opposite signs. The ensuing effect may be expressed in terms of a shift of the distance coordinate at which a particular concentration prevails. Obviously, the activity corrections cause a contraction of the double layer (the effective excess molarity, i.e. the osmotic pressure, being reduced for a given value of the local volume density of countercharge). As was shown by Bolt (1955) this "shrinkage" effect may be calculated. For the curve of $y(x) \equiv \ln u(x)$ it amounts only to about 2 Å contraction for every tenfold increase of $u^{z_+} \equiv \exp(z_+ y)$, i.e. the Boltzmann factor for the (locally) dominant cation. Perhaps redundantly it is pointed out that the constancy of the shrinkage effect for each decade of u^{z_+} ensures that the effect is felt strongest in the regions of high ionic strength since the "width" on the distance axis per decade of u^{z_+} decreases, both with increasing κ_0 (and thus C_0 and z_+) and with increasing value of u^{z_+} (or U).

In effect the above implies that one may construct first the Gouy-Chapman curve according to the procedure discussed in the previous sections, yielding $\ln u (= y)$, as a function of $\kappa_0 x$. Taking as an example the monovalent system in equilibrium with 10^{-3} NaCl, as characterized in Fig. 1.4, one may now shift the point $u = 10$ two Å units to the right. For the chosen value of κ_0 this shift amounts to $\kappa_0 \Delta x = 0.02$. At $u = 100$ the shift equals another 2 Å units, or a total shift equal to 0.04; at $u = 1000$ the shift then equals 0.06 units on the distance axis. Noting that the distance between $u = 10$ and $u = 100$ equals about 0.45 distance units, as against 0.14 units between $u = 100$ and $u = 1000$, the "relative shrinkage" of the curve is about 5% between

$u = 10$ and $u = 100$ as against 15% between $u = 100$ and $u = 1000$. The ensuing corrected curve then represents the activity of the ions in the double layer. Applying now activity corrections according to the modified Davies equation suggested above, yields the ionic concentration as a function of position. Obviously the position of the charged surface must now also shift to the right, viz. to the point where $d \ln u/dx$ attains the correct value, equal to $\beta\Gamma/2$ (cf. (1.36)).

The situation for truncated double layers is treated in detail in Bolt (1955). As a first approximation one may use the following rule. Starting at x_t, the curve representing $z_+y = \ln u^{z+}$ is shrunk by 4 Å for the first tenfold increase of u^{z+}, and 2 Å for each following decade. As a result the value of $d_l = x_t - \delta$ corresponding to a given value of u_t (or U_t) has become smaller. Conversely the value of u_t corresponding to any given value of d_l has also decreased in comparison to the standard Gouy-Chapman curve. This effect is then, however, partly compensated if the relative activity u_t, is back translated to concentrations. For those homovalent systems where U_t exceeds 3, i.e. in the region where the "Langmuir" approximation (cf. (1.21)) applies, one may summarize the situation by plotting ${}^mC_t$ against $d_l = x_t - \delta$. The corresponding "Langmuir" approximation of the Gouy equation then reads:

$$\sqrt{\beta C_0 z_+} x_t \sqrt{u_t^{z+}/2} = \pi/\sqrt{2} \tag{1.40}$$

With $C_0 u_t^{z+}/z_+ \equiv {}^mC_t$, this yields:

$${}^mC_t = \pi^2/z_+^2\beta(d_l + \delta)^2 \tag{1.41}$$

In homovalent systems this total molarity at the point of truncation thus becomes independent of the equilibrium concentration, as is implicit in the Langmuir approximation. Applying now the activity correction as suggested before, one may plot the corrected curve together with the Gouy-Langmuir curve as shown in Fig. 1.6. The overall effect is indeed very small. No data are available for the mixed systems but it seems likely that the effect should be of the same order of magnitude.

Turning now to the surface region one finds the situation on first sight literally crammed with deviations from the assumed ideal behavior. Some of these, viz. dielectric saturation effects and image effects due to a sudden change in dielectric constant from solution to solid, may still be estimated. For the given surface density of charge (1—$3 \cdot 10^{-9}$ keq/m^2) they should be small (cf. Bolt, 1955). Next, the ion-specific effects due to the ionic size (or polarizability) and the electrical and geometrical unevenness of the surface are of concern. Together these effects are formally equivalent to the existence of a "Stern layer", where an ion-specific energy of interaction is added to the Boltzmann exponent, and/or a crowding term is introduced. Starting with the latter, it appears that for clays with a *fixed* surface charge, this effect is hardly of concern. In contrast to colloids reversible to a particular potential determining ion, where a given surface potential may easily lead to lateral crowding if the electrolyte level is increased, the clay surface will, as

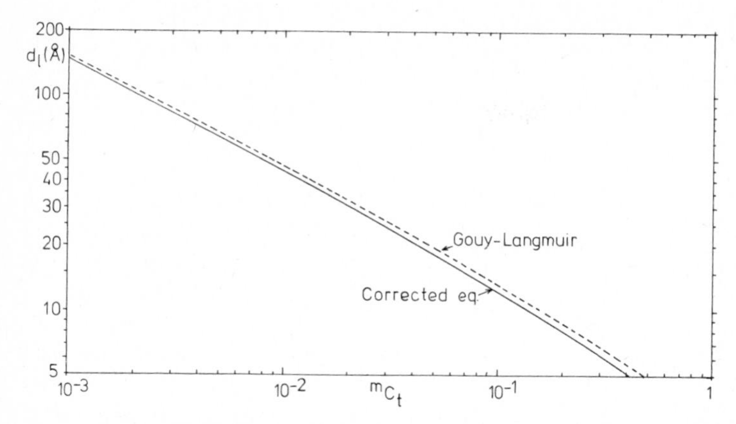

Fig. 1.6. The influence of corrections for ion activities, ion polarizability and dielectric saturation on the ionic distribution in a diffuse double layer (from Bolt, 1955). Shown is the concentration at the plane of truncation, ^mC_t, as a function of the thickness of the liquid layer, d_l, for a homoionic double layer, with $\Gamma = 10^{-9}$ keq/m^2, $z_+ = 1$. The dashed line represents the uncorrected Gouy-Langmuir approximation valid for $U_t > 3$ (cf. linear part of U_t in Fig. 1.1).

a rule, not accumulate a larger number of counterions than corresponds to its surface charge. Remembering the spacing of the surface charges at about 10 Å, lateral crowding is hardly of concern: if all counterions were stored in the first 10 Å of the double layer this would only amount to a concentration of 1—2 molar for monovalent ions and half that for the divalent ones. In this respect it is recollected that the constant molarity at the clay surface calculated by (1.39), i.e. 2—3 molar, leads to the same conclusion: *at* the plane parallel to the surface where the electric field strength reaches the value corresponding to the given surface density of charge, the concentration should reach about this value. Averaged over a finite region outward from this plane the concentration will be less than this absolute maximum.

Having thus dismissed lateral crowding effects as within the range of the activity corrections taken care of already, there remains the finite distance of approach towards the surface. If the cations remain predominantly hydrated, the situation becomes simple for homoionic clays. All counterions have the same (hydrated) size, and will never come closer to the surface than the distance of their hydrated radius. In effect this means that the point $x = \delta$ corresponds with a plane at some (small) distance from the surface. Using the value of å as used in the D-H theory, this should amount to 2—3 Å for the common cations. Fortunately this minimum spacing should be subtracted from any given value of $d_l = (x_t - \delta)$, the thickness of the truncated double layer, if one wants to find the value of x_t belonging to the system under consideration. Thus $x_t = (d_l - å) + \delta$. Where both å and δ are only a few Å, it appears that they will cancel within the limits of accuracy attainable.

Recollecting now that the surface charges are actually buried 2—5 Å within the clay plate, it is also clear that as long as the counterions remain hydrated, the ion specific "Stern" energy term should probably be very small. In conclusion it appears justified to assume that for homoionic clays saturated with common ions, if hydrated, the Stern layer will be an "empty" Stern layer according to the terminology of Grahame (1947). Accordingly the potential in such a double layer would follow the Gouy-Chapman curve — if necessary corrected for activity coefficients — from $x = x_t$ up to $x = \delta$. From there on $|\psi|$ would increase linearly with a slope corresponding to an appropriate value of the dielectric constant for the empty Stern layer, over a distance of å (a few Å units).

Recalling (1.36), one finds $(d\psi/dx)$ at $x = \delta = 4$ Å (pertaining to a Na-montmorillonite with $\Gamma = 10^{-9}$ keq/m^2) at roughly a hundred million Volt per m, i.e. 10 mV per Å-unit. Thus for a dielectric constant of 80 in the empty Stern layer the surface potential would be 20—30 mV larger than the value found for $x = \delta$, and for lower values of ϵ_r proportionally higher. Where at an equilibrium concentration of, say, $2.5 \cdot 10^{-3}$ normal NaCl the potential at $x = \delta$ should be about 180 mV, the actual surface potential should then be at least 200 mV.

In case of a mixture of ions with different ionic radii å one may distinguish the empty zone between the surface and $å_1$, a following zone with thickness $(å_2 - å_1)$, occupied only by the smallest ion, and finally the Gouy-Chapman region beyond $å_2$, occupied by the mixture. A calculation-example shows for two monovalent species differing 1 Å in radius that about 20% of the counterionic charge is located in the "middle" zone (cf. (1.37), using $\delta = 4$ for ion (1) and $\delta = 5$ for ion (2)). At a ratio of 1:1 of both ions in the equilibrium solution one would thus find a preference factor of about 1.5 for the smallest cation. This rather rough manipulation of the Gouy-Chapman theory shows that notwithstanding the fact that the Stern corrections as foreseen have probably a minor effect on the overall shape of the double layer, more specifically on its extent, the uncertainty in the precise structure of the "first" layer of counterions will reflect fully (actually in an exaggerated manner) on any attempt to predict the cationic composition of the entire double layer in mixed systems (cf. chapter 3).

If (partial) dehydration of counterions occurs, the influence would be along the same line as suggested before. That is, the extent would perhaps be influenced only moderately, but ionic preferences might show a major shift. In this respect, it should be pointed out that the ions K, Rb and Cs would probably become (partly) dehydrated in close contact with clays with predominantly tetrahedral substitution (illites!). They would then presumably locate themselves sunken partly into the hexagonal holes on the surface.

Finally it is noted that the Stern effects give rise to a decreased capacity of the double layer (as mentioned in the introductory section), but unfortunately it appears impossible to measure the electric capacity of the double layer on clays.

Recently Ravina and Zaslavski (1972) pointed to the occurrence of electrostriction of the solvent water when exposed to electric fields of high intensity, F_e. The latter may reach values in excess of 10^8 V/m in the first few Å adjacent to a clay surface (cf. (1.36)). The corresponding pressure increase calculated at about 10^3 bar by the above authors should be regarded as speculative, however, because of the use of the Clausius-Mosotti equation for this calculation. Nevertheless the liquid pressure very close to the clay surface will probably exceed (considerably) the value of about 100 bar induced locally by the osmotic effect of the accumulated counterions.

The influence of such high pressures on the distribution of the counterions in a homoionic double layer is likely to be minor in comparison with the effect of ionic interactions mentioned before. At the same time it will again be of much greater concern when considering ionic preferences in systems. It is then interesting to note that the proportion of the pressure increase due to electrostriction with F_e^2 will cause its effect to be closely correlated with the direct effect of the field intensity on the ionic polarisation, which is also proportional to F_e^2. This is particularly so because the ionic polarisation is roughly proportional to the ionic volume (cf. Ravina and De Bock, 1974).

1.7. SUMMARY OF THE PERSPECTIVES OF DL THEORY FOR CLAY COLLOIDS

The size and shape of the particles and the magnitude of the constant substitution charge of many clays offer interesting perspectives for the application of the (corrected) Gouy-Chapman theory towards calculation of the overall ionic distribution in the diffuse double layer. The shortcomings of the theory are often of minor concern in view of the accuracy needed. In that case a direct reading of the relative ionic strength at the terminal plane, U_t, as a function of the scaled distance parameter is amply sufficient (Fig. 1.1). For quick guesses eqn. (1.23) will be satisfactory, where replacing x_t by d_1 should often suffice. The actual course of the potential curve, or the Boltzmann factor, within the double layer may then be sketched with Fig. 1.3. This procedure should cover most types of ionic mixtures of interest. For freely developed double layers eqns. (1.32) and (1.33) and those given in Table 1.1 are easy to apply.

Although the construction of curves showing the distribution of ions in the double layer contributes to an overall picture, it seems often not warranted to do so for practical purposes. Usually approximating equations like (1.23) and (1.34) suffice to establish some limiting values.

As to the practical application, three aspects could be mentioned. Starting with the least promising one, it will be clear that DL theory as presented here offers little help in predicting exchange equations, except for the effect of valence (section 3.2 of chapter 3). Ion specificity is dominated to such a degree by the situation in the "first" layer that uncertainty prevails.

In contrast the calculation of co-ionic exclusion phenomena appears to be the most promising application. The virtual absence of co-ions in the first layer avoids the difficulties noted above. As long as the extent of the double layer is virtually unaltered, calculations will be very precise (cf. chapter 7).

Related to co-ionic exclusion are salt-sieving and other coupling phenomena (see chapter 11) which may also be tackled successfully with DL theory (cf. Kemper, 1960; Groenevelt and Bolt, 1969).

Finally the efforts to calculate swelling pressures in clays are mentioned (Schofield, 1946; Bolt and Miller, 1955; Bolt, 1956; Warkentin et al., 1957; Aylmore and Quirk, 1959, 1960a, b; Warkentin, 1960; Blackmore and Warkentin, 1960; Blackmore and Miller, 1961; Collis George and Bozeman, 1970). The results are somewhat ambiguous. Though the mechanism of the osmotic swelling of clays is explained, quantitative predictions are difficult in general. It appears that the uncertainty in this case is predominantly one of geometry in a clay paste: plate condensation, non-parallel arrangement in dense pastes make the use of an average value for d_1, based on moisture content and assumed surface area, questionable.

Returning to the application of co-ionic exclusion it is pointed out that the latter has been used to great advantage in determining the surface area carrying freely extended double layers (Schofield, 1947; Schofield and Talibuddin, 1948; Bolt and Warkentin, 1956; Bolt and Warkentin, 1958; Edwards and Quirk, 1962; De Haan, 1965; Edwards et al., 1965). This method appears rather dependable, particularly as one may often select a counterion not exhibiting reduction of the extent of the double layer because of extreme accumulation in the first layer.

REFERENCES

Aylmore, L.A.G. and Quirk, J.P., 1959. Swelling of Clay-Water systems. Nature, 183: 1752—1753.

Aylmore, L.A.G. and Quirk, J.P., 1960a. Domain or turbostratic structure of clays. Nature, 187: 1046—1048.

Aylmore, L.A.G. and Quirk, J.P., 1960b. The structural status of clay systems. Proc Natl. Clay Conf., 9: 104—130.

Bateman, H., 1953. In: A. Erdelyi (Editor), Higher Transcedental Functions, II. McGraw-Hill, New York, N.Y., 396 pp.

Blackmore, A.V. and Miller, R.D., 1961. Tactoid size and osmotic swelling in calcium montmorillonite. SSSA Proc., 25: 169—173.

Blackmore, A.V. and Warkentin, B.P., 1960. Swelling of calcium montmorillonite. Nature, 186: 823—824.

Bolt, G.H., 1955. Analysis of the validity of the Gouy-Chapman theory of the electric double layer. J. Colloid Sci., 10: 206—218.

Bolt, G.H., 1956. Physico-chemica analysis of the compressibility of pure clays. Geotechnique, 6: 86—93.

Bolt, G.H., and Miller, R.D. 1955. Compression studies of illite suspensions. SSSA Proc., 19: 285—288.

Bolt, G.H. and Warkentin, B.P., 1956. Influence of the method of sample preparation on the negative absorption of anions in montmorillonite suspensions. Proc. 6th Int. Congr. Soil Sci., Paris, B: pp. 33—40.

Bolt, G.H. and Warkentin, B.P., 1958. The negative adsorption of anions by clay suspensions. Kolloid Z., 156: 41—46.

Bresler, E., 1972. Interacting diffuse layers in mixed mono-divalent ionic systems. SSSA Proc., 36: 891—896.

Chapman, D.L., 1913. A contribution to the theory of electrocapillarity. Philos. Mag., 6 (25): 475—481.

Collis George, N. and Bozeman, J.M., 1970. A double layer theory for mixed ion systems as applied to the moisture content of clays under restraint. Aust. J. Soil Res., 8: 239—258.

De Haan, F.A.M., 1965. Determination of the specific surface area of soils on the basis of anion exclusion measurements. Soil Sci. 99: 379—386.

Edwards, D.G. and Quirk, J.P., 1962. Repulsion of chloride by montmorillonite. J. Colloid Sci., 17: 872—882.

Edwards, D.G., Posner, A.M. and Quirk, J.P., 1965. Repulsion of chloride ions by negatively charged clay surfaces. Trans. Faraday Soc., 61: 2808—2823.

Gouy, G., 1910. Sur la Constitution de la charge électrique à la surface d'un électrolyte. J. Phys., 9: 457—468.

Grahame, D.C., 1947. The electrical double layer and the theory of electrocapillarity. Chem. Rev., 41: 441—501.

Groenevelt, P. and Bolt, G.H., 1969. Permiselective properties of porous materials as calculated from diffuse double layer theory. IAHR Symp. Haifa, 241—255.

Kemper, W.D., 1960. Water and ion movement in thin filters as influenced by the electrostatic charge and diffuse layer of cations associated with clay mineral surfaces. SSSA Proc., 24: 10—16.

Kruyt, H.R., 1952. Colloid Science, vol. I. Elsevier, Amsterdam, 389 pp.

Langmuir, I., 1938. The role of attractive and repulsive forces in the formation of tactoids, thixotropic gels, protein crystals and coacervates. J. Chem. Phys., 6: 873—896.

Loeb, A., 1951. An interionic attraction theory applied to the diffuse layer around colloid particles. J. Colloid Sci., 6: 75—91.

Ravina, I. and Zaslavski, D., 1972. The water pressure in the electrical double layer. Isr. J. Chem., 19: 707—714.

Ravina, I. and De Bock, J., 1974. Thermodynamics of ion exchange. Soil Sci. Soc. Am. Proc., 38: 45—49.

Schofield, R.K., 1946. Ionic forces in thick films of liquid between charged surfaces. Trans. Faraday Soc., 42B: 219—228.

Schofield, R.K., 1947. Calculations of Surface Areas from measurements of negative adsorption. Nature, 160: 408—410.

Schofield, R.K. and Talibuddin, O., 1948. Measurement of the internal surface by negative adsorption. Discuss. Faraday. Soc., 3: 51—56.

Stern, O., 1924. Zur Theorie der elektrischen Doppelschicht. Z. Elektrochem., 30: 508—516.

Verwey, E.J.W. and Overbeek, J.Th.G., 1948. Theory of the Stability of Lyophobic Colloids. Elsevier, Amsterdam, 205 pp.

Warkentin, B.P., Bolt, G.H. and Miller, R.D., 1957. Swelling pressure of montmorillonite. SSSA Proc., 21: 495—497.

Warkentin, B.P. 1960. Swelling of Calcium montmorillonite. Nature, 186: 823—824.

CHAPTER 2

THERMODYNAMICS OF CATION EXCHANGE

G.H. Bolt

A cation exchange reaction involving the cations A and B, with valence z_A and z_B, respectively, may be formally presented as:

$$z_A^{-1}\,\bar{A} + z_B^{-1}\,B \rightleftharpoons z_B^{-1}\,\bar{B} + z_A^{-1}\,A \qquad (2.1)$$

in which the bar denotes the ion in the adsorbed phase. In this equation the reaction coefficients have been taken as being equal to the reciprocal valence of the ions, thus specifying the reaction per equivalent of the exchanger. Referring to the relevant discussion in chapter 2 of part A (cf. (A2.24)) one then finds that the "reduced" sum of the partial molar free enthalpies of reactants and products, $\Sigma_k' \bar{G}_k / z_k$, equals zero at equilibrium. Further working-out of this equilibrium condition necessitates the introduction of standard states of reactants and products, which then lead to the definition of the activities of these (cf. (A2.25)). In the present case, involving both a solution phase and an absorbed phase, this matter deserves some further attention.

2.1. SIGNIFICANCE OF ACTIVITIES AND ACTIVITY COEFFICIENTS OF CATIONS IN AN ADSORBED PHASE

The partial molar free enthalpy of a constituent k in an aqueous solution phase β is usually referred to as its chemical potential, μ_k^β. This chemical potential is (at least) a function of P, T and the phase composition. Separating out the effect of the solution composition one may put formally (cf. eqn. A2.25 in part A):

$$\mu_k^\beta \equiv \mu_k^\circ + RT \ln a_k^\beta \qquad (2.2)$$

where μ_k° is the value of μ_k in a chosen standard state, i.e. a system at the given pressure and temperature but with a specified composition. The activity a_k defined above is then a *relative activity* (cf. Guggenheim, 1967, § 4.12) covering the composition-dependent part of μ_k relative to the same one in the chosen standard state.

Introducing a suitable composition parameter, x_k, the activity is then formally related to this parameter via an activity coefficient $f_k \equiv a_k / x_k$, according to:

$$\mu_k^\beta = \mu_k^\circ + RT \ln f_k + RT \ln x_k \qquad (2.3)$$

The significance of f_k belonging to a given system thus depends upon the particular choice of standard state and on the composition variable employed.

For dissolved constituents the composition variable commonly used is their molal concentration. The use of the molar concentration is also satisfactory if pressure and temperature of the systems considered are constant. The standard state of the dissolved constituent is then specified as "a hypothetical unit molal (c.q. molar) solution exhibiting ideal behavior" which corresponds to a particular actual concentration. It should finally be pointed out that single ionic potentials are thermodynamically indeterminate. This is for the present discussion of no concern, however, as in case of ion exchange the application of (2.2) always leads to combinations of single ionic potentials corresponding to a zero transfer of charge.

Applying (2.2) to an adsorbed phase α some additions should be considered. While the aqueous solution phase β mentioned above was taken at a temperature and pressure which served to define the standard state of the solute, an adsorbed phase in contact with an aqueous solution may have a pressure which differs from the chosen standard pressure of the aqueous solution. Similarly, in case of ion adsorption, the electric potential in the adsorbed phase may differ from the assumed zero value in the aqueous solution. Indicating the partial molar free enthalpy of an ion in the adsorbed phase, subject to these phase-specific effects, with the symbol $\tilde{\mu}$ one could formally write something like (cf. Guggenheim, 1967, § 8.05):

$$\tilde{\mu}_k^\alpha \equiv \tilde{\mu}_k^{o\alpha} + RT \ln f_k^\alpha + RT \ln x_k^\alpha + z_k F \psi^\alpha + \bar{V}_k^\alpha P^\alpha + \ldots \qquad (2.4)$$

where the partial molar volume $\bar{V}_k$ has been taken as a constant. If the above adsorbed phase α is in equilibrium with a phase β, following (2.3), the equality $\tilde{\mu}_k^\alpha = \mu_k^\beta$ implies that the value of f_k^α defined by (2.4) depends (at least) on the specification of ψ^α, P^α, x_k^α and $\tilde{\mu}_k^{o\alpha}$. In general the complete information sought is not accessible for experimental determination and thus one takes recourse to a recombination of terms depending on the particular system. This then gives rise to many different definitions of the activity coefficient in the adsorbed phase. A few examples of such definitions will be given below, starting with a rather simple situation.

2.1.1. Homogeneous Donnan system

If phase α consists of an aqueous solution containing a fixed amount of a high molecular negatively charged material which is evenly distributed over α and prevented from entering phase β by a suitable membrane it seems practical and warranted to use the same standard state for a cation k in both phases. Using also the ionic concentration as the composition parameter in phase α, one then finds for (2.4):

$$\tilde{\mu}_k^\alpha = \mu_k^\circ + z_k F\psi^\alpha + \bar{V}_k P^\alpha + RT\ln f_k^\alpha + RT\ln {}^m c_k^\alpha \quad (2.4a)$$

where ${}^m c_k^\alpha$ indicates the molar (c.q. molal) concentration in phase α. The parameters P^α and ${}^m c_k^\alpha$ may be evaluated experimentally, while f_k^α could perhaps be approximated with e.g. a Davies-type equation. In that case the potential ψ^α for a phase α in equilibrium with an aqueous solution phase β may be estimated by using $\tilde{\mu}_k^\alpha = \mu_k^\beta$. Such a procedure then amounts to making a distinction between ionic interactions, characterized by f_k^α, and a phase-specific effect attributed to ψ^α and P^α. Its validity hinges on one's belief in the applicability of equations specifying single ionic activity coefficients in both phases α and β.

The reverse procedure, i.e. an attempt to determine experimentally the value of ψ^α by measuring the potential difference between reference calomel electrodes inserted in the phases α and β, was the basis of studies directed towards finding ion activities in clay and soil suspensions (Peech and Scott, 1950). As will be discussed in chapter 11, however, the potential difference determined in this manner should be interpreted in terms of ionic transport numbers within the phase α rather than in terms of the ratio of ionic activities in both phases.

Also a considerable number of publications by C.E. Marshall and collaborators, summarized in Marshall (1964), dealt with an attempt to determine $f_k^\alpha \, {}^m c_k^\alpha$ as defined by (2.4a). Similar to the standard determination of pH by insertion of a glass- and a calomel-electrode into a soil suspension, the value of pM was determined with clay-membrane electrodes reversible to different M-ions. The corresponding value a_M^α was then divided by the known total molar concentration of M-ions in the suspension to yield f_M^α, indicated as the "fraction active" of the M-ions in the suspension. Here, again, the presence of a KCl-junction of the calomel electrode in the suspension invalidates the interpretation of the results in terms of ion-activities in the suspension. In addition, the use of the total amount of M-ions present per unit volume of suspension to obtain the coefficient f^α obscures its interpretation as a measure of the interaction between exchanger and ion: dilution of a suspension with its own equilibrium solution will generally cause a change of f^α.

With reference to the relevant and clear discussion of the above system by Babcock (1963, pp. 461—463) it should be pointed out that the suggested distinction between intra-phase ionic interaction and the interphase effect in terms of ψ^α offers no perspectives for the direct evaluation of a_k^α. It only serves to explain that a_k^α could exceed a_k^β for a cation k if phases α and β are brought to equilibrium, but since ψ^α cannot be evaluated, the equality $\tilde{\mu}_k^\alpha = \mu_k^\beta$ yields no information about a_k^α / a_k^β.

The electric potential ψ^α must be associated with a minute separation of charges across the membrane separating phases α and β. Such a charge separation is established instantaneously on bringing phase α in contact with its equilibrium solution across a membrane preventing the crossing of the high molecular, negatively charged particles present in phase α. Since now the activity of the salt present, say KCl, is the same in phases α and β, while $c_K^\alpha > c_K^\beta$ and $c_{Cl}^\alpha < c_{Cl}^\beta$, a double layer is formed extending over some Å units within the membrane, containing a slight excess of K^+ over Cl^- on the side bordering phase β, and a slight excess of negative charge on the side bordering phase α.

Obviously the value of ψ^α becomes immaterial when the reduced ratio of the activities of two cation species is considered. If the above system contains two cation species, the distribution of these over the two phases may be regarded as an exchange equilibrium. Because of the chosen identity of the standard states of the ions in both phases, the standard free enthalpy of the exchange reaction following (2.1) equals zero (cf. part A, eqn. A 2.27). Constructing the expression for K°_{ex} then gives:

$$K^\circ_{ex} = 1 = \frac{(f^\alpha_B\, {}^m c^\alpha_B)^{1/z_B}}{(f^\alpha_A\, {}^m c^\alpha_A)^{1/z_A}} \; \frac{(f^\beta_A\, {}^m c^\beta_A)^{1/z_A}}{(f^\beta_B\, {}^m c^\beta_B)^{1/z_B}} \exp\left\{\frac{P^\alpha(\bar{V}_B/z_B - \bar{V}_A/z_A)}{RT}\right\} \qquad (2.5)$$

where the terms containing ψ^α cancel each other. Taking out the parameters which may in principle be determined experimentally, viz.:

$f^\beta_k\, {}^m c^\beta_k \equiv a^\beta_k$, ${}^m c^\alpha_k$ and $\Delta(\bar{V}_k/z_k) \equiv (\bar{V}_B/z_B - \bar{V}_A/z_A)$

one may express the reduced ratio of the activity coefficients in the absorbed phase as:

$$RT \ln \frac{(f^\alpha_B)^{1/z_B}}{(f^\alpha_A)^{1/z_A}} = - RT \ln K_c - P^\alpha \Delta(\bar{V}_k/z_k) \qquad (2.6)$$

in which K_c signifies a selectivity coefficient based on molar concentrations in the adsorbed phase, according to:

$$K_c \equiv \frac{({}^m c^\alpha_B)^{1/z_B} (a^\beta_A)^{1/z_A}}{({}^m c^\alpha_A)^{1/z_A} (a^\beta_B)^{1/z_B}} \qquad (2.7)$$

In summary, the present choice of the standard state of the ions in the adsorbed (Donnan) phase leads to $K^\circ_{ex} = 1$, while K_c approaches unity for ideally behaving ions, with a negligible pressure-term contribution. Although this formulation has also been used to describe the ion exchange equilibrium in soil (cf. Wiklander, in Bear, 1964), the necessary translation of "amounts adsorbed" into concentration in an adsorbed phase involves another assumption about the volume of the adsorbed phase, V^α. As the latter is neither known nor constant the application of (2.6) to soil systems offers little perspective with regard to predictions.

2.1.2. Resinous exchangers

If phase α consists of the interior of resin particles bathed in excess solution (phase β), one might still prefer to treat it as a modified aqueous solution, again accepting identical standard states for both phases. Eqn. (2.4a) is then retained, with the understanding that the pressure term may become considerable, while it escapes experimental determination. The same holds for the phase potential ψ^α. Finally the ionic composition inside the resin particle hardly resembles an aqueous salt solution (co-ions of the resin are likely to be virtually absent). Accordingly there is little sense in separating the effects of P^α, ψ^α and the ionic interactions. It then appears logical to combine these three effects in terms of a "total" activity coefficient, $\tilde{f}_k^\alpha$, and (2.4a) becomes:

$$\tilde{\mu}_k^\alpha = \mu_k^\circ + RT \ln \tilde{f}_k^\alpha + RT \ln {}^m c_k^\alpha \qquad (2.4b)$$

The corresponding total activity, $\tilde{a}_k^\alpha \equiv \tilde{f}_k^\alpha c_k^\alpha$, now equals the activity in the equilibrium solution, a_k^β, and is thus by itself of little interest. The selectivity coefficient K_c, defined in (2.7), then equals the reduced ratio of the coefficients $\tilde{f}^\alpha$. This ratio covers, in addition to the ionic interaction term, the pressure term, $\exp\{P^\alpha \Delta(\bar{V}_k/z_k)/RT\}$, and the effect of ion-specific interactions between ions and resin. As was the case with (2.6), the non-specific electrostatic term has canceled out.

In resin systems the total interior concentration is usually very high and rather insensitive towards the concentration in the equilibrium solution. In that case P^α, which corresponds to the difference in osmotic pressure between the phases α and β (because of the solvent equilibrium), will vary with the exterior concentration in a roughly predictable manner; this variation should be reflected in the variation of K_c with exterior electrolyte level. It may be noted that V^α, here the interior volume of resin particles, is usually known (roughly) enabling one to relate the interior concentration to the amount adsorbed.

2.1.3. Adsorbed phases that show little resemblance to aqueous solutions

This group may be considered as the left-over from the previous systems and comprises e.g. non-aqueous liquid phases and solid phases with surface adsorption. Actually it is often a matter of personal insight whether or not the adsorbed phase is thought to resemble a (modified) aqueous solution. Perhaps the basic contention here is the fact that the thermodynamic potential of the ion in the phase is determined chiefly by a direct interaction between the ion and some exchanger matrix in competition with other ion species and not, as in the previous group, by its concentration in an aqueous phase subject to adsorber derived effects like the presence of an electric field and/or excess hydrostatic pressure.

The above description indicates the direction in which (2.4) should be modified. If exchanger sites combined with an adsorbed ion are recognized as a chemical species, the single ionic potential of the adsorbed ion must be regarded as an indeterminate part of the partial molar free enthalpy of this uncharged species, $\bar{G}_{\bar{k}}$. So if one writes this part as:

$$\bar{\mu}_k \equiv \bar{\mu}_k^\circ + RT \ln \bar{a}_k \qquad (2.4c)$$

the present potential $\bar{\mu}_k$ should be distinguished from the previous one, $\tilde{\mu}_k^\alpha$, which derived its significance from a comparison between two aqueous solutions residing in the phases α and β, respectively, taking $\tilde{\mu}_k^\alpha = \mu_k^\beta$. In the present case $\bar{\mu}_k^{\alpha}$ is indeterminate with respect to μ_k^{β}. This is of no concern, however, if one wants to describe the exchange equilibrium between adsorbed phase and solution. Thus the difference $\bar{\mu}_B/z_B - \bar{\mu}_A/z_A$ is thermodynamically determinate for a given adsorbed phase, both ions A and B being associated with the same exchanger. Applying this to the exchange equilibrium following (2.1) gives:

$$\bar{\mu}_B/z_B - \bar{\mu}_A/z_A = \mu_B^{\beta}/z_B - \mu_A^{\beta}/z_A \tag{2.8}$$

Eqn. (2.4c) now requires the specification of a standard state for the ions in the adsorbed phase. Having abandoned the description in terms of an aqueous solution it seems logical to select the homoionic exchanger for this purpose. If the adsorbed phase also contains solvent molecules it is necessary to specify their activity as well. Following Gaines and Thomas (1953), the standard state of the adsorbed ions is conveniently taken as the homoinic exchanger in equilibrium with solvent at unit activity. In practice this is the homoionic exchanger in equilibrium of an infinitely dilute solution of a salt of the same cation.

In a discussion of the appropriate standard state for adsorbed ions Babcock (1963, p. 501) pointed to the fact that the finite solubility of silicate exchangers prevents the physical realization of e.g. a Na-clay in equilibrium with an infinitely dilute solution of NaCl. As was indicated by Gaines and Thomas (1953) this does not impede the definition of the standard state as such (cf. the discussion of the "solvent term" in the following section and in chapter 3). The difficulties inferred by Babcock (1963) when suggesting the use of a finite salt concentration for the definition of the standard state may be traced to the inappropriate use for the present system of an equality $\bar{\mu}_k \equiv \mu_k^{\beta}$.

The appropriate composition variable for a mixed exchanger should then be the mole fraction, M_k, or equivalent fraction adsorbed, N_k. Taking the latter as a more convenient one, the corresponding activity coefficient in the adsorbed phase becomes:

$$\bar{f}_k \equiv \bar{a}_k/N_k \tag{2.9}$$

Applying the above to the exchange equilibrium specified by (2.8) one finds:

$$-(\bar{\mu}_B^{\circ}/z_B - \bar{\mu}_A^{\circ}/z_A) + (\mu_B^{\circ}/z_B - \mu_A^{\circ}/z_A)$$
$$= RT \ln \frac{(N_B)^{1/z_B}(a_A)^{1/z_A}}{(N_A)^{1/z_A}(a_B)^{1/z_B}} \frac{(\bar{f}_B)^{1/z_B}}{(\bar{f}_A)^{1/z_A}} \tag{2.10}$$

where β has been omitted, as the phases are unambiguously indicated. The LHS of this equation may be identified by $-\Delta G_{ex}^{\circ}$, specified per equivalent of the exchanger, in the direction A → B.

This standard free enthalpy of the exchange reaction could be interpreted as the difference between the free enthalpy of one equivalent of B-ions and one equivalent of A-ions when both ions are in the adsorbed state at unit equivalent fraction, compared with the same difference when both ions are present in "ideally behaving unit molar electrolyte solutions", i.e. in their standard state in solution (cf. also section 2.3 of this chapter). The RHS of (2.10) contains the empirically accessible selectivity coefficient on equivalent fraction basis, K_N, multiplied by the "reduced" ratio of the activity coefficients in the adsorbed phase, so:

$$-\Delta G^{\circ}_{ex}/RT = \ln\left[K_N \frac{(\bar{f}_B)^{1/z_B}}{(\bar{f}_A)^{1/z_A}}\right] \tag{2.11}$$

Whereas on first sight the activity coefficients $\bar{f}$ might appear to be purely hypothetical parameters inaccessible for experimental evaluation, it was shown by Gaines and Thomas (1953) that a combination of (2.11) with the Gibbs-Duhem equation for the adsorbed phase allows in principle their calculation from a complete set of experimental data (cf. following section). At the same time it will be clear that the coefficients f^{α}_k (and the fraction active as used by Marshall (1964)), $\tilde{f}^{\alpha}_k$ and $\bar{f}_k$ have a totally different meaning and should not be compared numerically with each other.

A final word may be said about the experimentally accessible selectivity coefficient K_N, its counterpart K_M defined in terms of mole fractions adsorbed, and K_c. The first two are obtained directly from the composition of the exchanger and the equilibrium solution, the third one requires knowledge of the volume of the adsorbed phase. For resins the latter may usually be estimated, provided the co-ion penetration into the resin particle is negligible (cf. also p. 52) and will be indicated here as V^{α} per equivalent. The three coefficients are then related according to:

$$K_c = K_N \frac{(z_A V^{\alpha})^{1/z_A}}{(z_B V^{\alpha})^{1/z_B}} \tag{2.12}$$

$$K_M = K_N \frac{(N_A + N_B z_A/z_B)^{1/z_A}}{(N_B + N_A z_B/z_A)^{1/z_B}} \tag{2.13}$$

and become identical for the homovalent exchange equilibrium. For $z_A = 1$ and $z_B = 2$ these relations become $K_c = K_N\sqrt{V^{\alpha}/2}$ and $K_M = K_N\sqrt{(N_A + 1)/4}$, respectively.

2.2. CALCULATION OF THE STANDARD FREE ENTHALPY OF THE CATION EXCHANGE REACTION FROM EXPERIMENTAL DATA

In the previous section a selectivity coefficient on equivalent fraction basis was introduced for the exchange reaction (2.1):

$$K_N \equiv \frac{(N_B)^{1/z_B}(a_A)^{1/z_A}}{(N_A)^{1/z_A}(a_B)^{1/z_B}} \tag{2.14}$$

The reduced ratio of the solution phase activities, a, in this equation may be expressed in terms of the relevant molar concentrations with the help of mean activity coefficients of the corresponding salts, according to:

$$\frac{(a_A)^{1/z_A}}{(a_B)^{1/z_B}} = \frac{({}^m c_A)^{1/z_A}}{({}^m c_B)^{1/z_B}} \cdot \frac{(f_{\pm,A})^{\Sigma\nu/\nu_A z_A}}{(f_{\pm,B})^{\Sigma\nu/\nu_B z_B}} \tag{2.15}$$

in which $\Sigma\nu$ indicates $(\nu_+ + \nu_-)$ for the A-salt and B-salt, respectively (cf. chapter 2 of part A, section A 2.3). As the equivalent fraction adsorbed N_k, is also experimentally accessible (cf. below), K_N may be evaluated from experimental data on exchange equilibria. As a rule K_N will vary with the value of N_k and often it will also depend on the electrolyte level of the equilibrium solution.

Introducing the thermodynamic equilibrium constant of the exchange reaction as $\ln K^\circ_{ex} \equiv -\Delta G^\circ_{ex}/RT$, one finds according to (2.11):

$$K^\circ_{ex} = K_N \frac{(\bar{f}_B)^{1/z_B}}{(\bar{f}_A)^{1/z_A}} \tag{2.16}$$

The principle of the procedure used by Gaines and Thomas (1953) to derive K°_{ex} from a complete set of values of K_N, is easiest demonstrated with a simplified example. For this purpose an adsorbed phase is visualized containing solely a mixture of solvated monovalent cations to a fixed total amount. The ions are subject to binding forces of the adsorber and their activity in the adsorbed phase, $\bar{a}_k$, as defined according to (2.4c), will be a function of N_k, according to:

$$\bar{a}_k = \bar{f}_k N_k \tag{2.17}$$

with $\bar{f}_k \to 1$ for $N_k \to 1$, in accordance with the chosen homoionic standard state.

If now the value of K_N is determined for a complete range of equilibrium systems, for the present case stretching from $N_B \approx 0$ to $N_B \approx 1$, a plot of $\ln K_N$ against N_B may be made (cf. Fig. 2.1). From (2.16) one may ascertain that the slope of $\ln K_N$ with N_B must correspond to the divergence of the lines presenting $\ln \bar{f}_A$ and $\ln \bar{f}_B$ against N_B (for the present homovalent case):

$$d \ln K_N = d \ln \bar{f}_A - d \ln \bar{f}_B \tag{2.18}$$

Furthermore the activity coefficients are subject to the Gibbs-Duhem equation for the adsorbed phase, which for the present system containing only two species of solvated monovalent cations reads:

$$N_A\, d \ln \bar{f}_A N_A + N_B\, d \ln \bar{f}_B N_B = 0 \tag{2.19}$$

With $N_A + N_B = 1$ this gives:

$$N_A\, d \ln \bar{f}_A + N_B\, d \ln \bar{f}_B = 0 \tag{2.20}$$

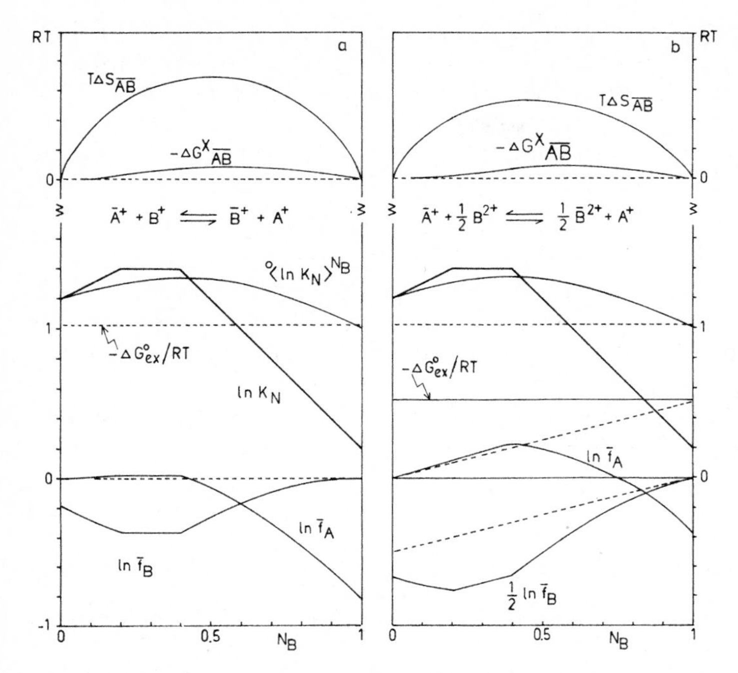

Fig. 2.1. Construction of $\ln \bar{f}_A$, $\ln \bar{f}_B$, ΔG°_{ex}, $\Delta G^x_{\overline{AB}}$ pertaining to a given (arbitrary) course of $\ln K_N$ (as a function of N_B), in the absence of a significant solvent term.
a. For mono-monovalent exchange.
b. For mono-divalent exchange.

Combination of this equation with (2.18) allows elimination of either activity coefficient. Using $N_A = 1 - N_B$ yields the differential equations:

$$d \ln \bar{f}_A = N_B \, d \ln (\bar{f}_A / \bar{f}_B) = N_B \, d \ln K_N \tag{2.21a}$$

$$d \ln \bar{f}_B = - N_A \, d \ln K_N \tag{2.21b}$$

subject to the condition $\bar{f}_k = 1$ for $N_k = 1$.
Integration of (2.21a) from $N_B = 0$ to $N_B = 1$ then gives:

$$\ln \bar{f}_A (N_B = 1) = \ln K_N - \int_{N_B=0}^{1} \ln K_N \, dN_B \tag{2.22}$$

Remembering that at $N_B = 1$, $\bar{f}_B = 1$, one finds with (2.16):

$$\ln K^\circ_{ex} = \ln K_N - \ln \bar{f}_A (N_B = 1) = \int_0^1 \ln K_N \, dN_B \tag{2.23}$$

For the present simplified case $\ln K^\circ_{ex}$ thus equals the *mean* value of $\ln K_N$ over the complete range of N_B, i.e. $\langle \ln K_N \rangle$. Furthermore, integration of (2.21) over part of the range of N_B gives:

$$\ln \bar{f}_A = N_B \ln K_N - \int_0^{N_B} \ln K_N \, dN_B \tag{2.24a}$$

$$\ln \bar{f}_B = -(1-N_B)\ln K_N + \int_{N_B}^{1} \ln K_N \, dN_B \qquad (2.24b)$$

In Fig. 2.1a the above "translation" of an arbitrarily chosen "curve" for $\ln K_N$ into the corresponding curves for $\ln \bar{f}_k$ is shown. As follows from (2.20), both curves must have slopes of opposite sign everywhere. At the homoionic endpoints the limiting slope of the corresponding ion must always approach zero; at $N = 0.5$ the slopes are equal and opposite. The slopes of $\ln \bar{f}_k$ for $N_k \to 0$ must approach that of $\ln K_N$, where for the chosen direction of $A \to B$, $\ln \bar{f}_A$ follows $\ln K_N$, whereas the slope of $\ln \bar{f}_B$ is always opposite to that of $\ln K_N$ (cf. (2.16)). When $\ln K_N$ intersects its mean value, representing $\ln K^{\circ}_{ex}$, the same equation indicates that at this point the activity coefficients of both ions are equal. If finally K_N is constant throughout, its value corresponds to K°_{ex} and $\ln \bar{f}_k$ will be zero for both ions over the entire range (cf. the dashed lines in Fig. 2.1a).

This situation might be termed ideal homovalent exchange; it implies that the energy of interaction between neighboring ions $\bar{A}$ and $\bar{B}$ in the adsorbed phase equals the mean value of that for the pairs $\bar{A}$—$\bar{A}$ and $\bar{B}$—$\bar{B}$ (cf. chapter 4). The free enthalpy of mixing of the ions in the adsorbed phase then equals $-T\Delta S_{mix} = \{(N_A/z)\ln N_A + (N_B/z) \ln N_B\} RT$, so the excess free tnthalpy of mixing is zero over the entire range. Of course the energy of interaction between the ions and the adsorber will still be different if $K_N = K^{\circ}_{ex} \neq 1$ (cf. section 2.3).

Extending the above treatment to heterovalent exchange one finds:

$$\ln K^{\circ}_{ex} = -\left(\frac{1}{z_A} - \frac{1}{z_B}\right) + \int_{0}^{1} \ln K_N \, dN_B \qquad (2.25)$$

The activity coefficients are then given by:

$$\ln (\bar{f}_A)^{1/z_A} = N_B \left[\left(\frac{1}{z_A} - \frac{1}{z_B}\right) + \ln K_N\right] - \int_{0}^{N_B} \ln K_N \, dN_B \qquad (2.26a)$$

$$\ln (\bar{f}_B)^{1/z_B} = -(1-N_B)\left[\left(\frac{1}{z_A} - \frac{1}{z_B}\right) + \ln K_N\right] + \int_{N_B}^{1} \ln K_N \, dN_B \qquad (2.26b)$$

The term corresponding to the difference between the reciprocal valencies arises from the Gibbs-Duhem equation, which for heterovalent exchange gives:

$$N_A d \ln (\bar{f}_A)^{1/z_A} + N_B d \ln (\bar{f}_B)^{1/z_B} = \left(\frac{1}{z_A} - \frac{1}{z_B}\right) dN_B \qquad (2.27)$$

It should be noted that because of the presence of this term, $\ln \bar{f}_k$ remains a function of the exchanger composition even if K_N is constant over the entire range of N (cf. also chapter 3, Figs. 3.2 and 3.5).

In Fig. 2.1b the relevant curves of $\ln (\bar{f}_k)^{1/z_k}$, calculated from the arbitrary curve for $\ln K_N$ as used in Fig. 2.1a have been plotted, but taking $z_A = 1$ and $z_B = 2$. These curves deviate from their "ideal" ones in the same way as in Fig. 2.1a. These "ideal" curves, however, are now varying linearly with N_B, in accordance with the first term of (2.26a) and (2.26b). This feature is the result of the particular choice of activity coefficients coupled with equivalent fractions in conjunction with the definition of ideal behaviour of heterovalent exchange corresponding to a constant value of K_N. In section 2.3 and in chapter 4 this aspect wil be discussed in greater detail.

As was pointed out by Gaines and Thomas (1953), the above approach still suffers from oversimplification. Treating the exchanger as a phase containing only solvated cations is not realistic: at least it must contain also solvent molecules. Then the Gibbs-Duhem equation, at constant P and T, must be augmented to:

$$\frac{N_A}{z_A} d \ln \bar{f}_A + \frac{N_B}{z_B} d \ln \bar{f}_B + n_s d \ln a_s = \left(\frac{1}{z_A} - \frac{1}{z_B}\right) dN_B \tag{2.28}$$

in which n_s is the number of moles of solvent present in one equivalent of exchanger. At the same time the definition of the standard state of the cations in the absorbed state must now be specified with regard to the state of the solvent molecules, which varies with the electrolyte concentration in the equilibrium solution of the homoionic exchanger. Gaines and Thomas (1953) suggested the definition of the standard state of the absorbed cation as the one corresponding to a homoionic exchanger in equilibrium with an infinitely dilute solution of that ion, i.e. solvent molecules at unit activity. Combining (2.28) with the differential of (2.16) yields in this case:

$$d \ln (\bar{f}_A)^{1/z_A} = N_B d \ln K_N + \left(\frac{1}{z_A} - \frac{1}{z_B}\right) dN_B - n_s d \ln a_s \tag{2.29a}$$

and a similar extension of (2.21b) for $d \ln \bar{f}_B$. In order to arrive at the value of $\ln \bar{f}_A$ at the standard state of the adsorbed ion B (i.e. $N_B = 1$, $a_s = 1$) one may now integrate (2.29a) in three steps, viz. (a) from $N_A = 1$, $a_s = 1$ to $N_A = 1$ at finite salt level of the equilibrium solution; (b) from $N_A = 1$ to $N_B = 1$, preferably at constant value of $a_s \neq 1$; (c) from $N_B = 1$, $a_s \neq 1$ to $N_B = 1$, $a_s = 1$ (cf. Fig. 2.2). For the chosen integration path the last term of (2.29a) drops out in step (b), which then results in $\Delta \ln \bar{f}_A$ in that step in accordance with (2.26), $\ln K_N$ now to be specified for the chosen value of a_s. In addition one finds for step (a): $\Delta \ln (\bar{f}_A)^{1/z_A} = -\int_1^{a_s} n_s(A) d \ln a_s$ where $n_s(A)$ refers to the homoionic A-exchanger when the solvent activity is decreased from unity to a_s. Similarly step (c) gives $\Delta \ln (\bar{f}_A)^{1/z_A} = -\int_1^{a_s} n_s(B) d \ln a_s$, where $n_s(B)$ indicates the solvent content of the homoionic B-exchanger during the increase of the solvent activity from a_s to unity. Reconstructing (2.25) for this case then gives:

$$\ln K^\circ_{ex} = -\left(\frac{1}{z_A} - \frac{1}{z_B}\right) + \int_0^1 \ln K_N (a_s) dN_B + \int_1^{a_s} \{n_s(A) - n_s(B)\} d \ln a_s \tag{2.30}$$

Equations analogous to (2.26a) and (2.26b) may also be constructed,

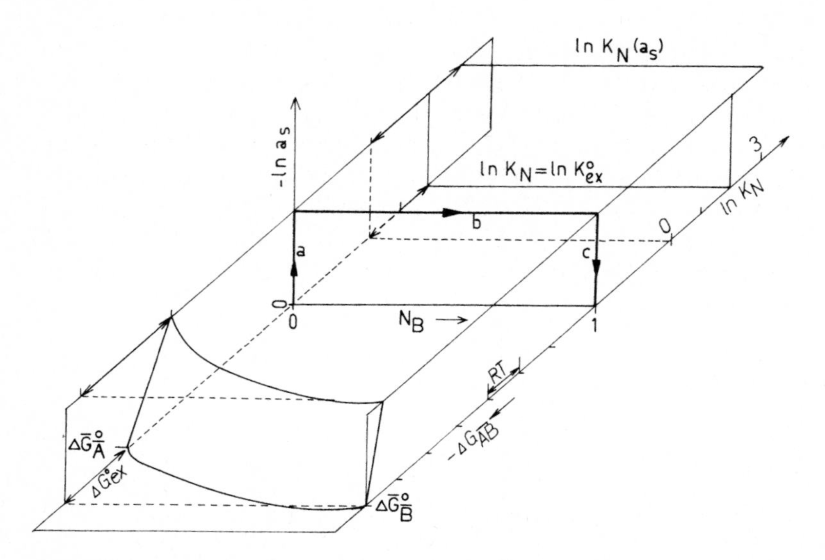

Fig. 2.2. Perspective view of the relation between $\ln K_N$ and $\Delta G_{\overline{AB}}$ in the case of (ideal) mono-monovalent exchange involving a solvent term of excessive magnitude. Base plane: $\Delta G_{\overline{AB}}$ and $\ln K_N = \ln K^{\circ}_{ex}$ belonging to exchange at vanishing concentration of the equilibrium solution ($\ln a_s \approx 0$). Vertical plane at centre: pathway of the line integal $-\oint n_s \, d \ln a_s$.

containing in addition a solvent term of the type $\int n_s \, d \ln a_s$ over the relevant range, according to:

$$\ln \bar{f}_A^{1/z_A} = N_B \left[\left(\frac{1}{z_A} - \frac{1}{z_B} \right) + \ln K_N \right] - \int_0^{N_B} \ln K_N \, dN_B - \int_{1,0}^{a_s, N_B} n_s \, d \ln a_s \tag{2.31a}$$

$$\ln \bar{f}_B^{1/z_B} = -(1 - N_B) \left[\left(\frac{1}{z_A} - \frac{1}{z_B} \right) + \ln K_N \right] + \int_{N_B}^{1} \ln K_N \, dN_B - \int_{1,1}^{a_s, N_B} n_s \, d \ln a_s \tag{2.31b}$$

where the last integral of (2.31a) follows step (a) above and then part of step (b), while the last integral of (2.31b) follows step (c) and part of step (b) in reverse direction.

The above procedure for obtaining $\ln K^{\circ}_{ex}$ from (2.30) is demonstrated for ideal homovalent exchange in the upper part of Fig. 2.2 (to be discussed in more detail in section 2.3). The integration path is indicated with the arrows (a), (b) and (c). Since, here, $\ln K_N(a_s)$ was taken as a constant, the integral of step (b) results in $\langle \ln K_N(a_s) \rangle = \ln K_N(a_s)$, at a value of 3, while the third term vanishes at constant a_s. Conversely the second term disappears in steps (a) and (c), while the third term may be calculated if $n_s(A)$ and $n_s(B)$ are known (cf. the model theories treated in the following chapter). Here the relevant integrals were arbitrarily put at -1.5 for step (a) and $+0.5$ for step (c), for demonstration purposes. This then yields $\ln K^{\circ}_{ex}$ at $3 - 1 = 2$.

In practice it may be inconvenient to determine K_N over the full range of N_B while keeping a_s constant. In this case the last term of (2.29a) must be retained and integrated also over the range of N_B. The line integral covering this full path will be referred to at later stages with the symbol $\oint n_s \, d \ln a_s$. Foregoing, at this point, a discussion about the expected magnitude of the solvent term (cf. chapter 3), it is obvious from (2.30) that it may be assessed experimentally from the change of $\ln K_N$ with $\ln a_s$, as $\ln K^{\circ}_{ex}$ is unaffected by a_s. Turning the argument around, the often observed shift of $\ln K_N$ with electrolyte level may be explained in terms of the above solvent term, indicating that $n_s(A) \neq n_s(B)$.

As was shown by Laudelout and Thomas (1965), the expression $\ln K_N \, dN_B + n_s \, d \ln a_s$ is a total differential, and thus:

$$\left[\frac{\partial \ln K_N}{\partial \ln a_s}\right]_{N_B} = \left[\frac{\partial n_s}{\partial N_B}\right]_{a_s} \tag{2.32}$$

Visualizing the trajectory of K_N at a particular value of a_s for a system in which $n_s(A)$ at a_s exceeds $n_s(B)$ at the same value of a_s, it seems likely that $n_s(A, B)$ in the mixture might decrease monotonically with increasing N_B. The RHS of (2.32) is then negative throughout the range $N_B = 0$ to $N_B = 1$, and thus $\ln K_N$ will increase with increasing electrolyte level (i.e. decreasing a_s). If it is further assumed that in case of homovalent exchange the A-ion, giving rise to the larger amount of solvent in the adsorbed phase, is likely to be the lesser preferred ion species, then $\ln K_N$ would in this case be positive. This would lead to the general rule that the homovalent selectivity of an exchanger will increase with increasing electrolyte concentration (cf. the example of Fig. 2.2).

The above treatment becomes more complicated if the adsorbed phase also contains anions (cf. Gaines and Thomas (1953)). A decision as to whether this is the case may depend on the definition of the extent of the adsorbed phase. As was indicated by Gibbs (1906), cf. Kruyt (1952, p. 116), some arbitrary decision must be made when defining the amounts of the different components present in a surface phase. For an interface between immiscible liquids it is customary to assign the major components of either liquid entirely to the bulk phase, leading to an experimental definiton of the amounts of other components present in the interfacial region according to:

$$S_i \Gamma_k = T_k - W^{\alpha} c^{\alpha}_{0,k} - W^{\beta} c^{\beta}_{0,k} \tag{2.33}$$

Here $S_i \Gamma_k$ indicates the excess present in the interfacial region with surface area S_i separating two bulk phases, with liquid volumes equal to W^{α} and W^{β}, respectively, and containing together the amount T_k. When applied to a solid—liquid interface, the concentration of dissolved components in the bulk solid phase is taken at zero. If specified per unit amount of the solid this gives the excess of species k as used in part A of this text:

$$S \Gamma_k = T_k - W c_{0,k} \tag{2.34}$$

where S, T_k and W all refer to a unit amount of the solid phase. In case of ion adsorption by e.g. clays, the presence of anions in the bulk liquid phase then usually leads to a specification of the surface phase in terms of excesses of the cations plus a deficit of anions in this region (cf. chapter 4 of part A). Furthermore the amount of solvent in the surface region is put at zero by definition. Obviously such a definition of the surface phase is hardly attractive for the application of the thermodynamic description presented above. It is then convenient to delimit the bulk liquid phase on the basis of the constituent repelled from the surface, i.e. the anion. Using the latter as a "tracer" one finds for the extent of the bulk liquid phase a volume equal to $T_{an}/c_{0,an}$, the remainder of the liquid phase being assigned to the adsorbed phase, according to:;

$$S\Gamma_s = (W - T_{an}/c_{0,an})\, c_{0,s} \tag{2.35}$$

in which $c_{0,s}$ is the solvent concentration in the bulk liquid. As now the adsorbed phase is per definition free from anions, the excesses of the cations found by subtracting $c_{0,k}T_{an}/c_{0,an}$ from T_k will sum up to the cation exchange capacity as introduced in chapter 4 of part A, to be indicated here as $S \cdot \Gamma$. In terms of the equivalent fraction adsorbed and in the bulk solution, N_k and f_k, respectively, one thus finds:

$$N_k = T_k/S\Gamma - f_k T_{an}/S\Gamma \tag{2.36}$$

In similar manner the solvent content of the adsorbed phase may be specified in moles per equivalent of the adsorber as:

$$n_s \approx (W - T_{an}/c_{0,an})/\bar{V}_s S\Gamma \tag{2.37}$$

where the partial molar volume of the solvent, $\bar{V}_s$, is used as an approximation for the inverse of the solvent concentration in the adsorbed phase. Taking into account that the above definition actually needs further specification if more than one anionic species are present, it would seem that the limited accuracy obtainable when interpreting experimental data on cation exchange in e.g. clays, warrants the use of the total amount and total equilibrium concentration of the anions in (2.36) and (2.37).

While the above procedure tends to dismiss the need for the inclusion of anions in the description of the exchange equilibrium, it should be remembered that this approach is likely to lead to ambiguous results if an anion present is known to be adsorbed strongly by the exchanger (e.g. phosphate). Interpretation of surface activity coefficients of cations in systems containing phosphate would indeed be in need of an extension as suggested by Gaines and Thomas (1953). At the same time the fact that the phosphate ions are adsorbed specifically on the edges of clays, i.e. at a location separate from the cationic adsorption sites, would probably require a more involved treatment than the one suggested.

Gaines and Thomas's (1953) approach may be viewed as the basic thermodynamics of cation exchange phenomena. Except for the specification of the components present in the adsorbed phase it involves no assumptions. It serves the purpose of enabling one to translate experimental data in terms of recognized thermodynamic variables ΔG°_{ex} (or K°_{ex}) and activity coefficients of the adsorbed species, which may then be used for interpretation purposes. By their very nature, these variables contain no more information than that already present in the experimental data consisting of $K_N(N_k, a_s)$. Thus their use for predictive purposes in practical soil science is limited, the knowledge of K_N under given conditions being sufficient for such predictions. On the other hand the value of K°_{ex} as derived with this procedure allows a consistency check on a series of consecutive exchange reactions by comparing K°_{ex} (A → B), K°_{ex} (B → C) with K°_{ex} (A → C). A good check in this respect was obtained by Lewis and Thomas (1963) for the exchange reaction on a montmorillonite (Chambers, Arizona, No. 23 of the A.P.I. reference clay minerals) involving pairs of the ions Na, Cs and Ba. A similar agreement was found by Martin and Laudelout (1963), employing a montmorillonite from Maroc (Camp Berteau) with pairs of the ions NH_4, K, Rb, Na, Li and also Ba.

At the same time the extensive experimental data available on the Camp Berteau montmorillonite (Martin and Laudelout, 1963; Fripiat et al., 1965; Van Bladel and Laudelout, 1967; Laudelout et al., 1968) indicate that often incomplete reversibility of the exchange reaction occurs. In part this is no doubt associated with aggregation of the clay particles and then appears to be reduceable if a series of measurements at different electrolyte level are extrapolated to vanishing concentrations (Van Bladel and Laudelout, 1967). On the other hand the data by Fripiat et al. (1965) indicate in the same clay an ion-specific irreversibility which could be attributed to a hysteresis in dislodging NH_4-ions by ions with a larger hydrated volume (i.e. Li, Na and several divalent cations).

2.3. THE FREE ENTHALPY OF THE HETERO-IONIC EXCHANGER; EXCESS FUNCTIONS

In the previous section it was shown that a complete set of experimentally determined values of $K_N(N_k, a_s)$ may be "translated" into the value of ΔG°_{ex} plus a pair of curves representing the activity coefficients of the adsorbed ions. The same data may also be used to calculate the free enthalpy of the exchanger with a mixed composition relative to an appropriate reference system.

Taking as an example one equivalent of an exchanger containing N_A equivalents of A-ions and N_B equivalents of B-ions one may write:

$$G_{\overline{AB}} = (N_A/z_A)(\overline{G}^{\circ}_{\overline{A}} + RT\ln \bar{f}_A N_A) + (N_B/z_B)(\overline{G}^{\circ}_{\overline{B}} + RT\ln \bar{f}_B N_B) \quad (2.38)$$

where $\overline{G}^{\circ}_{\overline{k}}$ indicates the partial molar free enthalpy of the uncharged exchanger (plus its ions) in its homoionic standard state with $a_s = 1$ (cf. (2.4c)). This equation thus defines the free enthalpy of one equivalent of the mixed exchanger relative to the free energy of the appropriate amounts of the homoionic forms of the exchanger according to:

$$G_{\overline{AB}} - \Sigma_{\bar{A},\bar{B}} N_k \bar{G}^\circ_{\bar{k}}/z_k = RT[\Sigma_{\bar{A},\bar{B}} N_k \ln (N_k)^{1/z_k} + \Sigma_{\bar{A},\bar{B}} N_k \ln (\bar{f}_k)^{1/z_k}] \tag{2.39}$$

As will be mentioned in chapter 4, the first term of the RHS corresponds to the free enthalpy of ideal (athermal) mixing of ions which occupy an equal volume in the adsorbed phase *per equivalent* (and not per mole). Following Howery and Thomas (1965) this situation will be designated "ideal adsorption". Subtracting this first term from (2.39) then defines the excess free enthalpy of the mixed exchanger as:

$$G^X_{\overline{AB}}/RT = N_A \ln (\bar{f}_A)^{1/z_A} + N_B \ln (\bar{f}_B)^{1/z_B} \tag{2.40}$$

Making use of (2.26a) and (2.26b) gives:

$$G^X_{\overline{AB}}/RT = N_B \int_0^1 \ln K_N \, dN_B - \int_0^{N_B} \ln K_N \, dN_B$$

$$\equiv N_B [{}^\circ\langle \ln K_N \rangle^1 - {}^\circ\langle \ln K_N \rangle^{N_B}] \tag{2.41}$$

thus specifying $G^X_{\overline{AB}}$ in terms of the difference between the mean value of $\ln K_N$ over the full range of $N_B = 0 \rightarrow 1$ and its mean value over the range from 0 to N_B. It is noted that $G^X_{\overline{AB}}$ depends solely on the course of $\ln K_N$ and not on the valence of the participating ions. In Fig. 2.1 the excess function is shown as derived from the chosen arbitrary "curve" for $\ln K_N$ (where the solvent term was considered negligible). In contrast to $G^X_{\overline{AB}}$, the free enthalpy of ideal mixing — equal to $-T\Delta S_{\overline{AB}}$ — differs for the two cases shown in Figs. 2.1a and 2.1b, the total number of moles being smaller in the case of mono-divalent exchange.

In case the solvent term is significant, (2.40) may be combined with (2.31a) and (2.31b) to yield:

$$\frac{G^X_{\overline{AB}}(a_s)}{RT} = N_B \left[{}^\circ\langle \ln K_N (a_s) \rangle^1 - {}^\circ\langle \ln K_N (a_s) \rangle^{N_B} + \oint n_s \, d \ln a_s \right] - \int_{1,0}^{a_s, N_B} n_s \, d \ln a_s \tag{2.42}$$

It may be noted that in the case when $\ln K_N (a_s)$ is given for a constant value of $a_s < 1$, the two integral terms may be combined to give:

$$-\left(N_A \int_1^{a_s} n_s(A) \, d \ln a_s + N_B \int_1^{a_s} n_s(B) \, d \ln a_s \right)$$

i.e. the weighted mean value of the solvent terms for the homoionic end-points.

Obviously a plot of $G^X_{\overline{AB}}$ against N_B contains as much information as that of the pair of activity coefficients used before, the two different curves in the latter case arising from the predetermined condition of unit values in the standard states. Both types of plots reflect the *variation* of $\ln K_N$ *with respect to* $\ln K^\circ$, i.e. non-ideality. In addition to this information one may then derive the value of $\ln K^\circ$ or ΔG°_{ex} from the mean value of $\ln K_N$, either by correcting $\langle \ln K_N(a_s) \rangle$ for the solvent term, or by extrapolating $\langle \ln K_N(a_s) \rangle$, determined at different values of a_s, towards $a_s = 1$.

One may also attempt to combine the information above in one curve by plotting a quantity $\Delta G_{\overline{AB}}$ defined according to:

$$\Delta G_{\overline{AB}} \equiv G_{\overline{AB}} - N_A \mathring{\mu}_A / z_A - N_B \mathring{\mu}_B / z_B - \mathring{\mu}_{ads} \tag{2.43}$$

By analogy with the definition of the free enthalpy of formation of a proton via an appropriate half-reaction as discussed in part A of this text (cf. section A2.7.1) the above quantity could be termed the free enthalpy of formation of one equivalent of the exchanger following the half reaction:

$$N_A/z_A\, A + N_B/z_B\, B + \text{ads} \rightleftharpoons \overline{AB}\text{-ads} \tag{2.44}$$

where the reactants are taken in their standard state. Comparison with a half-reaction involving the formation of the homoionic exchanger in its standard state and loaded with a chosen reference ion — here taken as the A-ion — gives:

$$1/z_A\, A + \text{ads} \rightleftharpoons 1/z_A\, \bar{A}\text{-ads} \tag{2.45}$$

$$\text{with } \Delta \bar{G}^\circ_A / z_A \equiv \bar{G}^\circ_A / z_A - \mathring{\mu}_A / z_A - \mathring{\mu}_{ads} \tag{2.46}$$

This indeterminate value of the molar free enthalpy of formation of the homoionic exchanger loaded with the reference ion may now be taken as the reference level for $\Delta G_{\overline{AB}}$, according to:

$$\Delta G_{\overline{AB}} = \Delta \bar{G}^\circ_A / z_A + G_{\overline{AB}} - \bar{G}^\circ_A / z_A - N_B (\mathring{\mu}_B / z_B - \mathring{\mu}_A / z_A) \tag{2.47}$$

Making use of (2.38) gives:

$$\begin{aligned} \Delta G_{\overline{AB}} = {} & \Delta \bar{G}^\circ_A / z_A + N_B \{(\bar{G}^\circ_B / z_B - \bar{G}^\circ_A / z_A) - (\mathring{\mu}_B / z_B - \mathring{\mu}_A / z_A)\} \\ & + RT\{N_A \ln (\bar{f}_A N_A)^{1/z_A} + N_B \ln (\bar{f}_B N_B)^{1/z_B}\} \end{aligned} \tag{2.48}$$

where the second term of the RHS is recognized as ΔG°_{ex} (cf. (2.10)). Accordingly, the newly defined "free enthalpy of formation of the mixed exchanger" (per equivalent!) relative to the same of the homoionic standard state equals:

$$\Delta G_{\overline{AB}} - \Delta \bar{G}^\circ_A / z_A = G^X_{\overline{AB}} + RT \Sigma_{\bar{A},\bar{B}} N_k \ln (N_k)^{1/z_k} + N_B \Delta G^\circ_{ex} \tag{2.49}$$

Making use of (2.42) and (2.30) this expression may be condensed to:

$$\frac{\Delta G_{\overline{AB}} - \Delta \bar{G}^\circ_A / z_A}{RT} = \Sigma_{\bar{A},\bar{B}} N_k \ln (N_k)^{1/z_k} - N_B \left[{}^\circ\langle \ln K_N(a_s) \rangle^{N_B} - \left(\frac{1}{z_A} - \frac{1}{z_B} \right) \right] - \int_{1,0}^{a_s, N_B} n_s \, d \ln a_s \tag{2.50}$$

Applying this equation to the homoionic endpoints at a given value of a_s gives:

$$\frac{\Delta G_{\bar{A}} - \Delta \bar{G}^\circ_A / z_A}{RT} = - \int_1^{a_s} n_s \, d \ln a_s \tag{2.51a}$$

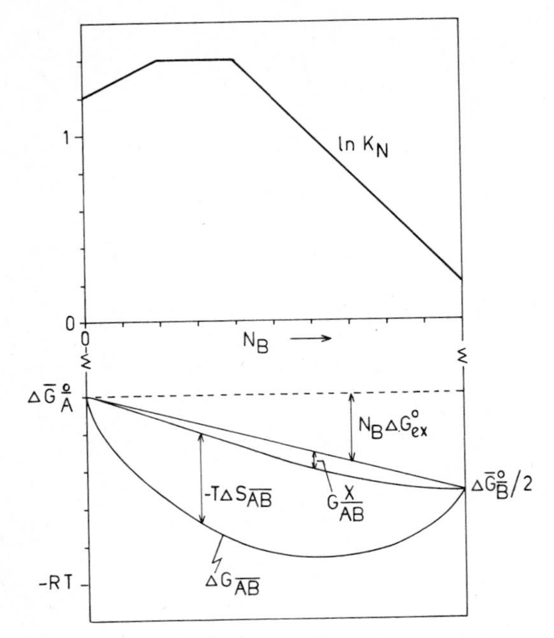

Fig. 2.3. Construction of $\Delta G_{\overline{AB}}$ corresponding to the system presented in Fig. 2.1b, involving mono-divalent exchange in the absence of a significant solvent term.

and

$$\frac{\Delta G_{\bar{B}} - \Delta\bar{G}^o_{\bar{A}}/z_A}{RT} = \left(\frac{1}{z_A} - \frac{1}{z_B}\right) - {}^o\langle \ln K_N(a_s)\rangle^1 - \int_{1,0}^{a_s,1} n_s \, d \ln a_s \tag{2.51b}$$

such that:

$$\frac{\Delta G_{\bar{B}} - \Delta G_{\bar{A}}}{RT} = \left(\frac{1}{z_A} - \frac{1}{z_B}\right) - {}^o\langle \ln K_N(a_s)\rangle^1 - \int_{N_B=0}^{1} n_s \, d \ln a_s \tag{2.52}$$

where $\Delta G_{\bar{A}}$ and $\Delta G_{\bar{B}}$ are defined again per equivalent of adsorber, in contrast to the *molar* quantity $\Delta\bar{G}^o_{\bar{A}}$.

In Fig. 2.2, lower part, the course of $\Delta G_{\overline{AB}}$ has been constructed as a perspective drawing for ideal exchange with $z_A = z_B = 1$ ($\ln K_N$ = constant!), using arbitrary (but excessive) values for $n_s(A) = 3 n_s(B)$. The curve in the base plane, corresponding to $a_s \approx 1$, follows the curve for ideal mixing which was shown before in Fig. 2.1a. It connects $\Delta\bar{G}^o_{\bar{A}}$ (situated at an arbitrary position) with $\Delta\bar{G}^o_{\bar{B}}$ (situated at a distance $-RT\langle \ln K_N\rangle = -RT \ln K^o$ from $\Delta\bar{G}^o_{\bar{A}}$). The increase of $\Delta G_{\bar{A}}$ with increasing electrolyte level (i.e. decreasing a_s) is depicted in the "vertical" plane through $N_B = 0$, and arbitrarily taken at 1.5 RT. Similarly the increase of $\Delta G_{\bar{B}}$ from $\Delta\bar{G}^o_{\bar{B}}$ with decreasing a_s is found in the vertical plane through $N_B = 1$ and taken here as 0.5 RT, implying that $n_s(A) = 3 n_s(B)$. Following the arrows marked a, b and c one finds the complete curve of $\Delta G_{\overline{AB}}$ according to (2.50). In Fig. 2.3 the construction of $\Delta G_{\overline{AB}}$ is given for mono-divalent exchange following the previous curve for $\ln K_N$, again neglecting the solvent term.

It is finally noted that for a constant, but arbitrary, value of a_s one finds:

$$\left[\frac{\partial G_{\overline{AB}}/RT}{\partial N_B}\right]_{a_s} = -\ln K_N + \ln \frac{N_B^{1/z_B}}{N_A^{1/z_A}} = \ln \frac{a_B^{1/z_B}}{a_A^{1/z_A}} \qquad (2.53)$$

REFERENCES

Babcock, K.L., 1963. Theory of the chemical properties of soil colloidal systems at equilibrium. Hilgardia, 34: 417—542.

Bear, F.E., 1964. Chemistry of the Soil. Reinhold, New York, N.Y., 373 pp.

Fripiat, J.J., Cloos, P. and Poncelet, A., 1965. Comparison entre les propriétés d'échange de la montmorillonite et d'une résine vis-á-vis des cations alcalins et alcalino-terreux. I. Réversibilité des processus. Bull. Soc. Chim. de France, 1965: 208—214.

Gaines, G.L. and Thomas, H.C., 1953. Adsorption studies on clay minerals. II. A formulation of the thermodynamics of exchange adsorption. J. Chem. Physics, 21: 714—718.

Guggenheim, E.A., 1967. Thermodynamics. North-Holland, Amsterdam, 390 pp.

Howery, D.G. and Thomas, H.C., 1965. Ion exchange on the mineral clinoptilolite. J. Phys. Chem., 69: 531—537.

Kruyt, H.R., 1952. Colloid Science. Elsevier, Amsterdam, 389 pp.

Laudelout, H. and Thomas, H.C., 1965. The effect of water activity on ion exchange selectivity. J. Phys. Chem., 69: 339—340.

Laudelout, H., Van Bladel, R., Bolt, G.H. and Page, A.L., 1968. Thermodynamics of heterovalent cation exchange reactions in a montmorillonite clay. Trans. Faraday Soc., 64: 1477—1488.

Lewis, R.J. and Thomas, H.C., 1963. Adsorption studies on clay minerals. VIII. A consistency test of exchange sorption in the systems sodium-cesium-barium montmorillonite. J. Phys. Chem., 67: 1781—1783.

Marshall, C.E., 1964. The Physical Chemistry and Mineralogy of Soils. Vol. I: Soil Materials. Wiley, New York, N.Y., 388 pp.

Martin, H. and Laudelout, H., 1963. Thermodynamique de l'échange des cations alcalins dans les argiles. J. Chim. Phys., 60: 1086—1099.

Peech, M. and Scott, A.D., 1950. Determination of ionic activities in soil-water systems by means of the Donnan membrane equilibrium. Soil Sci. Soc. Am. Proc., 15: 115—119.

Van Bladel, R. and Laudelout, H., 1967. Apparent irreversibility of ion-exchange reactions in clay suspensions. Soil Sci., 104: 134—137.

CHAPTER 3

THEORIES OF CATION ADSORPTION BY SOIL CONSTITUENTS: DISTRIBUTION EQUILIBRIUM IN ELECTROSTATIC FIELDS

G.H. Bolt

Different models have been used to describe the cation exchange equilibrium between the soil solution and the soil exchange complex. The primary goal of such models is to find exchange equations which check with reality and which may then be used for prediction purposes (cf. Bolt, 1967).

In view of the wide variety of materials contributing to the soil exchange complex and considering the differences between the adsorbed (occasionally liganded) cations in valence and size, it could hardly be expected that one reasonably simple model would explain the behavior of all systems of interest. Against this background it is of interest to review the different models, pointing out the underlying assumptions and the ensuing limitations. For this purpose it seems practical to distinguish broadly two classes of models, viz. those emphasizing the accumulation of cations in an electric potential field, against the ones stressing the presence of discrete adsorption sites.

The first class, to be treated in the present chapter, describes the charged adsorbent in terms of uniform surface (or volume) density of charge. The adsorbent charge then gives rise to a region in which either the electric potential varies in one direction only, or is constant throughout. The various cations are then assumed to be accumulated in that region, allowing for unconstrained mixing comparable with mixing in aqueous solutions. These theories provide a clear insight into the effect of adsorbent charge density and solution electrolyte level on the exchange equilibrium involving cations of different valence. Referring to chapter 2, these theories predict the *value* of K°_{ex} for hetero-valent exchange as a function of the charge density of the exchanger. They also predict the variation of K_N as a function of electrolyte level in solution and exchanger composition. Refinements, taking into account secondary effects like e.g. ionic polarizabilities and ionic interactions of the Debye-Hückel type could in principle be made. It remains questionable, however, if such refinements would be warranted in view of the assumed homogeneous distribution of the adsorbent charge.

The second class of models, to be discussed in the following chapter, emphasizes the interactions between individual adsorbent site and cation, where the presumed difference in energy of adsorption of different cation species on similar sites is usually taken as unknown. Taking into account the constraints of the mixing of ions adsorbed on a particular array of fixed sites, these models lead to a prediction of the variation of K_N with exchanger composition. If such predictions would check with experimental data for a certain group of systems, it should suffice to determine experimentally one point of an exchange isotherm pertaining to a particular member of that group to predict the course of K_N and thus K°_{ex}.

Comparing the assumptions underlying the two classes of models specified above, it should be expected that the adsorption of highly hydrated cations of the mono- and divalent type on surfaces with a high charge density approaches the characteristics predicted by the first class. As will be shown in section 3.2 the effect of electrolyte level on K_N should be particularly noticeable in dilute suspensions of e.g. Na-Ca-clays. In contrast, the adsorption of heavy metals with low hydration numbers should be much less sensitive to electrolyte level. A discrete-site model, particularly one recognizing at least two different types of sites then appears a more promising approach.

Finally it is mentioned that the two classes of models converge, even for heterovalent exchange, when applied to a system consisting of a monolayer of cations adsorbed on sites arranged in such a manner that each site has a large number of nearest neighbors. One then finds that at low electrolyte level both classes lead to a constant value of K_N, i.e. independent of the exchanger composition and electrolyte level (cf. the discussion on pp. 85—87).

3.1. ION ACCUMULATION IN A REGION WITH CONSTANT ELECTRIC POTENTIAL

The present model corresponds to the distribution of ions in an adsorbed phase of the type treated in sections 2.1.1 and 2.1.2 of the previous chapter. It could be indicated as a "micro-Donnan" system, each exchanger particle forming a separate compartment containing a large number of fixed charges distributed homogeneously throughout the particle. The idealized version of this model is discussed here at considerable length, because it lends itself very well to a fully verifiable application of the equations presented in chapter 2, thus providing a good understanding of the significance of the different terms of these equations.

In order to obtain a first estimate of the valence-selectivity in the present system one may visualize an adsorbed phase with a constant volume, V^{α}, per equivalent of adsorber (divided over many particles), containing ideally behaving cations A and B, with valence z_A and z_B, respectively.

The composition of the equilibrium solution may be given as:

$$\begin{aligned} c_{0,A} &= (1-f)C_0 \\ c_{0,B} &= fC_0 \end{aligned} \tag{3.1}$$

$$c_{0,-} = C_0$$

while the concentration of co-ions in the adsorbed phase equals:

$$c_-^\alpha = C_0/\bar{u} \qquad (3.2)$$

where for ideal behavior of the ions $\bar{u}$ equals the Boltzmann accumulation factor of monovalent cations in the adsorbed phase, $\exp(-e\psi^\alpha/kT)$. For the assumed ideal behavior, including a negligible effect of the pressure term mentioned in (2.4a), one finds the equivalent fraction adsorbed of A and B ions as:

$$N_A = (1-f)C_0(\bar{u}^{z_A} - 1/\bar{u})V^\alpha \qquad (3.3a)$$

$$N_B = fC_0(\bar{u}^{z_B} - 1/\bar{u})V^\alpha \qquad (3.3b)$$

Constructing K_N for the exchange equilibrium gives:

$$K_N = \frac{(\bar{u}^{z_B} - 1/\bar{u})^{1/z_B}}{(\bar{u}^{z_A} - 1/\bar{u})^{1/z_A}} \frac{(z_B V^\alpha)^{1/z_B}}{(z_A V^\alpha)^{1/z_A}} \qquad (3.4)^*$$

Using again the homoionic exchanger in equilibrium with a solution of the corresponding salt at vanishing concentration as the standard state, one finds — with $\bar{u} \rightarrow \infty$ for $C_0 \rightarrow 0$ — the standard free enthalpy of the exchange reaction according to (2.25) as:

$$-\Delta G^\circ_{ex}/RT = \ln K^\circ_{ex} = -(1/z_A - 1/z_B) + \ln \frac{(z_B V^\alpha)^{1/z_B}}{(z_A V^\alpha)^{1/z_A}} \qquad (3.5)^*$$

For orientation purposes it is pointed out that for $z_A = 1$ and $z_B = 2$, K_N equals $\sqrt{2^V\Gamma}$, where $^V\Gamma \equiv 1/V^\alpha$ indicates the volume charge density of the adsorbed phase in keq/m^3. For cation exchange resins with a porosity of about 0.5, $^V\Gamma$ has the order of 10 keq/m^3, yielding ΔG°_{ex} at about $-$ RT. Actually this estimate should be corrected at least for interionic interactions at such high concentrations in the adsorbed phase. For lack of information on the magnitude of the corresponding "modified solution phase" activity coefficients, f^α, it is pointed out that such coefficients would enter (3.4) as the reduced ratio $(f_A^\alpha)^{1/z_A}/(f_B^\alpha)^{1/z_B}$. For the chosen mono-divalent system this ratio is perhaps likely to exceed unity at the given value of the total concentration in the adsorbed phase and in that case K°_{ex} will increase correspondingly. Although also the pressure term should be taken into account its magnitude will be small in comparison to the uncertainty of the above ratio of activity coefficients f^α.

It is finally mentioned in passing that *because* of the chosen standard states, ΔG°_{ex} for the above "ideal" mono-divalent exchange approaches zero at $^V\Gamma \simeq 1.4$ keq/m^3.

Whereas the integration of $\ln K_N$ over N_B at vanishing concentration of the equilibrium solution produced $\ln K^\circ_{ex}$ directly because the solvent term (cf. (2.30)) remained absent, the integration at finite (constant) value of C_0 may be used to derive an expression for this particular term in the present system. Using for simplicity a system with $z_A = 1$ and $z_B = 2$, (3.4) may be written as:

---

* The chosen standard state in solution at unit molar concentration is reflected in the specification of K_N throughout this chapter in $(\text{kmol/m}^3)^{(1/z_A - 1/z_B)}$.

$$K_N = \sqrt{2/V^\alpha}\,\frac{\sqrt{(1-1/\bar{u}^3)}}{(1-1/\bar{u}^2)} \tag{3.6}$$

in which the first term of the RHS represents again the limiting value of K_N for $\bar{u} \to \infty$ at $C_0 \to 0$. Introducing the volume density of charge of the exchanger as ${}^v\Gamma \equiv 1/V^\alpha$ this term may also be written as $\sqrt{2^v\Gamma}$ (cf. small print section above). Reconstructing now (2.30) of the previous chapter, one finds:

$$\ln K^\circ_{ex} = -\tfrac{1}{2} + \ln\sqrt{2^v\Gamma} + \int_0^1 \ln \frac{\sqrt{(1-1/\bar{u}^3)}}{(1-1/\bar{u}^2)}\,dN_B + \oint n_s\, d\ln a_s \tag{3.7}$$

where according to (3.5) the first two terms of the RHS cancel the LHS. Thus the "solvent term" in case of the chosen system is found as:

$$-\oint n_s\, d\ln a_s = \underbrace{\int_0^1 \ln\sqrt{(1-1/\bar{u}^3)}\,dN_B}_{A} - \underbrace{\int_0^1 \ln(1-1/\bar{u}^2)\,dN_B}_{B} \tag{3.8}$$

Working out the RHS with the help of (3.3) gives:

$$A = \tfrac{1}{2} N_B \ln(1-1/\bar{u}^3)\Big]_0^1 - \tfrac{1}{2}\int_{N_B=0}^1 fV^\alpha C_0 \bar{u}^2\, d(1-1/\bar{u}^3)$$

$$= \tfrac{1}{2}\int V^\alpha C_0' \bar{u}^2\, d(1-1/\bar{u}^3) - \tfrac{1}{2}\int fV^\alpha C_0 \bar{u}^2\, d(1-1/\bar{u}^3) \tag{3.9a}$$

where the first integral covers the range $C_0' = 0$ (i.e. the standard state) to $C_0' = C_0$ (at $N_B = 0 = f$), while the second integral covers the range $N_B = 0 \to 1$, or $f = 0 \to 1$, at constant value C_0. Similarly:

$$B = -\int V^\alpha C_0' \bar{u}\, d(1-1/\bar{u}^2) - \int (1-f)V^\alpha C_0 \bar{u}\, d(1-1/\bar{u}^2) \tag{3.9b}$$

The two integrals of (3.9a) may be simplified by substituting $d(1-1/\bar{u}^3) = (3/\bar{u}^2)d(1-1/\bar{u})$ and combining them into one line integral of the form:

$$A = -\tfrac{3}{2} V^\alpha \oint fC_0'\, d(1-1/\bar{u}) \tag{3.10a}$$

where the integration path runs from $C_0' = 0$ at $f = 0$, to C_0 at $f = 0$ (zero value of the integral), next from $f = 0$ to $f = 1$ at $C_0' = C_0$ and then back from $C_0' = C_0$ to $C_0' = 0$, at $f = 1$. Similarly part B of (3.8) results in:

$$B = -2V^\alpha \oint (1-f)C_0'\, d(1-1/\bar{u}) \tag{3.10b}$$

The LHS (3.8) may be integrated by parts and then gives:

$$-\oint n_s\, d\ln a_s = -n_s \ln a_s\Big] + \oint \ln a_s\, dn_s \tag{3.11}$$

where the first term disappears for the chosen integration path from $a_s = 1$

at $N_B = 0$, to $a_s = 1$ at $N_B = 1$. Inserting the above results in (3.8) yields finally:

$$\oint \ln a_s \, d\, n_s = -\oint (2 - f/2) C_0' \, dV^\alpha (1 - 1/\bar{u}) \tag{3.12}$$

The integrand of the RHS of (3.12) now specifies the total molar concentration of the equilibrium solution, ${}^mC_0$. For the assumed ideal solution behavior the solvent activity is then found to be:

$\ln a_s \approx -\bar{V}_s {}^mC_0$; accordingly

$$n_s = V^\alpha (1 - 1/\bar{u})/\bar{V}_s \tag{3.13}$$

in which $\bar{V}_s$ is the partial molar volume of the solvent. This identifies n_s as the number of moles of solvent present in the anion-free volume associated with the adsorber (cf. the exclusion volume discussed in chapter 7). As to be expected this checks with the experimental definition given in (2.37).

With reference to (3.6) it is observed that K_N will *increase* with increasing concentrations of the equilibrium solution (which lead to a decrease of $\bar{u}$): for a given value of $c_{0,+}/\sqrt{c_{0,2+}/2}$ in the equilibrium solution (cf. the SAR-value mentioned in chapter 9 of part A) the amount adsorbed of divalent cations will increase upon increasing the electrolyte concentration. For the present model this effect is fully determined by the value of the Boltzmann factor $\bar{u}$ and becomes negligible if $\bar{u}$ exceeds 10 (i.e. less than 1%). In turn $\bar{u}$ is determined by the parameter ${}^v\Gamma/C_0 \equiv 1/V^\alpha C_0$ — which may be viewed as an overall accumulation factor — and by N_{2+} via (3.3). Thus it is found that this parameter varies from about 10 at $N_B = 0$ to about 100 at $N_B = 1$, for the above limit of $\bar{u}$ equal to 10. Accordingly the effect of a change in the total concentration is noticeable only if ${}^v\Gamma/C_0$ is less than about 20. In Fig. 3.1 the course of $\ln K_N$ is plotted as a function of N_B for several values of ${}^v\Gamma/C_0$, relative to its limiting value for $C_0 \to 0$ and indicated as $\Delta \ln K_N(C_0)$. Its absolute value is found by adding $\ln\sqrt{2^v\Gamma}$ to $\Delta \ln K_N(C_0)$.

The overall effect of a change in electrolyte concentration upon the valence-selectivity of the micro-Donnan system is conveniently expressed in $\langle \Delta \ln K_N(C_0) \rangle$, i.e. the mean shift of $\ln K_N$ relative to its position at vanishing concentration of the equilibrium solution. This quantity is found with the help of (3.12), as:

$$\langle \Delta \ln K_N(C_0) \rangle \equiv \int_0^1 \Delta \ln K_N(C_0) dN_B = \oint \ln a_s \, dn_s \left(= -\oint n_s \, d \ln a_s\right)$$

$$= -\ln\sqrt{\frac{{}^v\Gamma}{C_0}} - \ln\frac{\bar{u}_B}{\bar{u}_A} + \Delta\left[\frac{1}{2\bar{u}} + \frac{1}{4\bar{u}^2} + \ln(1 - 1/\bar{u}) + \frac{C_0}{{}^v\Gamma}\left\{\frac{2}{\bar{u}} + \frac{1}{4\bar{u}^2} + \frac{1}{6\bar{u}^3}\right\}\right]_A^B \tag{3.14}$$

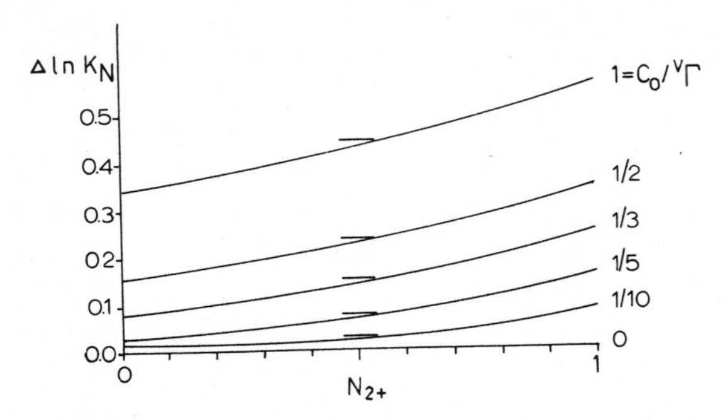

Fig. 3.1. The curve of $\Delta \ln K_N \equiv \ln(K_N/K^{\circ}_{ex})$ for mono-divalent exchange in an adsorbed phase with uniform electrostatic field, as a function of the relative electrolyte concentration, $C_0/^v\Gamma$. The mean values of $\Delta \ln K_N$ over the range $N_{2+} = 0 \rightarrow 1$ were calculated with (3.14).

where $\bar{u}_B$ and $\bar{u}_A$ are the values of $\bar{u}$ pertaining to the homoionic exchanger at a given value of $^v\Gamma/C_0$, according to:

$$\bar{u}_B^2 - 1/\bar{u}_B \;=\; ^v\Gamma/C_0 \;=\; \bar{u}_A - 1/\bar{u}_A \tag{3.15}$$

In Fig. 3.1 the corresponding values of $\langle \Delta \ln K_N (C_0) \rangle$ are indicated, showing that the increase of the valence-selectivity reaches a value of about $\sqrt{e}$ once $^v\Gamma/C_0$ falls below unity. Thinking in terms of $^v\Gamma$ equal to 1—10 keq/m^3 as may apply to soil organic matter polymer coils and closely spaced clay plates, it appears that for the present model the selectivity for divalent cations will hardly be influenced until the concentration in the equilibrium solution exceeds at least 0.1 *N*.

Returning to the assumption of ideal behavior inside the adsorbed phase it is pointed out that the reduced ratio of the D-H type activity coefficients, f^{α}, may remain fairly constant in the adsorbed phase, because its total ionic strength does not change very much for $\infty > {^v\Gamma}/C_0 > 10$. As the penetration of co-ions into the adsorbed phase will, if anything, be favored by non-ideal behavior it appears that n_s would be smaller, thus possibly decreasing the effect of the solvent term. As the latter determines the change of $\ln K_N$ with electrolyte level it is expected that the present model will probably not underestimate this change. Accordingly it is concluded that under conditions prevailing in soils a distribution of adsorbed cations according to the micro-Donnan model would lead to a very small influence of the electrolyte level on the value of K_N.

As will be discussed in the following section a diffuse distribution pattern of the adsorbed cations is more sensitive towards the electrolyte concentration. As calculations for such systems are rather involved it is of interest to look at another procedure available for estimating the effect of electrolyte level on K_N, using the present system for a check on its validity. For this purpose one may calculate the free enthalpy of the homoionic adsorber in equilibrium with a solution of the corresponding salt at concentration C_0 by means of a hypothetical discharging-charging procedure comparable to the one used in the derivation of the Debye-Hückel theory. As was shown by Bolt and Winkelmolen (1968) this leads to an equation of the type:

$$\frac{G_{\bar{K}}(C_0) - \bar{G}_{\bar{K}}^{\circ}/z_k}{RT} = \Delta\left[\frac{1}{\Gamma}\int_0^{\Gamma} \ln u' d\Gamma' + \frac{1}{z_k}\ln C_0/z_k\right]_0^{C_0} \quad (3.16)$$

where u' is the Boltzmann accumulation factor characteristic for the adsorbed phase during the charging process, when Γ' grows from zero to Γ. As is indicated by the symbol Δ at the beginning of the RHS, this charging process must be done twice, i.e. at the two concentration levels to be compared (here C_0 and 0, respectively), after which the resulting values of the bracketed portion of the equation are subtracted from each other. Referring to chapter 2, the LHS is recognized as the free enthalpy of formation of the homoionic adsorber at concentration C_0, specified by (2.48), (2.49) and (2.50).

In short this charging-discharging procedure runs as follows. For adsorbents with a constant charge density (per unit surface area, or per unit volume in the case of the present Donnan system) the surface or phase potential vanishes at very high electrolyte levels. The electrostatic part of the free enthalpy of the adsorbent with its counterions is thus conveniently compared with a hypothetical reference system at very high electrolyte level and is then found as a positive quantity equal to:

$$\Delta G_{el}(C_0) = \int_0^{\sigma} \psi' d\sigma' \quad (3.17)$$

In this equation ψ' indicates the phase or surface potential and σ' the charge density of the phase/surface during the process of charging, the discharging having been effected in the hypothetical reference system. It may be noted that (3.17) constitutes one part of the expression used for the calculation of the free enthalpy of e.g. the double layer on reversible surfaces with a fixed surface potential (cf. J.Th.G. Overbeek, in Kruyt, 1952). In the latter case the presence of a term equal to $-\psi\sigma$, correxponding to the chemical work due to adsorption of the potential determining ion, is combined with (3.17) to yield the well-known expression $-\int \sigma d\psi$. Referring to one equivalent of an adsorbed phase with a *volume* charge density (at the fully charged state) equal to ${}^{v}\Gamma$ keq/m^3, (3.17) may be written as:

$$\Delta G_{el}(C_0) = \frac{RT}{{}^{v}\Gamma}\int_0^{{}^{v}\Gamma} \ln \bar{u}' d\,{}^{v}\Gamma' \quad (3.18)$$

where $\bar{u}$ is the Boltzmann factor for the adsorbed phase, the primes indicating values during the charging process. By comparing the outcome of (3.18) for different values of C_0, one may compute the difference in free enthalpy of such systems, thus circumventing the actual use of the hypothetical reference state at very high electrolyte level. In addition to (3.18), covering the work associated with the charging, a second term must be added covering the change of free enthalpy of the uncharged counterions accompanying the dilution from $C_{0,ref}$ to C_0 equal to $(RT/z_k)\ln\{(C_0/z_k)/(C_{0,ref}/z_k)\}$ for one equivalent of ions. Addition of this term to (3.18) then yields (3.16) above.

Working-out (3.16) for e.g. mono- or divalent cations implies the substitution of (3.15), which also applies during the charging if primes are added to the variables $\bar{u}$ and ${}^{v}\Gamma$. Executing the above for a homoionic A-exchanger at C_0 and at vanishing electrolyte level then gives;

$$G_{\bar{A}}(C_0)/RT - \bar{G}^{o}_{\bar{A}}/RT = \ln \bar{u} C_0 \Big]_0^{C_0} - \frac{C_0}{{}^{v}\Gamma} \int_1^{\bar{u}} (1 - 1/\bar{u}'^2) d\bar{u}' \Big]_0^{C_0}$$

$$= \Big[\ln \bar{u} C_0 - 1 + 2(1 - 1/\bar{u}) C_0 / {}^{v}\Gamma \Big]_0^{C_0} \tag{3.19a}$$

Expanding the first term in order to separate its limit at $C_0 \rightarrow 0$, one finds:

$$\ln \bar{u} C_0 = \ln {}^{v}\Gamma - \ln (1 - 1/\bar{u}^2) \tag{3.20}$$

The first term of (3.20) constitutes the above limit, while the second term equals the first term of (3.9b). Writing out the latter again as an integral over $F(\bar{u})$ and dropping the constants of (3.19a) following application of the limits, the latter may be written as (omitting the primes on $\bar{u}$ and C_0):

$$G_{\bar{A}}(C_0)/RT - \bar{G}^{o}_{\bar{A}}/RT = - \int_0^{C_0} \frac{2C_0}{{}^{v}\Gamma} d(1 - 1/\bar{u}) + \left\{ \frac{2C_0}{{}^{v}\Gamma} (1 - 1/\bar{u}) \right\}$$

$$= + \int_0^{C_0} \left\{ (1 - 1/\bar{u}) / {}^{v}\Gamma \right\} d\, 2C_0 \tag{3.21a}$$

where the limits are expressed in C_0, but apply also to $\bar{u}$ via (3.15) as quoted above.

Eqn. (3.21a) may now be compared with (2.50) of chapter 2, because the *difference* between the free enthalpy of the homoionic exchanger at C_0 and at its standard state is independent of the reference point chosen for these parameters. Thus (2.50) immediately identifies n_s as $(1 - 1/\bar{u})/\bar{V}_s {}^{v}\Gamma$, since $\ln a_s = -2C_0 \bar{V}_s$, with $\bar{V}_s$ indicating the molar volume of the solvent.

The same calculation for a divalent B-ion gives, respectively:

$$G_{\bar{B}}(C_0)/RT - \bar{G}^{o}_{\bar{B}}/2RT = \Big[\tfrac{1}{2} \ln \bar{u}^2 C_0/2 - \tfrac{1}{2} + 1\tfrac{1}{2}(1 - 1/\bar{u}) C_0 / {}^{v}\Gamma \Big]_0^{C_0} \tag{3.19b}$$

$$= - \int_0^{C_0} \frac{3C_0}{2\,{}^{v}\Gamma} d(1 - 1/\bar{u}) + \frac{3C_0}{2\,{}^{v}\Gamma} (1 - 1/\bar{u})$$

$$= + \int_0^{C_0} \{(1 - 1/\bar{u}) / {}^{v}\Gamma\} d\, 3C_0/2 \tag{3.21b}$$

again identifying n_s as $(1 - 1/\bar{u})/\bar{V}_s {}^{v}\Gamma$, i.e. the salt-free volume per equivalent of exchanger, while this time $\ln a_s = -3C_0 \bar{V}_s/2$ for the chosen di-mono-valent equilibrium solution.

Finally the above procedure may be applied to the exchange reaction at a given level C_0. First one equivalent of the homoionic A-exchanger is discharged at C_0. One equivalent of B-salt at its standard state in solution is discharged (no work is involved in the case of the assumed ideal behavior), diluted to C_0 (involving $\frac{1}{2}$RT $\ln C_0/2$ for the cation) and the uncharged

B-ions are traded against the uncharged A-ions derived from the exchanger after which they are recharged on the exchanger. The uncharged A-ions are brought to their standard state in the presence of the uncharged anions of the B-salt and recharged. The total change of free enthalpy associated with this process equals:

$$(G_{\bar{B}} - \tfrac{1}{2}\mu_B^{\circ}) - (G_{\bar{A}} - \mu_A^{\circ}) = RT\Delta\left[\frac{1}{{}^{v}\Gamma}\int_0^{{}^{v}\Gamma} \ln \bar{u}' d^{v}\Gamma' + \frac{1}{z_k}\ln C_0/z_k\right]_{k=A}^{B} \tag{3.22}$$

i.e. the same expression as (3.16), with the understanding that now the bracketed expression must be calculated for both homoionic forms of the exchanger at C_0. The LHS may be recognized as $\Delta G_{\bar{B}} - \Delta G_{\bar{A}}$, i.e. the change of $\Delta G_{\overline{AB}}$ — as defined by (2.43) — when going from the homoionic A-exchanger at C_0 to the homoionic B-exchanger at C_0. Using the calculation as given in (3.19a) and (3.19b) one then finds:

$$\frac{\Delta G_{\bar{B}} - \Delta G_{\bar{A}}}{RT} = [\ln \bar{u}_B \sqrt{C_0/2} - \tfrac{1}{2} + 1\tfrac{1}{2}(1 - 1/\bar{u}_B)C_0/{}^{v}\Gamma]$$
$$- [\ln \bar{u}_A C_0 - 1 + 2(1 - 1/\bar{u}_A)C_0/{}^{v}\Gamma] \tag{3.23}$$

which is easily evaluated with the help of (3.15), either for a given value of C_0, or for a given value of a_s. In the latter case the answer obtained in this manner corresponds to $1/2 - {}^{0}\langle \ln K_N(a_s)\rangle^{1}$ (cf. (2.52) for a constant value of a_s). Accordingly the *mean* value of $\ln K_N(a_s)$ may be found directly by means of the present discharging—charging cycle as applied to both homoionic forms of the exchanger, i.e. without evaluation of the actual course of $\ln K_N(a_s)$. This fact may be used when invoking models that are more involved than the present Donnan distribution (cf. following section). For mono-divalent exchange this may be expressed in terms of a general equation of the type:

$${}^{0}\langle \ln K_N(a_s)\rangle^{1} = \tfrac{1}{2} - \Delta\left[\frac{F}{RT\sigma}\int_0^{\delta} \psi' d\sigma' + \frac{1}{z_k}\ln(a_k)\right]_{k=A_+}^{B_{2+}} \tag{3.24}$$

where again the bracketed expression is calculated at both homoionic endpoints, for a chosen value of a_s, and the corresponding values of the cationic activities a_k.

In contrast to the above, the mean value of $\ln K_N(C_0)$, i.e. the selectivity coefficient at constant electrolyte *concentration*, contains an additional term covering the change of the solvent term when going from $N_B = 0$ to $N_B = 1$ at constant C_0. As follows from (2.52) one finds in that case:

$${}^{0}\langle \ln K_N(C_0)\rangle^{1} = \tfrac{1}{2} - \frac{\Delta G_{\bar{B}} - \Delta G_{\bar{A}}}{RT} - \int_{n_B=0}^{1} n_s \, d\ln a_s \tag{3.25}$$

which still implies an integration from $N_B = 0$ to $N_B = 1$, though fortunately involving generally a rather small term. It may also be pointed out that $(\Delta G_{\bar{B}} - \Delta G_{\bar{A}})/RT$ already contains the same integral, but for the range from 0 to C_0 for both homoionic exchangers (cf. (3.21a) and (3.21b)). These components of the line integral $\int n_s \, d \ln a_s$ may actually be separated out by subtracting the limiting value of $(\Delta G_{\bar{B}} - \Delta G_{\bar{A}})/RT$ for $C_0 \to 0$, after which a reasonable guess may be made on its value in the connecting range from $N_B = 0$ to $N_B = 1$.

The above suggestion may be demonstrated for the present system as follows. Expanding the two bracketed expressions of the RHS of (3.23) (cf. (3.21)) gives:

$$
\begin{aligned}
{}^{\circ}\langle \ln K_N (C_0) \rangle^1 &= -\ln \sqrt{2^v\Gamma} + \int_0^{C_0} \{(1 - 1/\bar{u}_A)/{}^v\Gamma\} d\, 2C_0 \\
&\quad - \int_0^{C_0} \{(1 - 1/\bar{u}_B)/{}^v\Gamma\} d\, 3C_0/2 - \int_{n_B=0}^{1} n_s \, d \ln a_s \\
&\left(= \ln \underset{C_0 \to 0}{K_N} - \oint n_s \, d \ln a_s \right)
\end{aligned}
\qquad (3.26)
$$

Having thus identified $n_s(A) = (1 - 1/\bar{u}_A)/\bar{V}_s\, {}^v\Gamma$, i.e. the value of n_s for the homoionic A-exchanger at C_0, as well as $n_s(B)$, one may approximate the last integral by:

$$
- \int_{n_B=0}^{1} n_s \, d \ln a_s \approx - \frac{\{n_s(A) + n_s(B)\}}{2} \{\ln a_s(B) - \ln a_s(A)\} \qquad (3.27)
$$

Substituting this approximation into (3.25) and using the integral expression given in (3.23), one finds:

$$
\begin{aligned}
{}^0\langle \Delta \ln K_N (C_0) \rangle^1 &\equiv {}^0\langle \ln K_N (C_0) \rangle^1 - \ln \underset{C_0 \to 0}{K_N} \\
&\approx -\ln \sqrt{{}^v\Gamma/C_0} - \ln \frac{\bar{u}_B}{\bar{u}_A} + 2(1 - 1/\bar{u}_A) C_0/{}^v\Gamma \\
&\quad - 1\tfrac{1}{2}(1 - 1/\bar{u}_B) C_0/{}^v\Gamma - \tfrac{1}{4}\left(\frac{1}{\bar{u}_A} + \frac{1}{\bar{u}_B}\right) C_0/{}^v\Gamma
\end{aligned}
\qquad (3.28)
$$

In Table 3.1 the results of (3.28) are compared with those of the exact expression given in (3.14) as well as those that would be found from (3.25) neglecting completely the last term.

It appears from this Table that for the Donnan model the use of (3.28) is amply satisfactory, thus omitting the need for the actual integration over the range $N_B = 0$ to $N_B = 1$.

Aside from the course of $\ln K_N$ (and its mean value) as a function of the external electrolyte level as typified by $C_0/{}^v\Gamma$, it is also of interest to investigate the activity coefficients $\bar{f}_k$ for the present system. Using (2.31a) one finds for the present system:

TABLE 3.1

Exact and approximate calculation of the shift in $\langle \ln K_N(C_0)\rangle$ with increasing concentration of the equilibrium solution

${}^v\Gamma/C_0$	$\langle \Delta \ln K_N(C_0)\rangle$ (3.14)	$\langle \Delta \ln K_N(C_0)\rangle$ (3.28)	$\langle \Delta \ln K_N(C_0)\rangle$ (same, neglecting last term)
∞	0	0	0
20	0.013	0.012	0.034
10	0.033	0.031	0.071
5	0.083	0.080	0.148
3	0.153	0.148	0.245
2	0.238	0.232	0.353
1	0.447	0.440	0.596

$$\ln \bar{f}_A = \tfrac{1}{2} N_B + N_B \ln K_N - \int_0^{N_B} \ln K_N \, dN_B - \int_{0,0}^{C_0, N_B} n_s \, d\ln a_s \tag{3.29}$$

where the line integral is taken from the standard state via the homoionic A-exchanger at concentration C_0 to the point N_B at concentration C_0. Replacing $\ln K_N$ in the second and third term by $\Delta \ln K_N$, application of the integration procedure employed in (3.8)—(3.12) yields:

$$\ln \bar{f}_A = \tfrac{1}{2} N_B - \ln(1 - 1/\bar{u}^2) - \int_{0,0}^{C_0, N_B} [\ln a_s \, dn_s + n_s \, d\ln a_s]$$

$$= \tfrac{1}{2} N_B - \ln(1 - 1/\bar{u}^2) - n_s \ln a_s \tag{3.30a}$$

where the lower limit of the integral equals zero. In similar fashion one finds:

$$\tfrac{1}{2} \ln \bar{f}_B = -\tfrac{1}{2}(1 - N_B) - \ln\sqrt{1 - 1/\bar{u}^3} - n_s \ln a_s \tag{3.30b}$$

As $n_s = (1 - 1/\bar{u})/{}^v\Gamma\bar{V}_s$ and $\ln a_s \approx (2 - f/2)C_0\bar{V}_s$, the activity coefficients are solely dependent upon the relative concentration $C_0/{}^v\Gamma$ (and f). Accordingly the curves representing $\ln \bar{f}_A$ and $\tfrac{1}{2}\ln \bar{f}_B$ have been plotted against N_B in Fig. 3.2 for $C_0/{}^v\Gamma = 1$, $\tfrac{1}{2}$ and 0, respectively. In the last case the activity coefficients follow the "ideal" curve for mono-divalent exchange as discussed in the previous chapter in conjunction with Fig. 2.1b. It may be noted that the distance between the lines corresponding to $C_0/{}^v\Gamma = 1$ minus the distance between the ones corresponding to ideal behavior corresponds to the course of $\Delta \ln K_N$ for the same concentration in Fig. 3.1.

The discussion of the behavior of the "ideal Donnan system" may be summarized completely in one graph as shown in Fig. 3.3. Thus any deviation from "ideal exchanger behavior" — defined in chapter 2 as corresponding to K_N = constant — is caused solely by the penetration of co-ions (or salt) inside the exchanger compartment. This salt penetration is aptly reflected in the deviation of the parameter $(1 - 1/\bar{u})$ from unity (at $C_0/{}^v\Gamma = 0$). As $\bar{u}$ is fully determined by $C_0/{}^v\Gamma = V^\alpha C_0$ (and f, cf. (3.3),

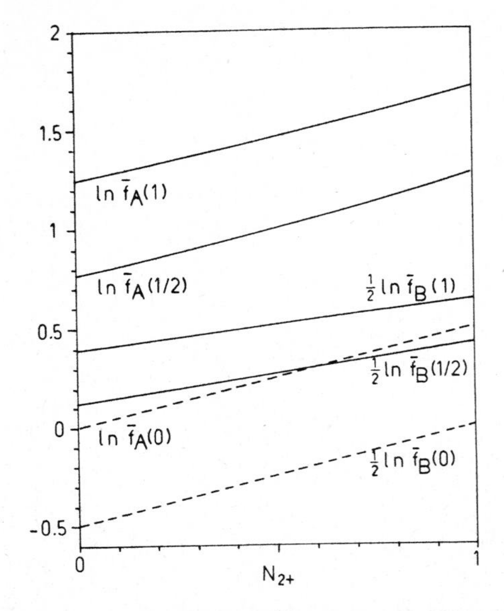

Fig. 3.2. Activity coefficients pertaining to an adsorbed phase with uniform electrostatic field containing monovalent A-ions and divalent B-ions, for different values of the relative electrolyte concentration, $C_0/^v\Gamma$, as indicated in parentheses.

while the latter parameters also define $\ln a_s/^v\Gamma\bar{V}_s = (2 - f/2)C_0/^v\Gamma$, the magnitude of all integrals of the type $\int n_s \, d\ln a_s$ and $\int \ln a_s \, dn_s$ are solely dependent upon $C_0/^v\Gamma$. In Fig. 3.3 the above dimensionless parameters are plotted against each other, for the monovalent A-exchanger marked $A(C_0/^v\Gamma)$ and for the divalent B-exchanger marked $B(C_0/^v\Gamma)$. Connecting these two curves are the lines corresponding to a constant value of $C_0/^v\Gamma$, as found from (3.3). The complete line integral $-\int n_s \, d\ln a_s$ then contains the corresponding surface area, giving a positive number (equal to $^0\langle\Delta\ln K_N\rangle^1$ as given in Fig. 3.1, e.g. 0.24 when going up to A(1/2)). The value of the activity coefficients is also implied in this figure. To read, for example, $\ln \bar{f}_A$ at the point marked M(1/2), i.e. a mixture at $C_0/^v\Gamma = \frac{1}{2}$, it may be established first that:

$$-\ln(1 - 1/\bar{u}^2) \equiv \int_\infty \frac{2}{(\bar{u} - 1/\bar{u})} \, d(1 - 1/\bar{u})$$

where the integrand equals $(\ln a_s)/^v\Gamma\bar{V}_s$ at $N_A = 1$ (or $N_B = 0$). This integral thus corresponds to the area (A(0), p, x) for the mixture M(1/2), in the present case equal to about 0.30. The last term of (3.30a) is represented by the rectangular area (O, x, M, y), and equal to 0.47. Together this gives about 0.75 for $\Delta \ln \bar{f}_A$, i.e. the deviation from the value of $\ln \bar{f}_A$ in the case of the ideal exchanger. In fact the given Figure could be completed by plotting the bundle of lines for constant N_B between the two extremes shown, in order

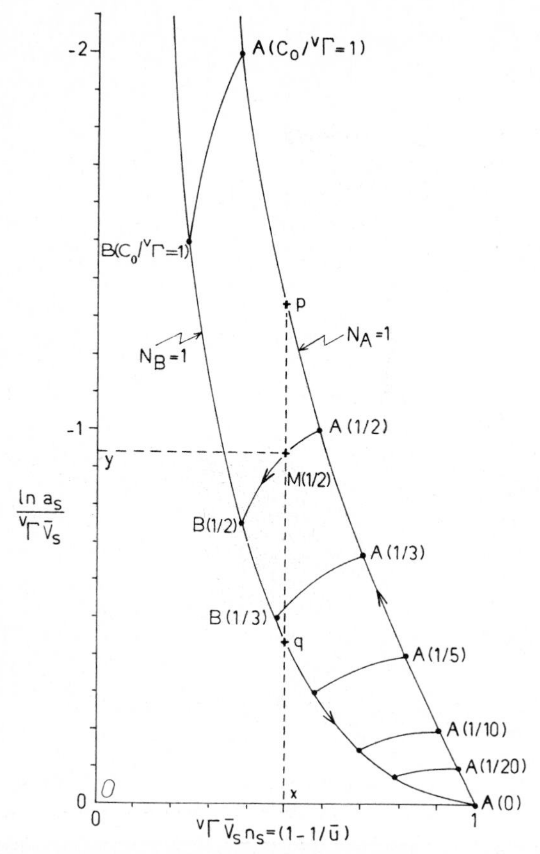

Fig. 3.3. Scaled diagram of the solvent content, n_s, as a function of the solvent activity, a_s, for an adsorbed phase with uniform electrostatic field. The line marked $A(C_0/{}^v\Gamma)$ indicates the homoionic adsorbed phase containing monovalent cations, the line marked $B(C_0/{}^v\Gamma)$ indicates the same but containing divalent cations. Connecting lines refer to mixed compositions at constant value of $(C_0/{}^v\Gamma)$.

to allow reading of the approximate value of N_B. Similarly, $\frac{1}{2}\ln \bar{f}_B$ is found from the areas (O, x, M, y) plus (B(0), q, x). Finally $\Delta \ln K_N$ is represented by the area (A(0), p, q), showing that $\Delta \ln K_N$ intersects its mean value slightly beyond the point M (which corresponds to $N_B = 0.44$) which may be verified from Fig. 3.1. For any level of $C_0/{}^v\Gamma$ one may thus estimate the deviation between the local value of $\Delta \ln K_N$ and its mean value from the area [(M, A, p) − (M, B, q)]. It is finally noted that $n_s \ln a_s$ goes from zero through positive values everywhere and back to zero. In contrast $-\int n_s \, d \ln a_s$ increases from zero to a maximum when leaving the A-line, after which it returns to lower values leaving the area [A(0), $A(C_0/{}^v\Gamma)$, $B(C_0/{}^v\Gamma)$, A(0)] as its value at the standard state of the B-exchanger. For the same cycle $-\int \ln a_s \, d n_s$ decreases from zero to negative values when moving along the A-line, falls off even more until the B-line is reached after which it returns. leaving the same area as above, but with a negative sign.

As has been made clear before, deviations from ideality remain rather small for the present system unless $C_0/{}^{v}\Gamma$ exceeds at least 0.1. In terms of soil colloids, with ${}^{v}\Gamma$ probably at least equal to $1\,N$, this would imply that the deviations due to the factors considered here should be small. Of course this leaves untouched the magnitude of $\ln K_N$ which should easily contain a (hopefully fairly constant) factor corresponding to non-ideal solution behavior inside the adsorbed phase.

3.2. ION ACCUMULATION IN THE DIFFUSE DOUBLE LAYER

The electric field present in the Gouy-type diffuse double layer may be considered as a contrasting extreme in comparison to the fully homogeneous field discussed in the previous section. In principle the distribution of mono- and divalent cation species follows the same rules as before, with the provision that now a distributed selectivity occurs throughout the adsorbed phase. Maximal preference for the divalent species is found close to the adsorber surface where the electric potential is highest.

With the help of the equations describing the distribution of the electric potential as a function of the position relative to the surface given in chapter 1, one may calculate the *excess* of any ion species via the equation:

$$\Gamma_k = f_k C_0 \int_{\delta}^{x_t} (u^{z_k} - 1)dx \tag{3.31}$$

in which δ and x_t indicate the limiting positions at the exchanger surface and plane of truncation of the double layer, u_k is the Boltzmann accumulation factor of the monovalent species and f_k indicates again the equivalent fraction in the equilibrium solution. Using a system composition as before, i.e.:

$$\begin{aligned} c_{0,A} &= (1-f)C_0 \\ c_{0,B} &= fC_0 \\ c_{0,-} &= C_0 \end{aligned} \tag{3.32}$$

with A and B representing mono- and divalent cation species, one finds:

$$N_A = \frac{(1-f)C_0}{\Gamma} \int_{\delta}^{x_t} (u - 1/u)dx \tag{3.33a}$$

$$N_B = \frac{fC_0}{\Gamma} \int_{\delta}^{x_t} (u^2 - 1/u)dx \tag{3.33b}$$

If u(x) is known (cf. chapter 1), (3.33) can be worked out after which the value of K_N is found for any given composition of the system. In practice this is done by replacing the integrals over x by the corresponding ones over u with the help of (1.9). In the present case this gives:

$$N_A = \frac{(1-f)C_0}{\Gamma} \frac{1}{\sqrt{\beta C_0}} \times$$

$$\int_{u_s}^{u_t} \frac{(u-1/u)du}{u\sqrt{\left\{\frac{f}{2}u^2+(1-f)u+1/u\right\}-\left\{\frac{f}{2}u_t^2+(1-f)u_t+1/u_t\right\}}} \tag{3.34}$$

and a corresponding expression for N_B containing $f(u^2-1/u)$ in the numerator of the integrand. At this stage it is clear that the calculation of N_A and N_B is much more involved than in the homogeneous field; nevertheless the outcome will follow the same overall pattern as found in the former case since the above integrals represent only the appropriate space average of the accumulation factors $(u-1/u)$ and (u^2-1/u). As will be shown later some special cases, particularly those pertaining to the freely extended double layer, are found fairly easily (see also Bolt and Page, 1965).

The easiest way to obtain an overall insight into the behavior of this system is to apply again the discharging-charging cycle to the homoionic endpoints. For this purpose one must first relate the surface density of charge, Γ, to the value of u_s, the Boltzmann factor at the exchanger surface. Applying (1.39) for this purpose one finds for the freely extended double layer:

$$\bar{\Gamma}^2 \equiv \frac{\beta\Gamma^2}{C_0} = 2(u_s-1)^2(f+2/u_s) \tag{3.35}$$

and thus for the homoionic endpoints:

$$\bar{\Gamma} = 2(u_{A,s}-1)/\sqrt{u_{A,s}} \tag{3.36a}$$

$$\bar{\Gamma} = (u_{B,s}-1)\sqrt{2(1+2/u_{B,s})} \tag{3.36b}$$

Applying (3.16) gives the free enthalpy of the homoionic exchanger at concentration C_0 as:

$$\frac{G_{\bar{A}}-\bar{G}_{\bar{A}}^{o}}{RT} = \Delta\left[\frac{1}{\bar{\Gamma}}\int_0^{\bar{\Gamma}} \ln u'_{A,s}\, d\bar{\Gamma}' + \ln C_0\right]_0^{C_0} \tag{3.37a}$$

Partial integration of the bracketed expression gives:

$$\begin{aligned}\left[\ \right] &= \ln u_{A,s} C_0 - \frac{2}{\bar{\Gamma}}\int_1 (\sqrt{u_{A,s}}-1/\sqrt{u_{A,s}})\, d\ln u_{A,s} \\ &= \ln u_{A,s} C_0 - 4(\sqrt{u_{A,s}}-1)/\bar{\Gamma} + 4(1-1/\sqrt{u_{A,s}})/\bar{\Gamma}\ * \\ &= \ln u_{A,s} C_0 - 2 + 8(1-1/\sqrt{u_{A,s}})/\bar{\Gamma}\end{aligned} \tag{3.38a}$$

It is noted in passing that the sum of the second and third term in this equation corresponds to $-(\int \Gamma' d\ln u')/\Gamma$, which is equivalent to the expression $-\int \sigma d\psi$ referred to on p. 53 as applicable to surfaces with a fixed surface potential controlled by a potential-determining ion. The above sum, if expressed per unit surface area, is also identical to eqn. (3.124) of Babcock (1963), though this identity is obscured somewhat by the latter's choice of parameters and his stated intent to find an expression for the free enthalpy applicable to surfaces with a given charge density rather than the one pertaining to surfaces with a fixed potential.

Expanding the first term with the help of (3.35) and (3.36a) gives:

$$\ln(u_{A,s}C_0) = \ln\frac{\beta\Gamma^2}{4} - 2\ln(1-1/u_{A,s})$$

$$= \ln\frac{\beta\Gamma^2}{4} - \int_\infty \frac{2}{1-1/u_{A,s}}\, d(1-1/u_{A,s})$$

$$= \ln\frac{\beta\Gamma^2}{4} - \int_\infty \frac{8}{\Gamma}\, d(1-1/\sqrt{u_{A,s}}) \qquad (3.39a)$$

Together this leads to:

$$[\;] = \left[\ln\frac{\beta\Gamma^2}{4} - 2\right] + \int_{C_0=0} 8(1-1/\sqrt{u_{A,s}})\, d\sqrt{C_0}/\Gamma\sqrt{\beta}$$

$$= \left[\ln\frac{\beta\Gamma^2}{4} - 2\right] - \frac{1}{\Gamma\bar{V}_s}\int_0 \{(1-1/\sqrt{u_{A,s}})2/\sqrt{\beta C_0}\}\, d(-2C_0\bar{V}_s) \qquad (3.40a)$$

where the first term presents the limiting value at $C_0 \rightarrow 0$, while the second one has been rearranged in such a manner that the integrator equals $\ln a_s$ as defined before in conjunction with (3.21a). Applying the limits of C_0 as required in (3.37a) makes this first term disappear and $(G_{\bar{A}} - G_{\bar{A}}^{o})/RT$ thus equals the second term. According to (2.51a) this term must then equal the solvent term, such that:

$$n_s(A)\, d\ln a_s = \frac{2(1-1/\sqrt{u_{A,s}})}{\sqrt{\beta C_0}}\, \frac{1}{\Gamma\bar{V}_s}\, d(-2C_0\bar{V}_s) \qquad (3.41a)$$

This then identifies the solvent associated with one equivalent of exchanger for the homoionic A-form as:

$$n_s(A) = \frac{2}{\sqrt{\beta C_0}}(1-1/\sqrt{u_{A,s}})\frac{1}{\Gamma\bar{V}_s} \qquad (3.42a)$$

As will be shown in chapter 7 this checks precisely with the volume of exclusion of the co-ions in the present mono-monovalent system.

The above procedure may be repeated for the homoionic B-exchanger yielding, respectively:

$$\frac{G_{\bar{B}} - \bar{G}_{\bar{B}}^{o}/2}{RT} = \Delta\left[\frac{1}{\bar{\Gamma}}\int_0^{\bar{\Gamma}} \ln u_{B,s}'\, d\bar{\Gamma}' + \tfrac{1}{2}\ln(C_0/2)\right]_0^{C_0} \qquad (3.37b)$$

with

$$\left[\;\right] = \tfrac{1}{2}\ln(u_{B,s}^2 C_0/2) - 1 + 3(\sqrt{6} - \sqrt{2 + 4/u_{B,s}})/\bar{\Gamma} \tag{3.38b}$$

and:

$$\tfrac{1}{2}\ln(u_{B,s}^2 C_0/2) = \tfrac{1}{2}\ln\frac{\beta\Gamma^2}{4} + \int_\infty \frac{3}{(u_{B,s} - 1)\sqrt{2 + 4/u_{B,s}}}\, d\sqrt{2 + 4/u_{B,s}}$$

$$= \tfrac{1}{2}\ln\frac{\beta\Gamma^2}{4} - \int_\infty \frac{3}{\bar{\Gamma}}\, d\{\sqrt{6} - \sqrt{2 + 4/u_{B,s}}\} \tag{3.39b}$$

such that:

$$\left[\;\right] = \left[\tfrac{1}{2}\ln\frac{\beta\Gamma^2}{4} - 1\right] - \frac{1}{\Gamma\bar{V}_s}\int_0 \frac{\sqrt{6} - \sqrt{2 + 4/u_{B,s}}}{\sqrt{\beta C_0}}\, d(-1\tfrac{1}{2}\, C_0 \bar{V}_s) \tag{3.40b}$$

Again the first term represents the limiting value at $C_0 = 0$, while the second term is recognized as $-\int n_s \, d\ln a_s$ for the homoionic B-exchanger, representing $(G_{\bar{B}} - \bar{G}_{\bar{B}}^{\circ}/2)/RT$. Thus:

$$n_s(B) = \frac{\sqrt{6} - \sqrt{2 + 4/u_{B,s}}}{\sqrt{\beta C_0}} \; \frac{1}{\Gamma\bar{V}_s} \tag{3.42b}$$

with checks again with the exclusion of monovalent anions in a di-monovalent system (see chapter 7). Finally the two homoionic exchangers may be connected in analogy with (3.22) and (3.23), which gives:

$$\frac{\Delta G_{\bar{B}} - \Delta G_{\bar{A}}}{RT} = \left[\tfrac{1}{2}\ln(u_{B,s}^2 C_0/2) - 1 + 3(\sqrt{6} - \sqrt{2 + 4/u_{B,s}})/\bar{\Gamma}\right]$$

$$- [\ln(u_{A,s} C_0) - 2 + 8(1 - 1/\sqrt{u_{A,s}})/\bar{\Gamma}] \tag{3.43}$$

This equation then determines again the mean value of $\ln K_N(a_s)$, as was already mentioned in connection with (3.23) for the homogeneous electric field. Again the mean value of $\ln K_N(C_0)$ requires the addition of the integral $-\int n_s \, d\ln a_s$ over the range $N_B = 0$ to $N_B = 1$. Using an averaging procedure to obtain n_s within that range (see (3.27)) it is thus possible to calculate the mean value of $\ln K_N(C_0)$ for different values of C_0 without prior calculations for mixed systems (which tend to be cumbersome in this case). As in the previous case it is convenient to separate out the limiting value of $\ln K_N$ at $C_0 \rightarrow 0$ from (3.43). Referring to the limiting values identified in (3.40a) and (3.40b) one finds for the Gouy system:

$$\frac{\Delta\bar{G}_{\bar{B}}^{\circ}/2 - \Delta\bar{G}_{\bar{A}}^{\circ}}{RT} \equiv \Delta G_{ex}^{\circ}/RT \equiv \ln K_{ex}^{\circ} \equiv \tfrac{1}{2} - \langle \ln K_N \rangle_{C_0 \rightarrow 0}$$

$$= 1 - \ln\sqrt{\frac{\beta\Gamma^2}{4}} \tag{3.44}$$

This gives the increment of $\langle \ln K_N(C_0) \rangle$ with respect to this limiting value as:

$$\langle \Delta \ln K_N(C_0) \rangle \approx -2\ln(1-1/u_{A,s}) + 8(1-1\sqrt{u_{A,s}})/\Gamma + \ln(1-1/u_{B,s}) + \tfrac{1}{2}\ln(1+2/u_{B,s}) - 3(\sqrt{6}-\sqrt{2+4/u_{B,s}})/\Gamma - \frac{\{n_s(A)+n_s(B)\}}{2}\Delta \ln a_s \tag{3.45}$$

In analogy with the homogeneous electric field treated in section 3.1, the influence of the concentration of the equilibrium solution on the "mean" valence selectvity in the Gouy layer may be summarized in one graph relating n_s to $\ln a_s$. In order to obtain such a single graph the solvent content is expressed in the scaled parameter $\Gamma n_s \bar{V}_s/(1/\beta\Gamma)$, of which the numerator specifies the solvent content in m^3 per m^2 adsorber surface (cf. the exclusion distance of co-ions discussed in chapter 7), while the denominator is a length parameter characterizing the surface charge density (much like the D-H length parameter characterizing the ionic strength of a solution). Dividing the solvent activity by the same scaling factor then gives $(\ln a_s)/\beta\Gamma^2\bar{V}_s$, which is a relative concentration of the equilibrium solution (for ideal behavior) in units of $C_0/\beta\Gamma^2$.

Disregarding for the time being the actual magnitude of the surface charge density occurring in soil clays one may first compare the present system with the previous one by plotting on the same scale as used in Fig. 3.3. As is immediately apparent from Fig. 3.4, the area covered by the line integral $\int n_s \, d\ln a_s$ is much larger than in the previous case, for any chosen value of the relative concentration of the equilibrium solution. This is caused by the growing extent of the co-ion exclusion zone with decreasing concentration (see chapter 7). Accordingly, the first conclusion is that the shift of $\langle \ln K_N \rangle$ with a change in electrolyte level is much more pronounced in the present case. A replot at one-tenth scale is given in the lower left hand corner of Fig. 3.4, indicating that $\langle \Delta \ln K_N(C_0) \rangle$ has already a magnitude exceeding about 0.5 at $C_0/\beta\Gamma^2 = 0.05$, as against a value of less than 0.02 for the homogeneous field pictured in Fig. 3.3. Although the present scale of plotting does not allow a graphical estimate of $\langle \Delta \ln K_N(C_0) \rangle$ of sufficient accuracy, it conveys the general picture to be expected, i.e. $\langle \ln K_N \rangle$ will start increasing with C_0 at fairly low values of $C_0/\beta\Gamma^2$. Sufficiently precise values may be found with (3.45), as given in Table 3.2.

Comparison with Table 3.1 shows the much greater influence of the electrolyte level on $\langle \ln K_N \rangle$ in the present case. Referring to Fig. 3.6, to be discussed later, the values tabulated above were used to indicate the mean position of $\ln K_N$, relative to its limiting value for $C_0 = 0$ situated at $\langle \ln K_N(C_0 = 0) \rangle = \tfrac{1}{2}\ln(\beta\Gamma^2/4) - \tfrac{1}{2}$.

In contrast to the homogeneous electrostatic field treated before, the present system does not exhibit a constant value of $\ln K_N$ at vanishing electrolyte level. This follows immediately from (1.39), in chapter 1, which

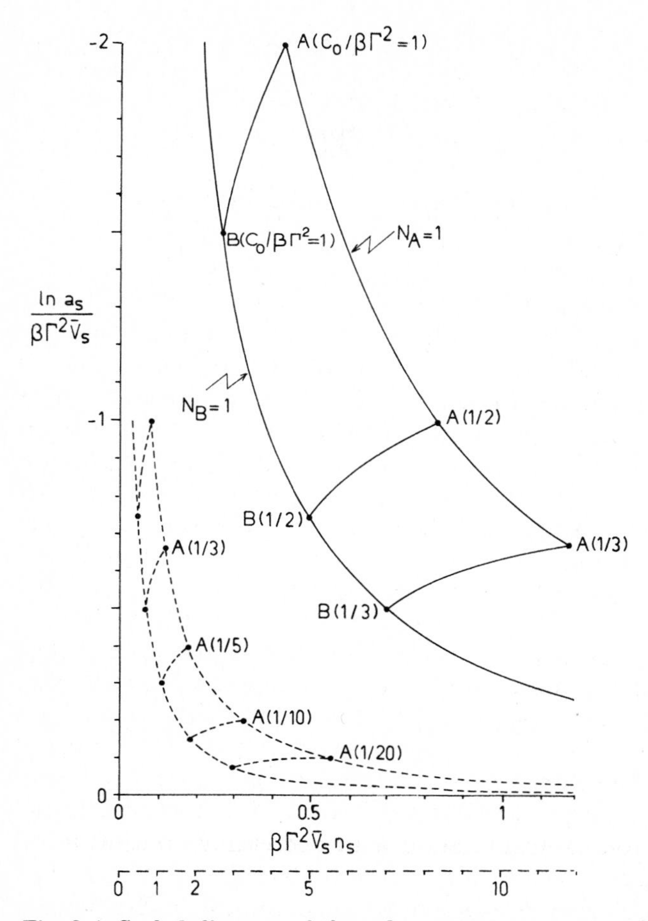

Fig. 3.4. Scaled diagram of the solvent content, n_s, as a function of the solvent activity, a_s, for an adsorbed phase consisting of a freely extended diffuse double layer (Gouy type). Dashed lines refer to the corresponding axis; cf. also Fig. 3.3.

TABLE 3.2

Shift of $\langle \ln K_N(C_0) \rangle$ for a Gouy distribution, as a function of $\beta\Gamma^2/C_0$

$\beta\Gamma^2/C_0$	$\langle \Delta \ln K_N(C_0) \rangle$	$\beta\Gamma^2/C_0$	$\langle \Delta \ln K_N(C_0) \rangle$
∞	0.00	30	0.64
10^4	0.04	20	0.75
$3 \cdot 10^3$	0.07	10	0.97
10^3	0.13	5	1.23
$3 \cdot 10^2$	0.23	3	1.44
10^2	0.38	2	1.61
		1	1.92

indicates that at very low values of C_0 the total *molar* concentration at the surface attains a fixed value equal to $\beta\Gamma^2/4$. With an increasing percentage of divalent ions in the diffuse double layer, the fractional neutralization of the surface charge by the ions present very close to the surface thus increases, giving rise to a shrinkage in the extent of the double layer field. Accordingly, the selectivity for divalent cations must decrease with increasing percentage of divalent ions present.

In order to find the actual course of $\ln K_N$ one must first calculate the distribution of the ions in the double layer. Using for this purpose (3.34) one finds for the freely extended double layer (cf. Bolt and Page, 1965):

$$\begin{aligned} N_A &= \frac{1-f}{\sqrt{f/2}}\,\frac{1}{\Gamma}\int_{u_s}^{1}\frac{(u+1)du}{u\sqrt{u^2+2f/u}} \\ &= \frac{1-f}{\sqrt{f/2}}\,\frac{1}{\Gamma}\left[\arg\sinh\left\{\Gamma\sqrt{f/2}\,\frac{u_s}{u_s-1}\right\} - \arg\sinh\sqrt{f(f+2)}\right] \\ &\quad + \frac{1}{\overline{\Gamma}}\left[\sqrt{2f+4}-\sqrt{2f+4/u_s}\,\right] \end{aligned} \tag{3.46}$$

in which the second bracketed term constitutes the co-ion deficit (cf. chapter 7). Together with (3.35), relating u_s to f for chosen values of Γ and C_0, one may then calculate N_A as a function of the reduced (molar) concentration ratio in the equilibrium solution, $r \equiv {}^m c_{0,+}/\sqrt{{}^m c_{0,2+}} = (1-f)\sqrt{2C_0/f}$. The corresponding value of K_N is then found as $r\sqrt{(1-N_A)}/N_A$. The course of $\ln K_N$ for the limiting value of $C_0 \to 0$ may actually be found from a simplified equation. In that case the second bracketed term obviously disappears with $\overline{\Gamma} \to \infty$ (co-ions being absent). At the same time the first term disappears for all finite values of f, leading to the conclusion that f must go to zero simultaneously with C_0 (at least for $N_A > 0$). In contrast, the reduced ratio r remains finite with f and C_0 going to zero. Introducing this limiting value as:

$$r_0 \equiv (1-f)\sqrt{2C_0/f}_{\;C_0\to 0,\, f\to 0} = \sqrt{2C_0/f} \tag{3.47}$$

one finds:

$$\lim_{C_0\to 0} N_A = \frac{\arg\sinh(\Gamma\sqrt{\beta}/r_0)}{\Gamma\sqrt{\beta}/r_0} \tag{3.48}$$

and:

$$\lim_{C_0\to 0}\ln K_N = \ln(\Gamma\sqrt{\beta}) + \tfrac{1}{2}\ln\left[\frac{1-\{\arg\sinh(\Gamma\sqrt{\beta}/r_0)\}/(\Gamma\sqrt{\beta}/r_0)}{\{\arg\sinh(\Gamma\sqrt{\beta}/r_0)\}^2}\right] \tag{3.49}$$

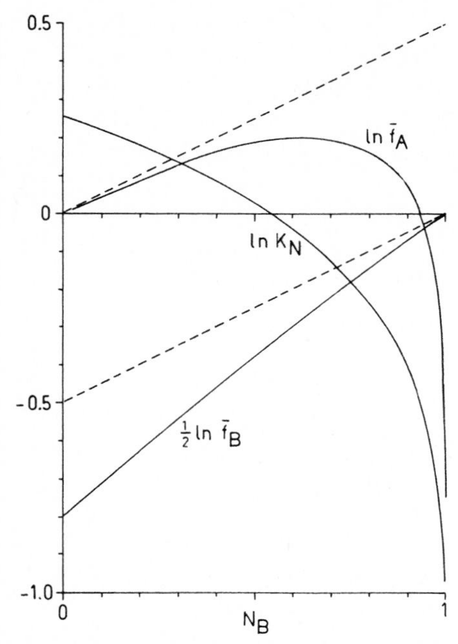

Fig. 3.5. Limiting trajectories of K_N, $\bar{f}_A$ and $\bar{f}_B$ as calculated for a Gouy-type diffuse double layer containing monovalent A-ions and divalent B-ions, in equilibrium with an infinitely dilute solution of the corresponding salts. Dashed lines refer to the ideal exchanger as defined in chapter 2 (p. 36).

For the homoionic A-exchanger the above expression gives, by expansion of the argument of the hyperbolic sine for $r_0 \to \infty$:

$$\lim_{C_0 \to 0,\, r_0 \to \infty} \ln K_N = \ln \Gamma\sqrt{\beta} + \tfrac{1}{2} \ln \left[\frac{1 - 1 + \beta\Gamma^2/6r_0^2}{\beta\Gamma^2/r_0^2}\right]$$

$$= \tfrac{1}{2} \ln (\beta\Gamma^2/6) \qquad (3.50)$$

which, compared with its mean value given in (3.44), gives:

$$\langle \ln K_N \rangle_{C_0 \to 0} = \tfrac{1}{2} \ln (\beta\Gamma^2/4) - \tfrac{1}{2} \qquad (3.51)$$

For the homoionic B-exchanger one finds $\ln K_N$ going to $-\infty$ corresponding $-\ln\ln(2\Gamma\sqrt{\beta}/r_0)$ with $r_0 \to 0$. In Fig. 3.5 the curve representing $\ln K_N$ ($C_0 \to 0$) has been plotted for $\beta\Gamma^2$ equal to $10\,kmol/m^3$ (corresponding roughly to the charge density of montmorillonite at about $0.1\,Coul/m^2$). The corresponding values of the activity coefficients in this limiting case are easily found from:

$$\ln \bar{f}_A = \tfrac{1}{2} N_B + \int_0^{N_B} N_B \, d \ln K_N$$

$$= N_B - \ln \operatorname{arg\,sinh} (\Gamma\sqrt{\beta}/r_0)\Big|_{\infty}^{r_0} + \int_1^{N_A} r_0 \, dN_A / r_0 \qquad (3.52a)$$

$$= N_B - \ln \operatorname{arg\,sinh} \Gamma\sqrt{\beta}/r_0 - \ln \{(1 + \sqrt{1 + \beta\Gamma^2/r_0^2})/2(\Gamma\sqrt{\beta}/r_0)\}$$

$$\tfrac{1}{2} \ln \bar{f}_B = -N_A - \tfrac{1}{2} \ln (1 - N_A) - \ln \{(1 + \sqrt{1 + \beta\Gamma^2/r_0^2})/(\Gamma\sqrt{\beta}/r_0)\} \qquad (3.52b)$$

As shown in Fig. 3.5, the activity coefficients are in the present case quite different from those pertaining to ideal behavior (dashed lines).

Using the complete equation (3.46), $\ln K_N$ may be plotted for chosen values of $\bar{\Gamma}$, as has been done in Fig. 3.6. The absolute value of $\ln K_N$ then hinges on the position of the limiting curve for $C_0 \rightarrow 0$, corresponding to $1/\bar{\Gamma}^2 \rightarrow 0$. To facilitate comparison with Fig. 3.4 the latter parameter has been indicated on the curves between parentheses.

As to the absolute value the vertical axes have been calibrated for $\beta\Gamma^2 = 10\,kmol/m^3$ (RHS, marked M, corresponding to, for example, montmorillonite with $\Gamma \approx 10^{-9}\,keq/m^2$) and for $\beta\Gamma^2 = 100\,kmol/m^3$ (LHS, marked I, corresponding to, for example, illitic clays). It is interesting to note that for plate adsorbers with a charge density comparable to that of montmorillonite, $\ln K_N$ would show a reversal in trend with increasing N_B in the concentration range of practical concern between 10^{-3} and $1\,N$ electrolyte. In contrast the illitic clays should be hardly sensitive to electrolyte levels up to $10^{-2}\,N$.

As the implicit character of the set of equations (3.46) and (3.35) necessitates computer assistance for the calculation of $\ln K_N$ (as a function of C_0 and N_B) and corresponding values of $\ln \bar{f}_k$, it is worth-while to use the much simpler equations pertaining to symmetric systems for further exploration of the system behavior. As introduced already in chapter 1 such a system is characterized by an equilibrium solution with symmetric anion—cation concentrations, according to:

$$c_0^+ = c_0^- = (1 - f) C_0$$

$$c_0^{2+} = c_0^{2-} = f C_0 \qquad (3.53)$$

In view of the variable composition of a soil solution with respect to the usually dominant anions Cl^- and SO_4^{2-} the assumed symmetric composition may be used correctly as an average situation. If in addition the simplifying convention is introduced to assign the deficit of monovalent anions fully to the monovalent cations in defining the equivalent fraction of the monovalent cations adsorbed, one finds:

$$N_A^* \equiv \frac{(1 - f) C_0}{\Gamma} \int_{\delta}^{x_t} (u - 1/u) dx \qquad (3.54a)$$

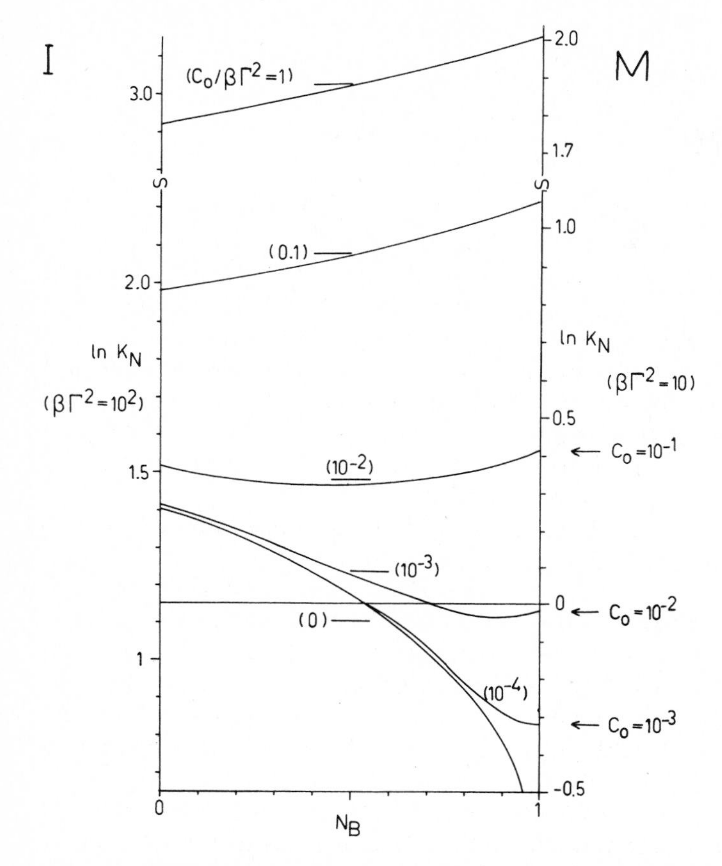

Fig. 3.6. The selectivity coefficient, K_N, for a fully extended Gouy-type diffuse double layer, containing monovalent A-ions and divalent B-ions at different values of the relative concentration of the equilibrium solution, $C_0/\beta\Gamma^2$, as indicated in parentheses. Mean values of $\ln K_N$ were calculated by (3.45) as given in Table 3.2. The absolute values of $\ln K_N$ as read on the RH-axis correspond to an exchanger surface with a charge density akin to that of montmorillonite (≈ 0.1 Coul/m^2), the LH-axis refers to illites.

$$N_B^* \equiv \frac{fC_0}{\Gamma} \int_{\delta}^{x_t} (u^2 - 1/u^2)\,dx \tag{3.54b}$$

With the help of (1.12) of chapter 1 this gives:

$$N_A^* = \frac{(1-f)}{\sqrt{f/2}} \frac{\sqrt{C_0}}{\Gamma\sqrt{\beta}} \int_{v_t}^{v_s} \frac{\sqrt{v^2-4}\,dv}{\sqrt{(v^2-4)(v-v_t)\{v+v_t+2(1-f)/f\}}} \tag{3.55}$$

in which $v = u + 1/u$, the "mixed" Boltzmann factor used before in chapter 1. Integration then gives:

$$N_A^* = \frac{r}{\Gamma\sqrt{\beta}} \operatorname{arg\,sinh} \frac{\sqrt{(v_s-v_t)\{v_s+v_t+2(1-f)/f\}}}{2\{v_t+(1-f)/f\}} \tag{3.56}$$

with $r \equiv {}^m c_{0,+}\sqrt{{}^m c_{0,2+}} = (1-f)\sqrt{2C_0/f}$. Expressing (1.39) in terms of v_s yields the equivalent of (3.35) which is applicable to the present symmetric system:

$$\bar{\Gamma}^2 = \frac{\beta\Gamma^2}{C_0} = 2f(v_s - v_t)\{v_s + v_t + 2(1-f)/f\} \tag{3.57}$$

Accordingly, (3.56) may be written:

$$N_A^* = \frac{r}{\Gamma\sqrt{\beta}} \arg\sinh \frac{\Gamma\sqrt{\beta}}{r + 2v_t C_0\sqrt{f/2C_0}} \tag{3.58}$$

which circumvents the calculation of u_s (or v_s) which was necessary in the case of (3.46). In fact (3.58) becomes an explicit relation of N_A^* in terms of f and C_0 (or r) in the case of the freely extended double layer as then $v_t \equiv u_t + 1/u_t = 2$.

Eqn. (3.58) was derived before by Eriksson (1952). For the freely extended double layer it may also be expressed in terms of the dimensionless system of parameters:

$$\bar{r}(f) \equiv (1-f)/\sqrt{f/2} = r/\sqrt{C_0}$$

$$\bar{\Gamma} \equiv \Gamma\sqrt{\beta}/\sqrt{C_0}$$

as

$$N_A^* = \frac{\bar{r}}{\bar{\Gamma}} \arg\sinh \left[\frac{\bar{\Gamma}}{\bar{r}}\,\frac{(1-f)}{(1+f)}\right] \tag{3.59}$$

In order to check the consequences of the simplifying convention with regard to N_A^* and the assumption of symmetric composition, Bolt and Page (1965) calculated the difference between N_A, following (3.4b), and N_A^* above. For the entire range of $\bar{\Gamma}$ between 10 (applying to clays in about 1 N electrolyte) and about 300 (applying to concentrations below 0.001 N) the difference was found to be less than 0.01. Obviously N_A always exceeds N_A^*, as the exclusion distance (cf. chapter 7) of divalent anions exceeds that of the monovalent anions. In view of the many shortcomings of the Gouy theory the above convenient simplification appears to be fully warranted.

With the help of (3.58) it is possible to investigate the effect of double layer truncation on the valence selectivity. It is immediately clear that the overall effect of truncation is in the same direction as that of the increase of the electrolyte level, viz. both will lead to an increase of K_N. Whereas an increase of C_0 leads to an increase in the solvent term $-\int n_s \, d\ln a_s$ (cf. Fig. 3.4 and the relevant discussion), double layer truncation leads to removal of its low potential tail-end in which particularly the monovalent cations reside. It may also be concluded beforehand that at high electrolyte level the effect of truncation will be minor, since then the fully extended double layer is still of limited width. Referring to Fig. 3.6, where the effect of C_0 in the fully extended double layer is exhibited, it should be expected that in systems containing truncated double layers (e.g. between "plate condensates", believed to exist particularly in Ca-clays, but also formed upon drying out of the soil) the severe deviations from "ideal behavior" at low electrolyte level will disappear.

In order to calculate the effect of truncation it is necessary to estimate the value of v_t. Referring to chapter 1, Fig. 1.5, this is conveniently done by first reading the appropriate value of $U_t \equiv I_t/I_0$ (with I = ionic strength) from the presumably known value of I_0 and x_t, the thickness of the liquid layer. Making use of (1.17) one may then calculate v_t, which is then used in (3.58) to calculate N_A^* and thus K_N. For this purpose it is convenient to write (3.58) as:

$$N_A^* = \frac{r}{\Gamma\sqrt{\beta}} \arg\sinh\left[\frac{\Gamma\sqrt{\beta}}{r}\,\frac{(1-f)}{(1+f)}\,T(U_t,f)\right] \tag{3.60}$$

with the truncation factor T equal to:

$$T(U_t,f) \equiv \frac{4(1+f)}{\sqrt{1+(16U_t-2)f+(16U_t+33)f^2}+3(1-f)} \tag{3.61}$$

An example may elucidate the procedure to be used. For an illitic clay with $\Gamma \approx 3\cdot10^{-9}$ keq/m^2, in equilibrium with 0.05 N Ca^{2+} and 0.05 N Na^+ one finds $\bar{\Gamma} \approx 30$ and $\bar{r} = 1$, $f = 0.5$. Accordingly, $N_A^* = (\arg\sinh 10)/30 = 0.10$, for the freely extended double layer. As $I_0 = 0.15$, one finds $1/\kappa_0$ is about 8 Å. If the clay is compressed to a moisture content of about 7%, the value of x_t may be guessed at 8 Å for a specific surface area of 10^5 m^2/keq, including $\delta = 1$ Å for the above clay largely loaded with Ca. This gives $\kappa_0 x_t$ as about 1 and U_t is estimated at 5 from Fig. 1.5. Equation (3.61) then yields T as 0.6, such that $N_A^*(U_t = 5)$ equals 0.084, i.e. a 15% decrease in the amount of Na adsorbed. As was indicated before the effect of truncation is minor at the chosen salt level of 0.1 N

As was shown in chapter 1, U_t may also be related directly to the thickness of the liquid layer by using the Langmuir approximation neglecting 1/u with respect to u. In that case one finds (cf. (1.21)):

$$\pi^2/2 \approx (\kappa_t x_t)^2 = \frac{\beta}{2}\{4C_0^2 v_t^2 f/2C_0 + (1-f)C_0 v_t - 4fC_0\}x_t^2 \tag{3.62}$$

which gives:

$$2v_t C_0\sqrt{f/2C_0} \approx \frac{\sqrt{r^2+16(4fC_0+\pi^2/\beta x_t^2)}-r}{4} \tag{3.63}$$

This gives the explicit relation between N_A^* and x_t as:

$$N_A^* \approx \frac{r}{\Gamma\sqrt{\beta}} \arg\sinh\left[\frac{4\Gamma\sqrt{\beta}}{3r+\sqrt{r^2+16(4fC_0+\pi^2/\beta x_t^2)}}\right] \tag{3.64}$$

(which gives $N_A^* = 0.083$ for the example discussed in the small print section above).

Using the above equation the effect of truncating the double layer to respectively 10, 20 and 40 Å was calculated for a system with $\beta\Gamma^2$ equal to 10 keq/m^3, comparable to montmorillonite clay (with a surface charge density of 0.1 Coul/m^2). As is shown in Fig. 3.7 the effect is quite considerable at low electrolyte levels (full lines corresponding to $C_0 \rightarrow 0$). At $C_0 = 10^{-2}\,N$ the effect becomes perceptible at a 40 Å truncation (dotted lines), while at $C_0 = 10^{-1}$ the curve for 20 Å truncation just about coincides with the curve for the freely extended double layer at that concentration. This implies that the above approximation becomes invalid, as is easily verified from the value of $\pi^2/\beta x_t^2 = 0.25$. According to (3.62) this term indicates the (estimated) ionic strength at the plane of truncation which must exceed the same in the equilibrium solution by a factor of, say, 3 in order to warrant the neglect of co-ions at the plane of truncation — as is implicit in the Langmuir approximation.

At 0.1 *N* I_0 equals $(1+f)C_0 = 0.1 - 0.2$, i.e. in this case the line shown for 40 Å truncation has no significance. Considering the fact that truncation must always lead to an increase of $\ln K_N$, Fig. 3.7 may be read in comparison with Fig. 3.6 to imply that at 10 Å truncation the system is insensitive towards a change in C_0 up to, say, 0.05 *N* electrolyte. In that range $\ln K_N$ is "truncation determined" and close to a constant.

As could be expected the value of $\ln K_N$ corresponding to a double layer formed on montmorillonite with $\delta = 4$ Å, and truncated at $x_t = 10$ Å corresponds closely to the one for the micro-Donnan system following (3.6). With ${}^v\Gamma$ approaching $\Gamma/(x_t - \delta)$ one finds for the latter system $\ln K_N = \ln\sqrt{2\,{}^v\Gamma} = \frac{1}{2}\ln 3.3 = 0.60$, which checks with the

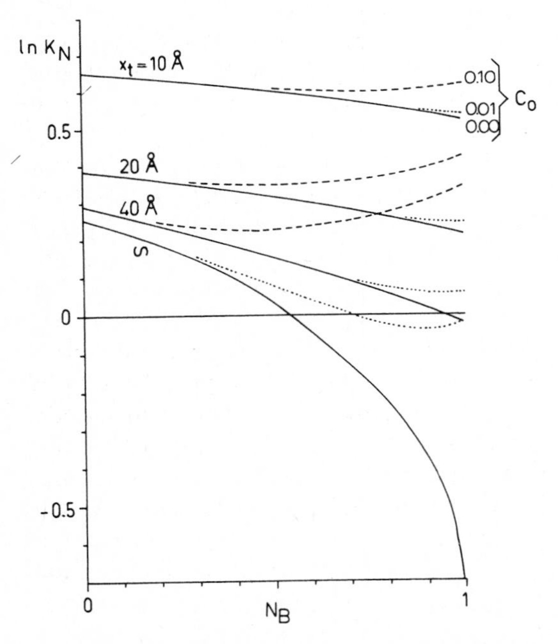

Fig. 3.7. Effect of double layer truncation, indicated as x_t in Å units, on the selectivity coefficient for mono and divalent cations in a Gouy layer, formed on a surface with a charge density akin to montmorillonite (≈ 0.1 Coul/m^2). Full lines refer to systems with vanishing electrolyte concentration, dotted and dashed lines refer to 0.01 and 0.1 *N* electrolyte, respectively.

line for $x_t = 10$ Å in Fig. 3.7. Severe truncation makes the DDL-model converge with the homogeneous Donnan model.

For 20 Å truncation the same holds up to 10^{-2} *N* electrolyte. Beyond these critical values the system very soon behaves like the fully extended systems as regards its reaction to an increase in electrolyte. In short, the truncation effect for $C_0 \to 0$ could be superimposed upon Fig. 3.6 for a quick estimate. Selecting the appropriate lines for a given x_t at $C_0 \to 0$ and for given C_0 at $x_t \to \infty$ (i.e. Fig. 3.6) the line highest on the $\ln K_N$ scale dominates the situation. If they intersect, the maximum values of the two will still be an underestimate and a more complete calculation according to (3.60) will be necessary.

In summary it is concluded that truncation of double layers, as is likely present in many Ca-clays, will delimit the electrolyte sensitivity of the system such that K_N remains fairly constant up to about 0.05—0.1 *N*. As the electrolyte sensitivity is particularly found near saturation with the divalent ions, it may be quickly assessed by comparing the value of $\pi/\sqrt{\beta x_t^2}$ with $\sqrt{8C_0}$, that is the value of the denominator of the hyperbolic sine term as it appears in (3.58) for the limiting case with $v_t = 2$ and $r = 0$. If the first number grossly exceeds the second one, the corresponding term in (3.64) becomes independent of C_0, and $\ln K_N$ is bounded by $\frac{1}{2}\ln(\beta\Gamma^2/6 + 2\pi^2/\beta x_t^2)$ at $N_A = 1$ and $[\frac{1}{2}\ln\beta\Gamma^2 - \ln \text{arg}\sinh(\beta\Gamma x_t/\pi)]$ at $N_A = 0$. The insensitivity of $\ln K_N$ in truncated double layers could, of course, also be demonstrated by constructing the integral $\int n_s \, d\ln a_s$ as was done before; the truncation length providing a cut-off value on the n_s-axis.

An interesting consequence of the above is the divergence in composition of external surfaces of polyplates (domains, cf. Aylmore and Quirk, 1959) as compared to the internal surfaces if both types of surfaces are in equilibrium with the same solution. Judging from Figs. 3.6 and 3.7 this effect is maximal at high saturation with divalent cations and a low total electrolyte level. For $x_t = 10$ Å, i.e. a half-spacing between the individual platelets of $x_t - \delta \approx 6$ Å, the effect on $\ln K_N$ may amount to 0.7 at $C_0 = 10^{-3}$ *N* and 10% saturation with monovalent cations. This amounts to about twice as much Na-ions on the external surface as compared to the internal ones. For even smaller values of the spacing within the polyplates as indicated by X-ray diffraction data of Ca-clays, this ratio of $N_+^*(\text{ext})/N_+^*(\text{int})$ could probably reach a value of 3 even if only the electrostatic fields were taken into account. As was pointed out by Bar-on et al. (1970) this may explain the observed rapid increase of the electrophoretic mobility of Ca-montmorillonite upon a small increase of the overall fraction of exchangeable Na-ions: at 10% exchangeable Na in the dominant interior sites, the external surface could already contain about 30% exchangeable Na.

The distribution equilibria discussed so far treat the cations as point charges and thus distinguish these solely on the basis of their valence. Taking recourse to computer assisted calculations the DDL theory could be refined by taking into account a number of ion specific characteristics, particularly size and polarizability. These will influence the ion-accumulation pattern in regions with high electric field strength and high concentrations (via D-H type ionic interactions). Obviously such refinements should be in balance with other needs for improvement of the DDL theory, taking into account for example, dielectric saturation, electrostriction, etc. Some efforts in this directions were made by Loeb (1951), Bolt (1955) and recently by Ravina and De Bock (1974) and Ravina and Gur (1978). As was pointed out in chapter 1 it is particularly the first layer of cations adjacent to the charged surface which is subject to all these secondary effects. In fact the multiplicity of these effects makes it unlikely that improvements of the DDL theory could be made which allow a sufficiently accurate prediction of the ionic distribution in that layer based on known ion characteristics. Moreover a detailed description of the first layer of counterions requires the recognition of the discrete nature of the adsorption sites, giving rise to the entropy effects to be discussed in the following chapter.

Against this background the broad approximation introduced by Heald et al. (1964) is of interest. In this approach, which was also used by Shainberg and Kemper (1967), the adsorbed phase is divided into a diffuse tail following regular Gouy theory plus a first ("Stern") layer of ions which are regarded as site-ion pairs. The equilibrium between these pairs and the "free" ions in the diffuse tail is then described in terms of a pair-formation reaction of the Bjerrum-type, characterized by:

$$\Gamma_S(k) = p_k c_k \Gamma_G \tag{3.65}$$

in which $\Gamma_S(k)$ indicates the number of site-ion pairs involving ion k, c_k is the concentration of that ion immediately outside the Stern layer and Γ_G signifies the non-paired sites in the Stern layer (which are equal to the total charge in the diffuse Gouy layer, Γ_G). The pair formation constant, p_k, thus tends to summarize all ion-specific effects of the first layer into one, hopefully reasonably constant, parameter for each ion species. The two constants p_k involved in a particular exchange equilibrium could then in principle be determined by suitable extrapolation of experimental data. Shainberg and Kemper (1966, 1967) actually estimated the value of p_k by considering the forces acting between a site and a cation in close proximity. Referring to Bolt (1967) and Bolt, Shainberg and Kemper (1967) for details of the derivation of an exchange equation based on the above model, it is mentioned that for the *homovalent* exchange equilibria studied by Shainberg and Kemper (1967) the agreement with experimental data was promising (cf. also chapter 5). No efforts have been made to apply the same model to heterovalent exchange. Notwithstanding the simplicity of the present model the solution of the relevant equations would then require computer assistance. As in this case K_N would be an implicit function of the two pair-formation constants p_k, the charge density Γ *and* the variables C_0 and f, the outcome of this model would have to be presented as an elaborate set of graphs or tables.

REFERENCES

Aylmore, L.A.G. and Quirk, J.P., 1959. Swelling of clay water systems. Nature, 183: 1752—1953.

Babcock, K.L., 1963. Theory of the chemical properties of soil colloidal systems at equilibrium. Hilgardia, 34: 417—542.

Bar-on, P., Shainberg, I. and Michaeli, I., 1970. Electrophoretic mobility of montmorillonite particles saturated with Na/Ca ion. J. Colloid Interface Sci., 33: 471—472.

Bolt, G.H., 1955. Analysis of the validity of the Gouy-Chapman theory of the electric double layer. J. Colloid Sci., 10: 206—218.

Bolt, G.H., 1967. Cation-exchange equations used in soil science — a review. Neth. J. Agric. Sci., 15: 81—103.

Bolt, G.H. and Page, A.L., 1965. Ion-exchange equations based on double layer theory. Soil Sci., 99: 357—361.

Bolt, G.H. and Winkelmolen, C.J.G., 1968. Calculation of the standard free energy of cation exchange in clay systems. Israel J. Chem., 6: 175—187.

Bolt, G.H., Shainberg, I. and Kemper, W.D., 1967. Discussion of the paper by I. Shainberg and W.D. Kemper entitled: "Ion exchange equilibria on montmorillonite". Soil Sci., 104: 444—453.

Eriksson, E., 1952. Cation-exchange equilibria on clay minerals. Soil Sci., 74: 103—113.

Heald, W.R., Frere, M.H. and De Wit, C.T., 1964. Ion adsorption on charged surfaces. Soil Sci. Soc. Am. Proc., 28: 622—647.

Kruyt, H., 1952. Colloid Science, I. Elsevier, Amsterdam, 389 pp.

Loeb, A., 1951. An interionic attraction theory applied to the diffuse layer around colloid particles. J. Colloid Sci., 6: 75—91.

Ravina, I. and De Bock, J., 1974. Thermodynamics of ion exchange. Soil Sci. Soc. Am. Proc., 38: 45—49.

Ravina, I. and Gur, Y., 1978. The application of electric double layer theory to prediction adsorption in mixed ionic systems. Soil Sci., 125: 204—209.

Shainberg, I. and Kemper, W.D., 1966. Hydration status of adsorbed ions. Soil Sci. Soc. Am. Proc., 30: 707—713.

Shainberg, I. and Kemper, W.D., 1967. Ion exchange equilibria on montmorillonite. Soil Sci., 103: 4—9.

CHAPTER 4

THEORIES OF CATION ADSORPTION BY SOIL CONSTITUENTS: DISCRETE-SITE MODELS

*K. Harmsen**

The present chapter deals with 'discrete-site' models: ion-exchange models that consider explicitly the occurrence of discrete and possibly non-uniform adsorption sites on the adsorbent surface. These models emphasize that the electric field is not uniform close to the adsorbent surface and that the dielectric constant of water may decrease significantly near the adsorbent surface. Small differences in distance to the solid surface between ions of different size may cause significant ion exchange selectivity. Configuration effects due to the arrangement of the sites on the adsorbent surface, interactions between ions in the adsorbed phase, specific interactions between ions and surface groups, heterogeneity of adsorption sites and the possible effect of charged or uncharged ligands on the adsorption of cations may all be described with discrete-site models.

The theories describing the distribution of ions in electric fields, treated in chapters 1 and 3, consider non-localized adsorption of point charges in laterally uniform electric fields provided by the fixed charges on an adsorbent. The application of the field theories would be favored by large hydrated radii and low charge numbers of the ions involved and a low ionic strength of the soil solution. However, even in dilute soil suspensions a large part of the adsorbed ions reside close to the adsorbent surface, whereas in soils at field capacity the liquid content is simply not high enough to allow an "extended" diffuse layer to develop freely. Therefore, when there are a low moisture content, a large proportion of divalent ions and a high electrolyte level, conditions that are commonly met in soils, most of the exchangeable ions will be at a distance from the adsorbent surface of the order of the hydrated radii of the ions. It may be noted that field models and discrete-site models are not mutually exclusive: they constitute two different approaches that both contribute to the understanding of ion exchange in soils. For example, in localized adsorption the equilibrium distance between counterions and adsorption sites may vary with surface charge density of the adsorbent, charge number of the counterions and electrolyte level of the soil solution, as predicted by the theory of the diffuse double layer.

* Present address: National Institute for Water Supply (RID), P.O. Box 150, Leidschendam, The Netherlands.

The preference of an ion for a particular adsorption site may be due to the arrangement of the sites (ideal mixtures), to interactions between neighboring ions in the adsorbed phase (regular mixtures), to ion-site interactions (two-site model) or to interactions between the ion in liganded form and the adsorption site (two-species model). The mixture models, treated in section 4.1 assume that there is only one type of adsorption sites and that the binding on any one site is independent of the binding on other sites. Such models could also be called "one-site" models. They describe the effect of the arrangement of adsorption sites (configuration effects) or interionic interactions on ion-exchange selectivity. The two-site and two-species models, treated in section 4.2, do not deal with configuration effects or interionic interactions. A "two-site" model describes the effect of the occurrence of different types of adsorption sites on ion exchange and a "two-species" model is concerned with the occurrence of ligands in solution that may form complexes with one of the cations involved in the ion-exchange reaction.

In the present chapter the attention is focused on the adsorption of cations on negatively charged adsorption sites, such as those on clay mineral surfaces, but generally the models presented in this chapter also hold for the adsorption of anions on positively charged surface sites, such as those associated with certain oxides at pH values below their points of zero charge.

4.1. THEORY OF MIXTURES APPLIED TO CATION EXCHANGE

The adsorption of cations on mineral surfaces or by organic matter may be described by the statistical thermodynamic theory of mixtures (Guggenheim, 1944a, b). Unlike the theory of the diffuse double layer, which assumes that the adsorbed ions can move freely in two dimensions, the theory of mixtures requires that the adsorption be localized: the ions being captured in potential "wells" associated with the fixed surface charges on the adsorbent surface.

The model considered here is a regular array of equivalent adsorption sites provided by a mineral or organic surface. Each site consists of a single negative charge, for example due to isomorphous replacement of Si^{4+} by Al^{3+} in a tetrahedral layer of a clay mineral or to dissociation of H^+ from a surface group. The adsorption sites are assumed to be equivalent: they do not have to be identical, but ions should not discriminate between sites. Each site is occupied by a monovalent cation or a "segment" of a polyvalent cation. The term "segment" (or: "element") stems from the theory of mixtures, which is mainly concerned with mixtures of large organic molecules and where it is assumed that molecules of unequal size occupy different numbers of sites in a hypothetical lattice. For monovalent cations this means that a surface charge is satisfied by one ion in direct contact with the surface or separated from it by one or more layers of water molecules, the equilibrium distance of an ion to the surface being constant for all ions of the same species. Polyvalent ions do satisfy two or more fixed surface charges at a time and may thus be located in between these sites, such that each site is "occupied" by a "segment" of a polyvalent cation.

The present model considers interactions between ions only and not between ions and the adsorbent surface. Ion—ion interactions are assumed to be confined to nearest neighbors and site—ion interactions are assumed to be constant for all ions of the same species and independent of the composition of the adsorbed phase, i.e. not affected by neighboring ions.

The adsorption sites are assumed to be distributed in a regular way over the adsorbent surface and the number of sites which are neighbors to any one site, denoted by $2l$, is used as a parameter to characterize the arrangement of sites. There may be some variation in the arrangement of the sites on the adsorbent surface and then the average value of $2l$ could be used instead. However, when the energy of mixing is non-zero, the condition of a single value of $2l$ is essential, since otherwise the ions would seek energetically favorable positions on the adsorbent surface and $2l$ would become a function of the adsorbent composition. The array of sites may be one-dimensional, as on certain organic chain molecules, two-dimensional, as on clay mineral surfaces, or three-dimensional, as in soil organic matter. The only condition imposed by the theory of mixtures is that the array of sites be regular, i.e. that the number of nearest neighbors be constant.

Application of the theory of mixtures to adsorption of ionic species amounts to assuming that ions of unequal charge occupy different numbers of adjacent sites on an adsorbent surface, not because of their ionic volumes, but because of their different charge numbers. It is thus assumed that the number of sites occupied by an ion of type k equals z_k, the charge number of the ion. The number of sites which are neighbors to an ion of species k then equals $2lh_k$, where h_k is determined from:

$$lh_k = (l-1)z_k + 1 \qquad (4.1)$$

The notion of h_k is fundamental in the statistical thermodynamic treatment of mixtures by Guggenheim (1944a, b), since it plays an essential role in determining the chance that if one particular site is occupied by an ion of species k, a given neighboring site is occupied by the same ion, by another ion of species k or by an ion of a different species.

The number of distinguishable ways of arranging a mixture of ionic species k and l on a regular array of sites (a lattice) is denoted by $g(m_k, m_l)$, where m_k denotes the number of ions of species k. When all configurations of ions on a lattice have the same interionic potential energy, the statistical weight of all configurations is equal and the arrangement of the particles is completely random. However, when the configurations differ in interionic potential energy, the statistical weight of any one configuration is also determined by the intermolecular energy. In the present treatment, following Guggenheim (1944b), it is assumed that the total interionic potential energy, E, may be regarded as a sum of terms, each pair of neighboring sites occupied by segments of different ions contributing one term. Hence only nearest neighbor interactions are considered and second or higher neighbor interactions are neglected. The contribution to the total potential energy by one pair of sites of which one is occupied by a segment of an ion of species k, the other by a segment of an ion of species l, is denoted by e_{kl}. The contribution of one pair of sites, one occupied by a segment of ion k and the other by a segment of ion l, to the energy of mixing of the system is defined by:

$$e_m \equiv e_{kl} - \tfrac{1}{2}(e_{kk} + e_{ll}) \tag{4.2}$$

whereas the energy of mixing of the system, E_{mix}, is defined by:

$$E_{mix} \equiv \Sigma_{kl}\, 2lX^*_{kl}\, e_m \tag{4.3}$$

where $2lX^*_{kl}$ denotes the number of pairs of neighboring sites occupied by different ions of which one is of species k and the other of species l, in the equilibrium state of the system.

The equilibrium state of a system consisting of a mixture of different types of ions on a lattice may be calculated from the condition of "quasi-chemical" equilibrium. Let (kl) denote a pair of sites occupied by segments of two different ions, one of species k, the other of species l. A rearrangement of adsorbed species such that two (kl)'s are created and one (kk) and one (ll) are destroyed can be written symbolically as:

$$(kk) + (ll) \rightleftharpoons 2(kl) \tag{4.4}$$

If this "reaction" represented a simple chemical reaction between molecules, the equilibrium condition would be:

$$\frac{X^2_{kl}}{4X_{kk}\,X_{ll}} = \exp(-2e_m/kT) \tag{4.5}$$

where k denotes the Boltzmann constant and T the temperature, and where the factor 4 arises from symmetry factors of $\tfrac{1}{2}$ in the molecular partition functions of kk and ll. If a formula such as (4.5) is applied to the number of pairs of neighboring sites occupied in several ways it may be referred to as the condition of quasi-chemical equilibrium (Guggenheim, 1944b).

For simplicity the discussion will be confined from here on to binary mixtures of two species of cations, A and B, with charge numbers z_A and z_B, respectively. The number of distinguishable ways of arranging a binary mixture of A and B ions on a lattice, all configurations having the same energy of mixing, is denoted by $g(m_A, m_B, X_{AB})$, where m_A and m_B denote the numbers of A and B ions and where E_{mix} is related to X_{AB} by (4.3). For a binary mixture, the condition of quasi-chemical equilibrium (4.5) becomes:

$$X_{AB}^2 = (h_A m_A - X_{AB})(h_B m_B - X_{AB}) \exp(-2e_m/kT) \quad (4.6)$$

where:

$$2X_{kk} = h_k m_k - X_{kl} \quad (4.7)$$

has been inserted for X_{AA} and X_{BB}. At this point it is convenient to define "adjusted" mole fractions according to:

$$M_A = \frac{h_A m_A}{(h_A m_A + h_B m_B)} \quad (4.8a)$$

$$M_B = \frac{h_B m_B}{(h_A m_A + h_B m_B)} \quad (4.8b)$$

and a quantity b by:

$$b = \{(M_B - M_A)^2 + 4M_B M_A \exp(2e_m/kT)\}^{1/2} \quad (4.9)$$

The solution of the quadratic equation (4.6) may now be written in the form:

$$X_{AB}^* = \frac{(h_A m_A + h_B m_B) 2M_A M_B}{(b+1)} \quad (4.10)$$

and this expression for X_{AB}^* may be inserted in $g(m_A, m_B, X_{AB}^*)$ to obtain the number of distinguishable configurations of the binary mixture under quasi-chemical equilibrium conditions. The expression for $g(m_A, m_B, X_{AB}^*)$ is rather involved and will not be given here; it may be obtained from Guggenheim (1944b).

The quantity b, defined by (4.9), is a measure for the interaction between different ionic species in the adsorbed state: if $e_m < 0$ it follows that $0 < b < 1$, if $e_m = 0$ it follows that $b = 1$, whereas $b > 1$ if $e_m > 0$. From $db/dM_A = 0$ it follows that b has an extremum for $M_A = M_B = \frac{1}{2}$, i.e. for $h_A m_A = h_B m_B$. This extremum is a minimum if $e_m < 0$ and a maximum if $e_m > 0$, in both cases the numerical value of the extremum is given by $\exp(e_m/kT)$. From (4.10) it follows that the number of pairs of sites occupied by segments of ions of different species decreases if $e_m > 0$ and increases if $e_m < 0$.

Commonly the ratio of the amount of substance of component k to the total amount of substance is expressed by a mole fraction, defined by:

$$M_k \equiv \frac{m_k}{\Sigma_l m_l} \quad (4.11)$$

where the summation is over all species. This mole fraction is based only on the number of moles of each species, and not on ionic properties, such as charge number, or properties of the adsorbent, such as the arrangement of fixed surface charges. In ion-exchange studies it is often convenient to define an "equivalent" fraction:

$$N_k \equiv \frac{z_k m_k}{\Sigma_l z_l m_l} \quad (4.12)$$

which takes into consideration the charge number of ions and where $\Sigma_l z_l m_l$ is constant during the exchange process. For the present application of the theory of mixtures to ion exchange, adjusted mole fractions, defined by (4.8) for binary mixtures, which take both the charge number of ions and the arrangement of fixed surface charges into consideration, seem to be the most convenient variables to describe the composition of the adsorbed phase. For l sufficiently large, the adjusted mole fraction becomes virtually equal to the equivalent fraction, as follows from:

$$\lim_{l \to \infty} h_k = z_k \quad (4.13)$$

In the trivial case of a binary mixture with $z_A = z_B$ it follows that:

$$N_k = M_k = \mathrm{M}_k$$

For $z_A \neq z_B$ and $l = 1$ it follows that $h_k = 1$ for all z_k, hence:

$$N_k \neq M_k = \mathrm{M}_k$$

However, in soils l will generally be greater than unity (which corresponds to a linear array of sites). For $l > 1$ it follows that:

$$N_k \neq M_k \neq \mathrm{M}_k$$

whereas M_k approaches N_k rapidly with increasing l.

With the aid of $g(m_A, m_B, X^*_{AB})$ several important thermodynamic functions can be derived for binary mixtures. For example, the entropy of a binary mixture of A and B ions is given by:

$$S = k \ln g(m_A, m_B, X^*_{AB}) \quad (4.14)$$

and the free enthalpy (or Gibbs function) follows from:

$$G = -kTm_A \ln \Omega_A - kTm_B \ln \Omega_B - kT \ln g(m_A, m_B, X^*_{AB}) + E_{mix} \quad (4.15)$$

where the energy of mixing is given by:

$$E_{mix} = 2lX^*_{AB}\, e_m \quad (4.16)$$

and where Ω_k denotes the partition function of an ion of species k attached to a set of z_k sites. The chemical potential of species k in the adsorbed state can be obtained by differentiating (4.15) with respect to m_k:

$$\frac{\partial G}{\partial m_k} = \bar{\mu}_k \quad (4.17)$$

where the bar refers to the adsorbed phase. The relative activity of species k in the adsorbed state, $\bar{a}_k$, is obtained from:

$$(\bar{\mu}_k - \bar{\mu}^0_k)/kT = \ln \bar{a}_k \quad (4.18)$$

where the superscript "0" refers to the standard state and where $\bar{a}_k^0 = 1$. The standard state for species k in the adsorbed state is here chosen as the homo-ionic adsorbent, i.e. all adsorption sites being occupied by ions of species k. Since the solvent or ions other than those occupying surface sites are not considered in this standard state, it is not necessary to specify the amount of solvent in the adsorbed phase or the electrolyte concentration in the solution phase. For a binary mixture of A and B ions the relative activities take the form:

$$\bar{a}_A = N_A \left[\frac{b + M_A - M_B}{N_A (b+1)} \right]^{lh_A} \tag{4.19a}$$

$$\bar{a}_B = N_B \left[\frac{b + M_B - M_A}{N_B (b+1)} \right]^{lh_B} \tag{4.19b}$$

where:

$$\lim_{m_A/m_B \to 0} \bar{a}_A = \lim_{m_B/m_A \to 0} \bar{a}_B = 0 \tag{4.20}$$

Activity coefficients may be defined in several ways: through the equivalent fraction, the adjusted mole fraction or the mole fraction as composition variable. Here the equivalent fraction seems to be the most suitable variable (cf. (4.19)), hence:

$$\bar{f}_k \equiv \bar{a}_k / N_k \tag{4.21}$$

It may be noted that the activity coefficients in the solution phase were defined on a molar basis:

$$f_k \equiv \frac{a_k}{({}^m c_k / {}^m c_k^0)} \tag{4.22}$$

where ${}^m c_k^0$ is a reference molarity, usually 1 mol/1. From (4.19) and (4.21) it follows that:

$$\lim_{m_A/m_B \to 0} \bar{f}_A^{1/z_A} = \{(z_B h_A / z_A h_B) \exp(2e_m/kT)\}^{lh_A/z_A} \tag{4.23a}$$

and:

$$\lim_{m_B/m_A \to 0} \bar{f}_B^{1/z_B} = \{(z_A h_B / z_B h_A) \exp(2e_m/kT)\}^{lh_B/z_B} \tag{4.23b}$$

An entropy of mixing may be defined according to:

$$\Delta S_{mix} = k \ln g(m_A, m_B, X_{AB}^*) - k \ln g(m_A, m_B = X_{AB}^* = 0) - k \ln g(m_B, m_A = X_{AB}^* = 0) \tag{4.24}$$

and the free enthalpy of mixing then becomes:

$$\Delta G_{mix} = -T \Delta S_{mix} + E_{mix} \tag{4.25}$$

Since also:

$$\Delta G_{mix} = m_A kT \ln \bar{a}_A + m_B kT \ln \bar{a}_B \tag{4.26}$$

it follows that the entropy of mixing may also be written as:

$$\Delta S_{mix} = E_{mix}/T - m_A k \ln \bar{a}_A - m_B k \ln \bar{a}_B \tag{4.27}$$

With:

$$\lim_{m_k/m_l \to 0} (m_k/m_l) \ln (m_k/m_l) = 0$$

it follows that:

$$\lim_{m_k/m_l \to 0} \Delta S_{mix} = 0 \tag{4.28}$$

Mixtures with zero energy of mixing ($e_m = 0$) are generally called ideal or perfect mixtures, whereas mixtures with small but non-zero energies of mixing may be referred to as regular mixtures. In the present treatment regular mixtures are not regarded as deviations from ideality, but as the general case: approximate equations for the regular mixture and formulae for the ideal mixture are derived as special cases from the general formulae for the regular mixture.

When the deviation from ideality is small, approximate equations, which are valid for small values of e_m only, may be derived for the regular mixture. For $2e_m/kT$ sufficiently small, b may be approximated by:

$$b \simeq 1 + 2M_A M_B (2e_m/kT) \tag{4.29}$$

Inserting this approximation for b in the exact formulae for $\ln \bar{a}_A$ and $\ln \bar{a}_B$ results in:

$$\ln \bar{a}_A \simeq \ln N_A + lh_A \ln (M_A/N_A) + lh_A M_B^2 (2e_m/kT) \tag{4.30a}$$

and:

$$\ln \bar{a}_B \simeq \ln N_B + lh_B \ln (M_B/N_B) + lh_B M_A^2 (2e_m/kT) \tag{4.30b}$$

Hence for an ideal mixture of equal-charged ions the relative activities reduce to the equivalent fractions of the ions in the adsorbed state.

The approximate formula for the entropy of mixing becomes:

$$\Delta S_{mix}/k \simeq - m_A \ln N_A - m_B \ln N_B - lh_A m_A \ln (M_A/N_A) - lh_B m_B \ln (M_B/N_B)$$
$$- l(h_A m_A + h_B m_B) M_A^2 M_B^2 (2e_m/kT)^2 \tag{4.31}$$

from which it can be seen that a non-zero energy of mixing always results in a lowering of the entropy: the number of accessible configurations decreases both for attraction and for repulsion between unlike ions in the adsorbed state. For $z_A = z_B$ the approximate formulae for the regular mixture become similar to those given by Lewis & Randall (1961) for regular solutions of equal-sized particles.

A mixture involving two components only may be referred to as a binary solution and when the two components form a solid phase, as in the case of a mixture of $ZnCO_3$ and $CaCO_3$, it may be called a binary solid solution. In soil science literature the theory of mixtures is sometimes referred to as the theory of "solid solutions". In the present case, however, where this theory is applied to different ionic species in aqueous solution, it seems more appropriate to use the expression mixture rather than (solid) solution. Nevertheless, the formulae derived in the present treatment would also apply to solid solutions.

The regular and ideal mixture models for ion exchange describe the adsorbed phase as a mixture of two species of cations only. The solvent may be of significance, for example through the dielectric constant, but is not considered explicitly in these models. This is because the sites, the fixed surface charges, can only be occupied by cations and not by uncharged solvent molecules, whereas the remaining part of the adsorbent surface is inaccessible for cations. For polyvalent cations the situation is not essentially different although the accessible positions on the adsorbent surface may not coincide with the fixed surface charges for these ions. For each ionic species the accessible positions on the adsorbent are determined by the geometry of the fixed surface charges. Therefore there is only one way in which $1/z_k$ moles of ionic species k and n_s moles of solvent can be distributed over one equivalent of adsorbent containing discrete adsorption sites and thus the entropy of mixing is zero for such a mixture. Application of the theory of mixtures to ion exchange results in the following expression for the entropy of mixing of two ionic species, A and B, for $e_m = 0$ and $l \to \infty$ (cf. (4.27) and (4.62)):

$$\Delta S_{mix}/R = -n_A \ln N_A - n_B \ln N_B \tag{4.32}$$

where n_k denotes the number of moles of ionic species k.

In applying the theory of mixtures to ion exchange it is assumed that there is only one way in which a particular site can be occupied by an ion of species k, i.e. the distance between the ion and the site is the same for all ions of the same species. If adsorption were in a monolayer, the site—ion interactions would probably be at maximum, but repulsive ion—ion interactions would also be at maximum. The osmotic pressure of such an adsorbed phase would be high too, such that solvent molecules would tend to enter the adsorbed phase. If adsorption were in a diffuse layer, interaction between ions would decrease and also the osmotic pressure of the adsorbed phase would decrease whereas entropy would increase. Unlike the mixture models, treated in this chapter, the "field" models for ion exchange allow for a spatial variation in ion—site distance. In the field models treated in chapters 1 and 3 it is assumed that the whole surface area of the adsorbent is accessible both for ions and for solvent molecules. If ions and solvent molecules were completely interchangeable without affecting the internal energy of the system, the entropy of mixing for a mixture of n_k moles of ion k and n_s moles of solvent would be:

$$\Delta S_{mix}/R = -n_k \ln\left[\frac{n_k}{n_k + n_s}\right] - n_s \ln\left[\frac{n_s}{n_k + n_s}\right] \tag{4.33}$$

Now the "sites" are not formed by the fixed surface charges, as in the discrete site models, but the positions of all particles at some instant are regarded as constituting an array of $(n_k + n_s)N_{Av}$ sites. Of course, ions and solvents molecules will generally differ in size, whereas the repulsive interactions between cations will result in a lowering of the entropy of mixing. For water as the solvent, the attractive interactions between ions and water molecules (permanent dipoles) will also lower the entropy of mixing, since they lower the number of accessible configurations for the system. The entropy of mixing given in (4.33) refers to $n_k = 0$ as standard state for the solvent and to $n_s = 0$ for the ions. Obviously these states cannot be reached without a simultaneous charging or discharging of the adsorbent surface. If the charge on the adsorbent is not altered, the number of moles of ion k per equivalent of adsorbent is constant $(1/z_k)$ and it follows that the entropy of mixing per equivalent of adsorbent becomes a function of n_s only. For water as the solvent and a surface charge density of the adsorbent of 0.16 Coul/m^2 (1 site per 100 Å^2) there would be about 14 moles of water and 1 mole of monovalent ions per equivalent of adsorbent, assuming that water molecules and ions are of the same size and that adsorption would be in a monolayer. Hence if the standard state for mixing of ions and water is chosen as $n_k^0 = 1/z_k$ and $n_w^0 = (15 - 1/z_k)$, the standard entropy of mixing, $\Delta S^0_{mix}/R$, according to (4.33) would become 3.67, 2.19 and 1.60, for mono-, di- and trivalent ions in the standard state of the system. If the adsorbed phase consisted of two, three or four layers of elementary particles on the adsorbent surface, the entropy of mixing per equivalent of adsorbent would be 0.71, 1.13 or 1.41, respectively, for monovalent ions relative to their standard state. Although these estimates do not take into consideration interactions between particles, they illustrate that entropy increases upon increasing the volume of the adsorbed phase. The volume of the adsorbed phase per equivalent of adsorbent, V^α, is defined by:

$$V^\alpha \equiv \frac{\Sigma n_k \bar{V}_k^\alpha}{\Sigma z_k n_k} \tag{4.34}$$

where $\bar{V}_k^\alpha$ is the partial molar volume of component k in the adsorbed phase (cf. (2.4)). Assuming that ions and solvent molecules are of equal size, the entropy of mixing per equivalent of adsorbent for a mixture of two ionic species, A and B, in water as the solvent, may be written:

$$\Delta S_{mix}/R = -n_A \ln N_A + n_A \ln (z_A V^\alpha/\bar{V}^\alpha) - n_B \ln N_B + n_B \ln (z_B V^\alpha/\bar{V}^\alpha) - n_w \ln (n_w \bar{V}^\alpha/V^\alpha) \tag{4.35}$$

where now $\bar{V}^\alpha$ denotes the partial molar volume of one mole of A, B or H_2O, i.e. $\bar{V}^\alpha \simeq 0.018$ l/mol. The assumption that all particles are of equal size may be a reasonable approximation for mixtures of H_2O and K^+, Rb^+ or Ba^{2+}. Also, the hydration numbers of these rather large cations will be low, such that adsorption in a monolayer may be possible. Since only the effect of mixing of ionic species on the entropy is considered, it is convenient to select the homoionic states (specified by $N_k = 1$) as the standard state for the ionic species. However, since (4.35) also contains n_w as an independent variable, it is necessary to specify also the water content of the adsorbed phase in the homoionic states. Assuming that V^α is a constant, it follows that:

$$n_w = (V^\alpha/\bar{V}^\alpha)(z_A n_A + z_B n_B) - n_A - n_B \tag{4.36}$$

is a function of n_A and n_B only. The condition that V^α be constant thus implies that n_w varies with surface composition if $z_A \neq z_B$, i.e. water molecules from the solution phase leave or enter the adsorbed phase. It may be noted that for a charge density of 0.16 Coul/m^2 the volume of the adsorbed phase would equal about 0.27 l/eq if adsorption were in a monolayer and for $\bar{V}^\alpha = 0.018$ l/mol. The volume of the adsorbed phase would increase to 1.08 l/eq if the adsorbed phase comprised 4 layers of elementary particles on the adsorbent surface, i.e. would have a thickness of about 10 Å. The entropy of mixing per equivalent of adsorbent for a mixture of ionic species A and B, relative to the homoionic states and for V^α = constant, becomes:

$$\Delta S_{mix}/R = -n_A \ln N_A - n_B \ln N_B$$

$$- n_A z_A \{(V^\alpha/\bar{V}^\alpha) - 1/z_A\} \ln \left[\frac{(V^\alpha/\bar{V}^\alpha) - (n_A + n_B)}{(V^\alpha/\bar{V}^\alpha) - 1/z_A} \right]$$

$$- n_B z_B \{V^\alpha/\bar{V}^\alpha) - 1/z_B\} \ln \left[\frac{(V^\alpha/\bar{V}^\alpha) - (n_A + n_B)}{(V^\alpha/\bar{V}^\alpha) - 1/z_B} \right] \qquad (4.37)$$

For $z_A = z_B$ (4.37) reduces to (4.32) since then the water content of the adsorbed phase does not change upon ion exchange. For $z_A \neq z_B$ it follows that:

$$\lim_{V^\alpha/\bar{V}^\alpha \to \infty} (\Delta S_{mix}/R) = -n_A \ln N_A - n_B \ln N_B$$

However, for a surface charge density of 0.16 Coul/m^2 (one site per 100 Å^2) (4.37) deviates less than 0.5% from (4.32) even if adsorption is in a monolayer. Therefore, the last two terms in (4.37) may well be omitted without major error.

It may be concluded from the considerations in the previous small print section that the entropy of mixing for a binary mixture of ionic species of equal size, obtained from the theory of mixtures for $e_m = 0$ and l sufficiently large, is approximately equal to the entropy of mixing obtained from the "field" model (Donnan system) for V^α sufficiently large and taking the homoionic states ($N_k = 1$) as the standard states for mixing of ions. This is because both types of models consider only the contribution to the entropy due to mixing of ionic species relative to their homoionic states, and not the possible contributions due to mixing of a single ionic species with the solvent. It may be noted that interionic interaction terms in the partial molar entropies of the individual ionic species would largely cancel when mixing of ions relative to homoionic states is considered. Only if energies of mixing were non-zero would interionic interactions contribute to the entropy of mixing (cf. section 4.1.2.).

4.1.1. Selectivity coefficients based on ideal mixing

Ideal or perfect mixtures are mixtures with zero energy of mixing. If $e_m = 0$ it follows that:

$$e_{kl} = \tfrac{1}{2}(e_{kk} + e_{ll}) \qquad (4.38)$$

Hence e_{kk} and e_{ll} may have values different from zero and may differ from each other. The only requirement is that e_{kl} be the arithmetic mean of e_{kk} and e_{ll}. For a binary ideal mixture the relative activities reduce to:

$$\bar{a}_A^{1/z_A} = N_A^{1/z_A} (M_A/N_A)^{lh_A/z_A} \qquad (4.39a)$$

$$\bar{a}_B^{1/z_B} = N_B^{1/z_B} (M_B/N_B)^{lh_B/z_B} \qquad (4.39b)$$

where:

$$lh_k/z_k = (l-1) + 1/z_k$$

The entropy of mixing for an ideal binary mixture becomes:

$$\Delta S_{mix} = -m_A k \ln N_A - lh_A m_A k \ln (M_A/N_A) - m_B k \ln N_B - lh_B m_B k \ln (M_B/N_B) \quad (4.40)$$

For the exchange reaction involving A and B ions (2.1):

$$(1/z_A)\bar{A}^{z_A+} + (1/z_B)B^{z_B+} \rightleftharpoons (1/z_B)\bar{B}^{z_B+} + (1/z_A)A^{z_A+}$$

where bars denote the adsorbed phase, a thermodynamic exchange constant, K^0_{ex}, was defined as (2.16):

$$K^0_{ex} = \frac{\bar{a}_B^{1/z_B}\, a_A^{1/z_A}}{\bar{a}_A^{1/z_A}\, a_B^{1/z_B}}$$

The selectivity coefficient based on equivalent fractions of the ions in the adsorbed phase, K_N, is related to K^0_{ex} by:

$$\ln K_N = \ln K^0_{ex} + (l-1)\ln(h_A z_B/h_B z_A) + (1/z_A)\ln(M_A/N_A) - (1/z_B)\ln(M_B/N_B) \quad (4.41)$$

and it follows that K_N varies with surface composition of the adsorbent. The selectivity coefficient based on mole fractions of the ionic species in the adsorbed phase, K_M, is related to K^0_{ex} by:

$$\ln K_M = \ln K^0_{ex} + (l-1)\ln(h_A z_B/h_B z_A) + (1/z_A)\ln(M_A/\mathrm{M}_A) - (1/z_B)\ln(M_B/\mathrm{M}_B) \quad (4.42)$$

and it can be seen that K_M varies with composition of the adsorbent too. Therefore, if the adsorbed phase behaves as an ideal mixture of ions, K_N and K_M are not the most suitable coefficients to describe selectivity between ionic species. It may thus be useful to define a selectivity coefficient based on adjusted mole fractions, K_{M}, according to:

$$K_{M} = \frac{M_B^{1/z_B}\, a_A^{1/z_A}}{M_A^{1/z_A}\, a_B^{1/z_B}} \quad (4.43)$$

which does not vary with surface composition, as follows from:

$$\ln K_{M} = \ln K^0_{ex} + (l-1)\ln(h_A z_B/h_B z_A) \quad (4.44)$$

For a binary mixture of ions of equal charge, i.e. for $z_A = z_B = z$, it follows that $h_A = h_B$ for all l such that $\mathrm{M}_k = M_k = N_k$. Hence for homovalent ion exchange, i.e. for ion exchange involving ions of equal charge only, the ratio of activity coefficients of ions in the adsorbed phase would equal unity and:

$$K_N = K_{M} = K_M = K^0_{ex}$$

for all l. The Kerr exchange "constant" (Kerr, 1928), which applies to homovalent exchange only, may be defined according to:

$$K\,(\text{Kerr}) = \frac{M_B\ {}^m c_A}{M_A\ {}^m c_B} \tag{4.45}$$

and it follows that for homovalent exchange:

$$K\,(\text{Kerr}) = (K'_N)^z = (K'_{\mathcal{M}})^z = (K'_M)^z \tag{4.46}$$

where the primes indicate that the selectivity coefficients are based on the ratio of molar concentrations of ionic species in solution rather than on their activities.

The application of the theory of ideal mixtures to heterovalent ion exchange, i.e. to ion exchange involving ionic species of unequal charge, results in relationships for the ratio of the activity coefficients in the adsorbed phase that vary with the composition of the adsorbent and that contain l and z_k as parameters. It is thus of significance that the number of nearest neighbors on a particular adsorbent be known in order to be able to choose the correct form of $K_{\mathcal{M}}$. For heterovalent exchange $\mathcal{M}_k$ will always be situated in between M_k and N_k and thus $K_{\mathcal{M}}$ in between K_M and K_N, as follows from:

$$1 \leqslant h_k \leqslant z_k \tag{4.47}$$

The limiting values of $K_{\mathcal{M}}$ are obtained for $l = 1$, when $K_M = K_{\mathcal{M}} \neq K_N$, and for $l \to \infty$, when $K_M \neq K_{\mathcal{M}} = K_N$.

Bloksma (1956) drew attention to the fact that in the calculation of the number of sites that are neighbors to an ion of species k (4.1) it is assumed that no site has two nearest neighbors that are nearest neighbors to each other at the same time. Unfortunately this assumption does not hold for most two-dimensional arrays of sites and for those arrays (4.1) would result in an overestimation of $2lh_k$. For example, Bloksma (1956) obtained a number of 10 nearest neighbors for a trivalent ion, visualized as a linear triatomic molecule, placed on a triangular array of sites ($2l = 6$), whereas (4.1) predicts a number of 14 nearest neighbors. In fact the situation is even more complicated since a trivalent ion does not really occupy 3 adjacent sites but will probably be situated in the center of the triangle formed by 3 adjacent sites and if so, a trivalent ion would have only 9 nearest neighbors (cf. Fig. 4.1e).

Equation (4.1) may be written in the more general form:

$$2lh_k = 2lz_k - a_k(z_k - 1) \tag{4.48}$$

where a_k represents an adjustable parameter that depends on the configuration of the sites, the number of nearest neighbors and the valency of the ionic species involved. For ion exchange in soils only $z_k = 1$, 2, 3 and possibly 4, are assumed to be of significance. In fact even most tri- and quadrivalent cations in soil solution will hydrolyze to form lower charged species. The hydrolysis products may combine to form polynuclear complexes of higher valency, but those complexes are not considered here. For a linear ($2l = 2$) and a hexagonal ($2l = 3$) array of sites parameter a_k equals 2, as in (4.1), since in these arrays no site has two neighbors that are also nearest neighbors to each other (see Fig. 4.1a, b). A square array ($2l = 4$) would also possess the same characteristics if only

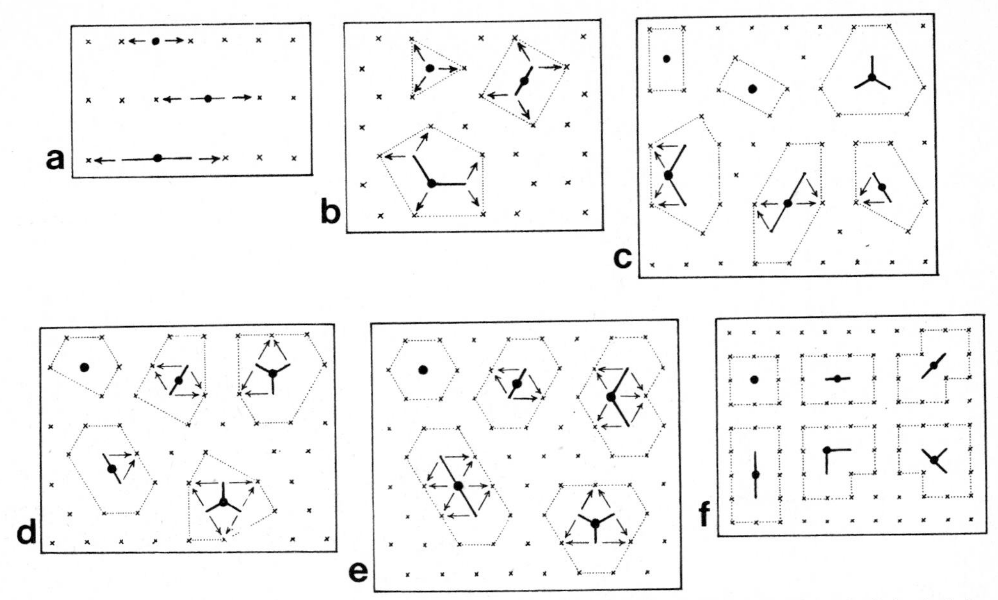

Fig. 4.1. Two-dimensional arrays of adsorption sites (crosses) with 2 (a), 3 (b), 4 (c), 5 (d), 6 (e), and 8 (f) nearest neighbors. Modes of occupation by mono-, di- and trivalent ions are indicated schematically. Arrows point to nearest-neighbor sites (a and b) or to those sites that are nearest neighbors to two sites at the same time (c—e). Dotted lines connect nearest neighbor sites (b—f).

first-nearest neighbors are considered to be "nearest" neighbors; if second-nearest neighbors are also considered as "nearest" neighbors each site in a square array would have 8 nearest neighbors (Fig. 4.1f). In an array of sites as pictured in Fig. 4.1c, where also $2l = 4$, each site has two pairs of nearest neighbors that are nearest neighbors to each other at the same time and it follows from Fig. 4.1c that parameter a_k equals 3 for all configurations. In an array as pictured in Fig. 4.1d ($2l = 5$) a divalent ion may have 6 ($a_k = 4$) or 7 ($a_k = 3$) nearest neighbors, whereas a trivalent ion may have 6 ($a_k = 4.5$), 7 ($a_k = 4$) or 8 ($a_k = 3.5$) nearest neighbors. Figure 4.1d also illustrates that although the array of sites may be regular the number of nearest neighbors to a pair of sites may differ with location on the lattice. Fig. 4.1e shows that for the triangular array parameter a_k equals 4 except for a trivalent ion situated in the center between 3 sites ($a_k = 4.5$), whereas $a_k = 4.67$ for a quadrivalent ion located in the center between 4 sites, corresponding to 10 nearest neighbors. If first and second-nearest neighbors are both considered to be "nearest" neighbors, a divalent ion on a square array of sites ($2l = 8$) would have 10 nearest neighbors ($a_k = 6$) if it were placed between two sites and 12($a_k = 4$) if it were placed between 4 sites, whereas a trivalent ion would have 12 ($a_k = 6$) nearest neighbors (Fig. 4.1f). A quadrivalent ion placed on the same lattice would also have 12 nearest neighbors ($a_k = 6.67$) if it were placed in the middle of the square formed by 4 adjacent sites. If first, second and third-nearest neighbors are all considered to be "nearest" neighbors, a site in a hexagonal array would have 12 nearest neighbors (Fig. 4.15a) and parameter a_k would equal 10 for $z_k = 1, 2, 3$ and 4.

It may be concluded that parameter a_k increases with increasing number of nearest neighbors that are also nearest neighbors to each other and for certain arrays also with increasing valency of the ions, in particular if polyvalent ions are placed between the sites.

If $a_k/2l$ approaches unity with increasing l, the adjusted mole fraction would approach the mole fraction rather than the equivalent fraction, in contrast to (4.13). A strictly two-dimensional array of sites is a rather poor model for a soil, however, and the inaccuracy of (4.1) would be smaller for a three-dimensional array (Guggenheim, 1944a). For the present treatment of ion exchange the exact form of (4.48) is not essential and for simplicity it is assumed throughout section 4.1 that (4.1) holds. If one wishes to remove this restriction it is merely necessary to replace (4.1) by (4.48) in the relevant formulae. The essential point in the present treatment is that a relation of the form of (4.48) holds for a certain array of sites and thus that inequality (4.47) holds.

The simplest possible array of sites that is of significance for heterovalent exchange is a linear array of sites, with $2l = 2$ (see Fig. 4.1a). Such an array of sites may be realized on an organic chain-molecule, the sites being formed by ionizable groups such as carboxyl groups. In soils, however, these chain molecules will generally be coiled and thus form three-dimensional structures such that $2l > 2$. Although a value of $2l = 2$ is unlikely to be representative for soils, it is of interest to investigate the limiting behavior of the relevant formulae for $2l = 2$, since in that case $h_k = 1$ for all z_k and thus the adjusted mole fraction reduces to the ordinary mole fraction. For $2l = 2$ and $z_A \neq z_B$ it follows that:

$$\ln K_N = \ln K^0_{ex} + (1/z_A)\ln(M_A/N_A) - (1/z_B)\ln(M_B/N_B) \tag{4.49}$$

whereas:

$$K_M = K_{\mathcal{M}} = K^0_{ex} \tag{4.50}$$

Hence K_N varies with the composition of the adsorbent surface, whereas $K_{\mathcal{M}}$ and K_M are both constants.

The Vanselow exchange constant (Vanselow, 1932), which applies to heterovalent exchange, may be defined as:

$$K(\text{Vanselow}) \equiv \frac{M_B^{z_A}\, a_A^{z_B}}{M_A^{z_B}\, a_B^{z_A}} \tag{4.51}$$

hence for the present model ($l = 1$):

$$K(\text{Vanselow}) = (K_M)^{z_A z_B} = (K^0_{ex})^{z_A z_B} \tag{4.52}$$

It thus follows that the Vanselow exchange constant may be derived from the theory of ideal binary mixtures for $l = 1$. Vanselow (1932) himself did not consider the configuration of adsorption sites on an adsorbent surface and disregarded the fact that each ion should occupy a number of sites ("surface area") proportional to its charge. The derivation of K(Vanselow) from the ideal mixture would cast some doubts upon the applicability of K(Vanselow) to soils, since a value of $l = 1$ is probably too low to be representative for soils. It is not impossible, however, that also for certain arrays of sites with higher numbers of nearest neighbors, $K_{\mathcal{M}}$ would be close to K_M such that K(Vanselow) would be approximately constant with adsorbent composition for those arrays.

Another exchange coefficient that is frequently used in soil science, is the Gapon exchange "constant" (Gapon, 1933), which may be defined according to (Ermolenko, 1966):

$$K'(\text{Gapon}) \equiv \frac{N_A \, {}^m c_B^{1/z_B}}{N_B \, {}^m c_A^{1/z_A}} \tag{4.53}$$

In general the application of K′(Gapon) is confined to ion exchange involving monovalent and divalent cations and therefore K′(Gapon) is commonly defined as (compare (A4.10) of part A):

$$K'(\text{Gapon}) \equiv \frac{N_A \, {}^m c_B^{1/2}}{N_B \, {}^m c_A} \tag{4.54}$$

Hence, if $z_A = 1$ and $z_B = 2$, it follows that:

$$K'(\text{Gapon}) = N_B^{-1/2} (K_N')^{-1} \tag{4.55}$$

or, if N_A is sufficiently small:

$$K'(\text{Gapon}) \simeq (1 + \tfrac{1}{2} N_A)(K_N')^{-1} \tag{4.56}$$

It follows that for N_A of the order of 0—20%, K′(Gapon) is almost equal to $(K_N')^{-1}$, which might explain in part why this coefficient is rather constant for a number of saline soils. It may be noted that the selectivity coefficients in this chapter are consistently defined on the basis of (2.1), with $z_A = 1$ and $z_B = 2$ whenever this applies, whereas K′(Gapon) as defined by (4.53) refers to the reverse of (2.1). The Gapon constant may be obtained by applying the law of mass action to the following equilibrium:

$$A + B_{1/2}X \rightleftharpoons AX + \tfrac{1}{2}B \tag{4.57}$$

where $z_A = 1$ and $z_B = 2$ and where X refers to the adsorbed phase. The physical significance, however, of the notation $B_{1/2}X$ instead of $\frac{1}{2}$ BX is somewhat obscure. It seems impossible to derive the Gapon constant from the theory of ideal mixtures, but it may be obtained from the theory of regular mixtures (cf. section 4.1.2.) as an approximation valid for a limited range of composition of the adsorbent. The Gapon constant may also be shown to follow as an approximation from the theory of the electric double layer (Eriksson, 1952; Lagerwerff and Bolt, 1959).

Although the Kerr, Vanselow and Gapon exchange constants are widely used in soil science, none of them takes explicitly the structure of the adsorbent into consideration. This was done by Davis (1950) and Davis and Rible (1950) who applied Guggenheim's (1944a) model to ion exchange. Davis (1950) introduced an exchange "constant" according to:

$$K(\text{Davis}) \equiv \frac{m_B^{z_A} (\Sigma_k h_k m_k)^{z_B - z_A} a_A^{z_B}}{m_A^{z_B} a_B^{z_A}} \tag{4.58}$$

hence:

$$K(\text{Davis}) = \left[\frac{h_A^{z_B}}{h_B^{z_A}}\right] K_M^{z_A z_B} \tag{4.59}$$

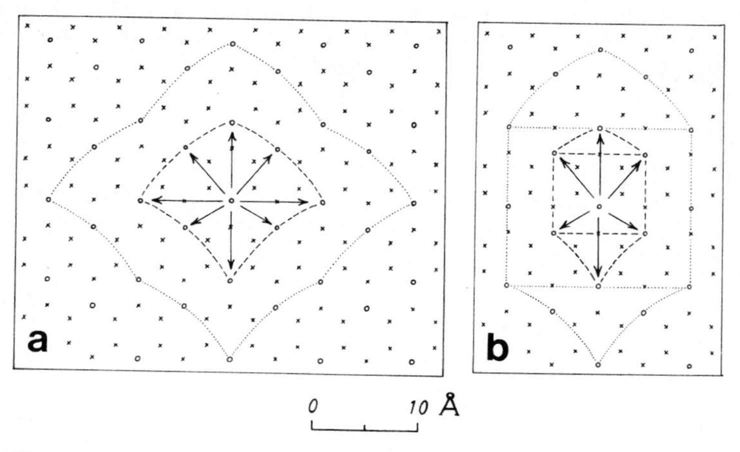

Fig. 4.2. Schematic representation of the pyrophyllite surface. Si^{4+} (crosses) → Al^{3+} (circles) substitution is 1/4 (muscovite) and the surface charge density is about 0.37 Coul/m^2. Arrows point to first-nearest neighbors, connected by a broken line; second-nearest neighbors are connected by a dotted line. The number of (first) nearest neighbors is 8 (a), 6, 5 or 4 (b).

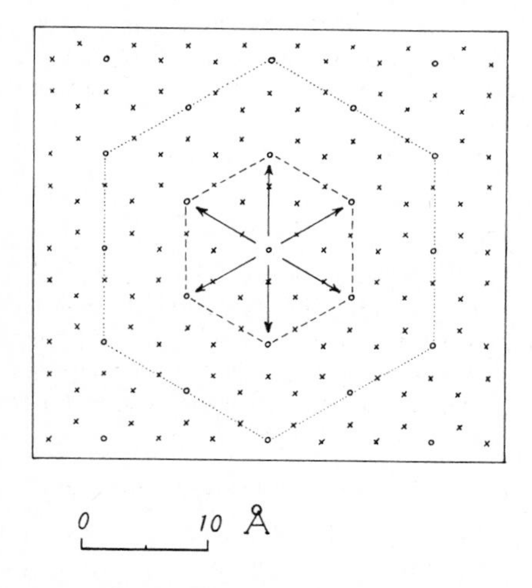

Fig. 4.3. Schematic representation of the pyrophyllite surface. Si^{4+} (crosses) → Al^{3+} (circles) substitution is 1/6 and the charge density is about 0.24 Coul/m^2. Arrows point to first-nearest neighbors, connected by a broken line; second-nearest neighbors are connected by a dotted line. The number of (first) nearest neighbors is 6.

and:

$$(1/z_A z_B) \ln K(\text{Davis}) = \ln K_{ex}^0 + (l-1) \ln \left[\frac{h_A z_B}{h_B z_A}\right] + (1/z_A) \ln h_A - (1/z_B) \ln h_B \qquad (4.60)$$

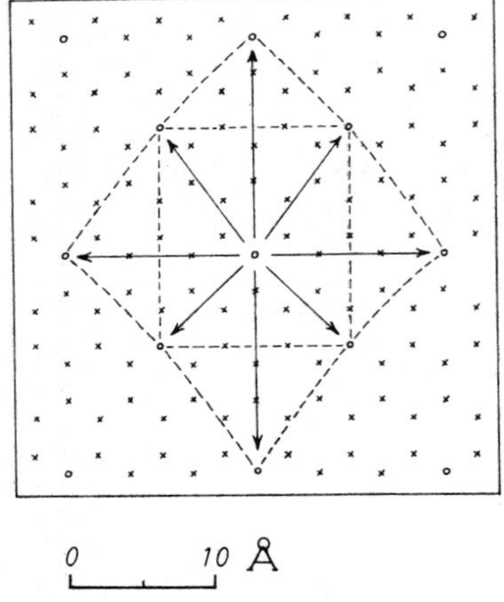

Fig. 4.4. Schematic representation of the pyrophyllite surface. Si^{4+} (crosses) → Al^{3+} (circles) substitution is 1/12 and the charge density is about 0.12 $Coul/m^2$. Arrows point to first-nearest neighbors, connected by a broken line. The number of (first) nearest neighbors is 4 or 8.

from which it can be seen that K(Davis) does not vary with the composition of the adsorbent. For the application of the Davis constant it is necessary to estimate the value of l. Krishnamoorthy and Overstreet (1950) proposed to use K(Davis) with $l = 3$. This seems to be a reasonable choice for ion exchange on clays, as is illustrated in Figs. 4.2—4.4.

The mineral pyrophyllite, $Al_2Si_4O_{10}(OH)_2$, may serve as a prototype for non-expanding three-layer clays. It has no net surface charge, but when part of the Si^{4+} in the tetrahedral layers is replaced by the lower charged Al^{3+}, a net negative surface charge results. Although the center of negative charge of a "site" coincides approximately with the nucleus of the Al atom, the excess negative charge is distributed over the surrounding O atoms. The probability of finding the excess electron on any of these O atoms is in first approximation equal for the 4 neighboring O atoms, whereas also the O atoms adjacent to the 4 coordinated O atoms will have a little increase in negative charge. Hence the excess negative charge tends to be spread out over the mineral surface. But when a cation approaches the surface the distribution of negative charge will change: the probability of finding an electron on one of the O atoms in the immediate proximity of the cation will increase.

Generally the negative surface charge of minerals that are derived from pyrophyllite is neutralized by K^+ ions embedded in hexagonal "holes" in tetrahedral layers of the minerals. The K^+ ions that are "sandwiched" in between the platelets are difficult to replace, but those adsorbed at the mineral—solution interface may be replaced by other cations.

If 1/4 of the Si^{4+} in both tetrahedral layers is replaced by Al^{3+}, the mineral muscovite, $KAl_2(Si_3Al_{10})(OH)_2$, is obtained. From Figs. 4.2a and 4.2b it can be seen that even if the configuration of the sites on the mineral surface is known, it may be difficult to determine the correct value of $2l$: 8 (Fig. 4.2a), 6, 5 or 4 (Fig. 4.2b). If less than one quarter of Si^{4+} is replaced by Al^{3+}, clay minerals of the illite group are obtained, $K_xAl_2(Si_{4-x}Al_x O_{10})(OH)_2$, where x ranges from 0.5 to 0.8, approximately. From Fig. 4.3 it follows that $2l = 6$ for illite clays with $x = 0.67$. From Fig. 4.4 it follows that for a lower charge density (x = 0.33) the choice is between $2l = 4$ and $2l = 8$. Hence $2l = 6$ seems to be a reasonable choice for the present group of (mica) minerals, but the correct value of $2l$ may have to be determined experimentally. In naturally occurring minerals the substitution charges will be more randomly distributed than in Figs. 4.2—4.4, but on electrostatic grounds it is unlikely that the sites would cluster together and therefore the substitution charges tend to be distributed evenly over the mineral surface. The distribution of sites may further be influenced by lattice imperfections or fractures.

It is interesting to investigate the limiting behavior of the present model for $l \to \infty$ and assuming that (4.1) holds, even though these conditions may not seem to be realistic for soils. For example:

$$\lim_{l \to \infty} (\bar{f}_A^{1/z_A}) = \exp\{N_B(1/z_A - 1/z_B)\} \tag{4.61a}$$

and:

$$\lim_{l \to \infty} (\bar{f}_B^{1/z_B}) = \exp\{N_A(1/z_B - 1/z_A)\} \tag{4.61b}$$

whereas the entropy of mixing becomes:

$$\lim_{l \to \infty} \Delta S_{mix} = m_A k \ln N_A - m_B k \ln N_B \tag{4.62}$$

where use has been made of:

$$\lim_{m \to \infty} (1 + x/m)^m = \exp(x)$$

From:

$$\lim_{l \to \infty} \left[\frac{\bar{f}_A^{1/z_A}}{\bar{f}_B^{1/z_B}}\right] = \exp(1/z_A - 1/z_B) \tag{4.63}$$

it follows that, if $l \to \infty$:

$$\ln K_M = \ln K_N = \ln K^0_{ex} + (1/z_A - 1/z_B) \tag{4.64}$$

and:

$$\ln K_M = \ln K^0_{ex} + (1/z_A - 1/z_B) + (1/z_A)\ln(N_A/M_A) - (1/z_B)\ln(N_B/M_B) \tag{4.65}$$

Lewis and Thomas (1963) proposed an expression for K_N similar to (4.64). They obtained the entropy of mixing, given by (4.62), from the theory of liquid mixtures of molecules of unequal size (see: Lewis and Randall, 1961, p. 287 ff.) by assuming that the ratio of the molal volumes of the pure liquids correspond to the charge ratio of the exchanging ions. For an ideal mixture ΔH_{mix} equals zero and relative activities of ionic species follow from:

$$\frac{\partial}{\partial m_k}(-T\Delta S_{mix}) = kT \ln \bar{a}_k \tag{4.66}$$

Thus (4.61a) and (4.61b) are obtained from (4.62) by partial differentiation with respect to m_A and m_B, respectively.

The question may arise as to whether the limit for $l \to \infty$ can be interpreted physically within the present model. Equation (4.1) is based on the assumption that no two nearest neighbors are nearest neighbors to each other at the same time. It is unlikely, however, that it would be possible to increase the number of nearest neighbors without violating that assumption. Therefore it seems more appropriate to turn to (4.48), which does allow for a simultaneous increase in pairs of nearest neighbors that are also each other's nearest neighbors.

The number of nearest neighbors to an adsorption site could be made to increase by increasing the number of fixed surface charges per adsorption site. If they no longer coincide it becomes necessary to distinguish between "fixed surface charges" and "adsorption sites". The meaning of "fixed surface charge" remains unaltered, but the charge of an individual surface charge would decrease from unit electronic charge initially to zero in the limit for $l \to \infty$. An "adsorption site" may then be defined as a collection of neighboring surface charges with a total charge equal to the charge of one electron (-1.6×10^{-19} Coul). It may further be assumed that when the charge of a fixed surface charge decreases from 1 to $1/d$, the number of fixed charges per "site" increases from 1 to d, such that the surface charge density on the adsorbent remains constant. If so, a "site" with a charge of -1.6×10^{-19} Coul would correspond to a constant unit of surface area on the adsorbent. This is illustrated in Fig. 4.5 where from the top downwards the surface charge density and thus the area of one adsorption site remains constant, but the number of surface charges per "site" increases from 1 (a), 7 (b), 19 (c) to "infinity" (d), whereas the charge per fixed surface charge decreases from 1 (a), 1/7 (b), 1/19 (c) to "zero" (d), such that the surface charge density remains constant: 1 electronic charge per 100 Å^2 or 0.16 Coul/m^2.

"Nearest neighbors" to a particular site may now be defined as all distinct sites that are accessible to one segment of a divalent ion when the other segment is attached to that particular site. Thus the number of nearest neighbors is still a lattice property, but the "sites" may partly coincide and the number of surface charges per site may vary. The number of surface charges per site, d, may be introduced in (4.48) according to:

$$2lh_k d = 2lz_k d - a_k(z_k d - 1) \tag{4.67}$$

For $d = 1$ the fixed surface charges and the sites coincide and (4.67) reduces to (4.48). But for $d > 1$ an ion of species k, with valency z_k, would occupy $z_k d$ surface charges at the time, each with charge $1/d$. Although the number of charges per site would increase, the configuration of the sites with respect to each other, characterized by $2l$, would not change. Therefore, z_k and l in (4.67) would be independent of d, whereas h_k and a_k would vary with d. From:

$$h_k = z_k - \left[\frac{a_k}{2l}\right](z_k - 1/d) \tag{4.68}$$

it follows that:

$$\lim_{d \to \infty}(h_k) = z_k - z_k \lim_{d \to \infty}\left[\frac{a_k}{2l}\right] \tag{4.69}$$

The limit for a_k for $d \to \infty$ is not easily evaluated since a_k is not known as a function of d. But when d increases, the nearest neighbor "sites" do overlap more and more and therefore it may be expected that a_k approaches $2l$ if $d \to \infty$. This is illustrated schematically in Fig. 4.5 where the number of nearest neighbors to a single site increases from top to bottom: 6 (a), 18 (b), 30 (c) and "infinity" (d), whereas h_k decreases from the top downwards: 1 (a), 0.43 (b), 0.26 (c) and "zero" (d) for $z_k = 1$, and 1.33 (a), 0.57 (b), 0.35 (c)

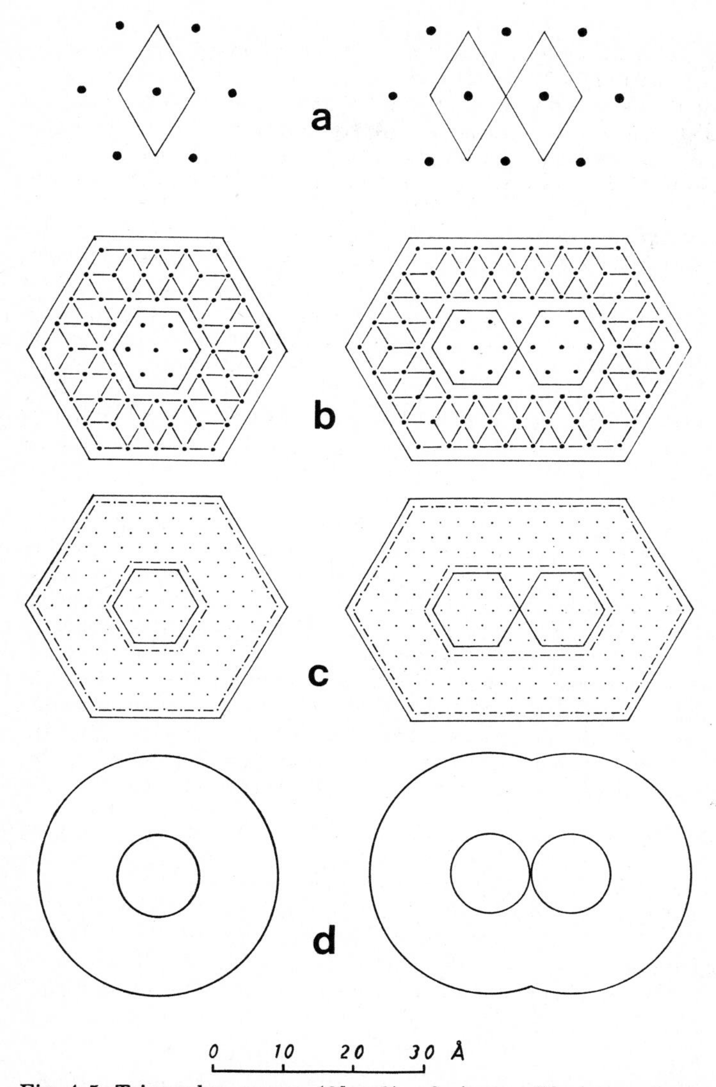

Fig. 4.5. Triangular arrays ($2l = 6$) of sites, with increasing numbers of fixed surface charges (dots) per site: 1 (a), 7 (b), 19 (c) and "infinity" (d). Nearest-neighbor sites to a single site (left column) and to a pair of sites (right column) are indicated schematically. The charge density is 0.16 Coul/m^2 (a—d).

and "zero" (d) for $z_k = 2$. Hence Fig. 4.5 illustrates that the limit of h_k for $d \rightarrow \infty$ (equation 4.69)) tends to zero, reflecting that the number of fixed charges per "site" increases more rapidly than the number of nearest neighbors to an ion of species k. But the limit of $2lh_k d$ also tends to infinity if d approaches infinity, although much slower than d itself, and thus it may be stated that 'the number of nearest neighbors to an ion of species k approaches infinity' if $d \rightarrow \infty$ at constant charge density on the adsorbent.

Fig. 4.5 further illustrates that although h_k tends to zero if $d \rightarrow \infty$, the ratio h_k/h_l, where k and l refer to ionic species of unequal valency, tends to a finite, non-zero value. For the situation pictured in Fig. 4.5 it follows that:

$$\lim_{d \to \infty} (h_A / h_B) = 3/4$$

where $z_A = 1$ and $z_B = 2$, and where it is assumed that, at constant charge density, the number of nearest neighbors to one or two sites remains proportional to the surface area neighboring one or two sites (Fig. 4.5d). This observation is of significance since it would imply that the adjusted mole fraction would be defined in the limit for $d \to \infty$ according to:

$$\lim_{d \to \infty} (M_k) = \frac{m_k}{m_k + m_l \lim\limits_{d \to \infty} (h_l / h_k)} \tag{4.70}$$

Hence the activity coefficient of species k in the adsorbed state would tend to unity:

$$\lim_{d \to \infty} (M_k / N_k)^{l h_k / z_k} = 1 \tag{4.71}$$

since the number of sites that are neighbors to an ion of valency z_k divided by the number of surface charges occupied by an ion of valency z_k tends to zero. Therefore the activity of species k in the adsorbed state would tend to its equivalent fraction, reflecting that in the field model ($d \to \infty$) there are no problems of placement for a polyvalent ion other than the need for a surface area proportional to the charge of the ion. These conclusions support the choice of the equivalent fraction as the most suitable composition variable for ions in an adsorbed phase (Gaines and Thomas, 1953; Lewis and Thomas, 1963). Finally it may be noted that the result of (4.71) would also be obtained if it were assumed that $z_k \to \infty$ at constant z_k / z_l, but this assumption would be physically less realistic than the assumption that the charge on the adsorbent be distributed evenly over the adsorbent surface, as in the field model.

The present ideal mixture model based on Guggenheim (1944a) comprises the Vanselow (1932) model, for $l = 1$, the Davis (1950) model, for $l = 3$, as proposed by Krishnamoorthy and Overstreet (1950), and the Lewis and Thomas (1963) model, for $l \to \infty$. The transition between these models is illustrated in Fig. 4.6, which shows that the difference between the ratio of the activity coefficients for $l = 3$ and $l \to \infty$ is rather small ($\leqslant 5\%$). Assuming that $l = 3$ is representative for soils, it may be expected that K_N (Lewis and Thomas, 1963) be more constant with adsorbent composition than K_M (Vanselow, 1932). It may be concluded that the Lewis and Thomas model, with K_N = constant, would provide a good first approximation to describe ion exchange in soils. If detailed information is available, an equation of the Davis (1950) type, with l, h_k and h_l as parameters, may be used.

It may be noted that a comparison between different selectivity coefficients (for example: Magistad et al. 1944; Krishnamoorthy and Overstreet, 1950; Bolt, 1967) should be performed with respect to the same amount of adsorbent. In the present treatment one equivalent of adsorbent has been used throughout the text for K_M, $K_{\mathcal{M}}$ and K_N. In contrast, K(Vanselow) and K(Davis) refer to $z_A z_B$ equivalents of adsorbent, whereas K(Kerr) refers to z equivalents of adsorbent.

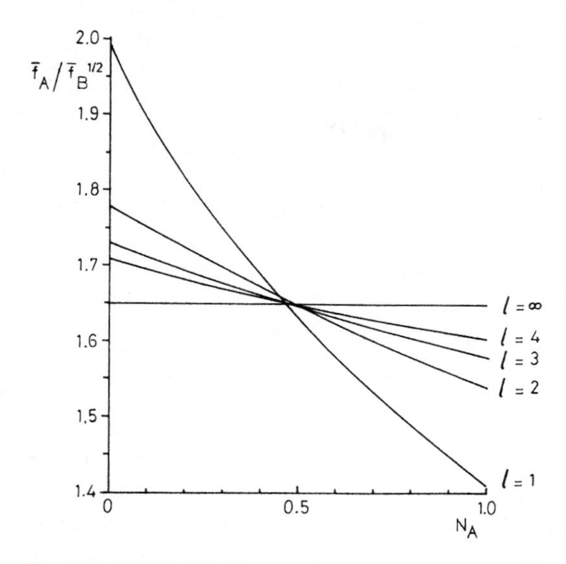

Fig. 4.6. Ratio of activity coefficients for an ideal mixture of ions of species A ($z_A = 1$) and B ($z_B = 2$) calculated from (4.21) and (4.39) for $l = 1$ (Vanselow, 1932), $l = 2, l = 3$ (Davis, 1950; Krishnamoorthy and Overstreet, 1950), $l = 4$ and $l \rightarrow \infty$ (Lewis and Thomas, 1963).

4.1.2. Regular mixtures

Regular mixtures are mixtures with a non-zero energy of mixing, i.e. mixtures for which:

$$e_{kl} \neq \tfrac{1}{2}(e_{kk} + e_{ll}) \tag{4.72}$$

In ideal mixtures all configurations of k and l ions in an adsorbed phase have the same statistical weight: the arrangement of particles is completely random. In regular mixtures, however, the statistical weight of distinguishable configurations of ions is also determined by the intermolecular potential energy.

Cations adsorbed on a regular array of adsorption sites do repel each other because of their net positive charges. The energy of interaction, e_{kl}, between two cations, one of species k and the other of species l, is given in first approximation by Coulomb's Law:

$$e_{kl} = \frac{z_k z_l e^2}{4\pi\epsilon_0 \epsilon_w s_{kl}} \tag{4.73}$$

where ϵ_0 is the permittivity of vacuum, ϵ_w the relative dielectric constant of water, s_{kl} the distance between the two cationic species, e the charge of an electron, and z_k and z_l the charge numbers of the two cationic species. The ionic species involved may differ in an ionic property such as the polarizability, which is a measure of the ease with which a dipole moment can be induced in an elementary particle and which increases with increasing mass, size and number of electrons in the outer shell of the particle (Margenau, 1939). The ionic species may also differ in the number of water molecules or other uncharged ligands in the inner coordination sphere of the ion, which may alter the dielectric constant of water in the proximity of the ion. Thus the species may have "effective" charges, $\lambda_k z_k e$ and $\lambda_l z_l e$, with λ_k and λ_l close to unity, which may differ slightly from $z_k e$ and $z_l e$. Introducing these effective ionic charges in (4.73) and taking $\epsilon_0 = 8.85 \cdot 10^{-12}$ Coul^2 N^{-1} m^{-2}, $\epsilon_w = 80$ and $e = 1.60 \cdot 10^{-19}$ Coul, it follows that:

$$e_{kl} = 2.88 \cdot 10^{-20} z_k \lambda_k z_l \lambda_l / s_{kl} \tag{4.74}$$

where s_{kl} is in Angstrom (10^{-10} m) and where e_{kl} is in Joules. A value of 80 for ϵ_w at room temperature (293°K) has to be considered as a maximum value, and would apply to "free" water. Close to the adsorbent surface ϵ_w may decrease to values as low as 6 in case of complete dielectric saturation. The energy of mixing may be written as:

$$e_m = 2.88 \cdot 10^{-20} (z_k z_l \lambda_k \lambda_l / s_{kl} - \tfrac{1}{2} z_k^2 \lambda_k^2 / s_{kk} - \tfrac{1}{2} z_l^2 \lambda_l^2 / s_{ll}) \tag{4.75}$$

It appears to be difficult to predict the magnitude of e_m or even its sign. But a few qualitative remarks can be made. If $z_k = z_l = z$ and $s_{kl} = s_{kk} = s_{ll} = s$, (4.75) reduces to:

$$e_m = 2.88 \cdot 10^{-20} (\lambda_k \lambda_l - \tfrac{1}{2} \lambda_k^2 - \tfrac{1}{2} \lambda_l^2) z^2 / s \tag{4.76}$$

hence:

$$e_m = -1.44 \cdot 10^{-20} (\lambda_k - \lambda_l)^2 z^2 / s \leqslant 0 \tag{4.77}$$

It follows that if cationic species of equal valency differ in an ionic property that influences coulombic interactions between ions, the energy of mixing is always negative, provided interparticle distances are all equal.

If $z_k = z_l = z$ and $\lambda_k = \lambda_l = 1$ it follows that:

$$e_m = 2.88 \cdot 10^{-20} z^2 (1/s_{kl} - \tfrac{1}{2}/s_{kk} - \tfrac{1}{2}/s_{ll}) \tag{4.78}$$

hence $e_m \neq 0$ if:

$$2/s_{kl} \neq 1/s_{kk} + 1/s_{ll} \tag{4.79}$$

If equilibrium distances between cations and adsorbent surface differ for different species this may result in inequality (4.79) to hold. Fig. 4.7 illustrates that in that case $s_{kl} > s_{kk}$ even though $s_{kk} = s_{ll}$, and from:

$$2/s_{kl} < 1/s_{kk} + 1/s_{ll} \tag{4.80}$$

it follows that again $e_m < 0$. Hence for the situation pictured in Fig. 4.7 the energy of mixing would also be negative. So it may be expected that in general the energy of mixing for equal-charged ions will be zero or slightly negative.

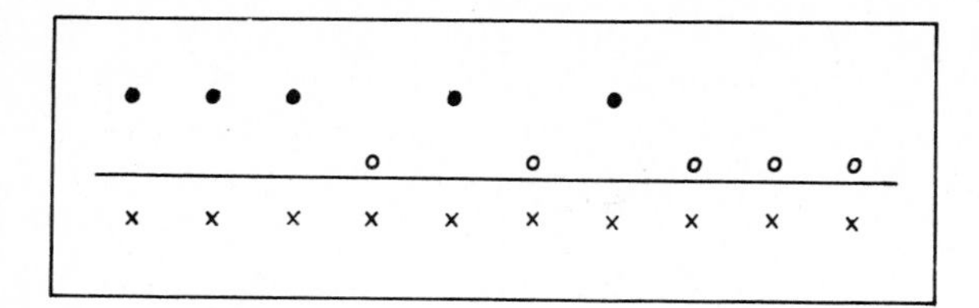

Fig. 4.7. Monovalent cations of species A (open circles) and B (dots) adsorbed on a linear array of sites. Equilibrium distances to the centers of negative charge (crosses) of ions of species A exceed those of species B, so that inequality (4.80) holds.

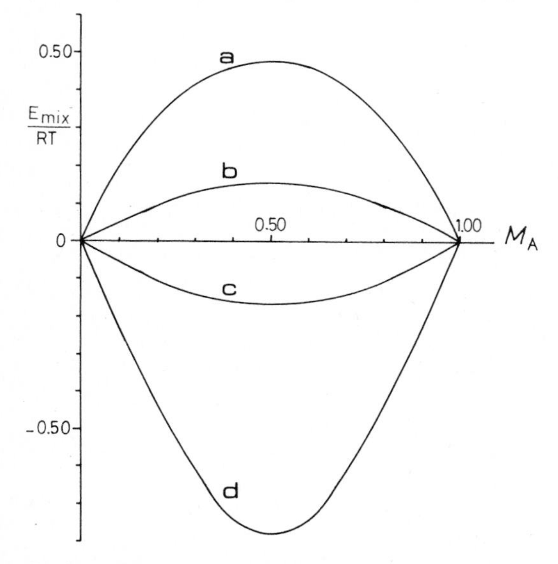

Fig. 4.8. Energy of mixing for a mixture of one equivalent of divalent ions of species A and B on a triangular array of sites ($l = 3$) calculated from (4.16) for $2e_m/kT = 1.00$ (a), 0.25 (b), −0.25 (c) and −1.00 (d).

The energy of mixing, E_{mix}, has a maximum ($e_m > 0$) or a minimum ($e_m < 0$) for $M_A = M_B = \frac{1}{2}$ as follows from (4.10) and (4.16). Figure 4.8 shows that for a binary mixture of one equivalent of divalent ions of species A and B the following inequality holds:

$$-E_{mix}(e_m = -a) > E_{mix}(e_m = a) \tag{4.81}$$

where a is a positive number. This is because for $e_m < 0$ the ions tend to mix intensively with each other to realize a minimum energy. These configurations are easy to achieve. For $e_m > 0$, however, unlike particles will avoid contact with each other, because of the unfavorable energy of mixing. Therefore, like particles will tend to form "islands" on the adsorbent surface in order to minimize contact between unlike particles. This tendency

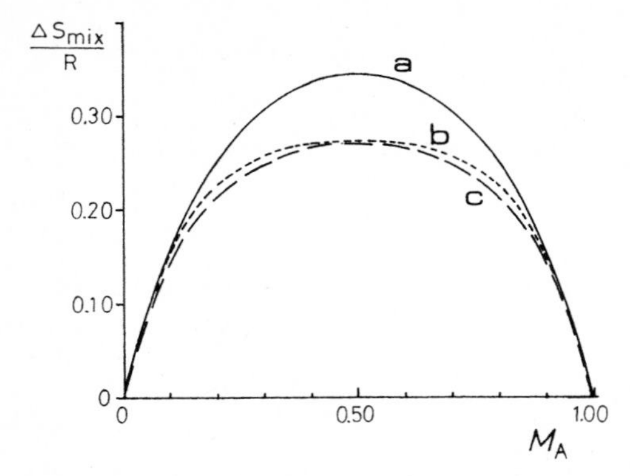

Fig. 4.9. Entropy of mixing for a mixture of one equivalent of divalent ions of species A and B on a triangular array of sites ($l = 3$) calculated from (4.27) for $2e_m/kT = 0$ (a), -1.00 (b) and 1.00 (c).

to demix is opposite to the tendency of the system to realize a maximum entropy and therefore these "low" energy configurations are less easily realized. Hence for $e_m \neq 0$ the tendency of the mixture to achieve a minimum energy always results in a lowering of the entropy of the system as compared to the ideal mixture (curve a, Fig. 4.9), but for $e_m < 0$ this lowering is less pronounced (curve b, Fig. 4.9) than for $e_m > 0$ (curve c, Fig. 4.9).

The entropy of mixing, ΔS_{mix}, for an ideal mixture reaches its maximum at $M_A = M_B = \frac{1}{2}$, indicating that the number of accessible configurations of a binary mixture is highest at that composition. For non-zero energies of mixing the number of accessible configurations decreases, but to a limited extent only: for $2e_m/kT = \pm 0.25$ the entropy curves practically coincide with the curve for ideal mixing (Fig. 4.9). This is because the energies of mixing considered here are rather low. For a binary mixture of one equivalent of A and B ions of valency z, the entropy of mixing reaches an extremum at $M_A = M_B = \frac{1}{2}$, given by:

$$\frac{\Delta S_{mix}}{R} = \left(\frac{lh}{z}\right)\left(\frac{e_m/kT}{1+\exp(e_m/kT)}\right) - \left(\frac{lh}{z}\right)\ln\left(\frac{\exp(e_m/kT)}{1+\exp(e_m/kT)}\right) - \left(\frac{lh-1}{z}\right)\ln 2 \tag{4.82}$$

This extremum may be a (relative) maximum or minimum depending on the value of e_m. The value of the entropy of mixing given by (4.82) is symmetrical around $e_m = 0$ (curve a, Fig. 4.10): $\Delta S_{mix}/R$ reaches a maximum of $(1/z)\ln 2$ at $e_m = 0$ and then decreases with e_m:

$$\lim_{e_m \to \pm\infty} (\Delta S_{mix}/R) = -\left(\frac{lh-1}{z}\right)\ln 2 \tag{4.83}$$

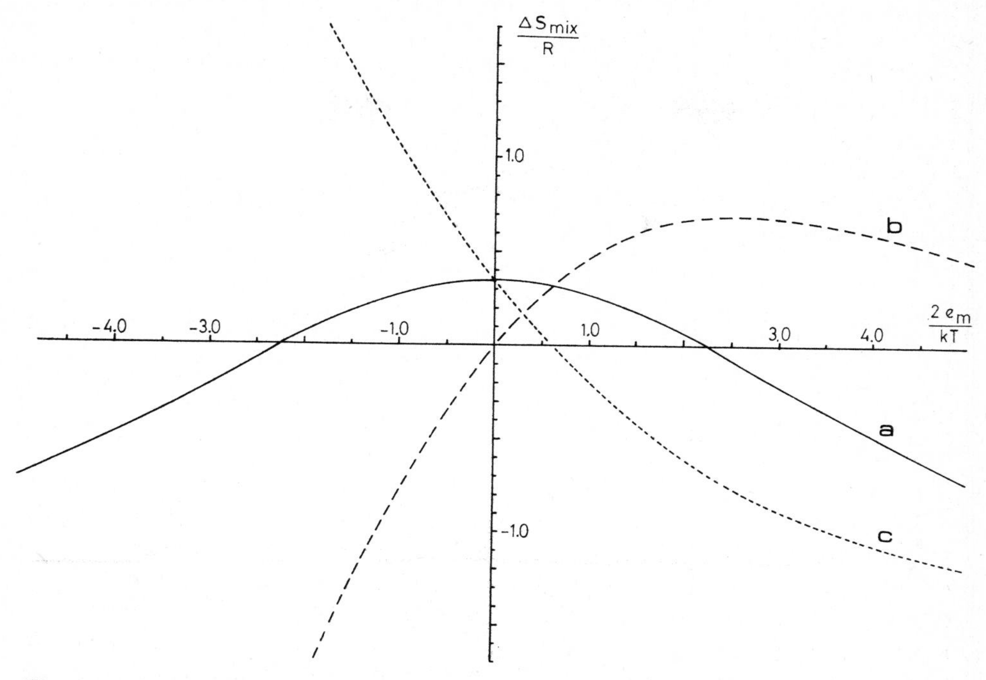

Fig. 4.10. A plot of the value of the entropy of mixing at $M_A = M_B = \frac{1}{2}$ (4.82) for a mixture of one equivalent of divalent ions of species A and B on a triangular array of sites ($l = 3$) as a function of $2e_m/kT$ (curve a). Curve b represents E_{mix}/RT and curve c the contribution to $\Delta S_{mix}/R$ due to the ionic activities in the adsorbed phase.

Hence if the absolute value of e_m is sufficiently large, mixing may result in a lowering of the entropy of the system. The first term in (4.82) is the energy of mixing, E_{mix}/RT, and the last two terms represent the contribution to the entropy of mixing due to ionic activities in the adsorbed phase (compare (4.27)). If the absolute value of e_m tends to infinity, the mixture rearranges into the energetically most favorable configuration. Hence, for $e_m \ll 0$ it follows that:

$$E_{mix}/RT \rightarrow \left(\frac{lh}{z}\right) \left(\frac{e_m}{kT}\right) \tag{4.84}$$

whereas:

$$\lim_{e_m \rightarrow +\infty} (E_{mix}/RT) = 0 \tag{4.85}$$

The energy of mixing (curve b, Fig. 4.10) reaches a maximum of $0.28(lh/z)$ at $2e_m/kT \simeq 2.56$: this maximum marks the point where the effect of demixing of the mixture starts to exceed the effect of the increase in e_m. In the limit for $e_m \rightarrow \infty$ the whole mixture would be separated in two homoionic halves such that contact between ions of different species would become negligible.

The decrease in entropy of mixing due to a non-zero energy of mixing can be measured by the "excess" entropy of mixing:

$$\Delta S^{X}_{mix} \equiv \Delta S_{mix}(e_m \neq 0) - \Delta S_{mix}(e_m = 0) \tag{4.86}$$

The excess entropy of mixing decreases with surface coverage until a minimum is reached at $M_A = M_B$ (Fig. 4.11). From definition (4.86) it follows that:

$$\Delta S^{X}_{mix} < 0 \tag{4.87}$$

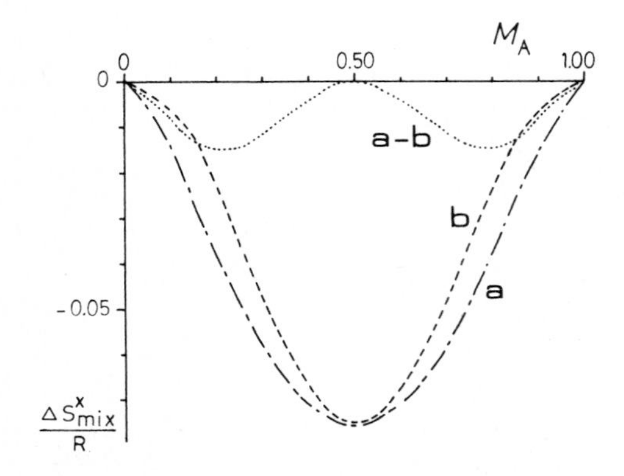

Fig. 4.11. Excess entropy of mixing for a mixture of one equivalent of divalent ions of species A and B on a triangular array of sites ($l = 3$) calculated from (4.86) for $2e_m/kT = 1.00$ (a) and -1.00 (b). Curve a—b represents the difference between curves a and b (4.88).

Figure 4.11 shows that ΔS^{X}_{mix} decreases more rapidly for $2e_m/kT = 1.0$ (curve a) than for -1.0 (curve b). Curve a—b in Fig. 4.11 represents the difference:

$$\Delta S^{X}_{mix}(2e_m/kT = 1.0) - \Delta S^{X}_{mix}(2e_m/kT = -1.0) \tag{4.88}$$

and it can be seen that at low coverage of the adsorbent with ions of one species the decrease in entropy of mixing is larger for $e_m > 0$ than for $e_m < 0$. But after a threshold coverage of about 20%, curve a—b approaches zero at $M_A = M_B = \frac{1}{2}$.

This is because at $M_A = M_B = \frac{1}{2}$ the distribution of frequencies of occurrence of configurations with equal energy of mixing is symmetrical around the most probable configuration. This is illustrated in Fig. 4.12. There are 6 ways in which 5 ions of species B and 1 ion of species A can be distributed over a linear array of 6 fixed sites (Fig. 4.12, situation I). Two of these configurations belong to category a (1 AB interaction) and 4 belong to category b (2 AB interactions). If $e_m > 0$ the mixture tends to avoid AB interactions and hence the probability of the 4 configurations that belong to category b

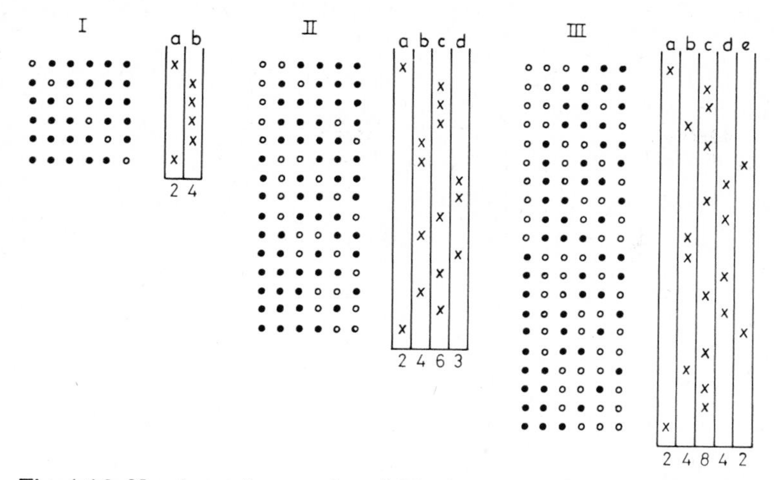

Fig. 4.12. Number of ways in which 1, 2 and 3 monovalent ions of species A (open circles) and 5, 4 and 3 of species B (dots) can be distributed over 6 sites in a linear array: 6 (I), 15 (II) and 20 (III), respectively. Distinguishable configurations are categorized according to the number of interactions between unlike ions: 1 (a), 2 (b), 3 (c), 4 (d) or 5 (e).

is decreased with respect to those belonging to category a. If $e_m < 0$ the probability of the 2 configurations belonging to category a is decreased with respect to the 4 belonging to category b. Of course the sum of the relative probabilities remains unity. The decrease in entropy is proportional to the number of distinguishable configurations which are lowered in probability of realization, i.e. which have a decreased accessibility. Hence the decrease in entropy is larger if $e_m > 0$ than if $e_m < 0$. A similar argument holds for situation II in Fig. 4.12: the probability of the configurations with a small number of AB interactions is decreased if $e_m < 0$ and increased if $e_m > 0$. When equal amounts of ions of species A and B are distributed over the sites (situation III) the distribution of frequencies of occurrence of configurations with equal numbers of AB interactions appears to be symmetrical around the most probable configuration (category c). Hence if $e_m < 0$ the configurations of category a and b are decreased in probability with respect to those of category d and e, whereas the reverse happens if $e_m > 0$, but the effect on the entropy of mixing is the same in both cases. It may be noted that when side effects are not taken into consideration in Fig. 4.12, the argument remains essentially similar. Then the configurations may be classified according to the number of neighboring sites occupied by ions of species A: no (class 1), 2 (class 2) or 3 (class 3). Hence if $e_m > 0$ the probability of configurations belonging to class 1 is raised with respect to those belonging to class 3, whereas the reverse happens if $e_m < 0$.

Figure 4.13 shows the influence of non-zero energies of mixing on the ratios of the activity coefficients of the ions in the adsorbed state calculated from (4.19) and (4.21). With the aid of these expressions for the activity coefficients it is possible to construct "exchange isotherms", i.e. plots of N_k as a function of f_k at constant temperature. Equation (2.16) can be written as:

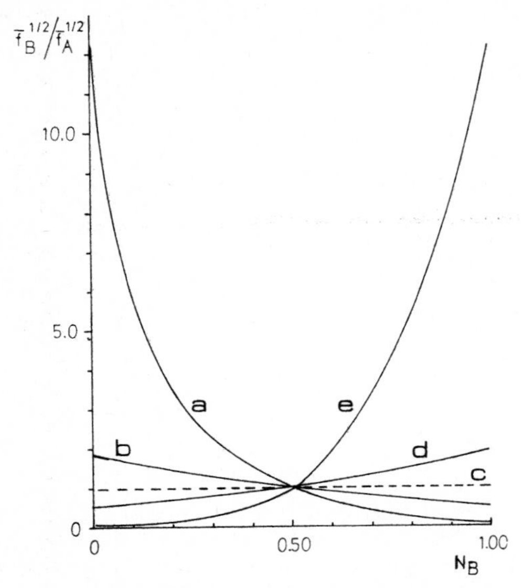

Fig. 4.13. Square root of the ratio of activity coefficients of divalent ions of species A and B on a triangular array of sites ($l = 3$) calculated from (4.19) and (4.21) for $2e_m/kT = 1.00$ (a), 0.25 (b), 0 (c), -0.25 (d) and -1.00 (e).

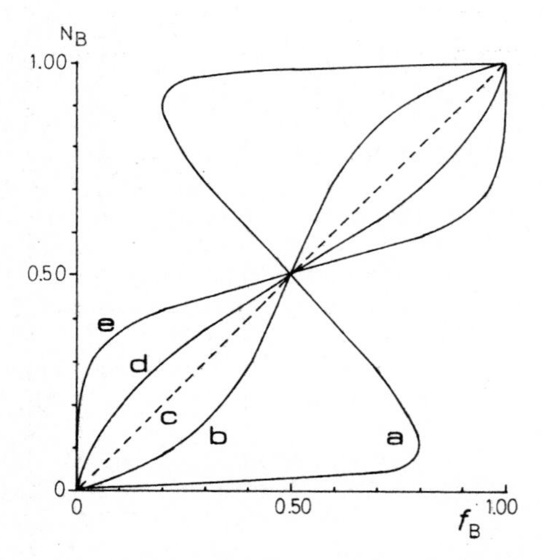

Fig. 4.14. Exchange isotherms for a mixture of divalent ions of species A and B on a triangular array of sites ($l = 3$) calculated from (4.90) for $2e_m/kT = 1.00$ (a), 0.25 (b), 0 (c), -0.25 (d) and -1.00 (e).

$$\left(\frac{\bar{f}_B N_B}{\bar{f}_A (1 - N_B)}\right)^{1/2} = K^0_{ex} \left(\frac{f_B f_B}{f_A (1 - f_B)}\right)^{1/2} \tag{4.89}$$

for a binary mixture of A and B ions with $z_A = z_B = 2$. Taking $K^0_{ex} = 1$ and $(f_B/f_A) = 1$ (4.89) reduces to:

$$\left(\frac{\bar{f}_B N_B}{\bar{f}_A (1 - N_B)}\right) = \left(\frac{f_B}{1 - f_B}\right) \tag{4.90}$$

The exchange isotherms in Fig. 4.14 clearly show the consequences of non-zero energies of mixing for ion exchange. Curve a ($2e_m/kT = 1.0$) shows that initially it is difficult to bring ions of species B on an adsorbent in the homoionic A form, since this state is energetically more favorable than a mixture. Entropy increases with mixing, however, and after the adsorbent has reached a critical composition, the adsorbent starts to take up ions of species B "spontaneously" thereby increasing N_B and decreasing f_B. All exchange isotherms for $K^0_{ex} = 1$ show what may be referred to as "selectivity reversal": for $f_B < 0.5$ one species of cation is preferred over the other by the adsorbent and for $f_B > 0.5$ the situation is reversed.

Since divalent or higher charged ions will be located in between two or more adsorption sites the notion of "nearest-neighbor" interactions becomes somewhat difficult to apply and average distances between ions in the adsorbed state depend on the configuration of ions on the adsorbent. In a hexagonal array of sites, for example, each site has 3 nearest neighbors at a distance of s from the central site, 6 second-nearest neighbors at a distance of $s\sqrt{3}$ and 3 third-nearest neighbors at 2s from the central site (Fig. 4.15a). For the configuration of divalent ions visualized in Fig. 4.15b the assumption that there are only 3 nearest neighbors would result in interactions between the central ion and 4 neighboring ions at a distance of $s\sqrt{3}$. However, interactions with two other ions at the same distance from the central ion would not be considered. Therefore it may seem more appropriate to assume that there are 12 "nearest" neighbors. Then interactions between

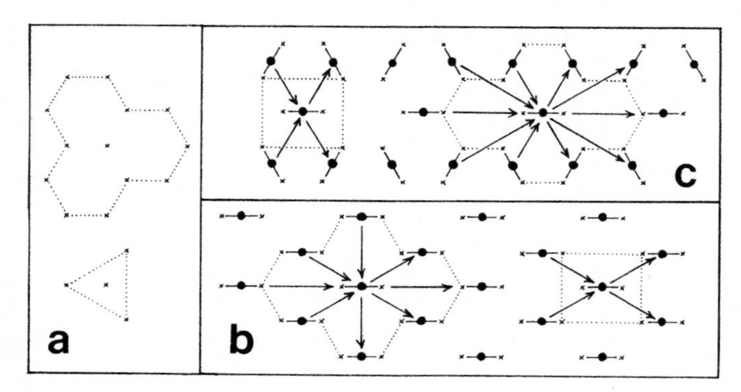

Fig. 4.15. Number of nearest neighbors to a site in a hexagonal array: 3 or 12 (a). Sites that are nearest neighbors to one site (a) or to a pair of sites (b and c) are connected by dotted lines. Interactions between divalent ions occupying sites that are nearest neighbors to a pair of sites and a central ion are indicated with arrows for two arrangements of ions (b and c).

the central ion and 6 neighboring ions at a distance of $s\sqrt{3}$ would be considered, but also with 2 ions at a distance of 3s. Since interactions with 4 other ions at a distance of 3s would again not be considered, the assumption that $2l = 12$ leads to an inconsistency as well. For the configuration of divalent ions pictured in Fig. 4.15c the assumption that $2l = 3$ results in interactions between the central ion and 4 ions at a distance of 1.5s. But the assumption that $2l = 12$ would result in interactions between the central ion and 10 surrounding ions: 4 of which are at a distance of 1.5s, 4 at $1.5s\sqrt{3}$ and 2 at 3s from the central ion, whereas interactions with 4 other ions at a distance of 3s would again not be considered. The best way to solve this problem might be to take $2l = 12$ and to consider interactions only with those neighboring ions that are situated in between two "neighboring" sites. For the configuration in Fig. 4.15b this would result in interactions with 6 ions at a distance of $s\sqrt{3}$, and for the configuration in Fig. 4.15c in interactions with 4 ions at 1.5s. The configuration in Fig. 4.15b would then be energetically less favorable ($e_{BB} = 2.88 \cdot 10^{-20} \times 6.93/s$) than the configuration of Fig. 4.15c ($e_{BB} = 2.88 \cdot 10^{-20} \times 5.33/s$), assuming that each pair of sites contributes one term to the intermolecular energy of the system. The assumption that only nearest neighbor interactions contribute to the intermolecular energy of the system probably represents a simplification in the case of adsorption of ions, where the forces of interaction may not be of sufficiently short range to rule out second or higher neighbor interactions. But the assumption is useful to demonstrate in a simple way the origin and nature of non-zero energies of mixing for binary mixtures of ions.

So far the discussion has been confined to binary mixtures of cations of equal charge. From here on the attention will be focused on binary mixtures of cations of unequal charge. For such mixtures the average distances between adsorbed ions depend on the configuration of the ions on the adsorbent and may vary with the composition of the adsorbent. For ions of unequal charge the parameters λ_k and λ_l probably represent a minor effect and therefore it will be assumed here that they equal unity. Hence (4.75) becomes:

$$e_m = 2.88 \cdot 10^{-20} \left(z_k z_l / s_{kl} - \tfrac{1}{2} z_k^2 / s_{kk} - \tfrac{1}{2} z_l^2 / s_{ll}\right) \qquad (4.91)$$

where $z_k \neq z_l$. The simplest possible case for which $z_k \neq z_l$, is a binary mixture of monovalent ($z_A = 1$) and divalent ($z_B = 2$) ions on a linear array of sites ($2l = 2$). The energy of mixing for this system becomes:

$$E_{mix} = 2X^*_{AB}\, e_m \qquad (4.92)$$

The contribution of one pair of sites, one occupied by A the other by a segment of B, to the energy of mixing of the system is estimated to be:

$$e_m = 2.88 \cdot 10^{-20} (2/15 - 1/20 - 1/10) = -4.81 \cdot 10^{-22} \qquad (4.93)$$

where it is assumed that the distance between the sites is 10 Å and that a divalent cation is situated midway in between two sites. Of course the position of an ion of species B need not be fixed at that position and may depend on whether neighboring sites are occupied by ions of species A or B. However, as a first approximation this assumption seems reasonable. Figure 4.6 (curve a) shows that the extremum of E_{mix} for a binary mixture of one equivalent of ions of unequal charge is not located at $M_A = \frac{1}{2}$. This is because the sum $m_A + m_B$ varies with surface composition. Figure 4.17 shows

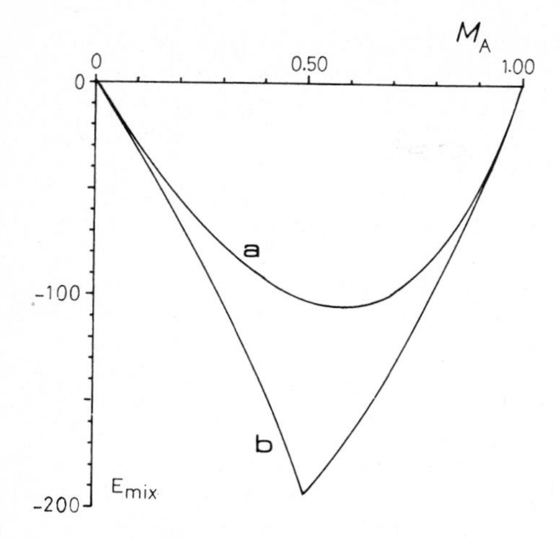

Fig. 4.16. Energy of mixing (J/mol) for a mixture of one equivalent of ions of species A ($z_A = 1$) and B ($z_B = 2$) on a linear array of sites calculated from (4.92) (curve a) and (4.95) (curve b).

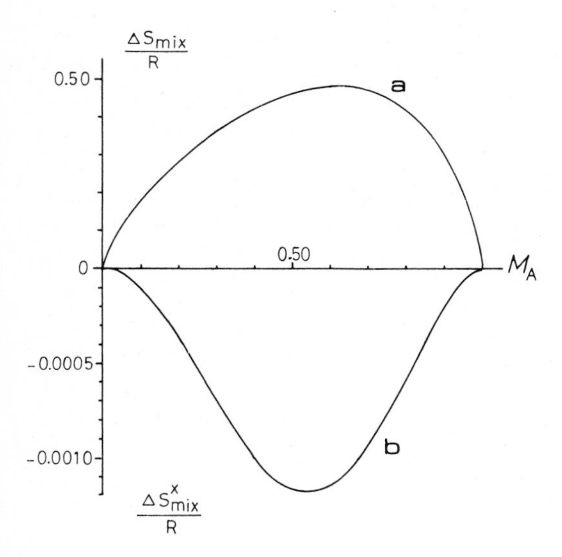

Fig. 4.17. Entropy of mixing (curve a) and excess entropy of mixing (curve b) for a mixture of one equivalent of ions of species A ($z_A = 1$) and B ($z_B = 2$) on a linear array of sites calculated from (4.27) for $2e_m/kT = 0$ and -0.237 (both curve a) and from (4.86) for $2e_m/kT = -0.237$ (curve b).

that the influence of the non-zero energy of mixing on the entropy of mixing is negligible: the deviation from ideality (curve b) is less than 0.3%. Hence the curves for $e_m = 0$ (ideal mixture) and $2e_m/kT = -0.237$ coincide (curve a) in Fig. 4.17. The energy of mixing of the system may also be estimated with the assumption that at any composition of the adsorbent the ions conform to the energetically most favorable configuration, i.e. disregarding the tendency of the mixture to realize a maximum entropy. When an ion of species A is placed on a site which is surrounded by ions of species B, the increase in the number of AB interactions is equal to $2lh_A$, i.e. equal to the number of nearest-neighbor sites occupied by an element of an ion of species B. This also follows from:

$$\lim_{m_k/m_l \to 0} (2lX_{kl}e_m) = 2lh_k m_k e_m \tag{4.94}$$

Hence for $e_m < 0$ the minimum energy of mixing at a given value of M_A is given by:

$$E_{mix}^{min} = 2lh_A m_A e_m \tag{4.95a}$$

until an absolute minimum is reached at $h_A m_A = h_B m_B$, and thereafter:

$$E_{mix}^{min} = 2lh_B m_B e_m \tag{4.95b}$$

in accordance with (4.94). For the present system ($z_A = 1$, $z_B = 2$ and $2l = 2$) the minimum energy of mixing becomes:

$$E_{mix}^{min} = 2m_A e_m \quad \text{for} \quad m_A \leqslant m_B ,$$

and:

$$E_{mix}^{min} = 2m_B e_m \quad \text{for} \quad m_A \geqslant m_B ,$$

whereas the minimum value of E_{mix}^{min} is reached at $m_A = m_B = N_{Av}/3$, i.e. at $M_A = \frac{1}{2}$, where e_m is given by (4.93). From Fig. 4.16 it follows that the deviation between E_{mix}^{min} (curve b) and the value predicted by (4.92) (curve a) is significant, notably near $M_A = \frac{1}{2}$.

This may be illustrated by the following considerations. There are three ways in which the energetically most favorable configuration of a mixture of $N_{Av}/3$ ions of species A and $N_{Av}/3$ ions of species B can be realized on a linear array of sites, if side effects are neglected. The total number of distinguishable configurations, $g(m_A, m_B)$, is given by:

$$g(N_{Av}/3, N_{Av}/3) = \frac{(2N_{Av}/3)!}{(N_{Av}/3)!(N_{Av}/3)!} \tag{4.96}$$

where N! denotes factorial N. Hence:

$$\ln g(N_{Av}/3, N_{Av}/3) \simeq (2N_{Av}/3) \ln 2 \tag{4.97}$$

where Stirling's approximation:

$$\ln N! \simeq N \ln N - N$$

has been used. The probability that one of the minimum energy configurations be formed in an ideal mixture is then $3 \times 2^{-2N_{Av}/3}$ which is an extremely small number. For comparison:

$\ln(3 \times 2^{-2N_{Av}/3}) \simeq -2.77 \cdot 10^{23}$

whereas: $\ln 10^{-99} \simeq -228$. Although the probability that one of these configurations be formed will be larger, because of $e_m < 0$, the total number of distinguishable configurations is so large that e_m is too small to influence the probability distribution of the configurations substantially.

In general, the energy of mixing of a binary mixture of ions of unequal valency on a linear array of sites may be expected to be negative, but for arrays of sites with higher numbers of nearest neighbors the situation is more complicated and both negative and positive energies of mixing may occur.

As an example, a triangular array ($2l = 6$) of one equivalent of sites will be considered, the distance between neighboring sites being 10 Å. It is convenient to define a quantity e'_{kl} by:

$$e'_{kl} = z_k z_l / s_{kl} \tag{4.98}$$

where s_{kl} is in Angstrom, and a quantity E'_{kk} by:

$$E'_{kk} = lh_k m_k e'_{kk} \tag{4.99}$$

where each pair of sites, one occupied by a segment of an ion of species k and the other by a segment of another ion of species k, contributes one term to E'_{kk}, and where side effects are not taken into consideration.

Hence the prime denotes that the energy is multiplied by a factor:

$$\frac{4\pi\epsilon_0\epsilon_w}{e^2 10^{10}} \simeq 3.47 \cdot 10^{19}\ \mathrm{J^{-1}\,Å^{-1}} \tag{4.100}$$

such that e'_{kl} and E'_{kk} have the dimension of $Å^{-1}$. There is only one way in which N_{Av} ions of species A ($z_A = 1$) can be distributed over N_{Av} sites and each ion interacts with its 6 neighbors in the same way (Fig. 4.18). Each ion then contributes $lh_A e'_{AA} = 0.300$ to E'_{AA}, and with $lh_A = 3$ it follows that $e'_{AA} = 0.100$. There are several ways in which $\frac{1}{2} N_{Av}$ ions of species B ($z_B = 2$) can be distributed over a triangular array of N_{Av} sites. Two of these configurations are visualized in Fig. 4.18 (i and j) and it follows that the distance between the ions, s_{BB}, is not a constant: s_{BB} takes the values 10, $10\sqrt{3}$ and 20 for the configuration in Fig. 4.18i and $10\sqrt{3}$ and $5\sqrt{7}$ for Fig. 4.18j. The contribution of one ion of species B to E'_{BB}, $lh_B e'_{BB}$, equals 0.831 (Fig. 4.18i) or 0.836 (Fig. 4.18j), and therefore e'_{BB} equals, by definition, 0.166 (Fig. 4.18i) or 0.167 (Fig. 4.18j). It follows that the configuration shown in Fig. 4.18i is energetically slightly more favorable ($E'_{BB} = 0.415 N_{Av}$) than the one in Fig. 4.18j ($E'_{BB} = 0.418 N_{Av}$) and therefore more likely to occur. When 2 ions of species A on a homoionic lattice are replaced by 1 ion of species B, the energy of the system is calculated to be (Fig. 4.18b):

$$E'_1 = E'_{AA} + 0.233$$

whereas:

$$E'_{mix} = E'_1 - (E'_{AA} - 2lh_A e'_{AA} + lh_B e'_{BB}) = +0.002$$

which is small but significant. Hence the energy of mixing is positive. When 4 ions of species A are replaced by 2 ions of species B, and when these 2 ions are placed on neighboring pairs of sites (Fig. 4.18c and d) the energy of the system is calculated to be:

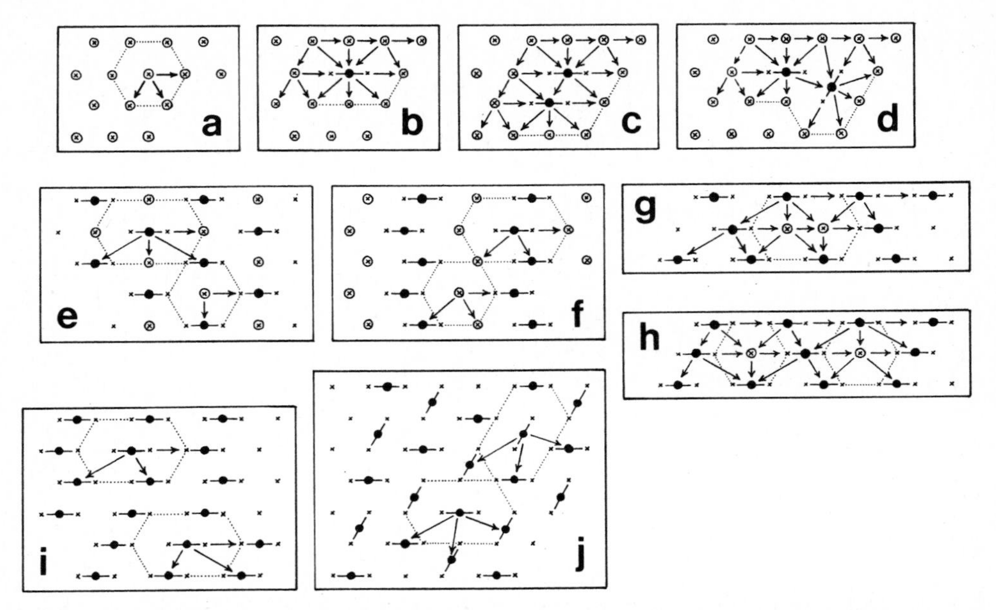

Fig. 4.18. Mixture of monovalent ions of species A (open circles) and divalent ions of species B (dots) on a triangular array of sites (crosses): only ions of species A (a), one ion (b) and two ions (c and d) of species B amidst ions of species A, equal amounts of ions of species A and B (e and f), two ions of species A amidst ions of species B (g and h) and only ions of species B (i and j). Nearest neighbor sites are connected by dotted lines and interactions between ions are indicated with arrows.

$$E_2' = E_{AA}' + 0.402$$

whereas:

$$E_{mix}' = E_2' - (E_{AA}' - 4lh_A e_{AA}' + 2lh_B e_{BB}') = -0.060$$

for both configurations. When ions of species B on a homoionic lattice are replaced by ions of species A the situation is reversed: when the two ions of species A are placed on sites which are not neighbors, the energy of mixing is negative (Fig. 4.18h):

$$E_{mix}' = E_{BB}' - 0.262 - (E_{BB}' - lh_B e_{BB}' + 2lh_A e_{AA}') = -0.031$$

whereas the energy of mixing becomes positive when the two ions are placed on neighboring sites (Fig. 4.18g):

$$E_{mix}' = E_{BB}' - 0.229 - (E_{BB}' - lh_B e_{BB}' + 2lh_A e_{AA}') = +0.002$$

So in this case the energy of mixing would increase upon mixing. The energy of mixing for a mixture of equal amounts of ions of species A and B may range from positive to negative, as is illustrated by Fig. 4.18e and f. For the configuration in Fig. 4.18e the energy of mixing is positive:

$$\begin{aligned} E_{mix}' &= (N_{Av}/3)0.826 + (N_{Av}/3)0.364 - \{(N_{Av}/3)lh_A e_{AA}' + (N_{Av}/3)lh_B e_{BB}'\} \\ &= (N_{Av}/3)0.059 \end{aligned}$$

whereas for the configuration in Fig. 4.18f the energy of mixing is negative:

$$E'_{mix} = (N_{Av}/3)0.685 + (N_{Av}/3)0.385 - \{(N_{Av}/3)lh_A e'_{AA} + (N_{Av}/3)lh_B e'_{BB}\}$$
$$= -(N_{Av}/3)0.061$$

Figure 4.18 shows that the calculation of E_{mix} depends crucially on whether a particular ion is considered as a nearest neighbor or not. The configuration shown in Fig. 4.18f is energetically more favorable than the one in Fig. 4.18e mainly because an ion of species B is assumed to interact with 2 ions of species B in the first configuration and with 4 such ions in the second. The question might be raised as to whether interactions between the central ion and its four neighbors of species B (Fig. 4.18e) should be considered as nearest-neighbor interactions or not (cf. also Fig. 4.15). In soils the situation is further complicated by the fact that the number of nearest neighbors is not a constant but an average value. Hence not all sites are energetically equivalent and since ions tend to occupy the energetically most favorable positions, the energy of mixing may become a function of the composition of the adsorbent. For soils it may be expected that e_m increases with increasing surface coverage by a particular ion, since the probability of occupation of positions with a low energy of mixing will be raised with respect to those with a higher energy of mixing. It may be concluded that e_m may vary with surface composition and that at any composition e_m will depend on the configuration of ions of species A and B on the adsorbent.

The thermodynamic exchange constant (2.16) for $z_A = 1$ and $z_B = 2$ may be written as:

$$\frac{\bar{f}_A N_A}{(\bar{f}_B N_B)^{1/2}} K^0_{ex} = \sqrt{2(^m c_A + 2\, ^m c_B)}\, \frac{f_A f_A}{(f_B f_B)^{1/2}} \qquad (4.101)$$

where $^m c_A$ and $^m c_B$ are in mol/l. The ratio of the activity coefficients of the ions in the adsorbed phase may be calculated from (4.19) and (4.21) for $l = 3$ and different values of e_m (Fig. 4.19). For comparison the curves for $e_m = 0$ and $l = 3$ and $l \rightarrow \infty$ are also shown in Fig. 4.19 (compare Fig. 4.6). For $^m c_A + 2\, ^m c_B = 0.01$ eq/l, $f_A / f_B^{1/2} = 1.1$ (Kielland, 1937) and $K^0_{ex} = 0.16$ it follows that:

$$\frac{\bar{f}_A N_A}{(\bar{f}_B N_B)^{1/2}} = \frac{f_A}{f_B^{1/2}} \qquad (4.102)$$

The choice for K^0_{ex} is arbitrary, of course, but not unreasonable for Ca^{2+}—K^+ or Ca^{2+}—NH_4^+ exchange on clays. Exchange isotherms may now be calculated from (4.102) using the values for $\bar{f}_A / \bar{f}_B^{1/2}$ shown in Fig. 4.19. The electrolyte level of the solution, $^m c_A + 2\, ^m c_B$, has been set at 0.01 eq/l for the exchange isotherms shown in Fig. 4.20, but it may be emphasized that this choice is arbitrary and that the electrolyte level has a pronounced influence on exchange isotherms describing heterovalent ion exchange.

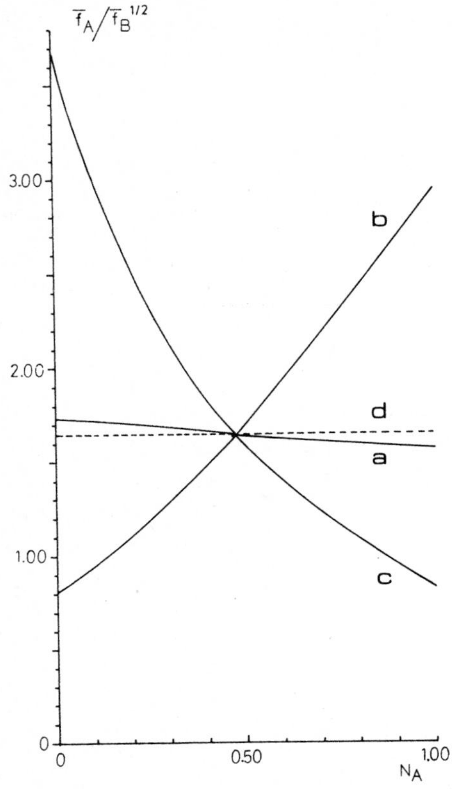

Fig. 4.19. Ratio of activity coefficients for a mixture of ions of species A ($z_A = 1$) and B ($z_B = 2$) calculated from (4.19) and (4.21) for $l = 3$ and $2e_m/kT = 0$ (a), -0.25 (b) and 0.25 (c), and from (4.63), i.e. for $l \rightarrow \infty$ and $e_m = 0$ (d).

Figure 4.19 shows that for $e_m < 0$ the ratio $\bar{f}_A/\bar{f}_B^{1/2}$ (curve b) is an increasing function of N_A which may be represented approximately by:;

$$\bar{f}_A/\bar{f}_B^{1/2} \simeq a(1 + nN_A) \tag{4.103}$$

where a and n are positive constants. Since for $N_B \ll N_A$:

$$aN_B^{-n} \simeq a(1 + nN_A) \tag{4.104}$$

it follows that for N_A sufficiently small:

$$\bar{f}_A/\bar{f}_B^{1/2} \simeq aN_B^{-n} \tag{4.105}$$

where a is determined from:

$$a = \lim_{m_A/m_B \rightarrow 0} (\bar{f}_A/\bar{f}_B^{1/2}) = \{(6/5)\exp(2e_m/kT)\}^3 \tag{4.106}$$

Inserting this expression for the ratio of the activity coefficients in (2.16) results in:

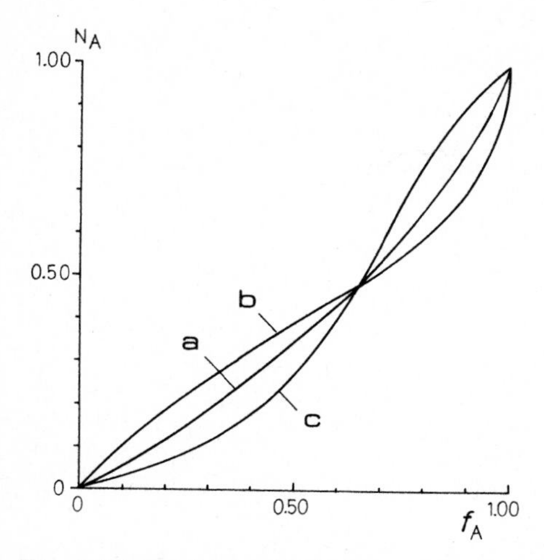

Fig. 4.20. Exchange isotherms for a mixture of ions of species A ($z_A = 1$) and B ($z_B = 2$) on a triangular array of sites ($l = 3$) calculated from (4.102) for $2e_m/kT = 0$ (a), -0.25 (b) and 0.25 (c).

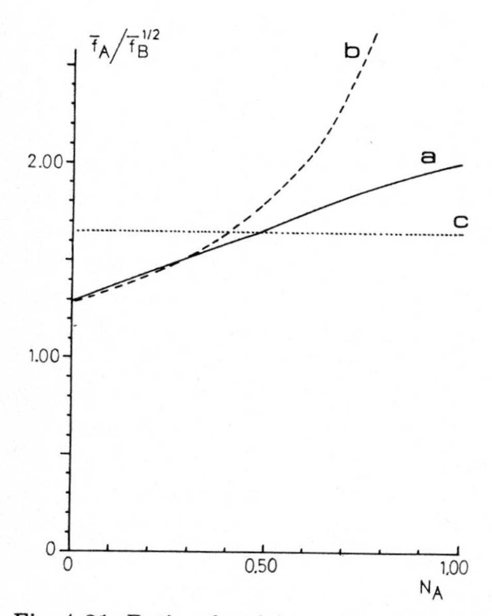

Fig. 4.21. Ratio of activity coefficients for a mixture of ions of species A ($z_A = 1$) and B ($z_B = 2$) on a triangular array of sites ($l = 3$) calculated from (4.19) and (4.21) for $2e_m/kT = -0.10$ (curve a), from (4.105) for $2e_m/kT = -0.10$ and $n = \frac{1}{2}$ (curve b) and from (4.63) (curve c).

$$\{(6/5)\exp(2e_m/kT)\}^3(N_A/N_B^{n+1/2}) \simeq \frac{1}{K_{ex}^0}\frac{f_A c_A}{(f_B c_B)^{1/2}} \qquad (4.107)$$

When the electrolyte level of the solution is kept constant, the ionic strength, $I = \frac{1}{2}{}^m c_A + 2\,{}^m c_B$, varies between rather narrow limits and it may well be assumed that $f_A/f_B^{1/2}$ will be nearly constant during the exchange process (Kielland, 1937). If it is assumed that $n = \frac{1}{2}$ and $f_A/f_B^{1/2}$ = constant, (4.107) becomes approximately equal to the Gapon exchange constant (equation (4.54)) with:

$$K'(\text{Gapon}) \simeq \frac{f_A/f_B^{1/2}}{K_{ex}^0\{(6/5)\exp(2e_m/kT)\}^3} \qquad (4.108)$$

Although there seems to be no general reason why n should equal exactly $\frac{1}{2}$ it may well be expected that in general for $e_m < 0$ (4.107) with $n = \frac{1}{2}$ will be a reasonable approximation for $\bar{f}_A/\bar{f}_B^{1/2}$ for small values of N_A. Figure 4.21 shows that (4.107) with $n = \frac{1}{2}$ (Gapon) is a good approximation for $f_A/f_B^{1/2}$ calculated from (4.19) and (4.21) for $2e_m/kT = -0.10$ for small values of N_A ($N_A \leqslant 0.3$). For comparison the relation $\bar{f}_A/\bar{f}_B^{1/2} = 1.65$ (equation (4.63)) obtained from the ideal mixture for $l \to \infty$, is plotted in Fig. 4.21 and it can be seen that this curve is approximating the regular mixture curve only near $N_A = \frac{1}{2}$. So if experimental data are described better by a Gapon exchange constant than by a selectivity coefficient based on equivalent fractions, i.e. if K'(Gapon) is more constant than K'_N, this may indicate that the adsorbed ions behave as a regular mixture with a small negative energy of mixing.

4.2. HETEROGENEITY OF ADSORPTION SITES AND IONIC SPECIES

Soils generally provide a range of adsorption sites with different bonding properties and contain a range of ionic and non-ionic aqueous complexes taking part in adsorption and possibly precipitation processes. Clay minerals, for example, provide at least 3 different types of negative adsorption sites: (1) planar sites, associated with isomorphous replacements in the mineral, (2) edge sites, formed by the partly dissociated Si-OH groups on the edges of the minerals, and (3) interlayer sites, in illites generally occupied by K^+ ions that become occluded ("sandwiched") between clay platelets. The planar sites may be further distinguished according to the number of nearest-neighbor sites, which may vary from 1 to 8 or more. At a given pH, the degree of dissociation of Si—OH groups on edges of clay minerals may depend upon the extent of isomorphous replacements in the tetrahedral layer "behind" a surface group. Phosphates or other anions that become specifically adsorbed on Al—OH groups on edges of clay minerals may also provide negative adsorption sites. Thus clay minerals alone provide a wide range of adsorption sites, a heterogeneity that is even augmented by the presence of lattice imperfections or fractures. Specifically adsorbed anions, oxides of Si and Mn, and, at high pH also oxides of Fe and Al, often present as coatings on other soil constituents, aluminosilicates other than clay minerals, and soil organic matter may all contribute to the cation adsorption capacity of a soil.

An ion may prefer a certain adsorption site because the interionic energy of mixing is negative for the particular configuration of ions around that site, or because the free enthalpy of adsorption for that site is smaller for the ion involved than for competing ions. In soils it may be expected that interionic energies of mixing will, at least locally, be non-zero. Since ions tend to occupy the energetically most favorable positions first, a negative energy of mixing would be observed as a selective uptake of a particular ion at low occupancy of the adsorbent, the selectivity decreasing with increasing occupancy. This behavior would be difficult to distinguish from ion-exchange selectivity caused by differences between ions in ion—site interactions. In the present section (4.2), however, ion—ion interactions are not considered, i.e. the energy of mixing is assumed to be zero for all configurations, and the preference of an ion for a certain site is assumed to be due to ion—site interactions only.

Contrary to the conditions prevailing in most cation-exchange experiments, where cations in solution are generally present as the simple aquocations, in soil solution cations often occur in complexed form. Heavy metals coordinate readily with organic ligands: up to about 90% of these metals in soil solution may be present as uncharged, anionic or cationic organic complexes. Coordination with organic ligands is especially important at high pH, where the solubility of aquo-ions is generally low.

It would be difficult to include all these factors in ion-exchange models, but fortunately this is not necessary, since in ion-exchange studies only differences between ionic species are considered and not absolute bonding energies. When the selectivity coefficient for a pair of ionic species is the same for all types of adsorption sites, no site heterogeneity would be observed at all on the basis of ion-exchange experiments.

The simplest way to introduce heterogeneity explicitly into ion-exchange equations is by assuming that a soil is a mixture of two types of sites: one type being selective for a certain cation and the other showing no preference for this cation. An ion-exchange model describing a system consisting of high-selectivity (h) and low-selectivity (l) sites, and two different cations (A and B), is hereafter referred to as a "two-site" model.

Ion-exchange studies have shown (see chapter 6) that sometimes not the simple cation (M), but an ionic complex ($L_n M$) is adsorbed preferentially, possibly by London-van der Waals forces, hydrogen bonding, coordinate bonding or due to enhanced electrostatic attraction. If there is a limited amount of ligand (L), which coordinates with one of the two cationic species, a situation may arise like the two-site system. An ion-exchange model describing a system with one type of site and two species of cations (M and B), one of which is present in two forms (M and $L_n M$) that both take part in ion exchange, is hereafter referred to as a "two-species" model. Here M is used instead of A to distinguish between cationic species that coordinate readily with ligands in solution, such as heavy metals (M), and species that do not, such as alkali and alkaline earth metals (A and B). Although the two-site and two-species models are simplified models, they provide a basis for understanding of specific ion-exchange behavior in soils.

4.2.1. Adsorption equations

a. Freundlich equation

The Freundlich adsorption equation is an empirical relation between the amount adsorbed of substance k per unit mass of adsorbent, ${}^{w}q_k$, and the aqueous concentration, ${}^{m}c_k$, of the form:

$$ {}^{w}q_k = k\,{}^{m}c_k^{1/n} \tag{4.109}$$

where k and n are positive constants, n being greater than unity in most cases of practical interest; see for example Freundlich (1922). The Freundlich equation is frequently used in soil science to describe the adsorption of a wide range of ionic and non-ionic species in soils. It can also be written in the following form:

$$\log {}^{w}q_k = \log k + (1/n)\log {}^{m}c_k \tag{4.110}$$

and it can be seen that when experimental data follow a Freundlich isotherm, a plot of $\log {}^{w}q_k$ against $\log {}^{m}c_k$ should yield a straight line with a slope equal to $1/n$ and an intercept equal to log k; such a plot is referred to as a "Freundlich plot".

Although the Freundlich equation is empirical, several authors have tried to give a theoretical derivation of the equation (see for example: Trapnell, 1955). Most of these derivations assume that the absolute values of the free enthalpy or the heat of adsorption decrease with increasing surface coverage, due to surface heterogeneity or to particle interactions.

The thermodynamic equilibrium condition for the adsorption of an uncharged molecule of species k from solution on an initially empty adsorbent surface, e.g. by London-van der Waals forces, may be written:

$$\mu_k^0 + RT\ln a_k = \bar{\mu}_k^0 + RT\ln \bar{a}_k \tag{4.111}$$

where bars refer to the adsorbed phase. With:

$$\Delta G^\circ_{ads} \equiv \bar{\mu}_k^0 - \mu_k^0 \tag{4.112}$$

it follows that:

$$\bar{a}_k/a_k = \exp(-\Delta G^\circ_{ads}/RT) \tag{4.113}$$

Assuming that ΔG°_{ads} varies with surface composition of the adsorbent, the Freundlich equation (4.109) may be obtained from (4.113). With:

$$ {}^{m}\bar{f}_k \equiv \bar{a}_k/M_k \tag{4.114}$$

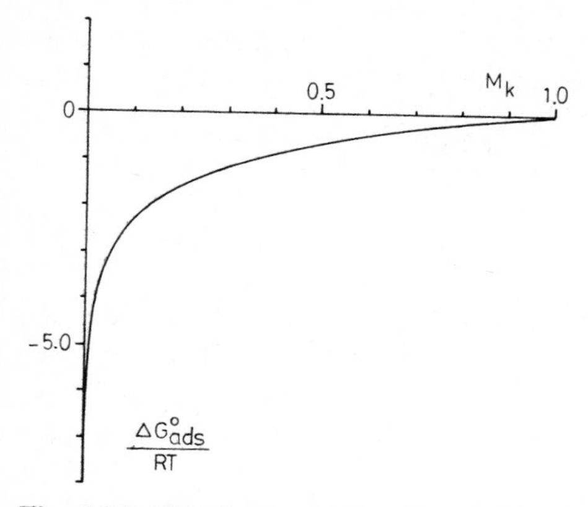

Fig. 4.22. The free enthalpy of adsorption of molecular species k according to (4.118) for $n = 2$ and $\ln \{(f_k/{}^m\bar{f}_k)({}^wQ_k/k)^2\} = 0$.

(4.113) can be written in the following form:

$$M_k = \{(f_k/{}^m\bar{f}_k)\exp(-\Delta G^\circ_{ads}/RT)\}\,{}^mc_k \tag{4.115}$$

The Freundlich equation can be written as:

$$M_k^n = (k/{}^wQ_k)^n\;{}^mc_k \tag{4.116}$$

where ${}^wQ_k$ denotes the adsorption maximum of species k per unit mass of adsorbent, which is reached at ${}^mc_k \doteq {}^mc_k^*$, and where:

$$M_k \equiv {}^wq_k/{}^wQ_k \tag{4.117}$$

Comparison of (4.115) and (4.116) shows that for ${}^mc_k \leqslant {}^mc_k^*$ (4.115) reduces to the Freundlich equation (4.116) if:

$$\Delta G^\circ_{ads}/RT = (n-1)\ln M_k + \ln\left[\frac{f_k}{{}^m\bar{f}_k}\left(\frac{{}^wQ_k}{k}\right)^n\right] \tag{4.118}$$

Hence for $n > 1$ it follows that ΔG°_{ads} would increase exponentially with increasing surface coverage by molecular species k. Figure 4.22 shows that if the second term in (4.118) equals zero, ΔG°_{ads} would tend to zero when the adsorbent surface approaches saturation with species k. It can readily be seen that a similar argument holds for adsorption of ionic species.

b. Rothmund-Kornfeld equation

The Rothmund-Kornfeld equation is an empirical equation that may be considered as a form of the Freundlich equation that applies to ion-exchange processes. In general form it can be written as:

$$\frac{{}^wq_A^{1/z_A}}{{}^wq_B^{1/z_B}} = k\left[\frac{c_A^{1/z_A}}{c_B^{1/z_B}}\right]^{1/n} \tag{4.119}$$

where A and B refer to ionic species. The equation was found to hold for

Ag^+—Na^+, K^+—Ag^+, Rb^+—Ag^+ and Li^+—Ag^+ exchange on permutite, for values of $1/n$ ranging from 0.39 to 0.69 (Rothmund and Kornfeld, 1918). The Rothmund-Kornfeld equation can also be written in logarithmic form (compare (4.110)) and it follows from (4.119) that a log-log plot of data that follow a Rothmund-Kornfeld isotherm should yield a straight line with slope $1/n$ and intercept log k; such a plot is referred to as a "Rothmund-Kornfeld plot".

For homovalent exchange, the Rothmund-Kornfeld equation may be derived as an approximation for ion-exchange equations based on the theory of regular mixtures (Garrels and Christ, 1965; Harmsen, 1977). The approximation would be valid for small (negative) values of $2e_m/kT$ and for values of M_A near $\frac{1}{2}$, i.e. near $M_A = M_B$.

When applied to adsorption of ionic species, the Freundlich equation may be considered as an approximate form of the Rothmund-Kornfeld equation valid for ${}^wq_B/c_B \simeq$ constant:

$$ {}^wq_A = k({}^wq_B/c_B^{1/n})^{z_A/z_B}(c_A^{1/n}) $$

Therefore, a derivation of the Rothmund-Kornfeld equation also applies to the Freundlich equation for a limited range of adsorbent composition.

c. Langmuir equation

The Langmuir equation as used in soil science may be represented as follows:

$$ {}^wq_k = \frac{k\,{}^wQ_k\,{}^mc_k}{(1 + k\,{}^mc_k)} \tag{4.120} $$

where the Langmuir constant, k, is a measure for the binding strength of the adsorption sites for molecular species k. The Langmuir equation (4.120) is often written in the linear form:

$$ \frac{{}^mc_k}{{}^wq_k} = \frac{{}^mc_k}{{}^wQ_k} + \frac{1}{k\,{}^wQ_k} \tag{4.121} $$

from which it can be seen that a plot of ${}^mc_k/{}^wq_k$ against ${}^mc_k$ ("Langmuir plot") yields a straight line with an intercept of $1/k\,{}^wQ_k$ and a slope of $1/{}^wQ_k$.

The Langmuir equation was originally derived (Langmuir, 1918) to describe the simple adsorption of gas molecules on a plane surface having only one kind of elementary space, each of which could hold only one adsorbed molecule. It was assumed that the binding of a molecule on any one elementary space was independent of the binding on the remaining elementary spaces. Brunauer et al. (1967) stated that the conditions of a uniform surface and no lateral interactions between adsorbed molecules are rarely, if ever, met in adsorption experiments. It is an experimental observation, however, that the Langmuir equation is obeyed in numerous instances. Brunauer et al. (1967) explained this observation by noting that for adsorption on energetically heterogeneous surfaces the heat of adsorption would decrease with increasing surface coverage, whereas the lateral interaction energies between physically adsorbed molecules would increase the heat of adsorption with increasing surface coverage. In certain cases the two opposing effects would compensate for each other, thus making the free enthalpy of adsorption approximately constant for a particular adsorption isotherm.

Although there seems to be no general theoretical justification for the use of the Langmuir equation to describe adsorption of molecular species in soils, it is easily shown that any type of adsorption approaching saturation with increasing aqueous concentrations can be approximated with a Langmuir isotherm, which is the simplest mathematical expression satisfying the following requirements:

$$ {}^{w}q_k = 0 \quad \text{at} \quad {}^{m}c_k = 0 \tag{4.122a} $$

$$ \lim_{{}^{m}c_k \to \infty} {}^{w}q_k = {}^{w}Q_k \tag{4.122b} $$

$$ \frac{d\,{}^{w}q_k}{d\,{}^{m}c_k} = k\,{}^{w}Q_k \quad \text{at} \quad {}^{m}c_k = 0 \tag{4.122c} $$

Generally (4.122a) will be satisfied for adsorption processes. Only if a molecular species is present in more than one chemical form in solution, one form taking part in the adsorption reaction and the others not, the "apparent" concentration of molecular species in solution may be non-zero when ${}^{w}q_k = 0$. Since in soils a wide range of adsorption sites may be expected to occur, k may decrease with increasing surface coverage. If so, the Freundlich equation would give a better fit than the Langmuir equation.

When the Langmuir equation is applied to adsorption of ionic species in soils, one must remember that (1) adsorption sites with a net charge are never "empty", but always associated with ionic species in solution to compensate their charge, (2) a wide range of adsorption sites and ionic species is to be expected in soils, and (3) interactions between ions in the adsorbed phase cannot be ruled out and for regular mixtures the heat of adsorption depends on the composition of the adsorbent. Therefore the Langmuir model does not apply to ion exchange in soils. This does not imply, however, that the Langmuir equation would never be obeyed by ionic species in soils. On the contrary, for homovalent exchange at constant total electrolyte concentration, a condition often met in ion-exchange experiments, the Langmuir equation becomes identical to the ion-exchange equation derived for a constant value of the selectivity coefficient K'_N (Bolt, 1967).

The selectivity coefficient K'_N (or K'_M) for homovalent ion-exchange may be written:

$$\frac{q_k}{Q - q_k} = (K_N')^z \frac{c_k}{C - c_k} \tag{4.123}$$

where z denotes the valency of the ionic species involved, Q the cation-exchange capacity of the adsorbent and C the total electrolyte concentration of the equilibrium solution. Equation (4.123) can be written in the form of a Langmuir equation, if C is constant:

$$\frac{c_k}{q_k} = \frac{((K_N')^z - 1)}{Q(K_N')^z} c_k + \frac{C}{Q(K_N')^z} \tag{4.124}$$

It can further be shown that at low coverage of the adsorbent, i.e. for q_k/Q and c_k/C sufficiently small, selectivity coefficients for both homovalent and heterovalent ion-exchange can be approximated by Langmuir equations (Harmsen, 1977).

An advantage of the Langmuir equation is that only c_k and q_k, calculated as loss from solution, have to be known, whereas for the application of ion-exchange equations also the concentrations of competing cations and the cation-exchange capacity have to be known. However, for the correct interpretation of the constants in the Langmuir equation in terms of binding strengths and adsorption maxima, the valency and concentration of the ion under consideration and of the major competing ions should be known.

4.2.2. Two-site model

The selectivity coefficient for homovalent ion exchange involving two ionic species, A and B, of valency z, may be written (compare with (4.123)):

$$\frac{{}^w q_B}{{}^w Q - {}^w q_B} = (K_N')^z \frac{c_B}{c_A} \tag{4.125}$$

where ${}^wQ - {}^wq_B = {}^wq_A$. From the definition of K_N' it follows that all concentrations are in mol/l, but the superscripts 'm' are omitted where possible, to simplify the notation. For homovalent exchange K_N' (or K_M') may be expected to be close to K_N, as follows from Kielland (1937). The superscripts w in (4.125) may be omitted without loss of generality, since this would amount to dividing the numerator and denominator on the left-hand side of (4.125) by ρ_s, the mass density of the adsorbent. In order to simplify the notation of the formulae in the following treatment, ${}^wq_B$ and $(K_N')^z$ in (4.125) will be denoted by q and K, respectively. For a system consisting of more than one type of adsorption site (4.125) may be written as:

$$q_i = \frac{K_i Q_i (c_B/c_A)}{1 + K_i Q_i (c_B/c_A)} \tag{4.126}$$

where subscript i refers to a particular type of adsorption site. An "average" selectivity coefficient, $\langle K \rangle$, for a system consisting of different types of site may be defined by the equation:

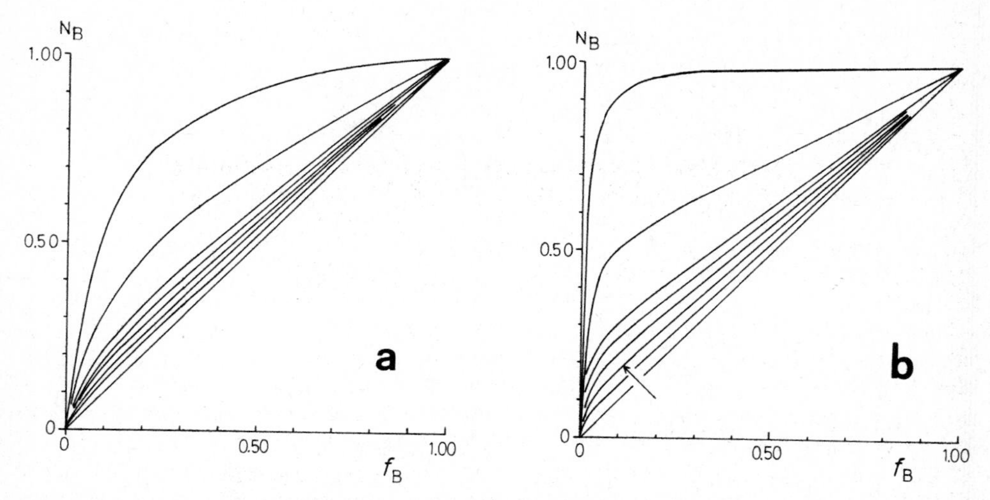

Fig. 4.23. A series of composite exchange isotherms derived from the two-site model (equation (4.128)) for $K_l = 1$, $K_h = 10$ (a) or 100 (b), and $Q_h/Q = 0$ (bottom curve), 0.05, 0.10, 0.15, 0.20, 0.25, 0.50 and 1.00 (top curve). (After Harmsen, 1977)

$$\langle K \rangle \equiv \frac{(c_A/c_B)\Sigma q_i}{\Sigma(Q_i - q_i)} \tag{4.127}$$

where the summation is over all types of site. In general $\langle K \rangle$ is taken to be a function of surface composition, but under equilibrium conditions $\langle K \rangle$ may also be considered as a function of c_B/c_A. For a system consisting of two types of site, low-selectivity (l) and high-selectivity (h) sites, it follows from (4.126) and (4.127) that:

$$\langle K \rangle \equiv \frac{K_l Q_l + K_h Q_h + K_h K_l (c_B/c_A)}{Q + (K_l Q_h + K_h Q_l)(c_B/c_A)} \tag{4.128}$$

where $Q = Q_l + Q_h$, the cation-exchange capacity of the soil. With this expression for $\langle K \rangle$, "composite" exchange isotherms can be constructed (Fig. 4.23). The exchange isotherm for $Q_h/Q = 0.10$ in Fig. 4.23b is indicated with an arrow: this isotherm will be treated in more detail in Figs. 4.24 and 4.25. Figure 4.23 illustrates that when K_h is sufficiently high, extrapolation of the straight part of the exchange isotherm towards the y-axis yields a reasonably accurate estimate for Q_h/Q. Taking the first derivative of $\langle K \rangle$ (4.128):

$$\frac{d\langle K \rangle}{d(c_B/c_A)} = \frac{-Q_l Q_h (K_h - K_l)^2}{\{(Q + (K_l Q_h + K_h Q_l)(c_B/c_A)\}^2} \tag{4.129}$$

it follows that for $K_h \neq K_l$:

$$\frac{d\langle K \rangle}{d(c_B/c_A)} < 0 \tag{4.130}$$

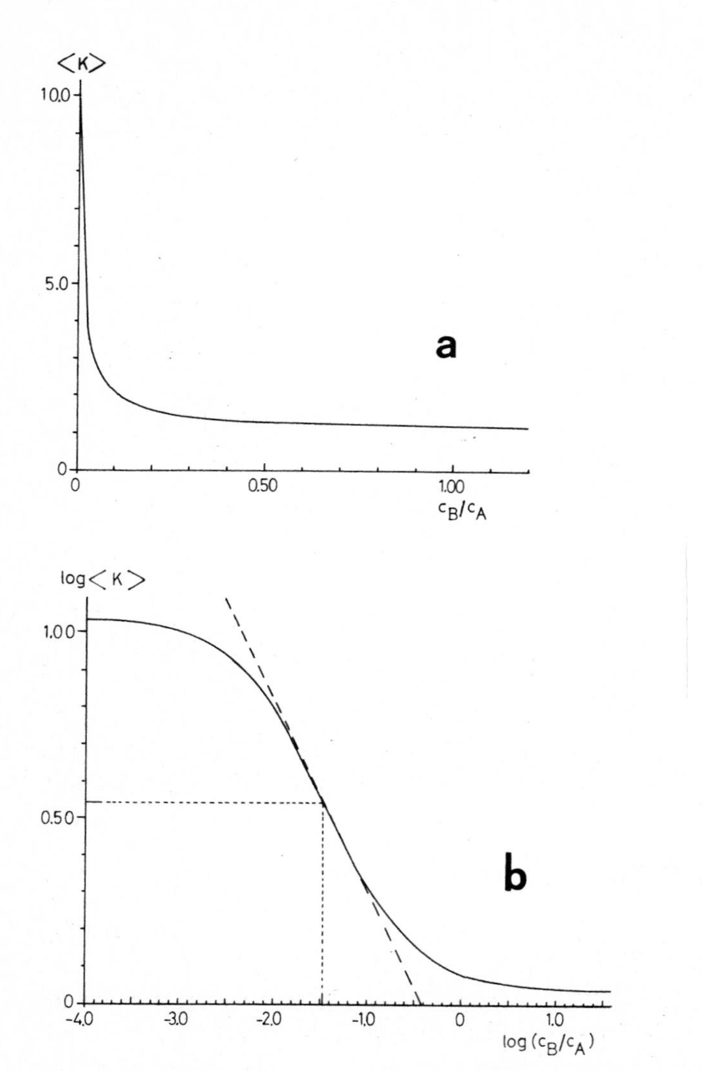

Fig. 4.24. Ordinary plot (a) and log-log plot (b) of the average selectivity coefficient derived from the two-site model (4.128) for $K_l = 1$, $K_h = 100$ and $Q_h/Q = 0.10$. The broken line (b) represents the tangent to the log⟨K⟩ curve at the point of inflection. (After Harmsen, 1977)

for all c_B/c_A. This implies that ⟨K⟩ decreases with increasing c_B/c_A, from a maximum at $c_B/c_A = 0$:

$$K_{max} = K_l (Q_l/Q) + K_h (Q_h/Q) \tag{4.131}$$

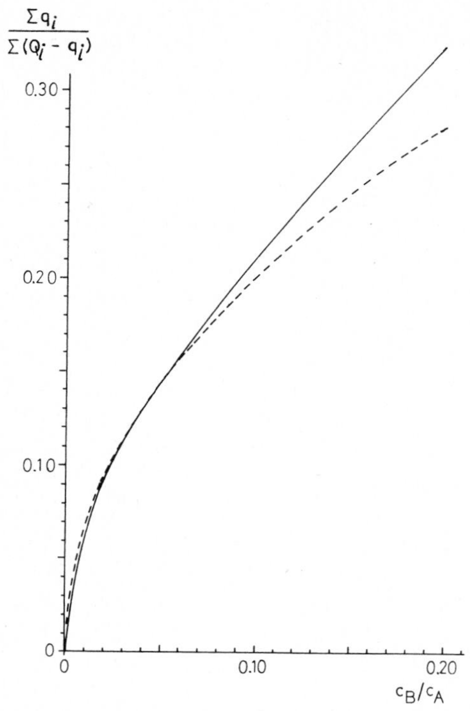

Fig. 4.25. Exchange isotherm (solid curve) calculated from (4.128) for $K_l = 1$, $K_h = 100$ and $Q_h/Q = 0.10$, and Rothmund-Kornfeld isotherm (broken curve). (After Harmsen, 1977)

to a minimum for $c_B/c_A \rightarrow \infty$:

$$K_{min} = \frac{K_h K_l}{K_h (Q_l/Q) + K_l (Q_h/Q)} \tag{4.132}$$

It may be noted that K_{max} and K_{min} depend on K_h, K_l and Q_i/Q, but not on the absolute value of Q. It is convenient to write (4.128) in the form:

$$\langle K \rangle = \frac{K_{min} (K_{max}/K_h K_l + c_B/c_A)}{K_{min}/K_h K_l + c_B/c_A} \tag{4.133}$$

The average selectivity coefficient is a decreasing function of c_B/c_A and thus can be approximated over a limited range by an equation of the form:

$$\langle K \rangle \simeq k (c_B/c_A)^{-(1-1/n)} \tag{4.134}$$

where $n > 1$. In doing so, a Rothmund-Kornfeld equation of the form:

$$\frac{\Sigma q_i}{\Sigma (Q_i - q_i)} \simeq k (c_B/c_A)^{1/n} \tag{4.135}$$

would be obtained. Hence, a composite exchange isotherm can be approximated by a Rothmund-Kornfeld isotherm over a limited range. Figure 4.23 shows that the composite exchange isotherms increase sharply at low concentration of species B in solution and then tend to conform to straight lines; this behavior is more pronounced for the curves in Fig. 4.23b ($K_h/K_l = 100$) than for the curves in Fig. 4.23a ($K_h/K_l = 10$). It is of interest to note that the location of the point where the tangent to the $\langle K \rangle$ curves in Fig. 4.23 equals unity, is not very sensitive to the choice of Q_h/Q: it is located in the low concentration region ($f_B \leqslant 0.2$) for all isotherms. The low concentration region is the range of interest for most trace elements, in particular heavy metals, and the adsorption of these elements can often be described successfully with Freundlich or Rothmund-Kornfeld equations. Therefore, it seems of interest to relate the constants k and n in (4.134) to $\langle K \rangle$ in such a way that the resulting Rothmund-Kornfeld isotherm does approximate the composite exchange isotherm of concern in the low concentration region. Considering (4.133) in more detail it appears that the tangent to the $\log\langle K \rangle$ curve, taken as a function of $\log(c_B/c_A)$, at the point of inflection is a rather good first approximation for $\log\langle K \rangle$ itself. Therefore it will be assumed that k and n in (4.134) and (4.135) are determined by the coordinates of the point of inflection and by the value of the first derivative at that point. The first derivative of $\log\langle K \rangle$ as a function of $\log(c_B/c_A)$ is given by:

$$\frac{d\log\langle K \rangle}{d\log(c_B/c_A)} = \frac{-(c_B/c_A)(K_{max} - K_{min})}{K_h K_l (K_{max}/K_h K_l + c_B/c_A)(K_{min}/K_h K_l + c_B/c_A)} \tag{4.136}$$

which is < 0 for all finite values of $\log(c_B/c_A)$ and $K_{max} > K_{min}$. From:

$$\frac{d^2 \log\langle K \rangle}{\{d\log(c_B/c_A)\}^2} = 0 \tag{4.137}$$

it follows that the point of inflection (pi) is at:

$$\log(c_B/c_A)_{pi} = \log\left[\frac{(K_{max} K_{min})^{1/2}}{K_h K_l}\right] \tag{4.138}$$

Figure 4.24a shows that $\langle K \rangle$ decreases rapidly at low c_B/c_A and then approaches asymptotically to K_{min}. The $\log\langle K \rangle$ curve, plotted as a function of $\log(c_B/c_A)$, has a pronounced point of inflection (Fig. 4.24b) and it can be seen that the tangent to the $\log\langle K \rangle$ curve would be a reasonable approximation for $\log\langle K \rangle$ itself. From (4.133) and (4.138) it may be deduced that the tangent at the point of inflection is a good first approximation for $\log\langle K \rangle$ itself for concentrations in the range:

$$-a + \log(c_B/c_A)_{pi} \leqslant \log(c_B/c_A) \leqslant a + \log(c_B/c_A)_{pi} \tag{4.139}$$

where constant a is close to unity. Of course the choice of a depends on the accuracy required for the approximation, but $a = 1$ seems reasonable, the deviation between the approximation and the exact solution being smaller than 25%. From the experimental point of view, the Rothmund-Kornfeld isotherm determined from the tangent to the $\log\langle K\rangle$ curve at the point of inflection may be a better approximation than would follow from inequality (4.139), notably in the low concentration region, since here the concentration of species B in solution will soon reach the detection limit of the element and thus be no longer measurable. If so, inequality (4.139) would simplify to:

$$\log(c_B/c_A) \leqslant a + \log(c_B/c_A)_{pi} \tag{4.140}$$

It is of interest to investigate the limiting behavior of $\log(c_B/c_A)_{pi}$ as a function of K_h/K_l and Q_l/Q_h. Equation (4.138) may be written as:

$$\log(c_B/c_A)_{pi} = \log K_l + \tfrac{1}{2}\log\left[\frac{K_h/K_l + Q_l/Q_h}{K_h/K_l + (Q_l/Q_h)(K_h/K_l)^2}\right] \tag{4.141}$$

From (4.141) it can be seen that $\log(c_B/c_A)_{pi}$ ranges from $\log K_l$ for $Q_l/Q_h \to 0$, to $\log(K_l^2/K_h)$ for $Q_l/Q_h \to \infty$, and from $\log K_l$ for $K_h = K_l$, to $-\infty$ for $K_h/K_l \to \infty$. Hence the location of the point of inflection moves to smaller values of c_B/c_A both with increasing K_h/K_l and with decreasing Q_h/Q_l.

The exponent in (4.134) is determined from the value of the first derivative at the point of inflection:

$$-(1 - 1/n) = \frac{-(K_{max} - K_{min})}{(K_{max}^{1/2} + K_{min}^{1/2})^2} \tag{4.142}$$

and since $K_{max} > K_{min}$, it follows that $n > 1$. The value of the constant k in (4.134) and (4.135) follows from the coordinates of the point of inflection:

$$\log k = \log\left[\frac{K_{min}\{K_{max} + (K_{max}K_{min})^{1/2}\}}{K_{min} + (K_{max}K_{min})^{1/2}}\right] + \frac{K_{max} - K_{min}}{(K_{max}^{1/2} + K_{min}^{1/2})^2}\log\left[\frac{(K_{max}K_{min})^{1/2}}{K_h K_l}\right] \tag{4.143}$$

For the values used in Fig. 4.24, $K_l = 1$, $K_h = 100$ and $Q_h/Q = 0.10$, the point of inflection is at $\log(c_B/c_A) = -1.46$ (equation (4.138)), the first derivative at that point is -0.52 (equation (4.142)) and $\log k = -0.22$ (equation (4.143)). Insertion of these values in (4.135) results in a Rothmund-Kornfeld isotherm of the form:

$$\frac{q_h + q_l}{Q - q_h - q_l} = 0.61\,(c_B/c_A)^{0.48}$$

In Fig. 4.25 this relation is plotted alongside the exact isotherm calculated from (4.127) and (4.128). Both curves practically coincide up to about 10% of B in solution, which is the range of interest for most trace elements in soils.

When ion-exchange data follow a Rothmund-Kornfeld isotherm (or a Freundlich isotherm) the values of the constant k and the exponent $1/n$ provide information about the values of K_h (when K_l is known) and the ratio Q_h/Q, if the specific exchange behavior is caused by the presence of high and low-selectivity sites. The selectivity coefficients for low-selectivity sites, e.g. planar sites on clay minerals, will be close to unity for a number of cations of equal charge, for example for Ca^{2+}, Mg^{2+}, Zn^{2+}, Cd^{2+} and Cu^{2+} at sufficiently low pH (Harmsen, 1977). When k and $1/n$ are known, K_h and Q_h/Q may be estimated from (4.142) and (4.143), if K_l is known, as it generally is. One may also construct "contour maps" of k and $1/n$, plotted as functions of two variables, K_h/K_l and Q_h/Q, where "contours" represent equal values of k or $1/n$. Examples of such contour maps are presented in Fig. 4.26. Combination of Figs. 4.26a and 4.26b in a single figure yields a figure from which K_h/K_l and Q_h/Q can be estimated if k and $1/n$ are known. In this way Q_h/Q and K_h for Zn^{2+}–Ca^{2+} exchange on a marine clay soil from The Netherlands (Winsum-Gr.) were estimated to be about 8% and 185, respectively, from a Rothmund-Kornfeld isotherm with $k = 0.45$ and $1/n = 0.42$, whereas K_l was equal to unity; the value obtained for Q_h was in good agreement with other estimates obtained in the same study (Harmsen, 1977).

When ion exchange is governed by more than two selectivity coefficients, the situation remains essentially like the two-site model, although the formulae describing the exchange behavior soon become cumbersome for those systems. For many mineral soils, however, a two-site model may be a good first approximation, notably for soils with a high clay content, if planar sites on clay minerals may be classified as low-selectivity sites, and dissociated surface groups, such as Si–OH groups on edges of clay minerals, as high-selectivity sites.

For heterovalent ion-exchange in a two-site system an average selectivity coefficient may be defined according to:

$$\langle K \rangle = \left[\frac{{}^{m}c_A^{1/z_A}}{{}^{m}c_B^{1/z_B}}\right] \frac{(q_h + q_l)^{1/z_B}}{(Q - q_h - q_l)^{1/z_A}} \tag{4.144}$$

For $z_A = 1$ and $z_B = 2$, $\langle K \rangle$ decreases from a maximum at ${}^{m}c_B^{1/2}/{}^{m}c_A = 0$:

$$K_{max} = \{K_h^2 (Q_h/Q)^2 + K_l^2 (Q_l/Q)^2\}^{1/2} \tag{4.145}$$

to a minimum for ${}^{m}c_B^{1/2}/{}^{m}c_A \to \infty$:

$$K_{min} = \frac{K_h K_l}{K_l (Q_h/Q)^{1/2} + K_h (Q_l/Q)^{1/2}} \tag{4.146}$$

whereas for $z_A = 2$ and $z_B = 1$, $\langle K \rangle$ decreases from a maximum at ${}^{m}c_B/{}^{m}c_A^{1/2} = 0$:

$$K_{max} = K_h (Q_h/Q)^{1/2} + K_l (Q_l/Q)^{1/2} \tag{4.147}$$

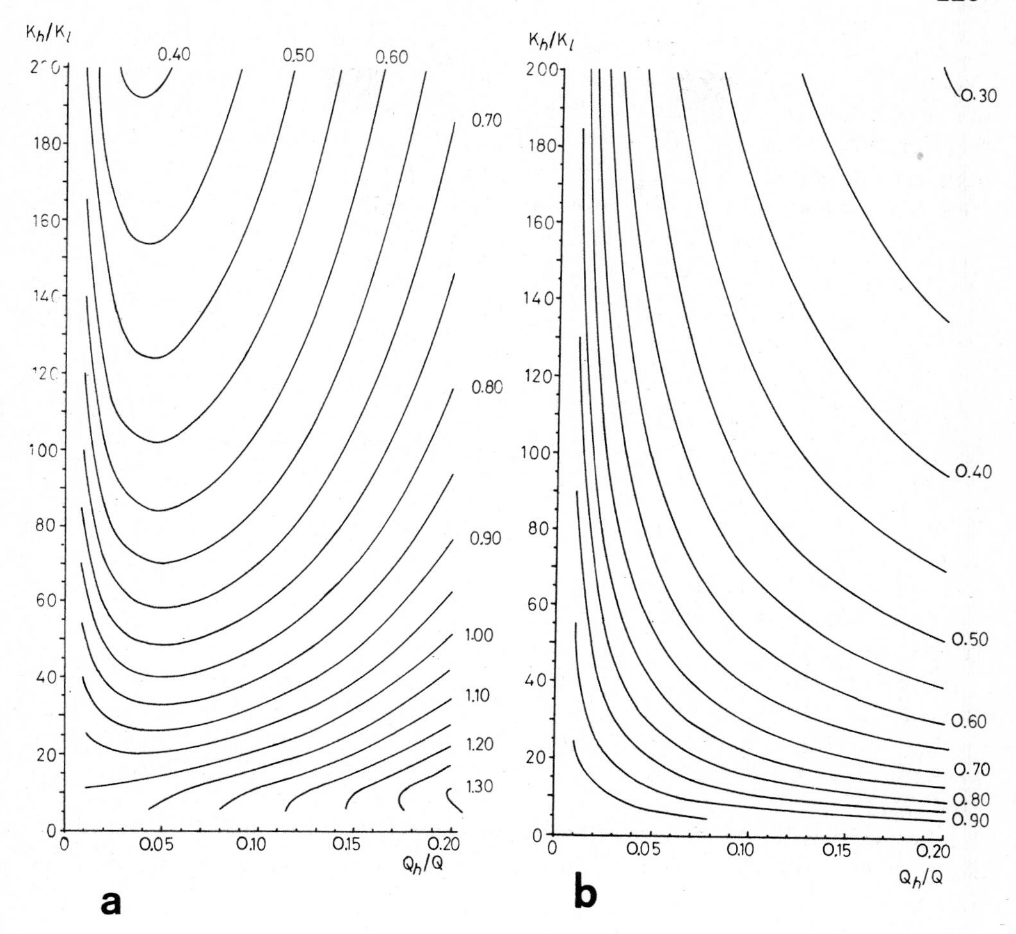

Fig. 4.26. Contour maps of k (a) and $1/n$ (b) calculated from (4.143) (a) and (4.142) (b) for $K_l = 1$. Solid curves represent equal values of k (a) and $1/n$ (b). Contours are drawn at intervals of 0.05. (After Harmsen, 1977)

to a minimum for ${}^{m}c_B/{}^{m}c_A^{1/2} \to \infty$:

$$K_{min} = \frac{K_h K_l}{\{K_l^2 (Q_h/Q)^2 + K_h^2 (Q_l/Q)^2\}^{1/2}} \tag{4.148}$$

Comparison of (4.145)—(4.148) with (4.131) and (4.132) shows that they are essentially similar. Hence, the two-site model also applies to heterovalent ion-exchange, but the formulae, notably for the first and second derivative of log⟨K⟩ as given by equation (4.144), soon become involved.

4.2.3. Two-species models

Cations in aqueous solution may coordinate with charged or uncharged ligands. For ligands other than water this implies that one or more water molecules from the inner coordination sphere of the cation are replaced by ligand molecules. Adsorption sites that are selective for liganded species may catalyze the formation of such complexes. When only a small amount of ligands is present in such a system, the situation may become comparable to the two-site system. If the liganded cations have the same net charge as the aquo ions, the only way to distinguish experimentally between a two-species and two-site system is to identify the ligand of interest and to measure its concentration in the course of the exchange experiment. When liganded cations have a different net charge, ion exchange will appear no longer stoichiometric. For example, apparent "superequivalent" (or "excess") adsorption of divalent cations may be caused by adsorption of monovalent complexes.

Detailed information about the thermodynamics of formation and adsorption of metal—(uncharged) ligand complexes can be found in chapter 6. The present section on two-species models is concerned mainly with the possible effect of ligands (and hydrolysis) on ion-exchange selectivity, in analogy with the two-site model treated in the previous section.

a. Uncharged ligands

A model describing selective adsorption of liganded cations (two-species model) has to deal with (a) regular ion-exchange involving the simple aquo cations, (b) coordination of one of the cationic species with one or more ligands, and (c) selective adsorption of liganded species. It is assumed that ligand molecules are neither adsorbed nor excluded by the adsorbent. For uncharged ligands these processes can be represented by the following equations:

$$M^{z+} + \bar{B}^{z+} \rightleftharpoons \bar{M}^{z+} + B^{z+} \qquad (4.149a)$$

$$M^{z+} + nL \rightleftharpoons L_n M^{z+} \qquad (4.149b)$$

$$L_n M^{z+} + \bar{B}^{z+} \rightleftharpoons \overline{L_n M}^{z+} + B^{z+} \qquad (4.149c)$$

where it is assumed that cationic species B and M are of equal charge (z +) and where n denotes the stoichiometric number of ligand molecules (L) involved in the complex formation. The equilibrium constants for equations (4.149a) through (4.149c) may be written as:

$$K_l = \frac{q_M c_B}{q_B c_M} \qquad (4.150a)$$

$$\mathscr{B}_n = \frac{c_{L_n M}}{c_M {}^m c_L^n} \qquad (4.150b)$$

$$K_h = \frac{q_{L_nM}\, c_B}{q_B\, c_{L_nM}} \qquad (4.150c)$$

where $\mathscr{B}_n$ may be referred to as a cumulative or gross constant (Sillén and Martell, 1971) and where subscripts l and h denote that M^{z+} and L_nM^{z+} are adsorbed with low and high selectivity, respectively. From (4.150a) and (4.150c) it follows that again (compare section 4.2.2.) the short notation K_i is used for $(K'_N)^z_i$.

The equilibrium composition of a system containing cationic species B^{z+} and M^{z+} and ligand species L may be estimated from the initial concentrations of these species in solution, and the cation-exchange capacity of the system. For simplicity c_k and q_k will both be expressed in equivalents per unit volume; if one wishes to remove this restriction, the mass density of the adsorbent and the liquid content of the system have to be specified. Now let it be assumed that initially the system contains only ions of species B and ligand species L. Addition of an amount of M^{z+}, corresponding to an initial concentration of $c_{i,M}$, is followed by formation of L_nM^{z+}, adsorption of M^{z+} and L_nM^{z+}, and desorption of equivalent amounts of B^{z+}. When equilibrium is reached, concentrations may be expressed by means of three composition variables, x_1, x_2 and x_3 (eq/l):

$$c_M = c_{i,M} - x_1 - x_2 \qquad q_M = x_2$$

$$c_B = c_{i,B} + x_2 + x_3 \qquad q_B = Q - x_2 - x_3$$

$$c_{L_nM} = x_1 - x_3 \qquad q_{L_nM} = x_3$$

$${}^{m}c_L = {}^{m}c_{i,L} - nx_1/z$$

where subscript i refers to the inital state and where Q denotes the cation-exchange capacity per unit volume. The equilibrium concentrations may be inserted in (4.150a—(4.150c) to yield a system of 3 equations with 3 unknowns. These equations can be solved for x_1, x_2 and x_3, but the formulae soon becomes cumbersome.

An average selectivity coefficient for the present system may be defined by the equation:

$$\frac{q_M + q_{L_nM}}{q_B} = \langle K \rangle \frac{{}^{m}c_M + {}^{m}c_{L_nM}}{{}^{m}c_B} \qquad (4.151)$$

which can be transformed into:

$$\langle K \rangle = \frac{K_l + K_h \mathscr{B}_n ({}^{m}c_{i,L} - nx_1/z)^n}{1 + \mathscr{B}_n ({}^{m}c_{i,L} - nx_1/z)^n} \qquad (4.152)$$

from which it can be seen that $\langle K \rangle$ may be considered as a function of nx_1/z, i.e. the decrease in ligand concentration upon addition of ionic species M. The first derivative:

$$\frac{d\langle K \rangle}{dx_1} = \frac{-(n^2/z)\, \mathscr{B}_n ({}^{m}c_{i,L} - nx_1/z)^{n-1} (K_h - K_l)}{\{1 + \mathscr{B}_n ({}^{m}c_{i,L} - nx_1/z)^n\}^2} \qquad (4.153)$$

is negative for all $0 \leqslant nx_1/z < {}^m c_{i,L}$ and $K_h > K_l$. Hence $\langle K \rangle$ decreases from a maximum at $x_1 \to 0$ (or ${}^m c_L \to {}^m c_{i,L}$), i.e. for ${}^m c_M \to 0$:

$$K_{max} = \frac{K_l + K_h \mathscr{B}_n {}^m c_{i,L}^n}{1 + \mathscr{B}_n {}^m c_{i,L}^n} \tag{4.154}$$

to a minimum for $nx_1/z \to {}^m c_{i,L}$ (or ${}^m c_L \to 0$), i.e. for ${}^m c_M \gg {}^m c_L$:

$$K_{min} = K_l \tag{4.155}$$

as for the two-site model.

The present two-species model would apply to adsorption of Cu^{2+}, Ni^{2+}, Cd^{2+}, Zn^{2+}, Hg^{2+} or Ag^+ on clay minerals initially saturated with alkali or alkaline earth metals in the presence of uncharged ligands such as ethylenediamine, cyclohexane-1,2-diamine or propane-1,2-diamine. The model would also apply to Ag^+—Na^+ exchange on montmorillonites and other clay minerals in the presence of thiourea (TU), the selectivity coefficient being of the order of unity for Ag^+—Na^+ exchange (K_l) and of the order of 10^3 for $AgTU^+$—Na^+ exchange (K_h). More details are discussed in chapter 6.

b. Monovalent ligands

Cations in aqueous solution may also coordinate with ligands that alter the net charge of the ionic complex. If so, the number of cations that become adsorbed would differ from the number of cations released, through simultaneous adsorption of ligand ions. For a system of two divalent cationic species and one monovalent negative ligand species, the relevant reactions may be written as:

$$M^{2+} + \bar{B}^{2+} \rightleftharpoons \bar{M}^{2+} + B^{2+} \tag{4.156a}$$

$$M^{2+} + L^- \rightleftharpoons LM^+ \tag{4.156b}$$

$$2LM^+ + \bar{B}^{2+} \rightleftharpoons 2\overline{LM}^+ + B^{2+} \tag{4.156c}$$

and the equilibrium constants become:

$$K_l = \frac{q_M c_B}{q_B c_M} \tag{4.157a}$$

$$\mathscr{B}_1 = \frac{c_{LM}}{{}^m c_M c_L} \tag{4.157b}$$

$$K_h = \frac{q_{LM}^2 \, {}^m c_B}{q_B c_{LM}^2 Q} \tag{4.157c}$$

For the present system an average selectivity coefficient may be defined by the equation:

$$\frac{q_M + q_{LM}}{q_B} = \langle K \rangle \frac{{}^m c_M + {}^m c_{LM}}{{}^m c_B} \tag{4.158}$$

where concentrations are expressed per unit volume. It may be noted that it might be more appropriate to express q_k in moles rather than equivalents since adsorption is commonly measured as loss from solution of one cationic species and ligand concentrations are generally not determined. Equation (4.158) may be transformed into:

$$\langle K\rangle = \frac{K_l + \mathscr{B}_1 K_h ({}^m c_{i,L} - x)(c_{LM}/q_{LM})Q}{1 + \mathscr{B}_1 ({}^m c_{i,L} - x)} \tag{4.159}$$

where ${}^m c_L = {}^m c_{i,L} - x$ has been inserted and where composition variable x is in mol/l. The derivative of (4.159) with respect to x is rather difficult to evaluate since the dependence of c_{LM}/q_{LM} on x is not known. It may be obtained by solving (4.157a)—(4.157c) for x_1, x_2 and x_3, as for uncharged ligands. The limiting behavior of $\langle K\rangle$ for $x \to 0$ may be estimated with the help of (4.157c) in the form:

$$q_{LM}/c_{LM} = (K_h Q q_B / {}^m c_B)^{1/2} \tag{4.160}$$

Hence $\langle K\rangle$ decreases from a maximum at $x \to 0$ (or ${}^m c_L \to {}^m c_{i,L}$):

$$K_{max} = \frac{K_l + \mathscr{B}_1 K_h^{1/2}\, {}^m c_{i,L}\, {}^m c_{i,B}^{1/2}}{1 + \mathscr{B}_1\, {}^m c_{i,L}} \tag{4.161}$$

to a minimum for $x \to {}^m c_{i,L}$ (or ${}^m c_L \to 0$):

$$K_{min} = K_l \tag{4.162}$$

as for uncharged ligands.

The present model would apply to adsorption of Zn^{2+} by ion-exchange resins and clays in the presence of glycine (Gly). The selectivity coefficient (K_N') for the $ZnGly^+$—Zn^{2+} exchange reaction may be estimated to be about $1.6 \cdot 10^3$ from data by Siegel (1966). An explanation for the enhanced uptake of $ZnGly^+$ would be the action of London-van der Waals forces on the organic ligand. If so, one would expect the selectivity for the metal amino-acid complexes to increase with increasing chain length of the amino acid.

c. Selective adsorption of hydrolysis products

The adsorption of heavy metals by soils is often accompanied by a decrease in pH. Under slightly alkaline conditions this may be due to precipitation of heavy metals as hydroxides, carbonates or silicates. But also at near-neutral or lower pH, when solutions are generally undersaturated with respect to known solubility products of hydroxides, carbonates or silicates, a decrease in pH upon addition of heavy metals may be observed (Harmsen, 1977). When precipitation is unlikely to occur, a release of H^+ may be due to (a) selective uptake of heavy metals on sites that are initially occupied by H^+, (b) selective uptake of hydrolysis products of heavy metals on regular sites, or (c) selective uptake of hydrolysis products on sites that are initially occupied by H^+. Since it is often difficult to distinguish experimentally between these mechanisms of adsorption, the role of hydrolysis in adsorption of heavy-metal cations is still subject to discussion.

The adsorption of hydrolyzable metal ions may be described in terms of electrostatic ion—solvent and ion—solid interactions (see for example: James and Healy, 1972). The free enthalpy of adsorption of an ionic species is made up of a positive contribution due to dehydration, which is unfavorable to adsorption, and a negative contribution due to coulombic attraction, which is favorable to adsorption. The free enthalpy of adsorption may further include contributions due to London-van der Waals interactions, hydrogen bonding or ligand exchange (or: coordinate bonding), which are generally negative and thus favorable to adsorption. The free enthalpy of (secondary) hydration is a function of the distance from the nucleus of the cation to the adsorbent surface and depends quadratically on the charge of the ion. Other parameters are the dielectric constants of the solid and the solvent. The change in coulombic free enthalpy is a function of distance to the surface and depends linearly on the charge of the ion, and further on the surface potential of the adsorbent and the ionic strength of the solution. For oxide surfaces, the influence of pH is both through the surface potential, which becomes more negative with increasing pH, and through the activity of first hydrolysis products, which increases with increasing pH. It may be concluded that under certain conditions, in particular at high pH or high surface charge density on the adsorbent, adsorption of hydrolysis products may be favored over the unhydrolyzed species, because of the lower hydration energy of the lower charged species.

Selective adsorption of first hydrolysis products may be described with a modified two-species model, which has to account for (a) regular ion-exchange on low-selectivity sites, (b) hydrolysis of one of the cations, (c) specific adsorption of first hydrolysis products, and (d) buffering of pH, i.e. neutralization of H^+ released upon hydrolysis. When all exchange reactions take place on one type of site, the model can be represented by the following set of equations:

$$M^{2+} + \bar{B}^{2+} \rightleftharpoons \bar{M}^{2+} + B^{2+} \tag{4.163a}$$

$$M^{2+} \rightleftharpoons MOH^+ + H^+ \tag{4.163b}$$

$$2MOH^+ + \bar{B}^{2+} \rightleftharpoons 2\overline{MOH}^+ + B^{2+} \tag{4.163c}$$

$$2H^+ + \bar{B}^{2+} \rightleftharpoons 2\bar{H}^+ + B^{2+} \tag{4.163d}$$

Equations (4.163a, c and d) form a complete description of all exchange reactions that can take place on one type of site. If there were two or three distinct types of site involved, the number of independent exchange equations required to describe the model would be doubled or tripled, since all exchange reactions would take place on all types of site. The formulae for the one-site system, obtained by solving the 4 equilibrium constants belonging to (4.163a—d) for 4 unknowns, become rather cumbersome and therefore it seems to be more appropriate to simplify (4.163c and d) further to:

$$MOH^+ \rightleftharpoons \overline{MOH}^+ \tag{4.163c'}$$

$$H^+ \rightleftharpoons \bar{H}^+ \tag{4.163d'}$$

which are valid for small values of q_{MOH} and q_H only. It is assumed that the sites that are selective for MOH^+ or H^+ are not selective for unhydrolyzed cations. If H^+ were adsorbed on neutral oxide surfaces, for example iron oxides, (4.163d′) would represent a reaction of the type:

$$H^+ + FeOOH \rightleftharpoons FeOOH_2^+$$

which would thus increase the adsorption capacity of the soil for anions. The simplified set of equations (4.163a, b and c′, d′) then yields the following set of equilibrium constants:

$$K_l = \frac{q_M c_B}{q_B c_M} \tag{4.164a}$$

$$*\mathscr{B}_1 = \frac{c_{MOH} c_H}{{}^m c_M} \tag{4.164b}$$

$$K_h = q_{MOH} / c_{MOH} \tag{4.164c'}$$

$$a = q_H / c_H \tag{4.164d'}$$

where $*\mathscr{B}_1$ denotes the first hydrolysis constant of species M and where a is a constant related to the buffer capacity of the soil. In general the buffer capacity is a function of pH and therefore the assumption that a is a constant would apply to a limited pH interval only. An essential feature of the present model is that H^+ produced upon hydrolysis lowers the pH and inhibits the hydrolysis to proceed further. Therefore only a limited amount of MOH^+ is formed and the average selectivity coefficient for the overall M—B exchange reaction decreases with increasing c_M / c_B, as for the two-site system. The extent to which $\langle K \rangle$ decreases with increasing c_M / c_B depends upon the selectivity of the adsorbent for the hydrolyzed species, the initial pH and the buffer capacity of the system.

The behavior of $\langle K \rangle$ as a function of c_M / c_B will be illustrated with an example, taking Zn^{2+} (M^{2+}) and Ca^{2+} (B^{2+}). It is assumed that an amount of Zn^{2+}, corresponding to an initial concentration of $c_{i,Zn}$, is added to a system containing Ca^{2+} and H^+ as cations. The equilibrium concentrations (eq/l) for this system may be written as follows:

$$c_{Zn} = c_{i,Zn} - x_1 - x_2 \qquad q_{Zn} = x_2$$

$$c_{Ca} = c_{i,Ca} + x_2 \qquad q_{Ca} = Q - x_2$$

$$c_{ZnOH} = \frac{x_1}{(K_h + 1)} \qquad q_{ZnOH} = \frac{x_1 K_h}{(K_h + 1)}$$

$$c_H = c_{i,H} + \frac{x_1}{(a + 1)} \qquad q_H = a c_{i,H} + \frac{x_1 a}{(a + 1)}$$

where K_h (4.164c′) and a (4.164d′) already have been introduced. Inserting these expressions for the concentrations in (4.164a and b), a system of two equations with two unknowns is obtained:

$$K_l = \frac{(c_{i,Ca} + x_2)x_2}{(Q - x_2)(c_{i,Zn} - x_1 - x_2)} \tag{4.165a}$$

$$*\mathscr{B}_1 = \frac{\{c_{i,H}(a+1) + x_1\}x_1}{(K_h + 1)(a+1)(c_{i,Zn} - x_1 - x_2)/2} \tag{4.165b}$$

which can be solved for x_1 and x_2 rather easily. Thus a cubic equation in x_1 is obtained, which can be solved analytically (see for example: Abramowitz and Stegun, 1968).

An average selectivity coefficient for the present system may be defined by the equation (compare with (4.158)):

$$\frac{{}^{m}q_{Zn} + q_{ZnOH}}{{}^{m}q_{Ca}} = \langle K \rangle \frac{{}^{m}c_{Zn} + c_{ZnOH}}{{}^{m}c_{Ca}} \tag{4.166}$$

where ${}^{m}q_k$ is in mol/l and not in eq/l, since it is difficult, if not impossible, to distinguish experimentally between adsorption of Zn^{2+} and $ZnOH^+$ together with H^+. In Fig. 4.27 $\log\langle K\rangle$ is plotted as a function of $\log\{({}^{m}c_{Zn} + c_{ZnOH})/{}^{m}c_{Ca}\}$ for $K_l = 1, Q = 10^{-2}$ eq/l, $*\mathscr{B}_1 = 2 \cdot 10^{-9}$ mol/l, $(K_h + 1) = 100$, $(a+1) = 100$, $c_{i,H} = 10^{-7}$ mol/l, ${}^{m}c_{i,Ca} = 5 \cdot 10^{-3}$ mol/l, and for ${}^{m}c_{i,Zn}$ ranging from $5 \cdot 10^{-8}$ to $5 \cdot 10^{-3}$ mol/l. The $\log\langle K\rangle$ curve in Fig. 4.27 behaves like the one in Fig. 4.24b for the two-site model. A detection limit for Zn is not included in Fig. 4.27, but it may be near $\log\{({}^{m}c_{Zn} + c_{ZnOH})/{}^{m}c_{Ca}\} = -4.0$, corresponding to a Zn concentration of about $5 \cdot 10^{-7}$ mol/l, which is about the

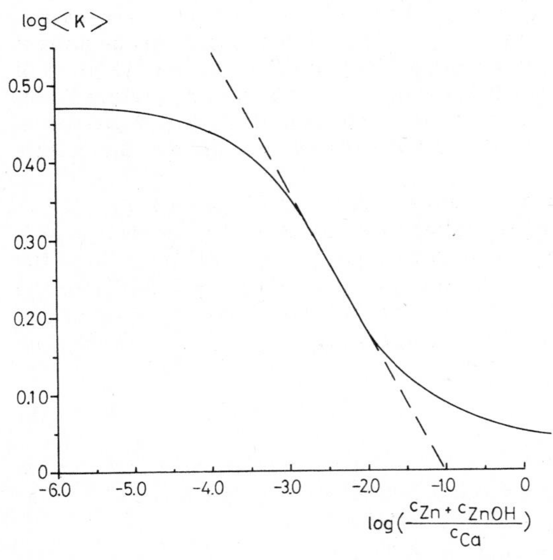

Fig. 4.27. Average selectivity coefficient for the Zn^{2+}—$ZnOH^+$—Ca^{2+}-system, calculated from (4.165a, b) and (4.166). The broken line represents the tangent to the $\log\langle K\rangle$ curve at the point of inflection. Concentrations are in mol/l. (After Harmsen, 1977)

limit for ordinary atomic absorption spectrophotometry. Figure 4.27 shows that the tangent to the log⟨K⟩ curve at the point of inflection is a good approximation for log ⟨K⟩ itself for $\log\{(^{m}c_{Zn} + c_{ZnOH})/^{m}c_{Ca}\}$ below -1.0, assuming that the detection limit for Zn is indeed near -4.0.

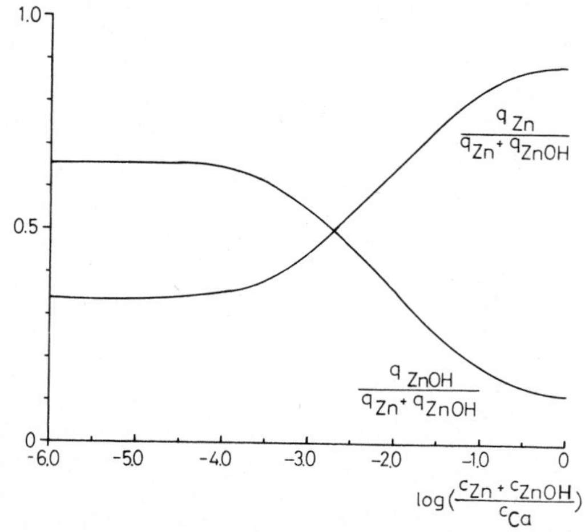

Fig. 4.28. Fractional contributions of Zn^{2+} and $ZnOH^{+}$ to the total amount of Zn adsorbed, calculated from (4.165a and b). Both q_k and c_k are in mol/l. (After Harmsen, 1977)

Figure 4.28 shows that the fractional contribution of q_{ZnOH} to the total amount of Zn adsorbed is highest at low Zn concentrations and decreases with increasing Zn concentration. At $\log\{(^{m}c_{Zn} + c_{ZnOH})/^{m}c_{Ca}\} = 0$ only about 5% of the total cation-exchange capacity is occupied by $ZnOH^{+}$ and another 5% by H^{+} which means that for the present example the errors made by neglecting the changes in $^{m}c_{Ca}$ and $^{m}q_{Ca}$ upon adsorption of $ZnOH^{+}$ and H^{+}, i.e. by taking (4.163c$'$ and d$'$) instead of (4.163c and d), are small over the whole range of concentrations. If H^{+} is adsorbed on uncharged hydroxyl groups on oxide surfaces, the errors are even smaller, since then (4.163d$'$) would give the correct picture and (4.163d) would be in error.

Figure 4.29 shows that $-\log c_H$ is a decreasing function of $\log(^{m}c_{Zn} + c_{ZnOH})$, where $^{m}c_{Zn} + c_{ZnOH}$ represents the total Zn concentration. For the present example c_{ZnOH} is less than 2% of $^{m}c_{Zn}$ and thus $^{m}c_{Zn} + c_{ZnOH}$ is approximately equal to $^{m}c_{Zn}$. Since the activity coefficient of H^{+} in aqueous solution ranges from about 0.90 to 0.87 for the present system (Kielland, 1937), it follows that $-\log c_H$ is approximately equal to pH. Hence pH decreases almost linearly with increasing $\log(^{m}c_{Zn} + c_{ZnOH})$ and the broken line in Fig. 4.29, which represents the relationship:

$$2\text{pH} + n\log(^{m}c_{Zn} + c_{ZnOH}) = \text{k} \tag{4.167}$$

for $n = 0.95$ and $\text{k} = 8.81$, is a good approximation for the $-\log c_H$ curve over the whole range of concentrations, assuming that the detection limit for Zn in aqueous solution is at $\log(^{m}c_{Zn} + c_{ZnOH}) = -6.3$.

A relationship of the form:

$$2\text{pH} + \log {}^{m}c_{Zn} = \text{k} \tag{4.168}$$

at constant ionic strength of the solution, would point to the presence of a hydroxide or,

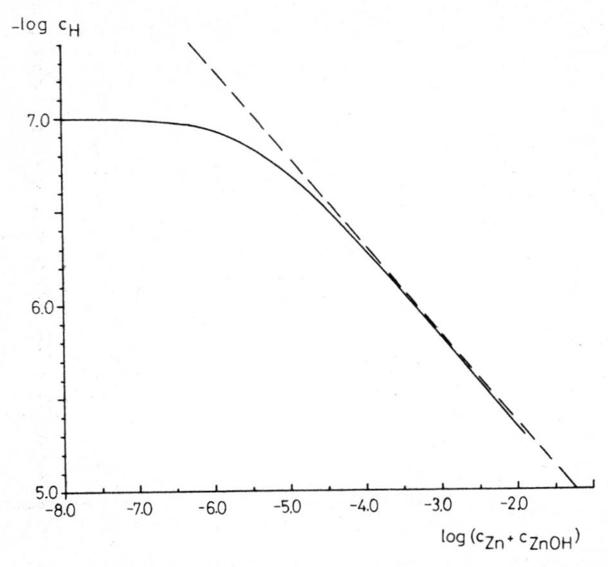

Fig. 4.29. Negative logarithm of H^+ concentration, calculated from (4.165a and b), as a function of total Zn concentration (mol/l). The broken line represents (4.167) for $n = 0.95$ and $k = 8.81$. (After Harmsen, 1977)

if equilibrium with atmospheric CO_2 or solid SiO_2 exists, to the presence of a carbonate or silicate. But from the present treatment it may be concluded that a relationship of the form of (4.168) may also be due to selective adsorption of first hydrolysis products. The coefficient n in (4.167) may be close to unity, as in Fig. 4.29, but in general n depends upon the initial pH and the buffer capacity of the system involved: a higher value of n would point to a lower buffer capacity and a smaller value to a higher buffer capacity.

REFERENCES

Abramowitz, M. and Stegun, I.A., 1968. Handbook of Mathematical Functions. Dover, New York, N.Y.

Bloksma, A.H., 1956. On a hidden assumption in the statistical theory of ion exchange of Davis and Rible. J. Colloid Sci., 11: 286—287.

Bolt, G.H., 1967. Cation-exchange equations used in soil science. A review. Neth. J. Agric. Sci., 15: 81—103.

Brunauer, S., Copeland, L.E. and Kantro, D.L., 1967. The Langmuir and BET theories. In: E.A. Flood (Editor), The Solid—Gas Interface, 1. Marcel Dekker, New York, N.Y., pp. 77—103.

Davis, L.E., 1950. Ionic exchange and statistical thermodynamics. I. Equilibria in simple exchange systems. J. Colloid. Sci., 5: 71—79.

Davis, L.E. and Rible, J.M., 1950. Monolayers containing polyvalent ions. J. Colloid Sci., 5: 81—83.

Eriksson, E., 1952. Cation-exchange equilibria on clay minerals. Soil Sci., 74: 103—113.

Ermolenko, N.F., 1966. Trace Elements and Colloids in Soils. Minsk. Transl. from Russian by Israel Program Sci. Transl., Jerusalem, 2nd ed., 1972.

Freundlich, H., 1922. Kapillarchemie. Akademische Verlagsgesellschaft, Leipzig, 2nd ed.
Gaines, G.L. and Thomas, H.C., 1953. Adsorption studies on clay minerals. II. A formulation of the thermodynamics of exchange adsorption. J. Chem. Phys., 21: 714—718.
Gapon, E.N., 1933. On the theory of exchange adsorption in soil. J. Gen. Chem. U.S.S.R., 3: 144—152. (In Russian)
Garrels, R.M. and Christ, C.L., 1965. Solutions, Minerals, and Equilibria. Freeman, Cooper & Co., San Fransisco, Calif.
Guggenheim, E.A., 1944a. Statistical thermodynamics of mixtures with zero energy of mixing. Proc. R. Soc. Lond., Ser. A., 183: 203—212.
Guggenheim, E.A., 1944b. Statistical thermodynamics of mixtures with non-zero energies of mixing. Proc. R. Soc. Lond., Ser. A., 183: 213—227.
Harmsen, K., 1977. Behaviour of Heavy Metals in Soils. Agric. Res. Rep. (Versl. Landbouwk. Onderz.), 886. Also: Thesis, Agric. Univ., Wageningen, Netherlands.
James, R.O. and Healy, T.W., 1972. Adsorption of hydrolyzable metal ions at the oxide-water interface. III. A thermodynamic model of adsorption. J. Colloid Interface Sci., 40: 65—81.
Kerr, H.W., 1928. The nature of base exchange and soil acidity. J. Am. Soc. Agron., 20: 309—335.
Kielland, J., 1937. Individual activity coefficients of ions in aqueous solutions. J. Am. Chem. Soc., 59: 1675—1678.
Krishnamoorthy, C. and Overstreet, R.., 1950. An experimental evaluation of ion-exchange relationships. Soil Sci., 69: 41—53.
Lagerwerff, J.V. and Bolt, G.H., 1959. Theoretical and experimental analysis of Gapon's equation for ion exchange. Soil Sci., 87: 217—222.
Langmuir, I., 1918. The adsorption of gases on plane surfaces of glass, mica and platinum. J. Am. Chem. Soc., 40: 1361—1403.
Lewis, G.N. and Randall, M., 1961. Thermodynamics. 2nd ed., revised by Pitzer, K.S., and Brewer, L. McGraw-Hill, New York, N.Y.
Lewis, R.J. and Thomas, H.C., 1963. Adsorption studies on clay minerals. VIII. A consistency test of exchange sorption in the systems sodium-cesium-barium montmorillonite. J. Phys. Chem., 67: 1781—1783.
Magistad, O.C., Fireman, M. and Mabry, B., 1944. Comparison of base-exchange equations founded on the law of mass action. Soil Sci., 57: 371—379.
Margenau, H., 1939. Van der Waals forces. Rev. Mod. Phys., 11: 1—35.
Rothmund, V. and Kornfeld, G., 1918. Der Basenaustausch im Permutit. Z. Anorg. Allg. Chem., 103: 129—163.
Siegel, A., 1966. Equilibrium binding studies of zinc-glycine complexes to ion-exchange resins and clays. Geochim. Cosmochim. Acta, 30: 757—768.
Sillén, L.G. and Martell, A.E., 1971. Stability Constants of Metal-Ion Complexes. Chem. Soc. Lond., Suppl. 1. Spec. Publ., 25.
Trapnell, B.M.W., 1955. Chemisorption. Butterworths, London.
Vanselow, A.P., 1932. Equilibria of the base-exchange reactions of bentonites, permutites, soil colloids and zeolites. Soil Sci., 33: 95—113.

CHAPTER 5

SURVEY OF EXPERIMENTAL INFORMATION ON CATION EXCHANGE IN SOIL SYSTEMS

M.G.M. Bruggenwert and A. Kamphorst

Referring to chapter 4 of part A of this volume the interaction between cations and soil constituents comprises different types, as follows:

(a) Reversible cation exchange, where different cations compete for a fixed amount of exchange sites (CEC), the exchanger exhibiting low to moderate preference for one cation species compared to the other.

(b) Highly selective and often only partly reversible adsorption of certain cation species, releasing equivalent amounts of other cations (or hydrogen ions) originally occupying the exchange sites concerned (cf. section A4.5.1 in part A).

(c) Adsorption reactions bordering on chemical bonding, implying in principle a modification of the surface charge of the adsorbent (cf. section A4.5.2 in part A).

(d) The adsorption of hydrolysable cations, where exchange equivalence is difficult to prove and non-Coulombic bonding may play a role (cf. section A4.6.1.2 in part A).

The transitions between these types may be gradual, even for one type of adsorbent.

The total amount of experimental information available on cation adsorption by soil is tremendous: thousands of publications, ranging from a simple report about the exchange behavior of common cations in a particular soil to, for example, studies of the temperature-dependence of the exchange equilibrium involving fairly well characterized clay minerals, may be found in many different journals, though predominantly in soil science and clay mineralogy journals. In an attempt to summarize here at least part of this information, the main effort of this chapter is directed towards tabulation of data which characterize *briefly* the exchange properties of soil materials with respect to different cations. Such data then concern, by necessity, particularly those exchange equilibria that may be summarized in terms of some types of selectivity coefficients, i.e. the equilibria of type (a) above.

For this purpose some 2500 titles were collected from the leading soil science journals covering the last 20 years and scrutinized for probable quantitative information. About 300 of these appeared sufficiently promising. After reading the corresponding papers some 200 more were added from references cited, covering in part journals outside the soils world. With these 500 papers the coverage of the relevant literature is obviously still far from complete. Inasmuch as the information lent itself to presentation in the form of selectivity coefficients pertaining to relatively well defined systems, these have been presented in the following section 5.3. It is hoped that the information provides an insight into the magnitude of the selectivity coefficients for soils and for soil constituents.

Before the presentation of these data in section 5.3 some information concerning materials, experimental methods, and modes of presentation of the data as used in the publications cited is given in the sections following. The presentation particularly needs some attention if a comparison between data from different sources is considered to be the primary goal.

5.1. MATERIALS AND METHODS EMPLOYED

A large share of the exchange studies reported here is concerned with comparatively "pure" systems, viz. suspensions of clay minerals obtained from known sources or the clay fraction of soils known to contain predominantly a particular clay mineral. Partly because of its high exchange capacity montmorillonite, e.g. montmorillonite-dominated sediments such as Wyoming bentonite, plays a major role in investigations directed towards obtaining detailed information about the mechanism of cation adsorption, covering almost complete ranges of composition for many different ion pairs. Less frequently the clay minerals vermiculite, kaolinite and illite (hydrous mica) have been used. The information about cation exchange involving the organic fraction of soils is scarce, no doubt because of the difficulties of isolation and identification. In some cases organic soils (peat soils) have been used (Ehlers et al., 1967; Bache, 1974). Cation exchange properties of soil-borne oxides have recently received attention because of the increasing interest in adsorption of heavy metals (Grimme, 1968; McKenzie, 1972; Bar-Yosef et al., 1975; Kinniburgh et al., 1976).

Aside from these (often detailed) investigations on specific soil constituents, samples of natural soils have also been used, usually for the purpose of obtaining an estimate of the expected behavior in the field. Here the fairly large-scale effort to characterize the sodication process in irrigated soils should be mentioned (Bower, 1959; United States Salinity Laboratory Staff, USDA Handbook 60, 1969), while a considerable amount of information is probably also hidden in reports of local soil laboratories. Obviously it is difficult to generalize about the exchange characteristics of natural soils, because the material description often does not go beyond the local name of the soil involved. Attempts to relate the exchange behavior of a soil to that of its components are not always successful because a mineralogical analysis may forego the effect of other factors like, for example, coatings present on certain minerals.

The cation species used for exchange studies clearly depend on the purpose of the investigation. In the early period the cations commonly investigated were particularly those of primary importance in the agricultural use of soils, viz. Na, K, NH_4, Ca, Mg (and H). When radioactive fall-out became of concern, shortly after the second world war, the cations Cs and Sr received much attention (Coleman et al., 1963a, b; Sawhney, 1965, 1966, 1967). Lately, concern about soil pollution has lead to an increase in the investigations on the adsorption behavior in soil of heavy metals (Banerjee and Mukherjee, 1972; Bergseth and Stuanes, 1976; DeMumbrum and Jackson, 1956, 1957; Hodgson et al., 1964, 1966; Khan, 1969; Mukherjee et al., 1973; Schnitzer and Skinner, 1967; Tiller, 1968; Verloo and Cottenie, 1972). Several publications are devoted to one specific ion in particular, e.g. Zn (Benson, 1966; Udo et al., 1970; Reddy and Perkins, 1974; Shuman, 1975), Cu (Muller, 1960; Davies et al., 1969; Du Plessis and Burger, 1971; Singhal and Singh, 1973), Ni (McLean et al., 1966), U (Szalay, 1969), V (Szalay and Szilagyi, 1967), Co (Zende, 1954; Hodgson, 1960; Hodgson et al., 1964; Taylor, 1968; Tiller et al., 1969). Reference is also made to the relevant sections of Tables 5.1—5.5.

Studies directed towards the elucidation of adsorption mechanisms, particularly on clay minerals, often include the less common members of the alkali and alkaline earth series. The adsorption behavior of hydrogen ions has been studied mainly via pH-titration curves of soils and soil components, later on in conjunction with the behavior of aluminum ions (cf. section A4.6 in part A), while Fe has received attention particularly in studies concerning redox equilibria.

The characterization of an exchange equilibrium requires the composition of the exchange complex and of its equilibrium solution. In the literature several methods have been described to obtain this information, which can be grouped in different ways. First one may distinguish the relatively fast "material balance" methods, in which one derives the change in composition of the adsorbed phase from the observed change in the equilibrium solution. The simplest example of such a method is the determination of the disappearance from solution of a cation added to a known amount. In that case one finds an "adsorption" (or "disappearance") isotherm of the particular cation involved, without necessarily knowing the amount and type of the competing cations. Except at the trace level this information is not sufficient for generalization of the observed behavior. If the original system has been made homoionic, one may in principle derive a description of the exchange equilibrium for the two cations involved, granted that equivalent exchange is involved. The interpretation of the data obtained by this method is further complicated by the usually (though not necessarily) implied change in electrolyte level upon progressive additions. A further improvement requires the determination of the amount of the resident cation released during adsorption, which provides a check on the constancy of the CEC.

In contrast to the "material balance" methods one may distinguish the "double analysis" methods. Here the composition of the adsorbed phase is also determined, either by means of isotopic dilution methods, or by some "stripping" technique meant to release the cations involved. In this manner one may gain an insight into the reversibility of the adsorption and a check on the accuracy of the data obtained. Furthermore this method lends itself better to the determination of cation exchange at a constant total electrolyte level.

Another way of grouping experimental methods is according to the procedure of equilibration used. Percolation methods allow "impregnation" with a solution of chosen composition with respect to a given pair of cation species, starting from any composition with regard to resident cations on the original sample. "Double analysis" is then always necessary, because excess leaching precludes the use of a material balance.

In contrast to the percolation method one may distinguish batch equilibration methods. Here one may choose to equilibrate repeatedly with a solution of a given composition (necessitating "double analysis") or use a one-time equilibration (allowing for a "material balance" if the original sample is homoionic). The latter tends to give the most accurate results, as presumably the amount of exchanger is not subject to losses that may occur upon repeated equilibration. The batch methods require a separation between exchanger and bulk solution, which is done either by centrifugation (fast, but difficult if the clay peptizes) or by means of dialysis (often slow but the most accurate method, unless trace amounts of a cation are involved that could be adsorbed by the membrane).

Finally, the method of preparation of the sample before exchange equilibration is of concern. Often a homoionic form of the exchanger is the simplest starting point, although the removal of excess salt may constitute a problem. Of importance is the fact that the exchange properties may depend on the history of the sample. The chemical instability of H-clays is an extreme example in this case, which is the reason why the use of H-clays as a starting point for exchange studies has now been given up completely.

In conclusion, it may be stated that the diversity of methods used in the literature, and their often incomplete specifications, render a comparison of the data as regards accuracy and dependability sometimes rather difficult.

5.2. PRESENTATION OF THE DATA

The "raw data" acquired in ion exchange experiments involving cation species A and B are usually presented in the form of exchange isotherms relating the amount adsorbed of, for example, species B to its concentration in the equilibrium solution for a given total electrolyte level. Assuming that for each measuring point the exchange capacity (comprising the sum of the amounts adsorbed of the ion species A and B) is known, these data may be presented conveniently as a normalized exchange isotherm relating the equivalent fraction adsorbed, N_B, to the same in solution, f_B. Full information about the exchange behavior of a given material with respect to the ion pair A, B then consists of a family of curves $N_B(f_B)$, each curve corresponding to a specified level of the total electrolyte concentration, C_0, plus a similar family of curves relating the exchange capacity to both N_B (or f_B) and C_0. Fortunately the variability of the exchange capacity is often small compared to the accuracy of measurements, at least for a constant value of C_0 and pH. In that case a single family of curves $N_B(f_B)$ suffices together with the average value of the exchange capacity.

Notwithstanding the wealth of information available in the literature on cation exchange of soil materials, one rarely finds the complete information mentioned above. With, for example, ten measuring points per curve, five electrolyte levels, numerous pairs of cations of interest, and many different types of soils and soil materials, it is easy to see that such information would require too much effort to be executed on a routine basis. Accordingly, with the exception of a limited number of systems, the data available tend to be limited in scope, i.e. a few points at one or two electrolyte levels.

In order to facilitate the use of experimental information for predictive purposes it has been attempted all along to express such data in terms of certain selectivity coefficients which would, it was hoped, be satisfactorily constant for a fairly wide range of experimental conditions. Depending on the investigator's beliefs in particular theoretical models which suggest that certain coefficients should be fairly constant, most experimental data were "translated" in terms of such coefficients. Unfortunately the "raw data" underlying these are not always given, which then prevents a check on the degree of constancy and the range of measuring conditions.

The different types of selectivity coefficients used most widely have been discussed in the previous chapters and are restated briefly below. The experimental results to be presented in the following section (5.3) are sometimes given in terms of solution concentrations and in other cases in terms of solution phase activities. Definition equations used, written out in terms of the latter, are given by eqns. (5.1)—(5.6). Similar definitions in terms of molar solution *concentrations* are implied and will be indicated by a *prime* in the Tables given in section 5.3. Furthermore all coefficients are expressed as a function of the selectivity coefficient K_N used throughout this text (cf. (2.14)). So, with:

$$K_N \equiv \frac{N_B^{1/z_B}}{N_A^{1/z_A}} \frac{a_A^{1/z_A}}{a_B^{1/z_B}}; \quad K_N' \equiv \frac{N_B^{1/z_B}}{N_A^{1/z_A}} \frac{c_A^{1/z_A}}{c_B^{1/z_B}} \tag{5.1}$$

one finds for homovalent exchange, with $z_A = z_B = z$:

$$K(\text{Kerr}) \equiv \frac{N_B}{N_A} \frac{a_A}{a_B} (= K_N^z) \tag{5.2}$$

and for mono—divalent exchange, with $z_A = 1$, $z_B = 2$, (in accordance with the definitions given in the original literature):

$$K(\text{Vanselow}) \equiv \frac{M_A^2}{M_B}\frac{a_B}{a_A^2}\ [= 4K_N^{-2}/(2-N_B)] \tag{5.3}$$

$$K(\text{Eriksson}) \equiv \frac{N_A^2}{N_B}\frac{a_B}{a_A^2}\ [= K_N^{-2}] \tag{5.4}$$

$$K(\text{DKO}) \equiv \frac{M_A^2}{M_B(M_A + 3M_B/2)}\frac{a_B}{a_A^2}\ [= 4K_N^{-2}/(2-N_B/2)] \tag{5.5}$$

$$K(\text{Gapon}) \equiv \frac{N_A\sqrt{a_B}}{N_B a_A}\ [= (K_N\sqrt{N_B})^{-1}] \tag{5.6}$$

As follows from these definitions the "standard" selectivity coefficient K_N defined before, is related to the reciprocal of the other constants. Because of this the numerical values presented in Tables 5.1 and 5.2 refer to these reciprocals, which are there indicated as K_G [$\equiv K(\text{Gapon})^{-1}$], K_v [$\equiv K(\text{Vanselow})^{-1}$], respectively.

The presentation of ion exchange data in terms of the selectivity coefficients specified above may be considered as a first step towards the condensation of experimental information.

Whether such an attempt is successful depends obviously on the constancy of the chosen coefficient in the range of interest. Admitting that authors have, in some cases, been shown to be guided by wishful thinking with regard to the constancy of their preferred coefficient, this procedure is perfectly safe if the actual experimental points are given as $K(N_B)$ or $N_B(f_B)$, fully specifying C_0, pH, etc. In that case no experimental information is lost and the decision whether or not a particular coefficient is sufficiently constant over a chosen range of conditions depends on the use envisaged. At the same time a plot of $\ln K(N_B)$ tends to obscure the influence of the variability of K and once an "average" value is chosen on the basis of data covering a limited range its use is later often extended beyond this range.

An attempt towards further condensation of the experimental information has been described in chapter 2. A given family of $K_N(N_B)$ curves corresponding to different values of C_0 may be used to obtain the thermodynamic exchange constant, K_{ex}^0, characterizing the exchange reaction. Referring to (2.30) this implies that one determines the integral of $\ln K_N$ over the full range of N_B, K_N referring to a system at vanishing electrolyte level. In practice one could either extrapolate $\ln K_N$ values for each given value of N_B towards zero electrolyte and then integrate, or first integrate $\ln K_N$ at different values of C_0 and then attempt to extrapolate the integral to $C_0 = 0$. The accuracy of such a procedure is obviously rather low, because it implies extrapolation of $\ln K_N$ towards $C_0 \to 0$ as well as towards $N_B \to 0$, $N_B \to 1$. As the complete information needed for this is rarely available, one must often be satisfied to determine $\langle \ln K_N \rangle$ at one low electrolyte level, thus assuming that the so-called solvent term (cf. (2.30)) is sufficiently small. In the Tables to be discussed in the section following only a few systems are present where the authors have attempted to obtain the value of $\ln K_{ex}^0$ in the double extrapolation method.

Obviously the single number specifying $\ln K_{ex}^0$ no longer contains the information present in the family of $\ln K_N$ (N_B) curves. This extra information contained in the variation of $\ln K_N$ around $\langle \ln K_N \rangle$ could then be presented in plots of $\ln \bar{f}_A$ and $\ln \bar{f}_B$ or of the excess free enthalpy, ΔG^X, as a function of N_B and C_0 (see the extensive discussion in sections 2.2 and 2.3). While this manner of presentation of experimental data serves the interpretation of the exchange behavior in terms of the mode of interaction between cations and exchanger surface, care must be exercised in back-translating such computed data in terms of the expected exchange behavior, for two reasons. An obvious one is the loss in accuracy incurred in the smoothing out of the log-plots necessary to obtain $\ln K_{ex}^0$ and $\ln \bar{f}$. Granted, however, that the original data are usually of rather limited accuracy, it is still possible that the prediction on the basis of such smoothed-out data presents a fair picture of the actual behavior. A second reason is that the fairly involved treatment of data makes one lose sight of the conventions introduced by this treatment.

The second point made above is most easily demonstrated by restating (2.30) in a somewhat simplified form as:

$$-\frac{\Delta G_{ex}^0}{RT} \equiv \ln K_{ex}^0 = -\left(\frac{1}{z_A} - \frac{1}{z_B}\right) + \langle \ln K_N \rangle \qquad (5.7)$$

where $\langle \ln K_N \rangle$ is the average value of $\ln K_N$ as determined for a sufficiently low electrolyte level to warrant neglect of the solvent term. From this equation it follows that $K_{ex}^0 = 1$ (or $\Delta G_{ex}^0 = 0$) corresponds to $\langle \ln K_N \rangle = -\{(1/z_A) - (1/z_B)\}$. Returning now to the actual exchange behavior as apparent from the normalized exchange isotherm specified as $N_B(f_B)$, one may state that a preference of the exchanger for cation B as compared to cation A corresponds to $N_B/f_B > 1$. As follows from the definition equation (5.2), the ratio N_B/f_B may be related directly to K_N in the case of homovalent exchange, according to:

$$N_B/f_B = (K_N')^z(1 - N_B) + N_B \qquad (5.8)$$

in which $(K_N')^z = (K_N)^z \mathrm{f}_B/\mathrm{f}_A$, with $z_A = z_B = z$. As expected, exchange preference for cation B then coincides with $K_N' > 1$. Accordingly, as in this case $\Delta G_{ex}^0 = 0$ implies that $\langle \ln K_N \rangle = 0$, one finds that *negative* values of ΔG_{ex}^0 (specified in the direction of the formation of the B-exchanger) correspond with exchange preference for cation B (aside from the small effect of electrolyte level on the ratio of the solution phase activity coefficients). Much in contrast, one finds for mono-divalent exchange:

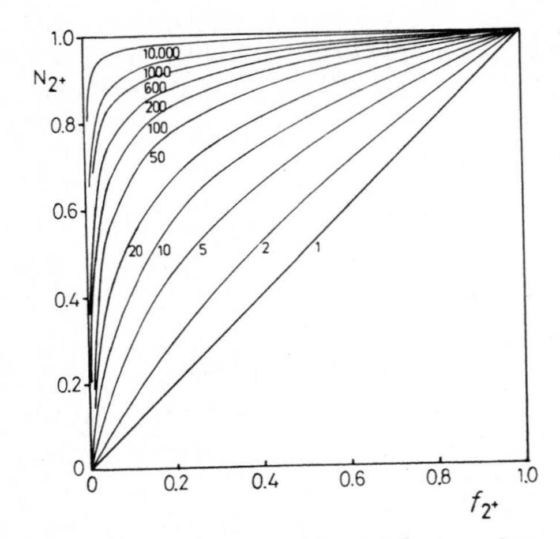

Fig. 5.1. N_{2+} as a function of f_{2+} at values for $(K_N)^2/2C_0$ as given in the Figure.

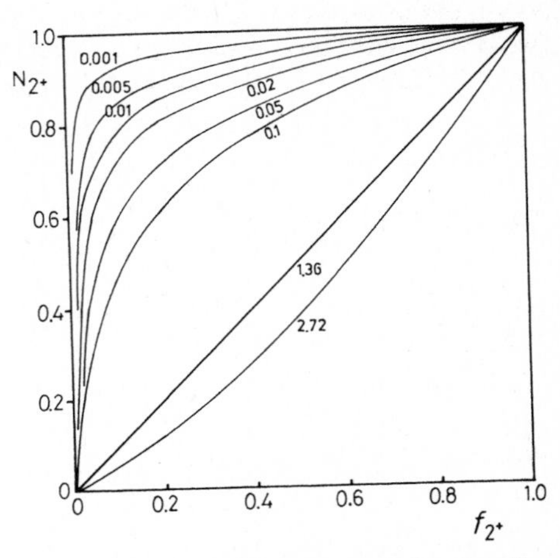

Fig. 5.2. N_{2+} as a function of f_{2+} for the case $K_N' = \sqrt{e}$ (i.e., $\Delta G^0_{ex} \simeq 0$), at values for C_0 as given in the Figure.

$$\frac{N_B}{f_B} = N_B \exp\left[2 \operatorname{arg\,sinh}\{(K_N'/\sqrt{2C_0}) \sinh(\ln N_B)^{-1/2}\}\right] \tag{5.9}$$

Exchange preference for the divalent cation B now coincides with $K_N'/\sqrt{2C_0} > 1$, and will thus depend strongly on the electrolyte level for any given value of K_N' or K_N. In Fig. 5.1 the relation $N_B(f_B)$ has been plotted for different values of $(K_N')^2/2C_0$, allowing one to reconstruct the normalized exchange isotherm pertaining to given values of K_N' and C_0.

The obvious consequence of the above is that even if ΔG^0_{ex} is zero for a given mono-divalent exchange reaction, the exchanger will still exhibit considerable preference for the divalent cation. The normalized adsorption isotherms in Fig. 5.2, plotted for $K^0_{ex} = 1$ or $K_N = \sqrt{e}$ (leaving solution phase activity coefficients out of consideration), indicate that the total electrolyte level must approach very high values before the preference for the divalent cations disappears.

Perhaps redundantly it is pointed out that with C_0 around, for example, 0.01 eq/l, a divalent cation remains preferred in comparison to a monovalent one for $\ln K^0_{ex}$ $(= -\Delta G^0_{ex}/RT) > -2.5$. Accordingly ΔG^0_{ex} (read in the direction of the formation of B^{2+}-exchanger) may have *positive* values up to 2.5 RT although the divalent ion is the preferred species.

This "apparent anomaly" (cf. Deist and Talibudeen, 1967) is the result of the convention adopted in defining the standard states of the species involved in the exchange reaction, viz. the homoionic exchanger, ersus the unit molar solution concentration. As has already been pointed out in discussing the micro-Donnan model of cation exchange (cf. chapter 3, p. 48, the condition $\Delta G^0_{ex} = 0$ then requires that the total electrolyte level *in* the accumulation zone equals about 1.4 *N*, which is of the same order of magnitude as the standard states in solution corresponding to 1 N monovalent and 2 N divalent cations, respectively (aside from activity corrections). Obviously exchange preference thus disappears if the totoal normality in the equilibrium solution reaches the same number (aside from activity corrections).

5.3. SELECTIVITY COEFFICIENTS OF REVERSIBLE EXCHANGE REACTIONS

Taking into account the observations made in the previous sections, it has been attempted to reproduce in Tables 5.1 and 5.2 the information available in a number of selected references, in such a manner that comparisons are possible. For this purpose the exchange selectivity has been specified in terms of a selectivity coefficient according to the definitions given in section 5.2.

In many cases the above implied that where the numbers given in the literature refer to one "mole" of exchanger, or to a reaction in the opposite direction, etc., recalculation was necessary. In cases where the experimental information was presented in a manner not related to any one of the coefficients mentioned, it was expressed in terms of K_N (or K_N').

Furthermore, to facilitate comparisons, all selectivity coefficients were calculated for a specified direction of the *reaction equation*, here chosen in the direction of the preferred cation being adsorbed for homovalent exchange, and in the direction of the cation of the highest valence for heterovalent exchange. The direction of the exchange reaction actually determined experimentally has been indicated separately with an arrow.

The normalized presentation of exchange behavior in terms of the selectivity coefficients defined in (5.1)—(5.6) does not yet allow a quick numerical comparison. For this purpose K_N (or K_N') was chosen as a centralizing concept and all other coefficients were converted to this coefficient.

With reference to Table 5.1, the following explanations are given per column specified.

The first column has been used to identify particular exchange experiments with a number.

The second column gives the relevant references abbreviated to "four letter words" plus the years of publication.

The third column specifies the reactants and the reaction involved in a chosen direction, the latter determining the numerical value of the selectivity coefficient (for homovalent exchange this implies generally $K_N > 1$, for heterovalent exchange $K_N > 1$ only if the valence effect is dominant, cf. section 5.2).

The fourth column contains one (or two) arrow(s) which specifies the direction of the exchange reaction studied, relative to the direction of the reaction equation in column three. A vertical arrow indicates that an "impregnation" technique was used (cf. section 5.1), the resident cation before impregnation being given if a homoionic clay was the starting point.

The column marked "material" specifies this in greater detail than was indicated previously in column three. The column "method" gives a sketchy description of the method used, as discussed in section 5.1; the abbreviation d.m. refers to "double analysis" and the term "microcal" refers to the determination of ΔH° by means of a microcalorimeter. The column "Exp. cond." gives the total electrolyte level at which the experiment was executed, as well as the temperature, if the latter was specified by the author beyond a general indication like "room temperature". In those situations where the range of compositions of the exchanger covered by the experiment was given or could be traced, the latter has been specified in terms of the corresponding values of N_B.

The "heart of the matter" is given in the column marked "Selectivity coefficients". Here a range of values corresponds with the range of N_B as specified; if an extremum occurred between these limiting values, this is given as such.

The last five columns are meant to facilitate the estimates of thermodynamic quantities of the reactions involved. Referring to chapter 2, the calculation of $\Delta G^\circ_{ex} \equiv RT \ln K^\circ_{ex}$ requires the mean logarithm of K_N over the full range of N_B, taking into account the effect of the so-called solvent term (cf. section 2.2). Here data were seldom fully supplied in the references cited; if so, the corresponding value of $\ln K^\circ_{ex}$ in the Table bears a superscript "0". Usually the solvent term was left out of consideration. If nevertheless the author produced a value of K°_{ex} (in certain cases by claiming constancy of K_N), such values have been entered without any superscript. Inasmuch as these were based on K'_N (i.e. without taking into account activity coefficients in the solution phase), the corresponding value of K°_{ex} has also been indicated by a prime.

The column marked "ΔG°_{ex}" must be regarded as a simple translation of the one containing $\ln K^\circ_{ex}$. The values of ΔH° and ΔS° were given only in a limited number of cases, as these involved either microcalorimeter measurements or the determination of the temperature coefficient of ΔG°_{ex}.

An extra column marked $\exp(\langle \ln K_N \rangle)$ was introduced in order to provide a link between the odd ranges of K_N values as presented in the column "Selectivity coefficients" and $\ln K^\circ_{ex}$. Thus $\exp(\langle \ln K_N \rangle)$ presents the logarithmically averaged value of K_N corresponding to $\ln K^\circ_{ex}$ as given, which should give a indication of how the tabulated values of K_N were situated with respect to this average value.

In Table 5.2 some information is given about exchange selectivity observed in natural soils, or in a specified size fraction of the latter. In this case only the selectivity coefficients are given, as the systems involved hardly ever warrant an attempt to specify ΔG°_{ex}.

The selection criterion for the particular systems chosen was the presence of information with regard to soil composition. In addition it was attempted to cover a fairly wide range of cation species as well as some geographical spreading.

The description given in the last column contains respectively: soil name as used by author, fraction used and composition data as given in the literature. The composition data comprise the percentage organic carbon (O.C.) and (sometimes as the percentage in a given size fraction) the predominant clays (and sometimes other minerals) like montmorillonite (M), kaolinite (K), illite (I), vermiculite (V), mica (Mi), biotite (B), hydrobiotite (HB), quartz (Q), chlorite (Cl), clinoptilolite (Clin), allophane (A) and calcite (Ca).

As was already apparent from the excellent review by Swartzen-Allen and Matijević (1974), it is not easy to generalize the exchange behavior of soils. Also the rather substantial amount of data given in the above Tables does not lend itself easily to further generalization. Obviously many different approaches could be chosen, like for example comparisons between different exchangers for a given ion pair or between different ion pairs for a given exchanger, construction of a preference order for e.g. alkali, alkaline earth, and other recognizable groups of cations, if possible with the appropriate range of K_N values. Furthermore one might endeavor to check the validity of certain theoretical models (see, for example, the chapter following, where the effect of surface charge density on exchange preference is discussed).

Systematic attempts in any of these directions have not been made here, mainly because the necessary insight into the accuracy of the data presented would require going back to the original papers, and because this was not the purpose of the present survey. Besides, in many cases such an attempt would be unsatisfactory because often not enough information is available. The main purpose of the Tables is to present data and their sources, leaving it to a user to select a limited number for further examination. In short, the Tables are primarily meant as a "table of contents" of existing research efforts of the last decades rather than a source for quantitative conclusions.

In certain cases, however, the need for information is of a fairly superficial nature. This is particularly so if factors other than those influencing directly the exchange equilibrium — while highly variable or even unknown — are likely to be of greater significance in determining the outcome of the process studied. An example of such a case is the movement of cations through soil as treated in chapter 9. Attempts to make a predictive model covering such a system usually have room only for the inclusion of a fairly simple exchange equation which covers roughly the range of the distribution ratio, R_D, to be expected in a usually ill-defined soil profile.

Seen from this standpoint, and accepting the data given as reasonably dependable, one might conclude from the observed variations of K'_N, that an exchange equation such as (5.1) might be used under the following limiting conditions: (a) the cations belong to the alkali or alkaline earth series; and (b) the cationic coverage exceeds that of trace amounts.

If one is satisfied to use a value of K_N accurate only within a factor of 2, it may often be possible to use the values given in Table 5.2, provided one has some insight into the relative preponderance of the different clay

TABLE 5.1

Selectivity coefficients and thermodynamic constants of different cation exchange reactions determined for clays and some

No.	Ref.	Equation A-clay + B		Material	Method	Exp. cond. C_0 (eq/1)	°C
Monovalent-monovalent exchange (alkali and NH_4 ions)							
1	Tabi60	Li-mont. + Na	→	Colony Wyom. Bent.; < 1 μ	batch	—	—
2	Tabi60	,,	←	,, ,,	,,	—	—
3	Mart63	,,	→	C. Berteau Maroc; < 0.2 μ	dial. d.m.	0.05	25
4	Gast69a	,,	←	Wyom. Bent.; < 0.2 μ	,,	0.001	,,
5	Gast 72	,,	←	Chambers API 23; 0.1—0.5 μ	dialysis	,,	,,
6	Gast72	Li-kaol. + Na	←	Georgia Kaol.	,,	,,	,,
7	Gast71	Li-verm. + Na	→	Transvaal S. Africa; 0.2—62 μ	batch	0.01	,,
8	Gast71	,,	←	,, ,,	,,	,,	,,
9	Gast71	,,	→	,, ,,	,,	,,	50
10	Kris50	Li-mont. + K	⇌	Utah Bent.; < 1 μ	,,	0.001—0.05	—
11	Shai67	,,	—	Wyom. Bent.; < 2 μ	batch d.m.	0.05	—
12	Gast69a	,,	→	,, < 0.2 μ	dial. d.m.	0.001	25
13	Gast72	,,	—	Chambers API 23; 0.1—0.5 μ	microcal	—	—
14	Kris50	Li-mont. + NH_4	⇌	Utah. Bent.; < 1 μ	batch	0.001—0.05	—
15	Mart63	,,	←	C. Berteau Maroc; < 0.2 μ	dial. d.m.	0.05	25
16	Frip65	,,	→	,, ; < 2 μ	dialysis	0.01	,,
17	Frip65	,,	←	,, ,,	,,	,,	,,
18	Mart63	Li-mont. + Rb	→	,, ; < 0.2 μ	dial. d.m.	0.05	,,
19	Mart63	Li-mont. + Cs	→	,, ,,	microcal.	—	—
20	Gast69b	,,	—	Wyom. Bent.; < 0.2 μ	,,	—	—
21	Gast72	,,	→	Chambers API 23; 0.1—0.5 μ	dialysis	0.001	25
22	Merr56	Li-atta. + Cs	↓	Attapulgite API 46; —	col. d.m.	0.02	30
23	Merr56	,,	↓	,, ; —	,,	,,	75
24	Vans32	Na-mont. + K	→	Bentonite 7; clay fr.	batch	0.1	25
25	Vans32	,,	→	Bentonite 5; ,,	,,	,,	,,
26	Kris50	,,	⇔	Utah Bent.; < 1 μ	,,	0.001—0.05	—
27	Tabi60	,,	→	Wyom. Bent.; < 1 μ	batch	—	—
28	Tabi60	,,	←	,, ,,	,,	—	—
29	Tabi60	,,	→	Belle Fourche, S.Dak.; < 1 μ	,,	—	—
30	Tabi60	,,	←	,, ,,	,,	—	—
31	Tabi60	,,	→	Plymouth, Utah; < 1 μ	,,	—	—
32	Tabi60	,,	←	,, ,,	,,	—	—
33	Mart63	,,	→	C. Berteau Maroc; < 0.2 μ	dial. d.m.	0.05	25
34	Shai67	,,	—	Wyom. Bent.; < 2 μ	batch d.m.	,,	—
35	Gast69a	,,	→	,, ; < 0.2 μ	dial. d.m.	0.001	25
36	Gast72	,,	→	Chambers API 23; 0.1—0.5 μ	dialysis	,,	,,
37	Kris50	Na-mont. + NH_4	⇌	Utah Bent.; < 1 μ	batch	0.001—0.05	—
38	Howe65	Na-clin. + NH_4	Na ↓	Clinoptilolite; —	col. d.m.	0.02	30
39	Mart63	Na-mont. + NH_4	←	C. Berteau Maroc; < 0.2 μ	dialysis	0.05	25
40	Frip65	,,	→	,, ; < 2 μ	,,	0.01	,,
41	Frip65	,,	←	,, ,,	,,	,,	,,
42	Vanb 66	,,	←	,, ; < 0.2 μ	,,	,,	,,
43	Vans 72	,,	→	,, ; ± 0.6 μ	,,	0.025	4
44	Vans72	,,	→	,, ,,	,,	,,	25
45	Vans72	,,	→	,, ,,	,,	,,	55

*1 ΔH° measured calorimetrically according to Calvet (Laudelout, 1965).
*2 Cf. Gast et al. (1969b).
*3 $K'_G \equiv 1/K'$ (Gapon), $K'_V \equiv 1/K'$ (Vanselow), as defined in (5.6) and (5.3), using solution phase concentrations.
*4 I = ionic strength.
*5 Oven-dried at 115° C.
*6 Thermally collapsed sample.

other soil components

No.	N_B-range of measurements	Selectivity coefficients in range specified	exp ($\langle \ln K_N \rangle$)	$\ln K^{\circ}_{ex}$	ΔG° (kcal/eq)	ΔH° (kcal/eq)	ΔS° (cal/eq°K)
1	—	K'_N = 0.98	0.98	−0.02′	0.01	—	—
2	—	K'_N = 0.90	0.90	−0.11′	0.07	—	—
3	—	—	1.27	0.24′	−0.14	-0.14^{*1}	0
4	0.15–0.9	K'_N = 1.2–1.0–1.1	1.08	0.08′	−0.05	$-0.15^{*1,2}$	−0.5
5	0.03–0.95	K'_N = 1.4–1.8–0.7	1.15	0.14′	−0.08	-0.11^{*1}	−0.1
6	0.2–0.95	K'_N = 2.0–1.3	—	—	—	—	—
7	0.03–0.92	K'_N = 6.3–22.2	11.48	2.44′	−1.44	−5.53	−13.7
8	0.4–0.9	K'_N = 10–20	11.5	2.44′	−1.44	−5.53	−13.7
9	0.05–0.9	K'_N = 3.5–7.3	5.2	1.64′	−1.07	—	—
10	—	K_N = 30.6	—	—	—	—	—
11	—	K'_N = 4.0	—	—	—	—	—
12	—	—	2.29	0.83′	−0.49	$-0.71^{*1,2}$	-0.7^{*2}
13	—	—	—	—	—	−1.16	—
14	—	K_N = 23.8	—	—	—	—	—
15	—	—	6.17	1.82′	−1.08	-1.24^{*1}	−0.5
16	0.3–1.0	K_N = 4–2.5–3	3.13	1.14	−0.67	—	—
17	0.4–0.95	K_N = 7–3	5.53	1.71	−1.01	—	—
18	—	—	33.5	3.51′	−2.08	-2.15^{*1}	−0.2
19	—	—	—	—	—	−3.7	−3.7
20	—	—	—	—	—	−2.55	—
21	—	—	43.1	3.76′	−2.22	-2.47^{*1}	−0.8
22	0–0.7	—	34.6	3.54′	−2.14	−4.08	−6.4
23	0–0.7	—	14.7	2.69′	−1.85	—	—
24	0.34–0.90	K'_N = 3.3–3.7–2.7	3.0	1.1′	−0.7	—	—
25	0.4–0.9	K'_N = 5.6–4.0	5.0	1.6′	−0.9	—	—
26	—	K_N = 6.2	—	—	—	—	—
27	0.23–0.89	K'_N = 1.74–1.76	1.76	0.57′	−0.34	—	—
28	0.24–0.82	K'_N = 1.78–1.73–1.79	1.76	0.57′	−0.34	—	—
29	—	—	2.23	0.80′	−0.48	—	—
30	—	—	2.27	0.82′	−0.49	—	—
31	—	—	3.09	1.13′	−0.68	—	—
32	—	—	3.27	1.19′	−0.71	—	—
33	—	—	4.06	1.40′	−0.83	-1.08^{*1}	−0.8
34	—	K'_N = 2.5	—	—	—	—	—
35	0.2–0.9	K'_N = 1.8–1.4	1.70	0.53′	−0.31	$-0.61^{*1,2}$	−1.0
36	0.03–0.95	K'_N = 2.6–3.3–2.8	3.45	1.24′	−0.73	$-1.16^{*1,2}$	−1.5
37	—	K_N = 4.76	—	—	—	—	—
38	—	—	8.60	2.15′	−1.29	−0.88	1.3
39	0.3–0.95	K'_N = 5.6–4.5–6.3	5.63	1.73′	−1.02	-0.93^{*1}	−0.3
40	0.3–1.0	K_N = 3.2–2.8–3.2	3.10	1.13	−0.67	—	—
41	0.35–0.95	K_N = 4.7–3.5	4.26	1.45	−0.86	—	—
42	0.1–0.9	K'_N = 5.8–6.7–4.9	5.64	1.73	−1.02	-1.23^{*1}	−0.7
43	0.3–0.9	—	3.78	1.33	−0.73	—	—
44	0.3–0.9	—	3.78	1.33	−0.79	0	2.6
45	0.3–0.9	—	3.80	1.34	−0.87	—	—

TABLE 5.1 (Cont.)

No.	Ref.	Equation A-clay + B		Material	Method	Exp. cond. C_0 (eq/1)	°C
46	Kris50	Na-mont. + Rb	⇌	Utah Bent.; < 1 μ	batch	0.001—0.05	—
47	Kuni68	"	→	" ; fine silt fr.	batch d.m.	0.0989	—
48	Kuni68	"	→	" "	"	0.0092	—
49	Kuni68	"	→	" "	"	0.00036	—
50	Kuni68	Na-kaol. + Rb	→	Hawthorne Florida; clay fr.	"	0.0996	—
51	Kuni68	"	→	" "	"	0.0096	—
52	Gast69a	Na-mont. + Rb	→	Wyom. Bent.; < 0.2 μ	dial. d.m.	0.001	25
53	Gast69a	"	→	" "	"	0.0075	"
54	Gast72	"	→	Chambers API 23; 0.1—0.5 μ	dialysis	0.001	"
55	Kuni68	Na-kaol. + Rb	→	Hawthorne Florida; clay fr.	batch d.m.	0.00058	—
56	Lewi63	Na-mont. + Cs	↓	Chambers API 23; 0.1—0.5 μ	column	0.04	30
57	Elia66	"	→	Bayard, New Mexico; 0.1—1 μ	"	0.05	25
58	Elia66	"	→	Chambers, Arizona; 0.1—0.5 μ	"	"	"
59	Crem68	"	→	C. Berteau Maroc; —	dial.	0.01	9.4
60	Crem68	"	→	" ; —	"	"	25.5
61	Gast69a	"	→	Wyom. Bent.; < 0.2 μ	dial. d.m.	0.001	25
62	Gast69a	"	→	" "	"	0.0075	"
63	Gast72	"	→	Chambers API 23; 0.1—0.5 μ	dialysis	0.001	"
64	Gast72	Na-kaol. + Cs	→	Georgia Kaol.; < 2 μ	"	"	"
65	Merr56	Na-atta. + Cs	↓	Attapulgite API 46; —	col. d.m.	0.02	30
66	Merr56	"	↓	" ; —	"	"	75
67	Frys62	Na-clin. + Cs	↓	Clinoptilolite; —	"	"	50
68	Frys62	"	↓	" ; —	"	"	75
69	Howe65	"	Na ↓	" ; —	"	"	30
70	Bolt63	Na-ill. + K	←	Winsum Ill., Neth.; < 2 μ	batch	0.5	—
71	Gast72	Na-kaol. + K	→	Georgia Kaol.; < 2 μ	dialysis	0.001	25
72	Merr56	Na-atta. + K	↓	Attapulgite API 46; —	col. d.m.	0.01	30
73	Kris50	K-mont. + NH_4	⇌	Utah Bent.; < 1 μ	batch	0.001—0.05	—
74	Mart63	"	←	C. Berteau Maroc; < 0.2 μ	dial. d.m.	0.05	25
75	Frip65	"	⇌	" ; < 2 μ	dialysis	0.01	"
76	Vanb 66	"	←	" ; < 0.2 μ	"	0.05	"
77	Fauc54	K-mont. + Cs	→	Chambers API 23; —	batch	0.01—0.4	20
78	Fauc54	"	←	" ; —	"	"	"
79	Shai67	"	—	Wyom. Bent.; < 2 μ	"	0.05	—
80	Gast69b	"	—	" ; < 0.2 μ	microcal	—	—
81	Gast72	"	—	Chambers API 23; < 2 μ	microcal	—	—
82	Merr56	K-atta. + Cs	↓	Attapulgite API 46; —	col. d.m.	0.02	30
83	Merr56	"	↓	" ; —	"	"	75
84	Kris50	NH_4-mont. + Rb	⇌	Utah Bent.; < 1 μ	batch	0.001—0.05	—
85	Mart63	"	→	C. Berteau Maroc; < 0.2 μ	dial. d.m.	0.05	25
86	Frip65	"	⇌	" ; < 2 μ	dialysis	0.01	25
87	Kris50	NH_4-mont. + Cs	⇌	Utah Bent.; < 1 μ	batch	0.001—0.05	—
88	Mart63	"	→	C. Berteau Maroc; < 0.2 μ	dial. d.m.	0.05	25
89	Frip65	"	→	" ; < 2 μ	dialysis	0.01	"
90	Frip65	"	←	" "	"	"	"
91	Howe65	NH_4-clin. + Cs	Na ↓	Clinoptilolite; —	col. d.m.	0.02	30
92	Kris50	Rb-mont. + Cs	⇌	Utah Bent.; < 1 μ	batch	0.001—0.05	—
93	Gast69a	"	→	Wyom. Bent.; < 0.2 μ	dial. d.m.	0.0075	25
94	Gast72	"	→	Chambers API 23; 0.1—0.5 μ	dialysis	0.001	25
Monovalent-divalent exchange (NH_4, alkali and earth alkali ions)							
95	Wild64	Na-verm. + Mg	←	Transvaal "World"; —	"	0.1	70
96	Wild64	"	⇌	" ; —	"	"	25
97	Schw62	Na-mont. + Ca	Na ↓	Clay Spur; < 2 μ	batch d.m.	0.01	—
98	Vanb72	"	→	C. Berteau Maroc; < 0.2 μ	dialysis	0.005	6.5
99	Vanb72	"	→	" "	"	0.01	"
100	Vanb72	"	→	" "	"	0.025	"

No.	N_B-range of measurements	Selectivity coefficients in range specified	exp (⟨ln K_N⟩)	ln K°_{ex}	ΔG° (kcal/eq)	ΔH° (kcal/eq)	ΔS° (cal/eq°K)
46	—	K_N = 15.4	—	—	—	—	—
47	0.001–0.97	K_N' = 24.8–15.5	—	—	—	—	—
48	0.001–0.97	K_N' = 21.4–10.3	—	—	—	—	—
49	0.007–0.67	K_N' = 25.2–1.6	—	—	—	—	—
50	0.003–0.80	K_N' = 66.0–9.6	—	—	—	—	—
51	0.005–0.82	K_N' = 62.2–11.7	—	—	—	—	—
52	0.1–0.9	K_N' = 3.3–2.2	2.92	1.07′	−0.63	−1.51*[1,2]	−2.9
53	0.03–0.95	K_N' = 5.7–2.5	3.82	1.34′	−0.79	—	—
54	0.03–0.95	K_N' = 17.8–5.2	9.86	2.29′	−1.35	−1.92*[1]	−1.9
55	0.002–0.78	K_N' = 48.0–17.4	—	—	—	—	—
56	0.1 – 0.95	K_N' = 40–12	36.00	3.58′	−2.15	—	—
57	0.051–0.998	K_N' = 53.5–176	47.67	3.86′	−2.28	—	—
58	0.044–0.998	K_N' = 46.3–18.4–45.5	31.74	3.46′	−2.04	—	—
59	0.036–0.98	K_N' = 52.6–18.8	49.90	3.91	−2.19	—	—
60	0.006–0.978	K_N' = 36.6–14.3	32.79	3.49	−2.07	−4.36	−7.7
61	0.05–0.95	K_N' = 12.2–2.0	6.23	1.83′	−1.08	−2.56*[1,2]	−4.9
62	0.03–0.95	K_N' = 14.8–6.4	7.92	2.07′	−1.22	—	—
63	0.03–0.95	K_N' = 39.8–7.9	24.61	3.20′	−1.89	−2.65*[1]	−2.6
64	0.5–0.95	K_N' = 18.1–1.1	—	—	—	—	
65	0.27–0.97	—	24.53	3.20′	−1.93	−3.75	−6.0
66	0.36–0.93	—	11.02	2.40′	−1.68	—	—
67	0.6–1.0	—	23.34	3.15	−2.02	−4.92	−9.0
68	0.5–0.9	—	13.20	2.58	−1.79	—	—
69	—	—	56.50	4.03′	−2.42	−2.70	−0.9
70	0–0.05	K_N' = around 400	—	—	—	—	—
71	0.5–0.95	K_N' = 7.8–2.7	—	—	—	—	—
72	0.3–0.95	—	9.20	2.22′	−1.34	—	—
73	—	K_N = 0.78	—	—	—	—	—
74	0.1–0.9	K_N' = 1.3–1.6–1.3	1.40	0.34′	−0.20	−0.20*[1]	0
75	0.1–0.95	K_N = 1.3–0.9–1.3	1.16	0.15	−0.09	—	—
76	0.1–0.9	K_N' = 1.3–1.5–1.2	1.39	0.33′	−0.20	−0.32*[1]	−0.4
77	0.35–1.0	K_N' = 18.3–10.8	13.07	2.57′	−1.53	—	—
78	0.35–1.0	K_N' = 18.3–10.8	13.07	2.57′	−1.53	—	—
79	—	K_N' = 13.3	—	—	—	—	—
80	—	—	—	—	—	−1.96*[1]	—
81	—	—	—	—	—	−2.05*[1]	—
82	0–1.0	—	7.61	2.03′	−1.22	−2.87	−5.4
83	0.1–0.9	—	4.11	1.41′	−0.97	—	—
84	—	K_N = 3.19	—	—	—	—	—
85	0.15–0.95	K_N' = 5.2–3.8–4.8	3.95	1.37′	−0.81	−1.10	−1.0
86	0.1–0.95	K_N = 1.8–2.3–2.5	2.69	0.99	−0.59	—	—
87	—	K_N = 14.60	—	—	—	—	—
88	0.3–1.0	K_N' = 16–10–25	13.60	2.61′	−1.54	−2.34	−2.7
89	0.5–1.0	K_N = 17.8–11.2–15.8	17.62	2.87	−1.69	—	—
90	0.5–1.0	K_N = 17.8–11.2–15.8	17.24	2.85	−1.69	—	—
91	—	—	5.42	1.69′	−1.02	−1.68	−2.2
92	—	K_N = 4.56	—	—	—	—	—
93	0.05–0.9	K_N' = 3.7–2.0	2.67	0.98′	−0.58	—	—
94	0–0.95	K_N' = 4.5–3.8	3.94	1.37′	−0.81	−1.08	−0.9
95	—	—	6.42	1.36	−0.92	4.8	16.6
96	0.2–0.85	K_N = 1.2–5.8	2.18	0.28	−0.16	4.8	16.4
97	0.80–0.996	$K_G'^{*3}$ = 1.67–3.52 (K_N' = 1.9–3.5)	—	—	—	—	—
98	0–1.0	K_N' = 1.5–4.2	—	—	—	—	—
99	0–1.0	K_N' = 1.6–3.3	1.75	0.06°	−0.03	—	—
100	0–1.0	K_N' = 1.4–5.0	—	—	—	—	—

TABLE 5.1 (Cont.)

No.	Ref.	Equation A-clay + B		Material	Method	Exp. cond. C_0 (eq/1)	°C
101	Vanb72	Na-mont. + Ca	→	C. Berteau Maroc; < 0.2 μ	dialysis	0.005	25
102	Vanb72	,,	→	,, ,,	,,	0.01	,,
103	Vanb72	,,	→	,, ,,	,,	0.025	,,
104	Wild64	Na-verm. + Ca	←	Transvaal "World"; —	batch	0.1	,,
105	Wild64	,,	←	,, ; —	,,	,,	70
106	Bolt55	Na-ill. + Ca	↓	Illite API 35; —	,,	0.001–0.4	—
107	Schw62	,,	Na ↓	Fithian Ill.; < 2 μ	batch d.m.	0.01	—
108	Elia66	Na-mont. + Sr	→	Bayard New Mexico; 0.1–1 μ	batch	0.05	25
109	Vanb69	,,	→	Wyom. Bent.; < 0.2 μ	dialysis	0.005	5.2
110	Vanb69	,,	→	,, ,,	,,	0.01	,,
111	Vanb69	,,	→	,, ,,	,,	0.025	,,
112	Vanb69	,,	→	,, ,,	,,	0.005	25
113	Vanb69	,,	→	,, ,,	,,	0.01	,,
114	Vanb69	,,	→	,, ,,	,,	0.025	,,
115	Wild64	Na-verm. + Sr	←	Transvaal "World"; —	batch	0.1	,,
116	Wild64	,,	←	,, ; —	,,	,,	70
117	Lewi63	Na-mont. + Ba	→	Chambers API 23; 0.1–0.5 μ	column	0.04	30
118	Laud68	,,	→	C. Berteau Maroc; < 0.2 μ	dialysis	0.05–0.005	25
119	Wild64	Na-verm. + Ba	←	Transvaal "World"; —	batch	0.1	,,
120	Wild64	,,	←	,, ; —	,,	,,	70
121	Vans63	K-ill. + Mg	←	Winsum Ill.; Neth.; < 2 μ	batch d.m.	0.03	—
122	Vans63	,,	←	,, ,,	,,	,,	—
123	Vans63	,,	←	,, ,,	,,	,,	—
124	Vans63	,,	←	,, ,,	,,	0.01	—
125	Vans63	,,	←	,, ,,	,,	0.003	—
126	Kris50	K-mont. + Ca	⇌	Utah Bent.; < 1 μ	—	—	—
127	Lage59	,,	H ↓	Volclay Bent.; < 2 μ	,,	0.0005	—
128	Lage59	,,	H ↓	,, ,,	,,	0.005	—
129	Lage59	,,	H ↓	,, ,,	,,	0.05	—
130	Schw62	,,	Na ↓	Clay Spur; < 2 μ	,,	0.005	—
131	Schw62	,,	Na ↓	,, ,,	,,	0.01	—
132	Schw62	,,	Na ↓	,, ,,	,,	0.05	—
133	Schw62	,,	Na ↓	,, ,,	,,	0.1	—
134	Schw62	,,	Na ↓	,, ,,	,,	0.5	—
135	Schw62	,,	Na ↓	,, ,,	,,	1.0	—
136	Schw62	,,	Na ↓	Wyom. Bent.; < 2 μ	,,	0.01	—
137	Schw62	,,	Na ↓	Clay Spur; < 2 μ	,,	,,	—
138	Schw62	,,	Na ↓	Amory; < 2 μ	,,	,,	—
139	Schw62	,,	Na ↓	Moosburg; < 2 μ	,,	,,	—
140	Schw62	,,	Na ↓	Gonterskirchen; < 2 μ	,,	,,	—
141	Schw62	,,	Na ↓	Jugoslavien; < 2 μ	,,	,,	—
142	Schw62	,,	Na ↓	Käste; < 2 μ	,,	,,	—
143	Schw62	,,	Na ↓	Putman; < 2 μ	,,	,,	—
144	Rich64	,,	Na ↓	Otay California; < 2 μ	,,	,,	—
145	Rich64	,,	Na ↓	Wyom. Bent.; < 2 μ	,,	,,	—
146	Hutc66	,,	→	Chambers API 23; —	batch	0.0113	30
147	Hutc66	,,	→	,, ; —	,,	,,	50
148	Hutc66	,,	→	,,	batch	0.0113	70
149	Dolc68	,,	Na ↓	Upton Wyoming; 0.2–0.8 μ	batch d.m.	0.01	—
150	Knib72	,,	←	Wyom. Bent.; < 2 μ	,,	0.055	—
151	Cars72	,,	←	Otay California API H24; 2–0.2 μ	,,	0.01	—
152	Cars72	,,	←	,, ; 0.2–0.08 μ	,,	,,	—
153	Cars72	,,	←	,, ; < 0.08 μ	,,	,,	—
154	Jens73	,,	K ↓	Clarsol FB5 CECA; —	col. d.m.	I^{*4} = 0.05	22
155	Jens73	,,	K ↓	,, ; —	,,	I^{*4} = 0.005	22
156	Lage59	K-ill. + Ca	H ↓	Grundite; < 2 μ	batch d.m.	0.0005	—
157	Lage59	,,	H ↓	,, ,,	,,	0.005	—
158	Lage59	,,	H ↓	,, ,,	,,	0.05	—
159	Schw62	,,	Na ↓	Fithian Ill.; < 2 μ	,,	0.005	—
160	Schw62	,,	Na ↓	,, ,,	,,	0.01	—

No.	N_B-range of measurements	Selectivity coefficients in range specified	exp ($\langle \ln K_N \rangle$)	ln K°_{ex}	ΔG° (kcal/eq)	ΔH° (kcal/eq)	ΔS° (cal/eq° K)
101	0—1.0	K'_N = 1.8—5.2	—	—	—	—	—
102	0—1.0	K'_N = 1.7—4.5	1.95	0.17°	−0.10	{0.92 {1.22°*1	0 —
103	0—1.0	K'_N = 1.7—2.7	—	—	—	—	—
104	0.2—0.8	K_N = 1.7—1.6	1.63	−0.01	0.01	4.7	15.7
105	—	—	4.62	1.03	−0.70	4.7	15.7
106	0.2—1.0	K'_G = 2.54(K'_N = 5.7—2.5)	—	—	—	—	—
107	0.8—0.99	K'_G = 1.62—1.86—1.50(K'_N = 1.8—1.5)	—	—	—	—	—
108	0.03—0.995	K'_N = 0.4—6.9	2.20	0.29′	−0.17	—	—
109	0—1.0	K'_N = 1.4—4.5	—	—	—	—	—
110	0—1.0	K'_N = 1.5—3.5	1.96	0.18°	−0.10	—	—
111	0—1.0	K'_N = 1.5—3.0	—	—	—	—	—
112	0—1.0	K'_N = 1.5—5.0	—	—	—	—	—
113	0—1.0	K'_N = 1.6—4.5	2.02	0.20°	−0.12	0.27	1.5
114	0—1.0	K'_N = 1.5—3.9	—	—	—	0.32°*1	—
115	0.2—0.8	K_N = 1.9—1.6	1.67	0.01	−0.01	4.4	14.7
116	—	—	4.44	0.99	−0.67	4.4	14.5
117	0.05—1.0	K'_N = 2.0—3.5—2.3	2.80	0.53	−0.32	—	—
118	0.4—0.9	—	1.79	0.08°	−0.05	0.68	2.5
119	0.2—0.8	K_N = 1.5—2.0	1.67	0.02	−0.01	5.3	17.8
120	—	—	5.53	1.21	−0.82	5.3	18.1
121	0.994—1.0	K'_G = $\ll$ 0.01	—	—	—	—	—
122	0.939—0.994	K'_G = 0.01(K'_N = 0.01)	—	—	—	—	—
123	0.939	K'_G = 0.45(K'_N = 0.45)	—	—	—	—	—
124	0.939	K'_G = 0.45(K'_N = 0.45)	—	—	—	—	—
125	0.939	K'_G = 0.45(K'_N = 0.45)	—	—	—	—	—
126	—	K_G = 0.21—0.25	—	—	—	—	—
127	0.48—0.943	K_G = 0.50—0.88(K_N = 0.72—0.91)	—	—	—	—	—
128	0.48—0.943	K_G = 0.65—0.98(K_N = 0.94—1.01)	—	—	—	—	—
129	0.48—0.943	K_G = 0.74—1.22(K_N = 1.07—1.26)	—	—	—	—	—
130	0.93	K'_G = 0.91(K'_N = 0.94)	—	—	—	—	—
131	0.90	K'_G = 0.91(K'_N = 0.96)	—	—	—	—	—
132	0.82	K'_G = 1.01(K'_N = 1.12)	—	—	—	—	—
133	0.76	K'_G = 1.00(K'_N = 1.15)	—	—	—	—	—
134	0.59	K'_G = 0.94(K'_N = 1.22)	—	—	—	—	—
135	0.42	K'_G = 0.72(K'_N = 1.11)	—	—	—	—	—
136	0.91	K'_G = 0.99(K'_N = 1.04)	—	—	—	—	—
137	0.62—0.98	K'_G = 0.67—0.98—0.83 (K'_N = 0.85—1.01—0.84)	—	—	—	—	—
138	0.84	K'_G = 0.58(K'_N = 0.62)	—	—	—	—	—
139	0.90	K'_G = 0.93(K'_N = 0.98)	—	—	—	—	—
140	0.82	K'_G = 0.46(K'_N = 0.51)	—	—	—	—	—
141	0.75	K'_G = 0.27(K'_N = 0.31)	—	—	—	—	—
142	0.87	K'_G = 0.72(K'_N = 0.77)	—	—	—	—	—
143	0.74	K'_G = 0.29(K'_N = 0.34)	—	—	—	—	—
144	0.79	K'_G = 0.39(K'_N = 0.44)	—	—	—	—	—
145	0.90	K'_G = 0.93(K'_N = 0.98)	—	—	—	—	—
146	0.2—0.9	K'_N = 0.5—0.6—0.4	0.51	−1.18′	0.70	—	—
147	0.25—0.9	K'_N = 0.6—0.7—0.6	0.63	−0.97′	0.62	—	—
148	0.3—0.9	K'_N = 0.7—0.8—0.7	0.79	−0.74′	0.51	—	—
149	0.90	K'_N = 0.90	—	—	—	—	—
150	0.94—1.0	K'_G = 0.95(K'_N = 0.98—0.95)	—	—	—	—	—
151	0.72—0.981	K'_G = 0.33—0.09(K'_N = 0.37—0.09)	—	—	—	—	—
152	0.74—0.985	K'_G = 0.37—0.12(K'_N = 0.43—0.12)	—	—	—	—	—
153	0.73—0.989	K'_G = 0.37—0.17(K'_N = 0.43—0.17)	—	—	—	—	—
154	0.15—0.87	K_V = 0.12—0.02(K_N = 0.5—0.3)	0.46	−1.28	0.75	—	—
155	0.49—0.93	K_V = 0.11—0.014(K_N = 0.5—0.3)	0.45	−1.30	0.76	—	—
156	0.465—0.939	K_G = 0.4—0.6—0.4(K_N = 0.6—0.4)	—	—	—	—	—
157	0.465—0.939	K_G = 0.5—0.25(K_N = 0.7—0.25)	—	—	—	—	—
158	0.465—0.878	K_G = 0.2—0.3—0.25(K_N = 0.4—0.3)	—	—	—	—	—
159	0.82	K'_G = 0.31(K'_N = 0.34)	—	—	—	—	—
160	0.77	K'_G = 0.34(K'_N = 0.39)	—	—	—	—	—

TABLE 5.1 (Cont.)

No.	Ref.	Equation A-clay + B		Material	Method	Exp. cond.	
						C_0 (eq/1)	°C
161	Schw62	K-ill. + Ca	Na ↓	Fithian Ill.; < 2 μ	batch d.m.	0.05	—
162	Schw62	,,	Na ↓	,, ,,	,,	0.1	—
163	Schw62	,,	Na ↓	,, ,,	,,	0.5	—
164	Schw62	,,	Na ↓	,, ,,	,,	1.0	—
165	Schw62	,,	Na ↓	Hungarian Ill.; < 2 μ	,,	0.01	—
166	Schw62	,,	Na ↓	Fithian Ill.; < 2μ	,,	,,	—
167	Schw62	,,	Na ↓	Morris Ill.; 2—0.2 μ	,,	,,	—
168	Vans63	,,	←	Winsum Ill., Neth.; < 2 μ	,,	0.03	—
169	Bolt63	,,	→	,, ; —	batch	0.5	—
170	Bolt63	,,	→	,, ; —	,,	,,	—
171	Schw62	K-Kaol. + Ca	Na ↓	Chodau Kaol.; < 2 μ	batch d.m.	0.01	—
172	Schw62	,,	Na ↓	Rosenthal Kaol.; < 2 μ	,,	,,	—
173	Jens73	,,	K ↓	Danish deposit; —	col. d.m.	$I^{*4} = 0.05$	22
174	Jens73	,,	K ↓	,, ,,	,,	$I^{*4} = 0.005$	,,
175	Schw62	K-verm + Ca	Na ↓	Kenya I; <2 μ	batch d.m.	0.01	—
176	Schw62	,,	Na ↓	Kenya II; ,,	,,	,,	—
177	Rich64	,,	Na ↓	Libby Montana; < 0.2 μ	,,	,,	—
178	Rich64	,,	Na ↓	,, ; 20—5 μ	,,	,,	—
179	Dolc68	,,	Na ↓	Transvaal S. Africa; < 5 μ	,,	,,	—
180	Schw62	K-musc. + Ca	Na ↓	Muscovite; < 2 μ	,,	,,	—
181	Rich64	,,	Na ↓	Ontario, Canada; < 0.2 μ	,,	,,	—
182	Dolc68	,,	Na ↓	Ward's H.S.E., N.Y.; 0.2—0.08 μ	,,	,,	—
183	Dolc68	,,	Na ↓	,, ; 2—0.2 μ	,,	,,	—
184	Dolc68	K-biot. + Ca	Na ↓	,, ; 2—0.2 μ	,,	,,	—
185	Dolc68	,,	Na ↓	,, ; 5—2 μ	,,	,,	—
186	Schw62	K-chlo. + Ca	Na ↓	Rimpfischwäng chlor.; < 2 μ	,,	,,	—
187	Frip65	NH_4 - mont. + Mg	→	C. Berteau Maroc; < 2 μ	dialysis	,,	25
188	Frip 65	,,	←	,, ,,	,,	,,	,,
189	Laud68	,,	→	,, ; < 0.2 μ	,,	0.05—0.005	7
190	Laud68	,,	→	,, ,,	,,	,,	25
191	Laud68	,,	→	,, ,,	,,	,,	37
192	Vans32	NH_4 -mont. + Ca	→	Bentonite 7; clay fr.	batch	0.1	25
193	Vans32	,,	←	,, ,,	,,	,,	,,
194	Schw62	,,	Na ↓	Clay Spur; < 2 μ	batch d.m.	0.01	—
195	Schw62	,,	Na ↓	Amory; ,,	,,	,,	—
196	Schw62	,,	Na ↓	Käste; ,,	,,	,,	—
197	Schw62	,,	Na ↓	Jugoslavien; < 2 μ	,,	,,	—
198	Schw62	,,	Na ↓	Putnam beidill.; < 2 μ	,,	,,	—
199	Frip65	,,	→	C. Berteau Maroc; < 0.2 μ	dialysis	,,	25
200	Frip65	,,	←	,, ,,	,,	,,	,,
201	Laud68	,,	→	,, ; < 0.2 μ	,,	0.05—0.005	7
202	Laud68	,,	→	,, ,,	,,	,,	25
203	Laud68	,,	→	,, ,,	,,	,,	37
204	Schw62	NH_4 -ill. + Ca	Na ↓	Fithian Ill.; 0.2—0.08 μ	batch d.m.	0.01	—
205	Schw62	,,	Na ↓	Morris Ill.; ,,	,,	,,	—
206	Schw62	NH_4 -kaol. + Ca	Na ↓	Chodau Kaol.; < 2 μ	,,	,,	—
207	Schw62	NH_4 -verm. + Ca	Na ↓	Kenya I; < 2 μ	,,	,,	—
208	Frip65	NH_4 -mont. + Sr	→	C. Berteau Maroc; < 2 μ	dialysis	,,	25
209	Frip 65	,,	←	,, ,,	,,	,,	,,
210	Laud68	,,	→	,, ; < 0.2 μ	,,	0.05—0.001	7
211	Laud68	,,	→	,, ,,	,,	,,	25
212	Laud68	,,	→	,, ,,	,,	,,	37
213	Frip65	NH_4 -mont. + Ba	→	,, ; < 2 μ	,,	0.01	25
214	Frip65	,,	←	,, ,,	,,	,,	,,
215	Laud68	,,	→	,, ; < 0.2 μ	,,	0.05—0.005	7
216	Laud68	,,	→	,, ,,	,,	,,	25
217	Laud68	,,	→	,, ,,	,,	,,	37

No.	N_B-range of measurements	Selectivity coefficients in range specified	exp ($\langle \ln K_N \rangle$)	$\ln K^{\circ}_{ex}$	ΔG° (kcal/eq)	ΔH° (kcal/eq)	ΔS° (cal/eq°K)
161	0.68	$K'_G = 0.48 (K'_N = 0.57)$	—	—	—	—	—
162	0.60	$K'_G = 0.47 (K'_N = 0.61)$	—	—	—	—	—
163	0.48	$K'_G = 0.64 (K'_N = 0.92)$	—	—	—	—	—
164	0.47	$K'_G = 0.89 (K'_N = 1.30)$	—	—	—	—	—
165	0.81	$K'_G = 0.42 (K'_N = 0.47)$	—	—	—	—	—
166	0.58–0.91	$K'_G = 0.58–0.14 (K'_N = 0.75–0.15)$	—	—	—	—	—
167	0.75	$K'_G = 0.30 (K'_N = 0.35)$	—	—	—	—	—
168	0.959	$K'_G = 0.47 (K'_N = 0.47)$	—	—	—	—	—
169	0.05–0.95	$K'_G = 0.5 (K'_N = 2.2–0.5)$	—	—	—	—	—
170	0.95–0.99	$K'_G = 0.5–0.025 (K'_N = 0.5–0.025)$	—	—	—	—	—
171	0.68	$K'_G = 0.21 (K'_N = 0.25)$	—	—	—	—	—
172	0.75	$K'_G = 0.30 (K'_N = 0.35)$	—	—	—	—	—
173	0.10–0.79	$K_V = 0.11–0.006 (K_N = 0.5–0.15)$	0.30	−1.71	1.00	—	—
174	0.35–0.91	$K_V = 0.09–0.006 (K_N = 0.5–0.15)$	0.28	−1.76	1.03	—	—
175	0.77	$K'_G = 0.33 (K'_N = 0.38)$	—	—	—	—	—
176	0.88	$K'_G = 0.74 (K'_N = 0.79)$	—	—	—	—	—
177	0.86	$K'_G = 0.56 (K'_N = 0.60)$	—	—	—	—	—
178	0.94	$K'_G = 0.22 (K'_N = 0.23)$	—	—	—	—	—
179	0.81	$K'_N = 0.48$	—	—	—	—	—
180	0.59	$K'_G = 0.14 (K'_N = 0.18)$	—	—	—	—	—
181	0.49	$K'_G = 0.10 (K'_N = 0.14)$	—	—	—	—	—
182	0.43	$K'_N = 0.12$	—	—	—	—	—
183	0.35	$K'_N = 0.10$	—	—	—	—	—
184	0.51	$K'_N = 0.15$	—	—	—	—	—
185	0.52	$K'_N = 0.15$	—	—	—	—	—
186	0.70	$K'_G = 0.25 (K'_N = 0.30)$	—	—	—	—	—
187	0.25–0.95	$K_N = 0.45–0.35$	0.39	−1.44	0.85	—	—
188	0.65–1.0	$K_N = 1.0–0.85$	1.06	−0.44	0.26	—	—
189	—	—	0.22	−2.02°	1.12	—	—
190	0.2–0.9	$K'_N = 0.3–0.2–0.4$	0.29	−1.73°	1.02	2.55*[1]	5.1
191	—	—	0.36	−1.53°	0.95	—	—
192	0.38–0.94	$K'_N = 0.51–1.15–0.57$	0.6	−1.1	0.7	—	—
193	0.28–0.91	$K'_N = 0.70–1.73–0.82$	0.8	−0.7	0.4	—	—
194	0.90	$K'_G = 0.90 (K'_N = 0.95)$	—	—	—	—	—
195	0.89	$K'_G = 0.81 (K'_N = 0.86)$	—	—	—	—	—
196	0.90	$K'_G = 0.86 (K'_N = 0.90)$	—	—	—	—	—
197	0.81	$K'_G = 0.41 (K'_N = 0.47)$	—	—	—	—	—
198	0.79	$K'_G = 0.37 (K'_N = 0.43)$	—	—	—	—	—
199	0.2–0.95	$K_N = 0.4–0.3–0.5$	0.38	−1.48	0.88	—	—
200	0.35–1.0	$K_N = 0.7–1.3$	0.73	−0.81	0.48	—	—
201	—	—	0.24	1.94°	1.09	—	—
202	0.2–0.9	$K'_N = 0.3–0.4$	0.31	−1.68°	1.00	2.50*[1]	5.1
203	—	—	0.38	−1.48°	0.91	—	—
204	0.79	$K'_G = 0.38 (K'_N = 0.43)$	—	—	—	—	—
205	0.84	$K'_G = 0.50 (K'_N = 0.54)$	—	—	—	—	—
206	0.75	$K'_G = 0.29 (K'_N = 0.35)$	—	—	—	—	—
207	0.82	$K'_G = 0.45 (K'_N = 0.51)$	—	—	—	—	—
208	0.3–0.95	$K_N = 0.5–0.45–0.7$	0.50	−1.20	0.71	—	—
209	0.5–1.0	$K_N = 0.9–0.85–1.0$	0.89	−0.62	0.37	—	—
210	—	—	0.24	−1.91°	1.06	—	—
211	0.2–0.9	$K_N = 0.3–0.5$	0.31	−1.67°	0.99	2.09*[1]	3.7
212	—	—	0.37	−1.49°	0.92	—	—
213	0.35–0.95	$K_N = 0.6–0.7$	0.58	−1.04	0.62	—	—
214	0.65–1.0	$K_N = 1.1–1.7$	1.09	−0.43	0.25	—	—
215	—	—	0.28	−1.77°	0.99	—	—
216	0.2–0.9	$K'_N = 0.3–0.6$	0.34	−1.58°	0.94	1.85*[1]	3.1
217	—	—	0.40	−1.42°	0.88	—	—

TABLE 5.1 (Cont.)

No.	Ref.	Equation A-clay + B		Material	Method	Exp. cond. C_0 (eq/1)	°C
218	Gilb70	NH_4-mont. + Mn^{2+}	→	C. Berteau Maroc; < 0.2 μ	dialysis	0.05—0.005	5
219	Gilb 70	,,	→	,, ; —	,,	,,	25
220	Gilb70	,,	→	,, ; —	,,	,,	37
221	Tabi60	Rb-mont. + Mg	→	Plymouth, Utah*5; < 1 μ	batch	0.001	—
222	Tabi60	,,	←	,, ,,	,,	,,	—
223	Tabi60	,,	→	Plymouth, Utah; < 1 μ	,,	,,	—
224	Tabi60	,,	←	,, ,,	,,	,,	—
225	Kris50	Cs-mont. + Ca	—	Utah Bent.; < 1 μ	batch	0.007—0.014	—
226	Gain55	Cs-mont. + Sr	↓	Chambers API 23; —	col. d.m.	0.05	5
227	Gain55	,,	↓	,, ; —	,,	,,	25
228	Gain55	,,	↓	,, ; —	,,	,,	50
229	Gain55	,,	↓	,, ; —	,,	,,	75
230	Elia66	,,	→	Bayard, New Mexico; 0.1—1 μ	column	,,	25
231	Lewi63	Cs-mont. + Ba	Na ↓	Chambers API 23; 0.1—0.5 μ	,,	0.04	30
232	Love65	,,	K ↓	,, ; —	col. d.m.	0.02	,,
233	Love65	,,	K ↓	,, *6 ; —	,,	,,	,,
Divalent-divalent exchange (earth alkali and Mn (II) ions)							
234	Kris50	Mg-mont. + Ca	←	Utah Bent.; < 1 μ	batch	0.001—0.05	—
235	Tabi60	,,	→	Colony Wyom. Bent.; < 1 μ	,,	0.001	—
236	Levy72a	,,	⇌	Wyom. Bent. API 26; < 2 μ	batch d.m.	0.06	—
237	Levy72a	,,	⇌	,, ,,	,,	0.01	—
238	Pete65	Mg-verm. + Ca	→	Libby, Montana; 50—74 μ	batch	—	—
239	Pete65	,,	←	,, ,,	,,	—	—
240	Pete65	,,	Na ↓	,, ,,	col. d.m.	—	—
241	Wild64	Mg-verm. + Ba	→	Transvaal "World"; —	dialysis	0.1	25
242	Wild64	,,	→	,, ,,	,,	,,	70
243	Kris50	Ca-mont. + Sr	→	Utah Bent.; < 1 μ	batch	0.001—0.05	—
244	Heal60	,,	→	,, ; —	,,	0.05—0.3	—
245	Heal60	,,	←	,, ; —	,,	,,	—
246	Juo69	,,	→	Wyoming Bent.; < 2 μ	,,	0.1	25
247	Heal60	Ca-ill. + Sr	→	Illinois Ill.; —	,,	0.05—0.3	—
248	Heal60	,,	←	,, ; —	batch	0.05—0.3	—
249	Heal60	Ca-kaol. + Sr	→	Florida Kaol.; —	,,	,,	—
250	Heal60	,,	←	,, ; —	,,	,,	—
251	Juo69	Ca-verm. + Sr	→	Africa Verm.; < 2 μ	,,	0.1	25
252	Heal60	,,	→	Montana Verm.; —	,,	0.05—0.3	—
253	Heal60	,,	←	,, ; —	,,	,,	—
254	Kuni68	Ca-cran. + Sr	↓	Crandallite; —	,,	0.002	—
255	Juo69	Ca-hum. + Sr	→	Humate, from Houghton muck	,,	0.1	25
256	Kris50	Ca-mont. + Ba	→	Utah Bent.; < 1 μ	,,	0.001—0.05	—
257	Wild64	Ca-verm. + Ba	→	Transvaal "World"; —	dialysis	0.1	25
258	Wild64	,,	→	,, ; —	,,	,,	70
259	Wild64	Sr-verm. + Ba	→	,, ; —	,,	,,	25
260	Wild64	,,	→	,, ; —	,,	,,	70
261	Gilb70	Ca-mont. + Mn	→	C. Berteau, Maroc; < 0.2 μ	dial. d.m.	0.005	25
262	Gilb70	,,	→	,, ,,	,,	0.01	,,
263	Gilb70	,,	→	,, ,,	,,	0.025	,,
Monovalent-monovalent exchange comprising H ions							
264	Gilb65	H-mont. + Na	←	C. Berteau, Maroc; < 0.2 μ	dialysis	0.04	5
265	Gilb65	,,	→	,, ; ,,	,,	,,	,,
266	Fosc69	,,	→	Otay California; < 1 μ	batch d.m.	0.005—0.025	—
267	Fosc69	,,	→	Colony Wyoming; < 1 μ	,,	,,	—
268	Fosc69	H-mont. + K	→	,, ,,	,,	,,	—
269	Fosc69	,,	→	Otay California; < 1 μ	,,	,,	—
270	Gilb65	H-mont. + NH_4	←	C. Berteau Maroc; < 0.2 μ	dialysis	0.04	5
271	Gilb65	,,	→	,, ; ,,	,,	,,	,,

No.	N_B-range of measurements	Selectivity coefficients in range specified	exp ($\langle \ln K_N \rangle$)	$\ln K^{\circ}_{ex}$	ΔG° (kcal/eq)	ΔH° (kcal/eq)	ΔS° (cal/eq°K)
218	0.1—0.9	K_N = 0.24—0.19—0.24	0.22	−2.01	1.11	—	—
219	0.1—0.9	K_N = 0.32—0.28—0.33	0.30	−1.70	1.00	2.74	5.83
220	0.2—0.9	K_N = 0.35—0.33—0.42	0.37	−1.50	0.92	—	—
221	0.09—0.47	K_N = 0.10—0.06	—	—	—	—	—
222	0.60—0.989	K_N = 0.27—0.23	—	—	—	—	—
223	—	K_N = 0.14—0.16	—	—	—	—	—
224	—	K_N = 0.14—0.16	—	—	—	—	—
225	0.32—0.80	K_N = 0.044—0.047	—	—	—	—	—
226	0.07—0.95	K'_N = 0.06—0.03	0.04	−3.85′	2.15	—	—
227	0.09—0.93	K'_N = 0.08—0.05	0.06	−3.32′	1.99	3.25	4.51
228	0.14—0.98	K'_N = 0.10—0.08—0.10	0.09	−2.81′	1.88	—	—
229	0.17—0.99	K'_N = 0.11—0.16	0.11	−2.67′	1.86	—	—
230	0.023—0.983	K'_N = 0.09—0.01—0.02	0.05	−3.42′	2.02	—	—
231	0—1.0	K'_N = 0.2—0.07	0.09	−2.92	1.75	—	—
232	0.05—1.0	K_N = 0.07—0.05—0.07	0.06	−3.30	2.01	—	—
233	0.1—1.0	K_N = 0.11—0.02	0.07	−3.17	1.93	—	—
234	—	K_N = 1.04	—	—	—	—	—
235	—	K'_N = 1.29	—	—	—	—	—
236	0.25—0.9	K_N = 1.3—1.1—1.3	1.20	0.18	−0.11	—	—
237	0.25—0.9	K_N = 1.3—1.1—1.3	1.20	0.18	−0.11	—	—
238	0.1—0.9	K'_N = 0.3—1.5	—	—	—	—	—
239	0—0.9	K'_N = 0.3—1.4	—	—	—	—	—
240	0—0.9	K'_N = 0.3—1.2	—	—	—	—	—
241	0.2—0.8	K_N = 0.3—1.0	0.73	−0.30	0.18	0.6	1.3
242	0.2—0.9	K_N = 0.6—2.0	0.84	−0.18	0.12	0.6	1.5
243	—	K_N = 1.05	—	—	—	—	—
244	0.05—0.95	K'_N = 1.14	—	—	—	—	—
245	0.2—0.95	K'_N = 1.14	—	—	—	—	—
246	0.01—0.07	K_N = 1.34—1.08	—	—	—	—	—
247	0.10—0.80	K'_N = 1.02	—	—	—	—	—
248	0.2—0.95	K'_N = 1.04	—	—	—	—	—
249	0.2—0.85	K'_N = 0.99	—	—	—	—	—
250	0.25—0.85	K'_N = 1.02	—	—	—	—	—
251	0.005—0.07	K_N = 1.12—1.07	—	—	—	—	—
252	0.1—0.9	K'_N = 1.17	—	—	—	—	—
253	0.1—0.95	K'_N = 1.15	—	—	—	—	—
254	—	K'_N = 2.1—2.8 (pH 4.7) K'_N = 1.4 (pH 9.3)	—	—	—	—	—
255	0.002—0.065	K_N = 0.95—0.81	—	—	—	—	—
256	—	K_N = 1.11	—	—	—	—	—
257	0.2—0.8	K_N = 0.8—1.2	0.94	−0.06	0.04	0.60	2.0
258	0.2—0.9	K_N = 1.0—1.3	1.08	0.08	−0.06	0.60	2.0
259	0.2—0.8	K_N = 1.0—0.9	0.99	−0.01	0.0	0.5	1.7
260	0.2—0.9	K_N = 1.1—1.4	1.1	0.09	−0.06	0.5	1.5
261	0.1—0.9	K'_N = 0.74—0.72—0.75	0.76	−0.27′	0.16	~0	—
262	0.1—0.9	K'_N = 0.66—0.69	0.70	−0.35′	0.21	—	—
263	0.1—0.9	K'_N = 0.74—0.63—0.69	0.73	−0.32′	0.19	—	—
264	0.2—0.75	K'_N = 0.8—0.4	0.51	−0.67′	0.37	—	—
265	0.1—0.9	K'_N = 0.7—0.5—0.8	0.65	−0.44′	0.24	—	—
266	0.30—0.86	K_N = 1.71—1.46	—	—	—	—	—
267	—	K_N = 1.27	—	—	—	—	—
268	—	K_N = 2.67	—	—	—	—	—
269	0.33—0.95	K_N = 5.15—4.94	—	—	—	—	—
270	0.7—0.95	K'_N = 9.6—2.3	8.85	2.18′	−1.20	—	—
271	0.4—0.95	K'_N = 4.5—1.6	3.39	1.22′	−0.67	—	—

TABLE 5.1 (Cont.)

No.	Ref.	Equation A-clay + B		Material	Method	Exp. cond. C_0(eq/1)	°C
Exchange reactions involving Al- and La-ions							
272	Fosc68	H-mont. + Al	←	Colony Wyoming; < 1 μ	batch	—	—
273	Fosc68	,,	←	Otay California; < 1 μ	,,	—	—
274	Fosc68	H-verm. + Al	←	Jefferite Verm.; —	,,	0.03	—
275	Coul 66	H-mont. + Al	←	Wyoming Bent.; < 2 μ	batch d.m.	0.01	5
276	Coul66	H-ill. + Al	←	Fithian Ill.; < 2 μ	,,	,,	,,
277	Clar65	Na-mont. + Al	↓	Wyom. Bent.; —	,,	0.05	—
278	Clar65	,,	↓	,, ; —	,,	1.0	—
279	Fosc68	,,	←	Colony Wyoming; < 1 μ	batch	—	—
280	Fosc68	,,	←	Otay California; < 1 μ	,,	—	—
281	Brug72	,,	→	Clay Spur; < 2 μ	dial. (pH3.5)	0.1	—
282	Brug72	,,	→	,, ,,	,,	0.05	—
283	Brug72	,,	→	,, ,,	,,	0.01	—
284	Brug72	,,	→	,, ,,	,,	0.005	—
285	Brug72	,,	→	,, ,,	dial. (pH 5.0)	0.05	—
286	Brug72	,,	→	,, ,,	,,	0.01	—
287	Fosc68	Na-verm. + Al	←	Jefferite Verm.; —	batch	0.03–0.05	—
288	Nye61	K-mont. + Al	↓	Montomorillonite; —	col. d.m.	0.05	—
289	Nye61	,,	↓	,, ; —	,,	0.00625–1.0	—
290	Clar65	,,	↓	Wyom. Bent.; —	batch d.m.	0.05	—
291	Clar65	,,	↓	,, ; —	,,	1.0	—
292	Fosc68	,,	←	Colony Wyoming; < 1 μ	batch	—	—
293	Fosc68	,,	←	Otay California; < 1 μ	,,	0.008–0.05	—
294	Nye61	K-kaol. + Al	↓	Kaolinite; —	col. d.m.	0.05	—
295	Fosc68	Mg-mont. + Al	←	Colony Wyoming; < 1 μ	batch	—	—
296	Fosc68	,,	←	Otay California; < 1 μ	,,	—	—
297	Fosc68	Mg-verm. + Al	←	Jefferite Verm.; —	,,	0.02	—
298	Clar65	Ca-mont. + Al	↓	Wyom. Bent.; —	batch d.m.	0.05	—
299	Clar65	,,	↓	,, ; —	,,	1.0	—
300	Fosc68	,,	←	Colony Wyoming; < 1 μ	batch	0.003–0.035	—
301	Fosc68	,,	←	Otay California; < 1 μ	,,	0.02	—
302	Coul68	,,	→	Wyom. Bent.; < 2 μ	,,	0.01	—
303	Col68	Ca-ill. + Al	→	Fouithian Ill.; < 2 μ	,,	,,	—
304	Coul68	Ca-verm. + Al	→	Montana Verm.; < 2 μ	,,	,,	—
305	Brug72	Ca-mont. + Al	→	Clay Spur; < 2 μ	dial. (pH 3.5)	,,	—
306	Brug72	,,	→	,, ,,	,,	0.05	—
307	Fosc68	Ca-verm. + Al	←	Jefferite Verm.; —	batch	0.02	—
308	Brug72	Na-mont. + La	→	Clay Spur; < 2 μ	dial. (pH 3.5)	0.1	—
309	Brug72	,,	→	,, ,,	,,	0.05	—
310	Brug72	,,	→	,, ,,	,,	0.01	—
311	Brug72	,,	→	,, ,,	,,	0.005	—
312	Brug72	,,	→	,, ,,	dial. (pH5)	0.05	—
313	Brug72	,,	→	,, ,,	,,	0.01	—
314	Brug72	Ca-mont. + La	→	,, ,,	dial. (pH3.5)	0.05	—
315	Brug72	,,	→	,, ,,	,,	0.01	—
316	Brug72	,,	→	,, ,,	,,	0.005	—
317	Brug72	,,	→	,, ,,	dial. (pH5.)	0.05	—
318	Brug72	,,	→	,, ,,	,,	0.01	—
319	Brug72	,,	→	,, ,,	,,	0.005	—

No.	N_B-range of measurements	Selectivity coefficients in range specified	exp ($\langle \ln K_N \rangle$)	$\ln K^{\circ}_{ex}$	ΔG° (kcal/eq)	ΔH° (kcal/eq)	ΔS° (cal/eq°K)
272	—	K_N = 5.9	5.9	1.2	− 0.7	—	—
273	—	K_N = 7.1	7.1	1.3	− 0.8	—	—
274	—	K_N = 3.4	—	—	—	—	—
275	—	—	6.3	—	—	—	—
276	—	—	2.8	—	—	—	—
277	0.85—1.0	K_N = 4.2	—	—	—	—	—
278	0.3—0.95	K'_N = 2.7	—	—	—	—	—
279	—	K_N = 5.9	5.9	1.1	− 0.7	—	—
280	—	K_N = 7.3	7.3	1.3	− 0.8	—	—
281	0.05—0.7	K_N = 1.8—5.2	2.12	0.08°	− 0.05	—	—
282	0.05—0.8	K_N = 1.3—5.9	2.05	0.05°	− 0.03	—	—
283	0.05—0.9	K_N = 0.3—3.5	1.31	− 0.40°	0.24	—	—
284	0.1—0.7	K_N = 0.2—1.1	0.64	− 1.11°	0.65	—	—
285	0.1—0.8	K_N = 0.8—6.8	2.07	0.06°	− 0.04	—	—
286	0.05—0.9	K_N = 0.1—4.3	1.23	− 0.46°	0.27	—	—
287	—	K_N = 1.6	1.6	− 0.2	0.1	—	—
288	0.3—0.95	K_N = 0.7—0.3	0.6	−1.2′	0.7	—	—
289	—	K'_N = 1.1—0.3	—	—	—	—	—
290	0.6—1.0	K_N = 1.6	1.6	− 0.2	0.1	—	—
291	0.05—0.9	K'_N = 1.1	—	—	—	—	—
292	—	K_N = 2.9	2.9	0.4	− 0.25	—	—
293	0.53—0.88	K_N = 0.9	0.9	− 0.8	0.5	—	—
294	0.2—0.95	K_N = 0.6—0.14	—	—	—	—	—
295	—	K_N = 1.7	1.7	0.4	− 0.2	—	—
296	—	K_N = 1.6	1.6	0.3	− 0.2	—	—
297	—	K_N = 4.1	4.1	1.2	− 0.7	—	—
298	0.45—0.95	K_N = 1.7	1.7	0.35	− 0.2	—	—
299	0.2—0.95	K'_N = 1.4	—	—	—	—	—
300	0.89—0.58	K_N = 1.6	1.6	0.3	− 0.2	—	—
301	—	K_N = 1.5	1.5	0.25	− 0.15	—	—
302	0.4—0.95	K_N = 2.0	2.03	0.54	− 0.33	—	—
303	—	K_N = 2.3	2.28	0.66	− 0.40	—	—
304	—	K_N = 3.4	3.35	1.04	− 0.63	—	—
305	0.05—1.0	K_N = 1.4—2.0	1.53	0.26°	− 0.16	—	—
306	0.1—0.9	K_N = 1.6—1.9	1.46	0.21°	− 0.13	—	—
307	—	K_N = 3.9	3.9	1.2	− 0.7	—	—
308	0.05—0.9	K_N = 2.5—12.4	2.48	0.24°	− 0.14	—	—
309	0.05—0.95	K_N = 2.4—14.6	2.75	0.34°	− 0.20	—	—
310	0.05—0.85	K_N = 1.3—4.7	2.44	0.22°	− 0.13	—	—
311	0.05—0.95	K_N = 0.9—3.4	1.46	− 0.29°	0.17	—	—
312	0.05—0.95	K_N = 2.6—23.3	3.06	0.45°	− 0.27	—	—
313	0.05—0.9	K_N = 1.1—6.4	2.51	0.25°	− 0.15	—	—
314	0.05—0.8	K_N = 1.7—2.3	1.63	0.32°	− 0.19	—	—
315	0.05—0.85	K_N = 2.1—2.4	1.97	0.50°	− 0.30	—	—
316	0.05—0.95	K_N = 2.2—6.3	2.22	0.63°	− 0.37	—	—
317	0.05—0.85	K_N = 1.8—2.3	1.68	0.35°	− 0.21	—	—
318	0.05—0.9	K_N = 2.0—2.7	2.03	0.54°	− 0.32	—	—
319	0.05—0.7	K_N = 1.8—2.5	2.15	0.60°	− 0.35	—	—

TABLE 5.1 (Cont.)

No.	Ref.	Equation A-clay + B		Material	Method	Exp. cond.	
						C_0 (eq/l)	°C
Exchange reactions comprising heavy metal ions							
320	Kris50	Ca-mont. + Cu	→	Utah Bent.; < 1 μ	batch	0.001—0.05	—
321	El-S70c	,,	→	Wyoming Bent.; < 2 μ	,,	0—0.02	25
322	El-S70c	,,	→	,, ,,	,,	,,	50
323	Maes73	Na-mont. + Cu	→	C. Berteau Maroc; < 0.5 μ	dial. d.m.	0.01	22
324	Sing73	Na-mont. + Co	→	Amori, Mississippi; < 2 μ	batch	± 0.003?	40
325	Sing73	,,	→	,, ,,	,,	,,	60
326	Maes73	,,	→	C. Berteau Maroc; < 0.5 μ	dial. d.m.	0.01	22
327	Bitt74	Ca-mont. + Cd	→	Wyoming API 25; < 2 μ	batch	0.002	—
328	Bitt74	,,	←	,, ,,	,,	,,	—
329	Bitt74	Ca-ill. + Cd	→	Beaver's Bend; < 50 μ	,,	,,	—
330	Bitt74	,,	←	,, ,,	,,	,,	—
331	Bitt74	Ca-kaol. + Cd	→	Sandersville Georgia; < 2 μ	,,	0.0005	—
332	Bitt74	,,	←	,, ,,	,,	,,	—
333	McLe66	Ba-mont. + Ni	Ca ↓	Otay California; —	batch d.m.	0.04	—
334	Maes73	Na-mont. + Ni	→	C. Berteau Maroc; < 0.5 μ	dial. d.m.	0.01	22
335	Maes73	Na-mont. + Zn	→	,, ; ,,	,,	,,	,,
336	Kris50	Ca-mont. + Pb	→	Utah Bent.; < 1 μ	batch	0.001—0.05	—
337	Bitt74	,,	→	Wyoming API 25; < 2 μ	,,	0.002	—
338	Bitt74	,,	←	,, ,,	,,	,,	—
339	Bitt74	Ca-ill. + Pb	→	Beaver's Bend; < 50 μ	,,	,,	—
340	Bitt74	,,	←	,, ,,	,,	,,	—
341	Bitt74	Ca-kaol. + Pb	→	Sandersville Georgia; < 2 μ	,,	0.0005	—
342	Bitt74	,,	←	,, ,,	,,	,,	—
343	Bitt74	Cd-mont. + Pb	→	Wyoming API 25; < 2 μ	,,	0.002	—
344	Bitt74	,,	←	,, ,,	,,	,,	—
345	Bitt74	Cd-ill. + Pb	→	Beaver's Bend; < 50 μ	,,	,,	—
346	Bitt74	,,	←	,, ,,	,,	,,	—
347	Bitt74	Cd-kaol. + Pb	→	Sandersville Georgia; < 2 μ	,,	0.0005	—
348	Bitt74	,,	←	,, ,,	,,	,,	—
349	Harm77	Ca-ill. + Zn	→	Grundite; —	,,	variable	—
350	Harm77	,,	←	,, ; —	,,	,,	—

No.	N_B-range of measurements	Selectivity coefficients in range specified	exp (⟨ln K_N⟩)	ln K_{ex}°	ΔG° (kcal/ eq)	ΔH° (kcal/ eq)	ΔS° (cal eq° K)
320	—	K_N = 0.94	—	—	—	—	—
321	0.1—0.9	K_N = 0.8—1.1	0.98	−0.02	0.01	−2.15	−7.0
322	0.1—0.9	K_N = 0.9—0.5	0.74	−0.30	0.19	—	—
323	0.05—0.95	K_N = 1.9—2.9	2.18	0.28	−0.17	0.47	2.1
324	0.33—0.66	K_N' = ±0.3?	—	—	—	—	—
325	0.36—0.60	K_N = ±0.3?	—	—	—	—	—
326	0.05—0.95	K_N = 1.8—6.1	2.29	0.33	−0.19	0.32	1.8
327	0.14—0.56	K_N' = 1.0—0.8	—	—	—	—	—
328	0.41—0.87	K_N' = 1.3—1.0	—	—	—	—	—
329	0.38—0.94	K_N' = 1.1—1.0	—	—	—	—	—
330	0.07—0.62	K_N' = 1.0—0.8	—	—	—	—	—
331	0.06—0.76	K_N' = 0.9—1.2	—	—	—	—	—
332	0.26—0.86	K_N' = 0.9—1.1	—	—	—	—	—
333	0.04—0.93	K_N' = 0.9—0.8—0.9	—	—	—	—	—
334	0.1—0.95	K_N = 1.8—2.9	2.18	0.28	−0.17	0.58	2.5
335	0.1—0.9	K_N = 1.9—4.5	2.20	0.29	−0.17	0.48	2.1
336	—	K_N = 1.18	—	—	—	—	—
337	0.11—0.65	K_N' = 1.5—1.1	—	—	—	—	—
338	0.47—0.96	K_N' = 1.4—1.3—1.4	—	—	—	—	—
339	0.13—0.75	K_N' = 1.9—1.4	—	—	—	—	—
340	0.50—0.98	K_N' = 1.7—1.3—2.0	—	—	—	—	—
341	0.10—0.87	K_N' = 2.4—1.8—2.2	—	—	—	—	—
342	0.44—0.92	K_N' = 1.7—1.3—1.6	—	—	—	—	—
343	0.09—0.79	K_N' = 1.1—1.1	—	—	—	—	—
344	0.38—0.96	K_N' = 1.5—1.4—1.6	—	—	—	—	—
345	0.10—0.69	K_N' = 2.1—1.1	—	—	—	—	—
346	0.48—0.95	K_N' = 1.6—1.3	—	—	—	—	—
347	0.16—0.83	K_N' = 2.1—1.5	—	—	—	—	—
348	0.47—0.97	K_N' = 1.8—2.8	—	—	—	—	—
349	0.1—1.0	K_N = 1.0	—	—	—	—	—
350	0.1—1.0	K_N = 1.0	—	—	—	—	—

TABLE 5.2
Selectivity coefficients as found for soils

No.	Ref.	Equation $A_s + B$		Exp. cond. C_0 (eq/l)	°C	N_B-range of measurements	Selectivity coefficients in range specified
Monovalent-monovalent exchange (alkali and NH_4 ions)							
1	Vans32	$Na_s + K$	→	0.1	25	0.4–0.9	$K'_N = 5.3–3.0$
2	Vans32	,,	→	,,	,,	0.45–0.95	$K'_N = 5.8–3.7$
3	Deis67	,,	←	0.01	,,	—	$K_N = 4.81$
4	Deis67	,,	←	,,	,,	—	$K_N = 4.53$
5	Deis67	,,	←	,,	,,	—	$K_N = 5.16$
6	Deis67	,,	←	,,	,,	—	$K_N = 5.58$
7	Deis67	,,	←	,,	,,	—	$K_N = 6.30$
8	Kris50	$Na_s + NH_4$	⇌	0.001–0.05	—	—	$K_N = 3.33$
9	Kris50	,,	⇌	,,	—	—	$K_N = 5.88$
10	Kris50	,,	⇌	,,	—	—	$K_N = 5.13$
11	Amph58	$Na_s + Cs$	↓	0.02–0.1	20	0.35–0.96	$K'_N = 26.5–41.8–9.0$
12	Kris50	$K_s + NH_4$	⇌	0.001–0.05	—	—	$K_N = 0.99$
13	Kris50	,,	⇌	,,	—	—	$K_N = 0.87$
14	Kris50	,,	⇌	,,	—	—	$K_N = 0.54$
15	Deis67	$K_s + Rb$	→	0.01	25	—	$K_N = 3.10$
16	Deis67	,,	→	,,	,,	—	$K_N = 2.14$
17	Deis67	,,	→	,,	,,	—	$K_N = 1.90$
18	Deis67	,,	→	,,	,,	—	$K_N = 2.10$
19	Deis67	,,	→	,,	,,	—	$K_N = 2.56$
20	Deis67	,,	→	,,	,,	—	$K_N = 2.23$
21	Deis67	,,	→	,,	,,	—	$K_N = 2.61$
22	Deis67	,,	→	,,	,,	—	$K_N = 2.39$
23	Kris50	$NH_{4s} + Rb$	⇌	0.001–0.05	—	—	$K_N = 2.34$
24	Kris50	,,	⇌	,,	—	—	$K_N = 3.30$
25	Kris50	,,	⇌	,,	—	—	$K_N = 3.60$
26	Amph58	$NH_{4s} + Cs$	↓	0.02–0.1	20	0.28–0.95	$K'_N = 7.5–8.6–5.7$
Monovalent-divalent exchange (NH_4, alkali and earth-alkali ions)							
27	Bakk73	$Na_s + Mg$	↓	0.2–0.01	—	0.9–1.0	$K'_G = 1.98 (K'_N = 2.1–2.0)$
28	Bowe59	$Na_s + Ca$	←	0.01–0.2	—	0.4–0.9	$K'_G = 2.4 (K'_N = 3.8–2.5)$
29	Bowe59	,,	←	,,	—	0.4–0.9	$K'_G = 2.1 (K'_N = 3.3–2.2)$
30	Bowe59	,,	←	,,	—	0.4–0.9	$K'_G = 1.9 (K'_N = 3.0–2.0)$
31	Bowe59	,,	←	,,	—	0.4–0.9	$K'_G = 1.8 (K'_N = 2.9–1.9)$
32	Bowe59	,,	←	,,	—	0.4–0.9	$K'_G = 1.4 (K'_N = 2.2–1.5)$
33	Bowe59	,,	←	,,	—	0.4–0.9	$K'_G = 1.9 (K'_N = 3.0–2.0)$
34	Schw62	,,	Na↓	0.01	—	0.82–0.992	$K'_N = 2.01–2.48–1.96$
35	Babc63	,,	↓	0.02(Cl)*2	—	0.78–0.89	$K_V = 3.60–3.36 (K_N = 3.4–3.5)$
36	Babc63	,,	↓	0.02(SO_4)*3	—	0.74–0.86	$K_V = 4.57–3.73 (K_N = 3.8–3.6)$
37	Babc63	,,	↓	0.08(Cl)*2	—	0.57–0.78	$K_V = 4.03–4.35 (K_N = 3.4–3.8)$
38	Babc63	,,	↓	0.08(SO_4)*3	—	0.48–0.70	$K_V = 5.08–4.13 (K_N = 3.7–3.6)$
39	Prat64	,,	↓	0.2	—	0.56	$K'_G = 3.96 (K'_N = 5.3)$
40	Prat64	,,	↓	,,	—	0.36	$K'_G = 1.78 (K'_N = 3.0)$
41	Prat64	,,	↓	,,	—	0.51	$K'_G = 3.33 (K'_N = 4.7)$

*1 Abbreviations (%): A = allophane, Al = hydrous oxides of Al, B = biotite, Ca = calcite, Cl = clorite, Clin = cline-pyroxen HB = hydrobiotite, I = illite, K = kaolinite, M = montomorillonite, Mi = mica, O.C. = organic carbon, Q = quartz, Sm = smectit V = vermiculite. Subscript "i" indicates "interstratified". – Al, – $CaCO_3$, – Fe, – O.C., – Si = Al resp. $CaCO_3$, Fe, O.C., Si remove *2 Cl^- as anion. *3 SO_4^{2-} as anion. *4 $CaCO_3$, Fe and organic carbon were removed from all samples indicated with Cars72. *5 N_B (M

No.	Method	Material*[1] Name; fraction; composition
1	batch	Soil Colloid 431; clay fraction; – O.C.
2	,,	Soil Colloid 5696; clay fraction; – O.C.
3	,,	Tedburn; whole soil;
4	,,	Bovey Basin; whole soil;
5	,,	Cegin; whole soil;
6	,,	Windsor; whole soil;
7	,,	Dunkeswick; whole soil;
8	,,	Yolo clay; $< 1\,\mu$; – O.C., M
9	,,	Hanford clay; $< 1\,\mu$; – O.C., I
10	,,	Aiken clay; $< 1\,\mu$; – O.C., K
11	col. d.m.	Lower Greensand, Oxon; < 40 mesh; M, I
12	batch	Yolo clay; $< 1\,\mu$; – O.C., M
13	,,	Hanford clay; $< 1\,\mu$; – O.C., I
14	,,	Aiken clay; $< 1\,\mu$; – O.C., K
15	,,	Dunkeswick; whole soil;
16	,,	Tedburn; whole soil;
17	,,	Bovey Basin; whole soil;
18	,,	Cegin; whole soil;
19	,,	Windsor; whole soil;
20	,,	Long Load; whole soil; $< 2\,\mu$: mixed M_i, V35
21	,,	Harwell; whole soil; $< 2\,\mu$: M65, Clin
22	,,	Newchurch; whole soil; $< 2\,\mu$: M35
23	,,	Yolo clay; $< 1\,\mu$; – O.C., M
24	,,	Hanford clay; $< 1\,\mu$; – O.C., I
25	,,	Aiken clay; $< 1\,\mu$; – O.C., K
26	,,	Lower Greensand, Oxon; < 40 mesh; M, I
27	col. d.m.	Shepperton subsoil; whole soil; $< 2\,\mu$: I75, K15, O.C. 0.5
28	batch d.m.	Billings silt loam, U.S.A.; whole soil; $\Gamma = 2.28 \cdot 10^{-7}$ me/cm^2
29	,,	Esquatzel loam, U.S.A.; whole soil; $\Gamma = 1.95 \cdot 10^{-7}$ me/cm^2
30	,,	Chino clay, U.S.A.; whole soil; $\Gamma = 1.76 \cdot 10^{-7}$ me/cm^2
31	,,	Chilcott silt loam, U.S.A.; whole soil; $\Gamma = 1.67 \cdot 10^{-7}$ me/cm^2
32	,,	Wyoming Bent.; whole sample; $\Gamma = 1.17 \cdot 10^{-7}$ me/cm^2
33	,,	Georgia Kaolin; whole sample; $\Gamma = 1.64 \cdot 10^{-7}$ me/cm^2
34	,,	Seemarsch; $< 2\,\mu$; M, I, K, Cl
35	col. d.m.	Yolo loam; whole soil;
36	,,	Yolo loam; whole soil;
37	,,	Yolo loam; whole soil;
38	,,	Yolo loam; whole soil;
39	batch d.m.	Organic soil (No. 29); whole soil; O.C. 15.1, pH 7
40	,,	Organic soil (No. 29); whole soil; – O.C., pH 7
41	,,	Organic soil (No. 30); whole soil; O.C. 9.5, pH 7

given as fraction of Cs-fixing sites. *[6] Referring to the discussion about two site models in chapter 4, these observations could be described in terms of highly selective sites covering 7% of the CEC with K_h around 185, plus the remainder of the CEC obeying $K_l = 1.0$ (see p. 123).

TABLE 5.2 (Cont.)

No.	Ref.	Equation $A_s + B$		Exp. cond. C_0 (eq/l)	°C	N_B-range of measurements	Selectivity coefficients in range specified
42	Prat64	$Na_s + Ca$	↓	0.2	—	0.37	$K'_G = 1.89(K'_N = 3.1)$
43	Tuck67	,,	←	0.07	—	0.92—1.0	$K'_G = 1$—$0.2(K'_N = 1.0$—$0.2)$
44	Tuck67	,,	←	,,	—	0.93—1.0	$K'_G = 0.5$—$0.2(K'_N = 0.5$—$0.2)$
45	Tuck67	,,	←	,,	—	0.9—1.0	$K'_G = 0.5$—$0.3(K'_N = 0.5$—$0.3)$
46	Tuck67	,,	←	,,	—	0.95—1.0	$K'_G = 1$—$0.4(K'_N = 1.0$—$0.4)$
47	Tuck67	,,	←	,,	—	0.94—1.0	$K'_G = 1$—$0.2(K'_N = 1.0$—$0.2)$
48	Levy68	,,	↓	0.042—0.059	—	0.61—0.97	$K_N = 3.6$—2.4
49	Levy68	,,	↓	,,	—	0.46—0.98	$K_N = 2.3$—2.2
50	Levy68	,,	↓	,,	—	0.55—0.97	$K_N = 3.0$—2.9
51	Nay169	,,	←	—	—	0.28—0.95	$K'_G = 5.68$—$5.87(K'_N = 10.7$—$6.1)$
52	Nay169	,,	←	—	—	0.35—0.96	$K'_G = 6.71$—$6.62(K'_N = 11.3$—$6.8)$
53	Unit69	$Na_s + (Ca + Mg)$	—	—	—	0.60—0.95	$K'_G = 2.15(K'_N = 2.8$—$2.2)$
54	Bakk73	$Na_s + Ca$	↓	0.2—0.01	—	0.9—1.0	$K'_G = 2.1(K'_N = 2.2$—$2.1)$
55	Ghey75	,,	←	0.02—0.21	—	0.71—0.98	$K_G = 7.0(K_N = 8.3$—$7.1)$
56	Ghey75	,,	←	,,	—	0.62—0.97	$K_G = 4.6(K_N = 5.8$—$4.7)$
57	Ghey75	,,	←	,,	—	0.55—0.96	$K_G = 3.4(K_N = 4.6$—$3.5)$
58	Ghey75	,,	←	,,	—	0.55—0.96	$K_G = 3.5(K_N = 4.7$—$3.6)$
59	Ghey75	,,	←	,,	—	0.56—0.96	$K_G = 3.7(K_N = 4.9$—$3.8)$
60	Ghey75	,,	←	,,	—	0.69—0.98	$K_G = 6.2(K_N = 7.5$—$6.3)$
61	Cohe62	$Na_s + Sr$	→	0.017	—	0—0.01	$K_N = 2.8$
				0.075	—	0—0.01	$K_N = 3.5$
				0.31	—	0—0.01	$K_N = 3.7$
62	Schw62	$K_s + Ca$	Na↓	0.01	—	0.71	$K'_N = 0.30$
63	Schw62	,,	Na↓	,,	—	0.74	$K'_N = 0.34$
64	Schw62	,,	Na↓	,,	—	0.77	$K'_N = 0.39$
65	Schw62	,,	Na↓	,,	—	0.76	$K'_N = 0.38$
66	Schw62	,,	Na↓	,,	—	0.70	$K'_N = 0.27$
67	Schw62	,,	Na↓	,,	—	0.76	$K'_N = 0.36$
68	Schw62	,,	Na↓	,,	—	0.70	$K'_N = 0.28$
69	Schw62	,,	Na↓	,,	—	0.53—0.95	$K'_N = 0.63$—0.25
70	Schw62	,,	Na↓	,,	—	0.71	$K'_N = 0.30$
71	Schw62	,,	Na↓	,,	—	0.74	$K'_N = 0.33$
72	Schw62	,,	Na↓	,,	—	0.72	$K'_N = 0.31$
73	Schw62	,,	Na↓	,,	—	0.68	$K'_N = 0.26$
74	Schw62	,,	Na↓	,,	—	0.72	$K'_N = 0.31$
75	Schw62	,,	Na↓	,,	—	0.71	$K'_N = 0.29$
76	Schw62	,,	Na↓	,,	—	0.71	$K'_N = 0.29$
77	Schw62	,,	Na↓	,,	—	0.68	$K'_N = 0.27$
78	Schw62	,,	Na↓	,,	—	0.66	$K'_N = 0.25$
79	Schw62	,,	Na↓	,,	—	0.67	$K'_N = 0.26$
80	Schw62	,,	Na↓	,,	—	0.64	$K'_N = 0.23$
81	Schw62	,,	Na↓	,,	—	0.64	$K'_N = 0.23$
82	Schw62	,,	Na↓	,,	—	0.65	$K'_N = 0.23$
83	Schw62	,,	Na↓	,,	—	0.68	$K'_N = 0.27$
84	Schw62	,,	Na↓	,,	—	0.59	$K'_N = 0.18$
85	Schw62	,,	Na↓	,,	—	0.61	$K'_N = 0.19$
86	Schw62	,,	Na↓	,,	—	0.62	$K'_N = 0.20$
87	Schw62	,,	Na↓	,,	—	0.57	$K'_N = 0.17$
88	Schw62	,,	Na↓	,,	—	0.66	$K'_N = 0.23$
89	Schw62	,,	Na↓	,,	—	0.66	$K'_N = 0.23$
90	Schw62	,,	Na↓	,,	—	0.57	$K'_N = 0.17$
91	Schw62	,,	Na↓	,,	—	0.58	$K'_N = 0.18$
92	Schw62	,,	Na↓	,,	—	0.56	$K'_N = 0.18$
93	Schw62	,,	Na↓	,,	—	0.58	$K'_N = 0.18$
94	Schw62	,,	Na↓	,,	—	0.64	$K'_N = 0.23$
95	Schw62	,,	Na↓	,,	—	0.31—0.83	$K'_N = 0.33$—0.07
96	Rich64	,,	Na↓	,,	—	0.66	$K'_G = 0.20(K'_N = 0.24)$
97	Rich64	,,	Na↓	,,	—	0.63	$K'_G = 0.17(K'_N = 0.21)$
98	Rich64	,,	Na↓	,,	—	0.53	$K'_G = 0.12(K'_N = 0.17)$

No.	Method	Material*[1] Name; fraction; composition
42	batch d.m.	Organic soil (No. 30); whole soil; − O.C., pH 7
43	batch	Willalooka; < 2 μ; pH 8.0
44	,,	Belalie; < 2 μ; pH 8.2
45	,,	McLaren Vale; < 2 μ; pH 8.0
46	,,	Urrbrae; < 2 μ; pH 8.0
47	,,	Claremont; < 2 μ; pH 8.2
48	batch d.m.	Dark Brown Residual Clay Loam; whole soil; O.C. < 2, < 2 μ: M60, K40
49	,,	Dark Brown Grumosolic Silty Clay; whole soil; O.C. < 2, < 2 μ: M80, K20
50	,,	Sandy Clay Loam Hamra; whole soil; O.C. < 2, < 2 μ: M50, K50
51	batch	Staten peaty muck, California; whole soil; O.C. 45—50
52	,,	Egbert muck, California; whole soil; O.C. 45—50
53	—	59 soil samples
54	col. d.m.	Shepperton subsoil; whole soil; < 2 μ: I75, K15, O.C. 0.5
55	batch d.m.	Mollisol, Nkheila, Morocco; whole soil; < 2 μ: M, I, M_i, I_i, K, Ca
56	,,	Mollisol, Nkheila, Morocco; whole soil; < 2 μ: I, M, K, Cl, V, Ca
57	,,	Mollisol, Nkheila, Morocco; whole soil; < 2 μ: I, K, Cl_i, V, Cl, M
58	,,	Vertisol, Gharb, Morocco; whole soil; < 2 μ: M, Cl, K, I, M_i, Cl_i
59	,,	Vertisol, Gharb, Morocco; whole soil; < 2 μ: M, Cl, K, I
60	,,	Mollisol, Nkheila, Morocco; whole soil; < 2 μ: M, K, Ca, Q
61	batch	Saclay mud, France; whole soil; < 2 μ: M45, K45, I10
62	batch d.m.	Auenboden, Germ.; < 0.08 μ; M80, I20, V < 10
63	,,	Seemarsch, Neth.; < 0.08 μ; M70, I30
64	,,	Seemarsch, Belg.; 0.2 − 0.08 μ; M70, I30
65	,,	Seemarsch, Belg.; < 0.08 μ; M85, I15
66	,,	Löss, Germ.; < 0.08 μ; M80, I20
67	,,	Pseudogley, Germ.; < 0.08 μ; M70, I30
68	,,	Mullrendzina (kro), Germ.; < 0.08 μ; M80, I20
69	,,	Seemarsch, Belg.; < 2 μ; M, I, K, Cl
70	,,	Seemarsch, Neth.; 2—0.2 μ; I60, K13, Cl10
71	,,	Seemarsch, Belg.; 2—0.2 μ; I35, M20, K12, Cl10
72	,,	Pseudogley, Germ.; 0.2—0.08 μ; I50, M40
73	,,	Mullrendzina (kro), Germ.; 2—0.2 μ; I30, Cl15, K11, M10
74	,,	Seemarsch, Neth.; 0.2—0.08 μ; I50, M30, (V + Cl)15
75	,,	Moräne, Schw.; < 0.08 μ; I40, M30, V10
76	,,	Pseudogley, Germ.; 2—0.2 μ; I35, M20, K20, V10
77	,,	Mullrendzina (kro), Germ.; 0.2—0.08 μ; M70, I20, V10
78	,,	Moräne, Schw.; 0.2—0.08 μ; I70, V10
79	,,	Moräne, Schw.; < 0.08 μ; I60, M20, V10
80	,,	Löss, Germ.; 0.2—0.08 μ; I50, M30, V20
81	,,	Moräne, Schw.; 0.2—0.08 μ; I55, M20, V20
82	,,	Moräne, Schw.; 2—0.2 μ; I60, V10
83	,,	Mullrendzina (mo), Schw.; < 0.08 μ; I60, M20, V20
84	,,	Auenboden, Germ.; 2—0.2 μ; I40, V20, K10
85	,,	Auenboden, Germ.; 0.2—0.08 μ; I40, M25, V25
86	,,	Löss, Germ.; 2—0.2 μ; I30, V20, K10
87	,,	Moräne, Schw.; 2—0.2 μ; I35, V20
88	,,	Mullrendzina (mo); Germ.; 2—0.2 μ; I50, V30
89	,,	Mullrendzina (mo); Germ.; 0.2—0.08 μ; I70, V20, M10
90	,,	Braunlehm, Germ.; 2—0.2 μ; I50, V30
91	,,	Braunlehm, Germ.; 0.2—0.08 μ; I50, V20, M10
92	,,	Braunerde, Germ.; 2—0.2 μ; Cl70, I10
93	,,	Braunerde, Germ.; 0.2—0.08 μ; Cl80; I < 20
94	,,	Braunerde, Germ.; < 0.08 μ; Cl80, I20
95	,,	Braunerde, Germ.; < 2 μ; Cl, I
96	,,	Putnam, Missouri; < 2 μ; − Fe, M
97	,,	Carrington Ap Iowa; 2—0.2 μ; − Fe, M, Mi20
98	,,	Berks Ap, Virginia; 2—0.2 μ; − Fe, Mi40

TABLE 5.2 (Cont.)

No.	Ref.	Equation $A_s + B$		Exp. cond. C_0 (eq/l)	°C	N_B-range of measurements	Selectivity coefficients in range specified
99	Rich64	$K_s + Ca$	Na↓	0.01	—	0.49	$K'_G = 0.10 (K'_N = 0.14)$
100	Rich64	,,	Na↓	,,	—	0.57	$K'_G = 0.13 (K'_N = 0.17)$
101	Beck64	,,	↓	0.04—0.06	—	0.87—1.0	$K_G = 0.23$—0.002
102	Deis67	,,	⇌	0.01—0.02	25	—	$K_N = 0.73$
103	Deis67	,,	⇌	,,	,,	—	$K_N = 0.81$
104	Deis67	,,	⇌	,,	,,	—	$K_N = 0.96$
105	Deis67	,,	⇌	,,	,,	—	$K_N = 0.63$
106	Deis67	,,	⇌	,,	,,	—	$K_N = 0.76$
107	Deis67	,,	⇌	,,	,,	—	$K_N = 0.34$
108	Deis67	,,	⇌	,,	,,	—	$K_N = 0.39$
109	Deis67	,,	⇌	,,	,,	—	$K_N = 0.52$
110	Deis67	,,	⇌	,,	,,	—	$K_N = 0.13$
111	Deis67	,,	⇌	,,	,,	—	$K_N = 0.47$
112	Ehle67	,,	↓	0.01	,,	0—0.81	$K_G = 1.20 (K_N \geqslant 1.33)$
						0.81—0.94	$K_G = 0.02 (K_N = 0.02)$
						0.94—1.0	$K_G = 8 \cdot 10^{-6} - 5 \cdot 10^{-6}$ $(K_N = 8 \cdot 10^{-6} - 5 \cdot 19^{-6})$
113	Ehle67	,,	↓	,,	,,	0—0.84	$K_G = 0.58 (K_N \geqslant 0.63)$
						0.84—0.988	$K_G = 0.01 (K_N = 0.01)$
						0.988—1.0	$K_G = 6 \cdot 10^{-5} (K_N = 6 \cdot 10^{-5})$
114	Ehle67	,,	↓	,,	,,	0—0.87	$K_G = 1.02 (K_N \geqslant 1.09)$
						0.87—0.98	$K_G = 0.02 (K_N = 0.02)$
						0.98—1.0	$K_G = 6 \cdot 10^{-4} (K_N = 6 \cdot 10^{-4})$
115	Ehle67	$K_s + Ca$	↓	0.01	,,	0—0.91	$K_G = 0.31 (K_N \geqslant 0.32)$
						0.91—0.955	$K_G = 0.02 (K_N = 0.02)$
						0.955—1.0	$K_G = 7 \cdot 10^{-4} (K_N = 7 \cdot 10^{-4})$
116	Ehle67	,,	↓	,,	,,	0—0.992	$K_G = 3.0 (K_N \geqslant 3.0)$
						0.992—0.999	$K_G = 0.04 (K_N = 0.04)$
						0.999—1.0	$K_G = 2 \cdot 10^{-5} (K_N = 2 \cdot 10^{-5})$
117	Tuck67	,,	←	0.07	—	0.85—0.99	$K'_G = 0.2$—$0.01 (K'_N = 0.2$—$0.01)$
118	Tuck67	,,	←	,,	—	0.9—0.98	$K'_G = 1$—$0.06 (K'_N = 1.05$—$0.06)$
119	Tuck67	,,	←	,,	—	0.85—1.0	$K'_G = 0.9$—$0.06 (K'_N = 1.0$—$0.06)$
120	Tuck67	,,	←	,,	—	0.85—0.98	$K'_G = 0.7$—$0.07 (K'_N = 0.8$—$0.07)$
121	Tuck67	,,	←	,,	—	0.85—0.99	$K'_G = 0.3$—$0.06 (K'_N = 0.3$—$0.06)$
122	Dolc68	,,	Na↓	0.01	—	0.79	$K'_N = 0.44$
123	Dolc68	,,	Na↓	,,	—	0.76	$K'_N = 0.36$
124	Dolc68	,,	Na↓	,,	—	0.76	$K'_N = 0.37$
125	Unit69	$K_s + (Ca + Mg)$	—	—	—	0.55—0.98	$K'_G = 0.30 (K'_N = 0.4$—$0.3)$
126	Andr70	$K_s + Ca$	↓	0.1	22	0.34—0.78	$K'_N = 0.71$—0.39
127	Andr70	,,	↓	,,	,,	0.41—0.76	$K'_N = 0.87$—0.37
128	Andr70	,,	↓	,,	,,	0.60—0.84	$K'_N = 1.56$—0.60
129	Cars72	,,	←	0.01	—	0.62—0.94	$K'_G = 0.21$—$0.03 (K'_N = 0.27$—$0.03)$
130	Cars72	,,	←	,,	—	0.66—0.97	$K'_G = 0.25$—$0.05 (K'_N = 0.31$—$0.05)$
131	Cars72	,,	←	,,	—	0.72—0.99	$K'_G = 0.36$—$0.14 (K'_N = 0.42$—$0.14)$
132	Cars72	,,	←	,,	—	0.65—0.94	$K'_G = 0.25$—$0.03 (K'_N = 0.31$—$0.03)$
133	Cars72	,,	←	,,	—	0.67—0.97	$K'_G = 0.26$—$0.06 (K'_N = 0.32$—$0.06)$
134	Cars72	,,	←	,,	—	0.71—0.98	$K'_G = 0.33$—$0.12 (K'_N = 0.40$—$0.12)$
135	Cars72	,,	←	,,	—	0.67—0.92	$K'_G = 0.26$—$0.02 (K'_N = 0.31$—$0.02)$
136	Cars72	,,	←	,,	—	0.70—0.98	$K'_G = 0.30$—$0.07 (K'_N = 0.36$—$0.07)$
137	Cars72	,,	←	,,	—	0.72—0.98	$K'_G = 0.33$—$0.10 (K'_N = 0.39$—$0.10)$
138	Cars72	,,	←	,,	—	0.56—0.94	$K'_G = 0.15$—$0.03 (K'_N = 0.20$—$0.03)$
139	Cars72	,,	←	,,	—	0.71—0.97	$K'_G = 0.32$—$0.06 (K'_N = 0.38$—$0.06)$
140	Cars72	,,	←	,,	—	0.68—0.98	$K'_G = 0.30$—$0.08 (K'_N = 0.37$—$0.08)$
141	Cars72	,,	←	,,	—	0.67—0.90	$K'_G = 0.22$—$0.02 (K'_N = 0.27$—$0.02)$
142	Cars72	,,	←	,,	—	0.67—0.97	$K'_G = 0.26$—$0.06 (K'_N = 0.32$—$0.06)$
143	Cars72	,,	←	,,	—	0.73—0.98	$K'_G = 0.37$—$0.10 (K'_N = 0.43$—$0.10)$
144	Cars72	,,	←	,,	—	0.76—0.96	$K'_G = 0.43$—$0.05 (K'_N = 0.49$—$0.05)$
145	Cars72	,,	←	,,	—	0.77—0.99	$K'_G = 0.45$—$0.18 (K'_N = 0.52$—$0.18)$
146	Knib72	,,	←	± 0.055	—	0.75—0.98	$K'_G = 0.20$—$0.05 (K'_N = 0.23$—$0.05)$
147	Knib72	,,	←	,,	—	0.77—0.98	$K'_G = 0.22$—$0.05 (K'_N = 0.25$—$0.05)$
148	Knib72	,,	←	,,	—	0.80—0.98	$K'_G = 0.25$—$0.05 (K'_N = 0.28$—$0.05)$

No.	Method	Material*[1] Name; fraction; composition
99	batch d.m.	Nason C, Virginia; 2—0.2 μ; — Fe, V46
100	,,	Nason A, Virginia; 2—0.2 μ; — Fe, V60, hydroxy Al-interlayers
101	batch	Lower Greensand soil; whole soil; < 2 μ: M, O.C. 1.7
102	,,	Tedburn, whole soil;
103	,,	Bovey basin; whole soil;
104	,,	Cegin; whole soil;
105	,,	Windsor; whole soil;
106	,,	Dunkeswick; whole soil;
107	,,	Sherborne; whole soil; < 2 μ: M, V
108	,,	Long Load; whole soil; < 2 μ: chloritized V40
109	,,	Denchworth; whole soil; < 2 μ: V20
110	,,	Harwell; whole soil; < 2 μ: M65, Clin
111	,,	Newchurch, whole soil; < 2 μ: M35
112	,,	Parabraunerde A, Germ.; whole soil; O.C. 20, < 2 μ: I + expand. clays
113	,,	Parabraunerde C, Germ.; whole soil; O.C. 0.7, < 2 μ: I + expand. clays
114	,,	Lehmrendzina A/B, Germ.; whole soil; O.C. 5.5, < 2 μ: M
115	,,	Braunerde B/C, Germ.; whole soil; O.C. 0.2, < 2 μ: K
116	,,	Carbonaatniedermoor, Germ.; whole soil; O.C. 34.5, < 2 μ: expand. clays
117	,,	Willalooka; < 2 μ; pH 8.0
118	,,	Belalie; < 2 μ; pH 8.2
119	,,	McLaren Vale; < 2 μ; pH 8.0
120	,,	Urrbrae; < 2 μ; pH 8.0
121	,,	Claremont; < 2 μ; pH 8.2
122	batch d.m.	Harpster B3; 0—0.2 μ; < 2 μ: M43, V12, Mi13, K10, Cl14, — O.C., — Fe
123	,,	Harpster B3; 2—0.2 μ;
124	,,	Triangle B2tb; < 5 μ; < 2 μ: M29, V29, Mi13, K18, A13, — O.C., — Fe, — $CaCO_3$
125	,,	59 soil samples
126	,,	Alluvial soil Rhine, France; whole soil; — $CaCO_3$
127	,,	Löss soil, France; whole soil; — $CaCO_3$
128	,,	Alluvial soil, France; whole soil; — $CaCO_3$
129	,,	Beaumont A11, Texas*[4]; 2—0.2 μ; K38, M21, Mi14, V9, Q18
130	,,	Beaumont A11, Texas*[4]; 0.2—0.08 μ; M52, K30, Mi9, V8
131	,,	Beaumont A11, Texas*[4]; < 0.08 μ; M71, K12, V12
132	,,	Houston Black A11, Texas*[4]; 2—0.2 μ; M33, K16, V14, Mi11, Q25
133	,,	Houston Black A11, Texas*[4]; 0.2—0.08 μ; M71, K12, V9, Mi8
134	,,	Houston Black A11, Texas*[4]; < 0.08 μ; M90, V7, Mi7
135	,,	Lake Charles A11, Texas*[4]; 2—0.2 μ; K18, M13, V12, Mi12, Q47
136	,,	Lake Charles A11, Texas*[4]; 0.2—0.08 μ; M64, K18, Mi10
137	,,	Lake Charles A11, Texas*[4]; < 0.08 μ; M79, V12, Mi7, K6
138	,,	Victoria C, Texas*[4]; 2—0.2 μ; Mi13, K11, M11, V7, Q58
139	,,	Victoria C, Texas*[4]; 0.2—0.08 μ; M73, Mi13, K7, V3
140	,,	Victoria C, Texas*[4]; < 0.08 μ; M79, Mi11, K7, V5
141	,,	Wilson B2, Texas*[4]; 2—0.2 μ; K14, Mi12, M11, V8, Q51
142	,,	Wilson B2, Texas*[4]; 0.2—0.08 μ; M63, Mi12, V17, K10
143	,,	Wilson B2, Texas*[4]; < 0.08 μ; M80, V8, Mi7, K7
144	,,	Guatamala soil AC*[4]; 0.2—0.08 μ; M73, K17, V7, Mi2, Q47
145	,,	Guatamala soil AC*[4]; < 0.08 μ; M85, K8, V6, Mi2
146	,,	Houston Black Clay, Texas; < 2 μ; — O.C., — $CaCO_3$, M, Mi19
147	,,	Houston Black Clay, Texas; < 0.2 μ; — O.C., — $CaCO_3$, — Fe, — Al, — Si, M, Mi18
148	,,	Montell Clay, Texas; < 2 μ; — O.C., — $CaCO_3$, M, Mi27

TABLE 5.2 (Cont.)

No.	Ref.	Equation $A_s + B$		Exp. cond. C_0 (eq/l)	°C	N_B-range of measurements	Selectivity coefficients in range specified
149	Knib72	$K_s + Ca$	←	± 0.055	—	0.75–0.98	$K'_G = 0.20–0.05 (K'_N = 0.23–0.05)$
150	Knib72	,,	←	,,	—	0.77–0.97	$K'_G = 0.25–0.03 (K'_N = 0.28–0.03)$
151	Knib72	,,	←	,,	—	0.80–0.99	$K'_G = 0.30–0.11 (K'_N = 0.31–0.11)$
152	Knib72	,,	←	,,	—	0.78–0.98	$K'_G = 0.22–0.08 (K'_N = 0.25–0.08)$
153	Kish74	,,	←	0.01	28	0.1–1.0	$\exp(\langle \ln K_N \rangle) = 0.39$
154	Kish74	,,	←	,,	,,	0.1–1.0	$\exp(\langle \ln K_N \rangle) = 0.42$
155	Kish74	,,	←	,,	,,	0.1–1.0	$\exp(\langle \ln K_N \rangle) = 0.45$
156	Kish74	,,	←	,,	,,	0.1–1.0	$\exp(\langle \ln K_N \rangle) = 0.49$
157	Scha75	$K_s + (Ca + Mg)$	↓	0.0084	—	0.91	$K'_N = 0.07$
158	Scha75	,,	↓	,,	—	0.89	$K'_N = 0.05$
159	Scha75	,,	↓	,,	—	0.74	$K'_N = 0.02$
160	Scha75	,,	↓	,,	—	0.985	$K'_N = 0.37$
161	Scha75	,,	↓	,,	—	0.74–0.96	$K'_N = 0.02–0.16$
162	Scha75	,,	↓	0.048	—	0.76	$K'_N = 2.34$
				0.03	—	0–0.01	$K_N = 0.68$
163	Cohe62	$K_s + Sr$	→	0.15	—	0–0.01	$K_N = 1.72$
				0.40	—	0–0.01	$K_N = 2.88$
164	Schw62	$NH_{4s} + Ca$	Na↓	0.01	—	0.82	$K'_N = 0.51$
165	Schw62	,,	Na↓	,,	—	0.75	$K'_N = 0.36$
166	Schw62	,,	Na↓	,,	—	0.77	$K'_N = 0.38$
167	Schw62	,,	Na↓	,,	—	0.76	$K'_N = 0.36$
168	Schw62	,,	Na↓	,,	—	0.72	$K'_N = 0.31$
169	Schw62	,,	Na↓	,,	—	0.71	$K'_N = 0.24$
				0.013	—	0–0.01	$K_N = 0.42$
170	Cohe62	$NH_{4s} + Sr$	→	0.10	—	0–0.01	$K_N = 0.70$
				0.50	—	0–0.01	$K_N = 0.76$
171	Cole65	$Cs_s + Mg$	→	0.1	—	—	$K'_N = 0.88 \cdot 10^{-4}–0.33 \cdot 10^{-4}$
172	Cole65	,,	→	,,	—	0.11–0.59*[5]	$K'_N = 0.11 \cdot 10^{-4}–0.19 \cdot 10^{-4}$
173	Cole65	,,	←	1.0	—	0.48–0.913*[5]	$K'_N = 0.45 \cdot 10^{-3}–0.88 \cdot 10^{-3}$
174	Cole65	,,	→	1.0	—	0.49–0.924*[5]	$K'_N = 0.21 \cdot 10^{-4}–0.47 \cdot 10^{-4}$
175	Amph56	$Cs_s + Sr$	↓	0.02	20	0.1–0.7	$K'_N = 0.14–0.05$
176	Amph56	,,	↓	0.05	,,	0.05–0.75	$K'_N = 0.14–0.05$
Divalent-divalent exchange (earth-alkali ions).							
177	Kris50	$Mg_s + Ca$	←	0.001–0.05	—	—	$K_N = 1.20$
178	Levy72b	,,	↓	0.06	—	0.15–0.85	$K'_N = 1.0–1.4$
179	Levy72b	,,	↓	0.01	—	0.15–0.85	$K'_N = 1.0–1.4$
180	Levy72b	,,	↓	0.06	—	0.25–0.80	$K'_N = 0.9–0.45$
181	Levy72b	,,	↓	0.01	—	0.25–0.80	$K'_N = 0.9–0.45$
182	Levy72b	,,	↓	0.06	—	0.25–0.90	$K'_N = 1.0–0.4$
183	Levy72b	,,	↓	0.01	—	0.25–0.90	$K'_N = 1.0–0.4$
184	Kris74	,,	→	0.05	—	0.42–0.91	$\exp(\langle \ln K_N \rangle) = 1.31$
185	Kris74	,,	←	,,	—	0.42–0.91	$\exp(\langle \ln K_N \rangle) = 1.19$
186	Kris74	,,	→	,,	—	0.25–0.72	$\exp(\langle \ln K_N \rangle) = 1.11$
187	Kris74	,,	←	,,	—	0.25–0.72	$\exp(\langle \ln K_N \rangle) = 1.10$
188	Kris74	,,	→	,,	—	0.38–0.95	$\exp(\langle \ln K_N \rangle) = 1.34$
189	Kris74	,,	←	,,	—	0.38–0.95	$\exp(\langle \ln K_N \rangle) = 1.20$
190	Ghey75	,,	←	0.02–0.21	—	—	$K_G = 1.52$
191	Ghey75	,,	←	,,	—	—	$K_G = 1.32$
192	Ghey75	,,	←	,,	—	—	$K_G = 1.24$
193	Ghey75	,,	←	,,	—	—	$K_G = 1.23$
194	Kris50	$Ca_s + Sr$	→	0.001–0.05	—	—	$K_N = 1.16$
				0.003	—	0–0.01	$K_N = 1.11$
195	Cohe62	,,	→	0.029	—	0–0.01	$K_N = 1.12$
				0.25	—	0–0.01	$K_N = 1.14$
196	Khas68	,,	—	—	—	—	$K'_N = 0.77–1.26$
197	Khas68	,,	↓	0.1	—	0.003	$K'_N = 0.76$
198	Khas68	,,	↓	,,	—	0.004–0.005	$K'_N = 0.81–0.92$
199	Khas68	,,	↓	,,	—	0.005–0.006	$K'_N = 0.93–1.02$
200	Khas68	,,	↓	,,	—	0.006–0.007	$K'_N = 1.03–1.12$
201	Juo69	,,	→	,,	25	0.005–0.055	$K_N = 1.26–0.79$
202	Juo69	,,	→	,,	,,	0.005–0.055	$K_N = 1.07–0.95$
203	Juo69	,,	→	,,	,,	0.005–0.075	$K_N = 1.24–1.02$
204	Kris50	$Ca_s + Ba$	→	0.001–0.05	—	—	$K_N = 1.34$

No.	Method	Material*[1] Name; fraction; composition
149	batch d.m.	Montell Clay, Texas; < 0.2 μ; − O.C., − $CaCO_3$, − Fe, − Al, − Si, M, Mi14
150	,,	Miller Clay, Texas; < 2 μ; − O.C., − $CaCO_3$, M, Mi26
151	,,	Beaumont Clay, Texas; < 2 μ; − O.C., − $CaCO_3$, M, Mi12
152	,,	Houston Clay, Mississippi; < 2 μ; − O.C., − $CaCO_3$, M, Mi10
153	,,	Alluvial soil I, Egypt; < 2 μ; oxidized, − $CaCO_3$, − O.C., M49, K18, V(5—10), Mi6
154	,,	Alluvial soil I, Egypt; < 2 μ; reduced, − $CaCO_3$, − O.C., M49, K18, V(5—10), Mi6
155	,,	Alluvial soil II, Egypt; < 2 μ; oxidized, − $CaCO_3$ − O.C., M41, K17, Mi9, V(5—8)
156	,,	Alluvial soil II, Egypt; < 2 μ; reduced, − $CaCO_3$, − O.C., M41, K17, Mi9, V(5—8)
157	,,	Trumao, Arrayan, Chile; O.C. 4.5, A > 50, pH 4.8
158	,,	Trumao, Temuco, Chile; O.C. 2.0, A > 50, pH 4.8
159	,,	Trumao, Santa Barbara, Chile, O.C. 1.0, A > 50, pH 4.8
160	,,	Trumao, Puerto Octay, Chile; O.C. 4.4, A > 50, pH 4.8
161	,,	Trumao, Santa Barbara, Chile; O.C. 1.0, A > 50, pH 4.8 → pH 7.5
162	,,	Trumao, Santa Barbara, Chile; O.C. 1.0, A > 50, pH 7.5
163	batch	Saclay mud, France; whole soil; < 2 μ: M45, I10
164	batch d.m.	Seemarsch, Belg.; 0.2—0.08 μ: M70, I30
165	,,	Moräne (2), Schw.; 0.2—0.08 μ; I70, V10
166	,,	Mullrendzina (kro), Germ.; 0.2—0.08 μ; M70, I20, V10
167	,,	Mullrendzina (mo), Germ.; 0.2—0.08 μ; I70, V20, M10
168	,,	Braunlehm, Germ.; 0.2—0.08 μ; I50, V20, M10
169	,,	Braunerde, Germ.; 0.2—0.08 μ; Cl80, I < 20
170	batch	Saclay mud, France; whole soil; < 2 μ: M45, K45, I10
171	column	San Joaquin — Hanford soil, California; 2—5 μ; V, HB, B
172	,,	San Joaquin — Hanford soil, California, 2—5 μ, V, HB, B
173	batch	San Joaquin — Hanford soil, California, 2—5 μ, V, HB, B
174	,,	San Joaquin — Hanford soil, California, 2—5 μ, V, HB, B
175	col. d.m.	Lower Greensand, Oxon; < 40 mesh; M, I
176	,,	Lower Greensand, Oxon; < 40 mesh, M, I
177	batch	Yolo clay; < 1 μ; − O.C., M
178	batch d.m.	Sandy clay loam, Israel; whole soil; M80
179	,,	Sandy clay loam, Israel; whole soil; M80
180	,,	Grumosolic clay, Israel; whole soil; M(40—50), Ca13, K7, V
181	,,	Grumusolic clay, Israel; whole soil; M(40—50), Ca13, K7, V
182	,,	Brown red grumosolic clay, Israel; whole soil; M40—50, Ca9, K9, V
183	,,	Brown red grumosolic clay, Israel; whole soil; M40—50, Ca9, K9, V
184	col. d.m.	Red sandy loam soil, Bangalore; whole soil; O.C. 0.4, < 2 μ: K, Fe, Al
185	,,	Red sandy loam soil, Bangalore, whole soil; O.C. 0.4, < 2 μ: K, Fe, Al
186	,,	Black clay soil, Dharwar; whole soil; O.C. 0.6, < 2 μ: M
187	,,	Black clay soil, Dharwar; whole soil; O.C. 0.6, < 2 μ: M
188	,,	Laterite sandy clay loam, Mangalore; whole soil; O.C. 1.9, K, Fe, Al
189	,,	Laterite sandy clay loam, Mangalore; whole soil; O.C. 1.9, K, Fe, Al
190	batch d.m.	Mollisol, Nkheila, Morocco; whole soil; < 2 μ: I, M, K, Cl, V, Ca
191	,,	Mollisol, Nkheila, Morocco; whole soil; < 2 μ: I, K, Cl_i, V, Cl, M
192	,,	Vertisol, Gharb, Morocco; whole soil; < 2 μ: M, Cl, K, I, M_i, Cl_i
193	,,	Vertisol, Gharb, Morocco; whole soil; < 2 μ: M, Cl, K, I
194	batch	Yolo clay; < 1 μ; − O.C., M
195	,,	Saclay mud, France; whole soil; < 2 μ: M45, I45, K10
196	—	63 Indiana soils (61 topsoils + 2 subsoils)
197	batch d.m.	Houghton muck; whole soil;
198	,,	3 Indiana soils
199	,,	15 Indiana soils
200	,,	2 Indiana soils
201	batch d.m.	Houghton muck; whole soil; O.C. 49.8
202	,,	Brookston surface soil; whole soil; O.C. 7.1, < 2 μ: M
203	,,	Sidell subsoil; whole soil; O.C. 21.2, < 2 μ: V, M
204	,,	Yolo clay; < 1 μ; − O.C., M

TABLE 5.2 (Cont.)

No.	Ref.	Equation $A_s + B$		Exp. cond. C_0 (eq/l)	°C	N_B-range of measurements	Selectivity coefficients in range specified
Exchange reactions involving Al ions							
205	Sing71	K_s + Al	↓	0.01	22	0.75–0.95	K_N = 1.1–0.4
206	Sing71	,,	↓	,,	,,	0.65–0.95	K_N = 1.4–0.3
207	Sing71	,,	↓	,,	,,	0.6–0.95	K_N = 1.2–0.3
208	Sing71	,,	↓	,,	,,	0.75–0.95	K_N = 1.1–0.6
209	Sing71	,,	↓	,,	,,	0.75–0.95	K_N = 1.7–0.9
210	Sing71	,,	↓	,,	,,	0.45–0.95	K_N = 0.7–0.25
211	Sing71	,,	↓	,,	,,	0.55–0.95	K_N = 1.4–0.45
212	Sing71	,,	↓	,,	,,	0.7–0.95	K_N = 1.9–0.8
213	Sing71	,,	↓	,,	,,	0.65–0.95	K_N = 1.1–0.45
214	Coul68	Ca_s + Al	→	,,	,,	0.3–0.95	K_N = 2.7(pH 4.3–3.0)
215	Coul68	,,	→	,,	,,	0.6–0.95	K_N = 2.5(pH 4.3–3.0)
216	Coul68	,,	→	,,	,,	0.5–0.9	K_N = 2.4(pH 2.4)
217	Coul68	,,	→	,,	,,	0.7–0.9	K_N = 2.4(pH 2.4)
218	Bach74	,,	←	,,	20	0.1–0.7	K_N = 3.97
219	Bach74	,,	←	,,	,,	0.2–0.8	K_N = 3.95
220	Bach74	,,	←	,,	,,	0.2–0.8	K_N = 3.45
221	Bach74	,,	←	,,	,,	0.02–0.5	K_N = 2.35
222	Bach74	,,	←	,,	,,	0–0.5	K_N = 2.26
223	Bach74	,,	←	,,	,,	0.05–0.8	K_N = 1.98
224	Bach74	,,	←	,,	,,	0–0.7	K_N = 1.67
225	Bach74	,,	←	,,	,,	0–0.4	K_N = 1.60
226	Bach74	,,	←	,,	,,	0.02–0.3	K_N = 1.18
227	Bach74	,,	←	,,	,,	0–0.2	K_N = 0.71
228	Bach74	,,	←	,,	,,	0–0.05	K_N = 1.62
Exchange reactions comprising heavy metal ions							
229	Lage72	Ca_s + Cd	→	0.0051	—	0.0023–0.015	K'_N = 1.8–1.5
230	Lage72	,,	→	0.0472	—	0.0014–0.007	K'_N = 2.4–1.7
231	Lage72	,,	→	0.0051	—	0.0021–0.016	K'_N = 4.2–4.6
232	Lage72	,,	→	0.0472	—	0.0018–0.012	K'_N = 5.0–3.3
233	Lage72	,,	→	0.0051	—	0.0023–0.016	K'_N = 7.7–4.3
234	Lage72	,,	→	0.0472	—	0.0019–0.011	K'_N = 4.7–3.1
235	Harm77	,,	→	variable	—	$<$0.1	$K_N > 2$
236	Harm77	,,	⇌	,,	—	0.1–0.3	K_N = 2–1.4
237	Harm77	,,	⇌	,,	—	0.3–1.0	K_N = 1.4–1.0
238	Harm77	Ca_s + Zn	→	,,	—	$<$0.1	$K_N > 2$*6
239	Harm77	,,	⇌	,,	—	0.1–0.3	K_N = 2–1.5*6
240	Harm77	,,	⇌	,,	—	0.3–1.0	K_N = 1.5–1.0*6
241	Harm77	Cd_s + Zn	→	,,	—	$<$0.1	K_N = 1.4
242	Harm77	,,	⇌	,,	—	0.1–0.3	K_N = 1.4–1.2
243	Harm77	,,	⇌	,,	—	0.3–1.0	K_N = 1.2–1.0
244	Lage73	Ca_s + Pb	→	0.0058	—	0.026–0.221	K'_N = 19.9–6.4
245	Lage73	,,	→	0.0494	—	0.026–0.208	K'_N = 31.9–9.0
246	Lage73	,,	→	0.0058	—	0.023–0.164	K'_N = 16.9–10.2
247	Lage73	,,	→	0.0494	—	0.023–0.149	K'_N = 27.0–6.9
248	Lage73	,,	→	0.0058	—	0.017–0.200	K'_N = 31.7–19.3
249	Lage73	,,	→	0.0494	—	0.017–0.197	K'_N = 31.3–20.1
250	Harm77	,,	→	variable	—	$<$0.3	$K_N > 6$
251	Harm77	,,	→	,,	—	0.3–0.5	K_N = 6–5
252	Harm77	Ca_s + Cu	→	,,	—	$<$0.2	$K_N > 4$
253	Harm77	,,	→	,,	—	0.2–0.4	K_N = 4–2
254	Lage72	Al_s + Cd	→	0.0058	—	0.0008–0.0024	K'_N = 1.2–0.7
255	Lage72	,,	→	0.0586	—	0.006–0.0008	K'_N = 2.1–0.8
256	Lage72	,,	→	0.0058	—	0.0006–0.0044	K'_N = 1.1–1.0
257	Lage72	,,	→	0.0586	—	0.0002–0.0024	K'_N = 1.2–1.5
258	Lage72	,,	→	0.0058	—	0.0007–0.0044	K'_N = 1.3–1.0
259	Lage72	,,	→	0.0586	—	0.0002–0.0026	K'_N = 1.3–1.6
260	Lage73	Al_s + Pb	→	0.0057	—	0.019–0.109	K'_N = 6.1–2.3
261	Lage73	,,	→	0.0539	—	0.017–0.080	K'_N = 7.6–3.2
262	Lage73	,,	→	0.0057	—	0.017–0.062	K'_N = 3.0–1.4
263	Lage73	,,	→	0.0539	—	0.011–0.022	K'_N = 3.6–1.5
264	Lage73	,,	→	0.0057	—	0.018–0.106	K'_N = 5.2–2.0
265	Lage73	,,	→	0.0539	—	0.015–0.059	K'_N = 5.9–2.6

No.	Method	Material*[1] Name; fraction; composition
205	batch d.m.	Rengam Latosol; whole soil; O.C. 2.0; $< 2\mu$: (1:1) 80, (Cl + V) < 5
206	"	Serdang Latosol; whole soil; O.C. 1.1, $< 2\mu$: (1:1) 50—65, (M + I) 35—50, (Cl + V) < 10
207	"	Selangor Gley; whole soil; O.C. 2.0, $< 2\mu$: (1:1) 50—65, (M + I) 30—50
208	"	Kuantan Latosol; whole soil; O.C. 2.4, $< 2\mu$: (1:1) 65—80
209	"	Prang Laterite; whole soil; O.C. 1.4, $< 2\mu$: (1:1) 65—80, (Cl + V) < 10
210	"	Segamat Latosol; whole soil; O.C. 2.1, $< 2\mu$: (1:1) 65—80
211	"	Batu Anam Latosol; whole soil; O.C. 1.6, $< 2\mu$: (M + I) 50—65, (1:1) 20—30
212	"	Chemor Latosol; whole soil; O.C. 1.2, $< 2\mu$: (1:1) 80, (M + I) 10—20
213	"	Ulu Tiram Latosol; whole soil; O.C. 0.9, $< 2\mu$: (1:1) 80, (Cl + V) < 5
214	"	Park Grass soil, Rothamsted, England; whole soil; — O.C., $< 2\mu$: M23, V12
215	"	Deerpark soil, Wexford, Ireland; whole soil; — O.C., $< 2\mu$: M17, V4.7
216	"	Park Grass soil, Rothamsted, England; whole soil; — O.C., $< 2\mu$: M23, V12
217	"	Deerpark soil, Wexford, Ireland; whole soil; — O.C., $< 2\mu$: M17, V4.7
218	"	Scottish subsoils, Insch; whole soil; O.C. 2.1, $< 2\mu$: V
219	"	Scottish subsoils, Foudland; whole soil; O.C. 0.8, $< 2\mu$: I, Cl_i, V_i
220	"	Scottish subsoils, Countesswells; whole soil; O.C. 0.3, $< 2\mu$: Cl, I
221	"	Shardlow subsoil; whole soil; O.C. 1.0, $< 2\mu$: I_i, V_i, I, K
222	"	Loamy topsoils, Insch; whole soil; O.C. 5.6, $< 2\mu$: V
223	"	Loamy topsoils, Countesswells; whole soil; O.C. 3.1, $< 2\mu$: V
224	"	Loamy topsoils, Shardlow; whole soil; O.C. 1.1, I_i, V_i
225	"	Sandy topsoil, Cawdor; whole soil; O.C. 5.3, $< 2\mu$: I_i, I, K
226	"	Sandy topsoil, Kennington; whole soil; O.C. 0.4, $< 2\mu$: I_i, I
227	"	Sandy topsoil, Wareham; whole soil; O.C. 0.9, $< 2\mu$: K, I
228	"	Sphagnum peat; whole soil
229	batch	Cecil sandy loam; whole soil; $< 2\mu$: K45, V25
230	"	Cecil sandy loam; whole soil, $< 2\mu$: K45, V25
231	"	Winsum clay loam; whole soil, $< 2\mu$: I80, Q10, K5
232	"	Winsum clay loam; whole soil, $< 2\mu$: I80, Q10, K5
233	"	Yolo silt loam; whole soil; $< 2\mu$: M47, Mi22, Cl13
234	"	Yolo silt loam; whole soil; $< 2\mu$: M47, Mi22, Cl13
235	"	Winsum soil, Neth.; whole soil; O.C. 0.6, $< 2\mu$: I, Sm
236	"	" " "
237	"	" " "
238	"	" " "
239	"	" " "
240	"	" " "
241	"	" " "
242	"	" " "
243	"	" " "
244	"	Cecil sandy loam; whole soil; $< 2\mu$: K45, V25
245	"	Cecil sandy loam; whole soil; $< 2\mu$: K45, V25
246	"	Winsum clay loam; whole soil; $< 2\mu$: I80, Q10, K5
247	"	Winsum clay loam; whole soil; $< 2\mu$: I80, Q10, K5
248	"	Yolo silt loam; whole soil; $< 2\mu$: M47, Mi22, Cl13
249	"	Yolo silt loam; whole soil; $< 2\mu$: M47, Mi22, Cl13
250	"	Winsum soil, Neth.; whole soil; O.C. 0.6, $< 2\mu$: I, Sm
251	"	Winsum soil; Neth.; whole soil; O.C. 0.6, $< 2\mu$: I, Sm
252	"	Winsum soil; Neth.; whole soil; O.C. 0.6, $< 2\mu$: I, Sm
253	"	Winsum soil; Neth.; whole soil; O.C. 0.6, $< 2\mu$: I, Sm
254	"	Cecil sandy loam; whole soil; $< 2\mu$: K45, V25
255	"	Cecil sandy loam; whole soil; $< 2\mu$: K45, V25
256	"	Winsum clay loam; whole soil; $< 2\mu$: I80, Q10, K5
257	"	Winsum clay loam; whole soil; $< 2\mu$: I80, Q10, K5
258	"	Yolo silt loam; whole soil; $< 2\mu$: M47, Mi22, Cl13
259	"	Yolo silt loam; whole soil; $< 2\mu$: M47, Mi22, Cl13
260	"	Cecil sandy loam; whole soil; $< 2\mu$: K45, V25
261	"	Cecil sandy loam; whole soil; $< 2\mu$: K45, V25
262	"	Winsum clay loam; whole soil; $< 2\mu$: I80, Q10, K5
263	"	Winsum clay loam; whole soil; $< 2\mu$: I80, Q10, K5
264	"	Yolo silt loam; whole soil; $< 2\mu$: M47, Mi22, Cl13
265	"	Yolo silt loam; whole soil; $< 2\mu$: M47, Mi22, Cl13

minerals as against organic matter. Obviously it would be of help to obtain one or two experimental check points for the soil concerned, but even then constancy of K_N is probably not better than within a factor of 1.5. For homovalent exchange involving divalent cations the variability of K_N is fortunately rather small, though organic exchangers exert substantial preferences for certain heavy metals and should thus be treated carefully.

For homovalent exchange involving monovalent cations, K_N varies considerably across the lyotropic series, even for substantial coverage of the exchanger. With this variation for different cation pairs one also finds a rather variable behavior as a function of type of clay mineral, and particularly towards the trace level for cations such as K, NH_4, Rb, Cs (see below). As is shown convincingly in the Tables, the same variability is found for mono-divalent exchange, except perhaps Na/Ca which appears to be predictable to within a factor of 2.

5.4. HIGHLY SELECTIVE ADSORPTION

The interaction between cations and soil constituents described in the previous sections was particularly of type (a) mentioned in the introductory section, i.e. reversible cation exchange where different cations compete for a fixed amount of exchange sites (CEC) and the exchanger exhibits a low-to-moderate preference for one cation species as compared to the other. In this section some attention will be given to highly selective adsorption of certain ion species, viz. the adsorption of heavy metal ions and potassium ions present at trace levels in solution, and to the adsorption of strongly hydrolysable cations (in particular Al^{3+}).

5.4.1. Heavy metals

It is well known that, in soil, heavy metal ions are adsorbed highly selectively when present at trace levels in solution. This is of particular interest because, for these ions, trace levels in solution may have great significance. Originally this interest centered on the role of such ions as micronutrients, where deficiencies could cause a significant yield depression (e.g. Zn, Co, Cu). Lately the presence of heavy metals in toxic amounts has become a matter of conern. In this context a rapidly increasing amount of data has been and is being accumulated in the literature, describing the adsorption of these ions by soil and soil components.

Inasmuch as heavy metal adsorption in soil comprises many different types of adsorption sites (organic constituents, oxides, edges and planar sites on clay minerals) one should expect a decrease of the selectivity with increasing saturation level. It is thus required to describe the adsorption behavior — particularly at low concentration levels — in terms of a selectivity coefficient as applicable to a limited (*but specified*) range. Admitting that a complete description might require the use of a series of selectivity coefficients covering consecutive ranges, it often suffices to recognize two or three ranges, i.e. a first range of specified capacity, with a highly selective adsorption, sometimes a fairly selective one in the range following, plus the remainder. This last range may still comprise excess adsorption or possible precipitation, but often the corresponding concentration levels in solution exceed that of practical concern. The first (or even second) range is conveniently derived from a Langmuir plot of adsorption data, according to (cf. (4.120) and (4.121)):

$$\frac{c}{{}^{w}q} = \frac{1}{ka} + \frac{c}{a} \qquad (5.10)$$

corresponding to an adsorption equation of the type:

$${}^{w}q = a\frac{kc}{1 + kc} \qquad (5.11)$$

In these equations ${}^{w}q$ indicates the amount adsorbed per unit mass of adsorbent (often written as x/m in the relevant literature), k is a measure of the binding strength of the sites (in the literature also given as b), while a designates the amount of sites subject to the chosen k value.

Depending on whether or not the adsorption of a particular heavy metal species is believed to be in competition with the common cations present in the soil system, say Ca ions, the value of k may depend on the concentration level of this competitor. Most authors appear to ignore this aspect when specifying a particular value of k characterizing the adsorption behavior of a heavy metal ion in a given soil.

In practice this may not be too serious a shortcoming. Thus at trace level adsorption the concentration of the competitor(s) will be a given fraction of the total electrolyte level, C, irrespective of the concentration of the heavy metal. Inasmuch as C depends mainly on soil moisture content and often varies around 0.01 N at e.g. field capacity, a particular value of k could thus indeed characterize the adsorption behavior of the heavy metal under field conditions, provided it was determined at a comparable level of total electrolyte.

For homovalent exchange (e.g. heavy metal against Ca) the connection between k and C is found by expressing the exchange equation (5.2) in terms of the corresponding Langmuir equation, as was shown in section 4.2.1c of the previous chapter. This gives:

$$k = (K_N')^2/C \qquad (5.12)$$

So, if the adsorption reaction follows an exchange equation with a constant value of K_N', k should be inversely proportional to C.

TABLE 5.3
Langmuir coefficients and K'_N-values for the adsorption of heavy metal ions by some soils and soil components

No.	Ref.	Ion	Material*[1]						Exp. cond.		Langmuir coeff.		K'_N	a/CEC
			Name	% clay	% O.M.	% $CaCO_3$	CEC (meq/100g)	pH	C_0 (eq/l)	pH	a (meq/100g)	^wk (l/mg)		
1	Shum75	Zn	Decatur cl, A, Georgia	28.1	2.37	—	12.43	5.41	—	—	1.32	7.94	—	0.11
											7.84	0.031	—	0.63
2	Shum75	Zn	Decatur, B2t, ,,	48.3	0.10	—	9.85	5.71	—	—	1.39	4.09	—	0.14
											8.01	0.027	—	0.81
3	Shum75	Zn	Cecil sl, A, ,,	7.6	1.60	—	4.44	6.67	—	—	1.32	6.04	—	0.30
											4.50	0.032	—	1.01
4	Shum75	Zn	Cecil sl, B2t, ,,	36.0	0.07	—	5.98	5.19	—	—	1.11	1.98	—	0.19
											5.39	0.013	—	0.90
5	Shum75	Zn	Norfolk ls, A, ,,	5.1	0.92	—	3.20	5.45	—	—	0.89	1.06	—	0.28
											2.19	0.065	—	0.68
6	Shum75	Zn	Norfolk ls, B2t, ,,	40.4	0.12	—	5.32	5.79	—	—	1.17	4.22	—	0.32
											3.60	0.041	—	0.68
7	Shum75	Zn	Leefield ls, A, ,,	2.4	1.14	—	3.13	5.27	—	—	0.74	1.70	—	0.24
											2.25	0.051	—	0.72
8	Shum75	Zn	Leefield ls, B2t, ,,	14.7	0.03	—	2.62	5.42	—	—	1.05	0.84	—	0.40
											1.57	0.184	—	0.60
9	Harm77	Zn	Winsum, Netherlands (< 2 μ: I, Sm)*[1]	50	0.6	—	20	7.0	—	—	1.6	—	13.6	0.08
10	Till69	Co	Prairie soil (F3) (< 2 μ: M)	41	4.1	—	40	6.0	0.1 $CaCl_2$	—	0.11	8.0	153	0.003
11	Till69	Co	,, (F4) (< 2 μ: M)	26	4.3	—	30	5.3	,,	—	0.6	1.12	57.5	0.002
12	Till69	Co	,, (F5) (< 2 μ: M, F)	42	4.0	—	44	5.9	,,	—	0.10	2.43	84.7	0.002
13	Till69	Co	,, (F6) (< 2 μ: M)	39	8.3	—	59	5.6	,,	—	0.10	2.16	79.8	0.002
14	Till69	Co	,, (F7) (< 2 μ: M)	36	4.8	—	43	5.6	,,	—	0.07	1.67	70.2	0.002
15	Till69	Co	Gleyed podzol (F8) (< 2 μ: I, Q, K)	26	3.1	—	14	5.2	,,	—	0.02	1.19	59.3	0.001
16	Till69	Co	,, (F9) (< 2 μ: Q)	18	3.5	—	12	5.0	,,	—	0.02	0.96	53.2	0.002
17	Till69	Co	,, (F10) (< 2 μ: Q)	9.2	2.4	—	8.8	5.3	,,	—	0.01	1.47	65.9	0.001
18	Till69	Co	,, (F11) (< 2 μ:Q)	19	2.6	—	12	5.0	,,	—	0.03	0.70	45.5	0.003
19	Till69	Co	,, (F12) (< 2 μ:Q)	11	2.7	—	11	4.6	,,	—	< 0.01	1.40	64.3	< 0.001
20	Till69	Co	Humus podzol (F13) (< 2 μ: Q)	4.4	2.9	—	11	4.6	,,	—	< 0.01	4.71	117.9	< 0.001
21	Till69	Co	,, (F14) (< 2 μ: Q)	4.9	3.5	—	12	4.0	,,	—	< 0.01	1.64	69.6	< 0.001
22	Till69	Co	,, (F15) (< 2 μ: Q)	4.2	2.9	—	10	6.9	,,	—	0.05	6.06	133.7	0.005
23	Till69	Co	,, (F16) (< 2 μ: Q)	4.6	3.5	—	14	4.0	,,	—	0.01	1.57	68.1	0.001
24	Till69	Co	,, (F17) (< 2 μ: Q)	5.2	3.7	—	14	3.9	,,	—	0.01	0.99	54.0	0.001

25	Till69	Co	Yellow podzolic soil (F18) (< 2 μ: Q)	16	3.8	—	15	4.5	0.1 $CaCl_2$	—	< 0.01	1.00	54.3	< 0.001
26	Till69	Co	,, (F19) (< 2 μ: Q, M)	29	2.8	—	16	4.9	,,	—	0.03	1.88	74.5	0.002
27	Till69	Co	,, (F20) (< 2 μ: Q, I)	33	3.7	—	19	4.3	,,	—	< 0.01	1.92	75.4	< 0.001
28	Till69	Co	,, (F21) (< 2 μ: Q, I)	29	4.5	—	22	5.2	,,	—	0.04	2.39	84.0	0.002
29	Till69	Co	,, (F22) (< 2 μ: Q)	18	3.9	—	14	4.2	,,	—	0.01	0.43	35.6	0.001
30	Till69	Co	Krasnozem (F23) (< 2 μ: V, C, K)	29	11	—	32	4.3	,	—	0.01	0.67	44.5	< 0.001
31	Till69	Co	Krasnozem (F24) (< 2 μ: V—C, K)	43	9.1	—	36	4.7	,,	—	0.04	1.33	62.6	0.001
32	Till69	Co	,, (F25) (< 2 μ: K, H)	38	8.3	—	34	5.2	,,	—	0.08	1.86	74.1	0.002
33	Till69	Co	,, (F26) (< 2 μ: K, H)	36	5.7	—	36	5.4	,,	—	0.12	3.94	107.8	0.003
34	Till69	Co	,, (F27) (< 2 μ: K)	26	5.8	—	24	4.8	,,	—	0.04	1.90	74.9	0.002
35	DuPl71	Cu	Fersiallitische-grond, Nelspruit, < 2 μ	—	—	—	—	—	,,	6.7	3.50	2.14	82.4	—
36	DuPl71		,, ,, — O.M.	—	—	—	—	—	,,	6.7	1.26 2.57	50.0 4.09	398 114	— —
37	DuPl71	Cu	,, ,, 2—50 μ	—	—	—	—	—	,,	6.2	1.34 0.27	9.4 4.06	173 114	— —
38	DuPl71	Cu	Ferrallitiese-grond, Witrivier, < 2 μ	—	—	—	—	—	,,	6.2	0.14 3.09	28.6 6.99	301 149	— —
39	DuPl71	Cu	,, ,, , < 2 μ, — O.M.	—	—	—	—	—	,,	6.6	1.10 2.85	360 3.81	1069 110	— —
40	DuPl71	Cu	,, ,, 2—50 μ	—	—	—	—	—	,,	6.0	0.94 1.26	167 6.76	728 147	— —
41	DuPl71	Cu	Mg-kaolinite	—	—	—	—	—	0.1 $MgCl_2$	6.4	0.35	1.27	63.5	—
42	DuPl71	Cu	Ca-kaolinite	—	—	—	—	—	0.1 $CaCl_2$	6.3	0.11 0.77	13.7 0.35	209 33.3	— —
43	DuPl71	Cu	Ba-kaolinite	—	—	—	—	—	0.1 $BaCl_2$	5.9	0.09 0.51	36.4 1.28	340 63.8	— —
44	DuPl71	Cu	Ca-montmorillonite, Clay Spur, Wyoming	—	—	—	—	—	0.1 $CaCl_2$	4.9	0.64	2.28	85.1	—
45	DupL71		Ca-montmorillonite, Clay Spur, Wyoming	—	—	—	—	—	,, ,,	7.1 7.1	0.33 0.93 0.32	38.6 51.92 3980	350 406 3555	— — —
46	McLa73	Cu	Astley Hall (Soil no. 1)	15.2	1.9	—	8.28	5.0	,,	5.5	1.83	1.06	580	0.22
47	McLa73	Cu	Fladbury (Soil no. 2)	60.2	9.1	—	13.80	6.1	,,	5.5	10.93	0.30	30.9	0.79
48	McLa73	Cu	Worcester (Soil no. 3)	37.5	0.4	—	11.87	4.9	,,	5.5	1.07	0.96	55.2	0.09
49	McLa73	Cu	Newport (Soil no. 4)	9.7	2.0	—	4.13	3.2	,,	5.5	2.65	0.71	47.5	0.64
50	McLa73	Cu	Blackwood (Soil no. 5)	9.8	2.8	—	11.46	5.8	,,	5.5	3.15	1.39	66.4	0.27
51	McLa73	Cu	Gilberdyke (Soil no. 6)	14.4	7.2	—	21.17	5.9	,,	5.5	5.13	1.47	68.3	0.24
52	McLa73	Cu	Banbury (Soil no. 7)	28.0	2.4	—	14.02	6.4	,,	5.5	3.94	0.83	51.3	0.28
53	McLa73	Cu	Gilberdyke (Soil no. 9)	6.7	9.1	—	28.03	5.9	,,	5.5	7.75	1.93	78.3	0.28
54	McLa73	Cu	Gilberdyke (Soil no. 10)	3.8	7.2	—	23.91	5.1	,,	5.5	7.43	1.65	72.4	0.31

TABLE 5.3. (Cont.).

No.	Ref.	Ion	Material*1 Name	% clay	% O.M.	% $CaCO_3$	CEC (meq/ 100g)	pH	Exp. cond. C_0 eq/l	Exp. cond. pH	Langmuir coeff. a (meq/ 100g)	Langmuir coeff. ${}^{w}k$ (l/mg)	K'_N	a/CEC
55	McLa73	Cu	Surface Water Gley (Soil no. 11)	6.4	1.4	—	7.91	6.2	0.1 $CaCl_2$	5.5	3.46	0.53	41.0	0.44
56	McLa73	Cu	Blackwood (Soil no. 12)	5.6	1.9	—	9.13	6.0	”	5.5	4.25	0.64	45.1	0.47
57	McLa73	Cu	Gilberdyke (Soil no. 13)	15.6	26.2	—	51.34	5.4	”	5.5	18.20	1.23	62.5	0.35
58	McLa73	Cu	Blackwood (Soil no. 14)	5.4	1.1	—	12.88	7.1	”	5.5	4.31	0.90	53.5	0.33
59	McLa73	Cu	Blackwood (Soil no. 15)	7.2	2.2	—	9.76	5.9	”	5.5	5.10	0.55	41.8	0.52
60	McLa73	Cu	Gilberdyke (Soil no. 16)	7.3	7.5	—	23.31	6.4	”	5.5	7.40	1.61	71.5	0.32
61	McLa73	Cu	Nordrach (Soil no. 17)	20.5	3.9	—	21.92	7.0	”	5.5	5.76	1.51	69.2	0.26
62	McLa73	Cu	Tynings (Soil no. 18)	23.5	2.9	—	13.76	5.3	”	5.5	7.69	0.46	38.2	0.56
63	McLa73	Cu	Maesbury (Soil no. 19)	19.1	5.6	—	22.49	5.7	”	5.5	6.83	1.20	61.7	0.30
64	McLa73	Cu	Ellick (Soil no. 20)	27.1	6.6	—	20.03	5.4	”	5.5	9.45	0.80	50.4	0.47
65	McLa73	Cu	Wrington (Soil no. 21)	27.4	3.6	—	19.18	6.4	”	5.5	7.69	0.65	45.4	0.40
66	McLa73	Cu	Spetchley (Soil no. 22)	50.5	3.4	—	33.39	6.8	”	5.5	6.17	1.05	57.7	0.18
67	McLa73	Cu	Chelwood (Soil no. 23)	12.5	2.2	—	12.88	6.8	”	5.5	4.28	0.58	42.9	0.33
68	McLa73	Cu	Allerton (Soil no. 24)	45.8	7.6	—	35.59	5.6	”	5.5	8.25	1.31	64.5	0.23
69	McLa73	Cu	Kaolinite, API 49—4, Georgia	—	—	—	—	—	”	5.5	0.38	0.47	38.6	—
70	McLa73	Cu	Halloysite, API 49—12, Indiana	—	—	—	—	—	”	5.5	2.55	0.42	36.5	—
71	McLa73	Cu	Montmorillonite, API 49—26, Wyoming	—	—	—	—	—	”	5.5	1.17	1.01	56.6	—
72	McLa73	Cu	Illite, API 49—36, Illinois	—	—	—	—	—	”	5.5	1.67	0.54	41.4	—
73	McLa73	Cu	Iron oxide	—	—	—	—	—	”	5.5	25.23	1.13	59.9	—
74	McLa73	Cu	Manganese oxide	—	—	—	—	—	”	5.5	215.12	3.71	108	—
75	McLa73	Cu	Ferro-manganese concretions	—	—	—	—	—	”	5.5	28.35	0.67	46.1	—
76	McLa73	Cu	Organic matter	—	—	—	—	—	”	5.5	36.91	1.49	68.8	—
77	Sold76	Pb	Casciana T., Tuscany (Soil no. 1)	6.8	7.8	25.7	32.5	7.7	—	—	34.9	—	—	1.07
78	Riff76	Pb	” ” ”	6.8	7.8	25.7	32.5	7.7	0.1 $CaCl_2$	—	22.3	0.25	50.7	0.68
79	Sold76	Pb	Parlascio, Tuscany (Soil no. 2)	4.0	0.9	12.4	8.8	8.2	—	—	8.3	—	—	0.97
80	Riff76	Pb	” ” ”	4.0	0.9	12.4	8.8	8.2	0.1 $CaCl_2$	—	3.3	0.36	61.3	0.37
81	Sold76	Pb	Colonnata, ” (Soil no. 3)	6.6	4.2	11.8	20.0	7.9	—	—	21.7	—	—	1.22
82	Riff76	Pb	” ” ”	6.6	4.2	11.8	20.0	7.9	0.1 $CaCl_2$	—	14.0	0.19	43.8	0.70
83	Sold76	Pb	Fornacette, ” (Soil no. 4)	7.6	2.1	5.3	16.2	8.3	—	—	15.6	—	—	0.96
84	Riff76	Pb	” ” ”	7.6	2.1	5.3	16.2	8.3	0.1 $CaCl_2$	—	8.6	0.10	31.5	0.53

85	Sold76	Pb	Bientina I, ,, (Soil no. 5)	23.1	4.4	5.8	31.2	8.1	—	—	47.7	—	—	1.55
86	Riff76	Pb	,, ,, ,,	23.1	4.4	5.8	31.2	8.1	0.1 $CaCl_2$	—	23.8	0.46	69.3	0.76
87	Sold76	Pb	Bientina II, ,, (Soil no. 6)	14.3	4.2	0.5	30.0	7.7	—	—	30.9	—	—	1.06
88	Riff76	Pb	,, ,, ,,	14.3	4.2	0.5	30.0	7.7	0.1 $CaCl_2$	—	15.9	0.26	51.6	0.53
89	Sold76	Pb	Subbiano, ,, (Soil no. 7)	10.7	2.0	1.2	16.5	8.6	—	—	16.8	—	—	1.01
90	Riff76	Pb	,, ,, ,,	10.7	2.0	1.2	16.5	8.6	0.1 $CaCl_2$	—	5.4	0.56	76.1	0.33
91	Sold76	Pb	Calci, Tuscany (Soil no. 8)	5.4	1.0	0.0	8.7	5.7	—	—	4.1	—	—	0.51
92	Riff76	Pb	,, ,, ,,	5.4	1.0	0.0	8.7	5.7	0.1 $CaCl_2$	—	1.5	0.13	36.1	0.17
93	Sold76	Pb	Cenaia, ,, (Soil no. 9)	12.0	3.1	15.0	16.2	8.1	—	—	21.3	—	—	1.32
94	Riff76	Pb	,, ,, ,,	12.0	3.1	15.0	16.2	8.1	0.1 $CaCl_2$	—	11.5	0.15	39.4	0.71
95	Sold76	Pb	Lavaiano, Tuscany (Soil no. 10)	3.8	1.2	0.0	14.0	5.8	—	—	9.7	—	—	0.71
96	Riff76	Pb	,, ,, ,,	3.8	1.2	0.0	14.0	5.8	0.1 $CaCl_2$	—	4.2	0.11	33.0	0.32
97	Sold76	Pb	Volterra, ,, (Soil no. 11)	7.0	1.8	10.6	23.7	7.6	—	—	33.3	—	—	1.43
98	Riff 76	Pb	,, ,, ,,	7.0	1.8	10.6	23.7	7.6	0.1 $CaCl_2$	—	11.2	0.17	42.1	0.47
99	Sold76	Pb	Fosdinovo; Tuscany (Soil no. 12)	1.7	1.4	41.6	8.7	8.1	—	—	17.1	—	—	2.15
100	Riff76	Pb	,, ,, ,,	1.7	1.4	41.6	8.7	8.1	0.1 $CaCl_2$	—	6.2	0.30	56.0	0.71
101	Riff75	Cd	Humic acid from a podzol	—	100	—	—	—	—	—	11.20	0.004	—	—
102	Riff75	Cd	Humic acid from a rendzina	—	100	—	—	—	—	—	18.73	0.015	—	—
103	Riff75	Cd	Humic acid from a brown Mediterranean soil	—	100	—	—	—	—	—	17.24	0.010	—	—
104	Levi76	Cd	Casciana T., Tuscany	7.0	7.8	25.7	32.5	7.7	—	—	18.1	0.22	—	0.56
105	Levi76	Cd	M. Spazzavento, Tuscany	16.1	2.7	0.8	27.5	7.9	—	—	15.7	0.07	—	0.57
106	Levi76	Cd	Parlascio, Tuscany	4.0	0.9	12.4	8.8	8.2	—	—	5.4	0.06	—	0.61
107	Levi76	Cd	Colonnata, Tuscany	6.8	4.2	11.8	20.0	7.9	—	—	13.8	0.16	—	0.69
108	Levi76	Cd	Torrenieri, Tuscany	1.8	1.7	18.0	15.0	8.1	—	—	7.5	0.13	—	0.50
109	Levi76	Cd	Santo Pietro, Tuscany	34.7	1.0	5.6	20.0	8.6	—	—	13.7	0.06	—	0.69
110	Levi76	Cd	Navacchio, Tuscany	10.9	3.1	8.0	17.5	8.5	—	—	9.4	0.14	—	0.54
111	Levi76	Cd	Fornacette, Tuscany	7.7	2.1	5.3	16.2	8.3	—	—	8.1	0.14	—	0.50
112	Levi76	Cd	Bientina I	24.9	4.5	5.8	31.2	8.1	—	—	18.9	0.21	—	0.61
113	Levi76	Cd	Bientina II	14.9	4.2	0.5	30.0	7.7	—	—	17.8	0.11	—	0.59
114	Garc76	Cd	Montmorillonite, Upton, Wyoming	—	—	—	—	—	0.0 $NaClO_4$	—	0.19	315	—	—
115	Garc76	Cd	,, ,, ,,	—	—	—	—	—	0.01 ,,	—	0.17	223	0.05	—
116	Garc76	Cd	,, ,, ,,	—	—	—	—	—	0.03 ,,	—	0.16	65	0.08	—
117	Garc76	Cd	,, ,, ,,	—	—	—	—	—	0.05 ,,	—	0.22	11	0.06	—
118	Garc76	Cd	,, ,, ,,	—	—	—	—	—	0.17 ,,	—	0.13	15	0.22	—
119	Garc76	Cd	,, ,, ,,	—	—	—	—	—	1.0 ,,	—	0.05	56	2.51	—

*[1] Dominant minerals in the fraction $< 2\,\mu$: F = feldspars, H = halloysite, I = illite, K = kaolinite, M = montmorillonite, Q = quartz, Sm = smectite, V—C = vermiculite-chlorite interstratified — O.M. indicates "organic matter removed".

For heterovalent exchange (i.e. divalent heavy metal against a monovalent competitor) one finds for $N_{2+} \leqslant 0.1$ (cf. Harmsen, 1977):

$$k = (K_N')^2/2C^2 \tag{5.13}$$

Obviously the above translations of k in terms of K_N' and C covers only those adsorption sites that are occupied in competition with other (dominant) cations. In those cases where, for example, heavy metal ions occupy additional sites (cf. excess adsorption) the influence of C on k will be minor. Against this background the experimentally observed variation of k with C should be regarded as an indication that the first type of adsorption is also involved (see reactions 114—119 in Table 5.3).

Table 5.3 gives the values of *a* and k as reported in the literature, covering a number of heavy metal species adsorbed on different soils or soil constituents. The second parameter is actually presented in units of l/mg and is indicated as ${}^{w}k(=2k/10^3\ M$, with M = molecular weight of heavy metal). Using (5.12) and (5.13), these values for ${}^{w}k$ are also expressed in terms of the corresponding value K_N' as given in Tables 5.1 and 5.2, if at least the total electrolyte level was given in the relevant publication.

As might be expected for highly selective adsorption at low coverage, the values of K_N' given in Table 5.3 vary greatly with ion species, with the range covered, and with the nature of the adsorbent. Thus an extensive set of data would be necessary to give a complete description of differences in selectivity between cations when heavy metal ions are involved. Nevertheless some general observations can be given. The K_N' values for the adsorption reactions of Zn, Cu, Co and Pb ions in calcium-saturated soils and soil components vary from 40 to 100 at low saturation ($<10\%$) with the heavy metal ions. Also the K_N' values for Ca-Pb exchange (Table 5.2, nos. 244—249) are within this range. A comparison with the K_N' values for the mutual exchange reactions between alkaline earth ions given in Table 5.1 (reactions 234—264) and Table 5.2 (nos. 177—205) shows that these values vary from 0.5 to 1.5. This indicates that the K_N' values for the adsorption reaction of the heavy metal ions in soils saturated with alkaline earth metals will be of the order of 40—100 as long as the heavy metal ions occupy not more than roughly 10% of the adsorption complex. The K_N' values for the adsorption of Cd ions in soils saturated with alkaline earth metals are in the range from 1 to 10 (Table 5.2, nos. 229—237), which is in accordance with the relatively low ${}^{w}q_{Cd}$ values of Table 5.3, nos. 101—119. Table 5.3 shows that K_N' generally varies with the degree of saturation of the heavy metal ion. An increase of this degree of saturation causes a decrease of K_N'. At high saturation with heavy metal ions the preference will be in the same order as the preference for alkaline earth ions (Table 5.1, nos. 320—322, 327—333, 336—350).

It is interesting to note that in the single case reported in Table 5.3 that covers more than one level of total electrolyte (nos. 114—119), K_N' appears to be fairly constant in the range of 0.01—0.05 N, in contrast to the strongly varying value of ${}^{w}k$ in that range. This may point to the need for more information at electrolyte levels akin to those expected under field conditions. In fact if K'_N were roughly constant, this would indicate that the value of ^wk applicable to an electrolyte level of 0.01 N might possibly be ten-times smaller than many values given. Finally it is pointed out that the ionic preference factor N_B/f_B as defined in (5.8) equals $(K'_N)^2$ in the present divalent systems. Thus a K'_N value of around 100 (e.g. Table 5.3, reaction no 42) indicates that N_B/f_B approaches 10^4 at very low levels of cobalt in solution.

As is done for the alkali and alkaline earth ions, some papers mention an order of preference for the heavy metal ions based on the results of the specific adsorption of different heavy metal ion species as measured in the same soils. It is difficult to generalize about these results because, as was mentioned above, the specific adsorption depends strongly on the type of adsorbent used. This is also the reason why in the literature so much attention is given to the influence of the composition of the soil on the specific adsorption. The factors investigated most frequently in this connection are the organic matter and clay content, the CEC, the specific surface area, the presence of oxides, and the pH. In order to establish the contribution of each of these factors to the "capacity" for specific adsorption (cf. the parameter a in (5.11)), multiple regression analyses have been carried out. The results of some of these analyses are given below.

(a) Adsorption of Cu ions in 24 soils (McLaren and Crawford, 1973; see Table 5.3, nos. 46—68), at pH 5, as a function of the percentage of organic matter, X_1, and the percentage of manganese oxide, X_2:

$$a_{Cu}\,(\mu g\,g^{-1}) = 759 + 198X_1 + 5820X_2 \quad (r = 0.93)$$

(b) Adsorption of Cd ions in 10 soils (Levi-Minzi et al., 1976; see Table 5.3, nos. 104—113) at 25°C as a function of the percentage of organic matter, X_1, and the CEC:

$$a_{Cd}\,(\text{meq per 100 g}) = -0.41 - 0.20X_1 + 0.63\,\text{CEC} \quad (r = 0.968)$$

(c) Adsorption of Zn ions in 10 soils (Udo et al., 1970) as a function of the carbonate equivalent and the percentage of organic matter, X_1:

$$a_{Zn} = 0.15 + 0.070\,(\text{carb. eq.}) + 0.88X_1$$

(d) Adsorption of Pb ions in 12 soils (Riffaldi et al., 1976); see Table 5.3, nos. 80—98) as a function of the percentage of organic matter, X_1, and the clay content:

$$a_{Pb}\,(\text{meq per 100 g}) = -0.81 + 2.65X_1 + 0.46\,(\text{clay content})$$

Apart from this type of analysis many studies have been undertaken to determine the adsorption behavior of the individual soil components. In these studies attention was paid to the differences in the preferences for different heavy metal ions and to the nature of the adsorption mechanisms.

The adsorption behavior of organic matter has been the subject of many investigations. Many authors have quantified the "binding strength" in terms of organometallic stability constants (Mangaroo et al., 1965; Randhawa and Broadbent, 1965; Schnitzer and Skinner, 1965, 1966, 1967; El-Sayed et al., 1970a, b; Ardakani and Stevenson, 1972; Stevenson et al., 1973; Stevenson, 1976). Szalay and Szilagyi (1967) and Szalay (1969) have calculated a so-called geological enrichment factor (G.E.F. value), which is a distribution ratio at equilibibrium between the amount of heavy metal ions adsorbed (on humic acid) and the amount present in an aqueous solution. Khan (1969) measured a decrease of the pH in solution of humic acids isolated from soils, after the latter had reacted with heavy metal ions. The magnitude of the decrease in pH is interpreted by this author as a qualitative measure of the complex formation. Verloo and Cottenie (1972) defined a stability index for the metal-organic matter complexes, as the ratio between the amount of the element present as soluble humate and the sum of the amounts precipitated as hydroxide and taken up by a resinous exchanger. Bergseth and Stuanes (1976) determined adsorption isotherms of heavy metals on organic matter. Schnitzer and Skinner (1965) and Davies et al. (1969) investigated the specific adsorption of heavy metals by certain "pretreated" organic materials. Such pretreatment was then meant to bloc particular functional groups (as e.g. carboxyl, hydroxyl, and quinone groups). These investigations could provide an insight into the relative contribution of such groups to the specific adsorption.

Generalization of these research findings is difficult, as the composition of soil organic matter varies with time, and between different locations. Apart from this, it is often impossible to make a quantitative comparison between the results reported by different authors, because they use different calculation methods. Owing to these circumstances, the summary in Table 5.4 gives only a qualitative rating of the relative binding strengths of certain organic materials for different metal ions. The order of preference as shown in Table 5.4 shows little consistency. This is hardly suprising as not only the organic materials are different but also the experimental conditions vary between the systems studied. In this context the effects of pH (Schnitzer and Skinner, 1966, 1967; Verloo and Cottenie, 1972) and of the electrolyte level (Stevenson, 1976) are mentioned.

Finally it should be mentioned that the organic constituents in soil profoundly influence the "mobility" of heavy metals in soil, which is of such pre-eminent importance in relation to pollution phenomena and deficiency symptoms in crops. It should thus be stressed that this mobility depends both on the binding strength between ion and soil components *and* on the mobility of the complex thus formed. The latter is determined mainly by the "size" of the organic constituents involved: strong binding of a heavy metal to low molecular organic material should increase rather than decrease its mobility in soil.

TABLE 5.4

Specific adsorption of metal ions onto organic matter

Ref.	pH	Material	Increasing complex stability c.q. binding strength
Schn66, 67	3.5	F.A.*1	$Mg^{2+} < Mn^{2+} < Zn^{2+} < Ca^{2+} < Co^{2+} < Pb^{2+} < Ni^{2+} < Fe^{2+} < Cu^{2+}$
Schn66, 67	5.0	"	$Mg^{2+} < Zn^{2+} < Ca^{2+} < Co^{2+} \simeq Mn^{2+} < Ni^{2+} < Fe^{2+} < Pb^{2+} < Cu^{2+}$
Szal69	4—4.6	H.A.*2	$Ni^{2+} < Fe^{2+} < Cu^{2+} < Co^{2+}$
Szal69	5.0	"	$Mn^{2+} < Zn^{2+} < Ba^{2+} \simeq UO_2^{2+} < La^{3+} < VO^{2+}$
Khan69	< 7.5	H.A.*3	$Mn^{2+} < Co^{2+} < Ni^{2+} < Zn^{2+} < Cu^{2+} < Al^{3+} < Fe^{3+}$
Verl72	4	H.A.*4	$Fe^{3+} < Mn^{2+} \leqslant Pb^{2+} < Cu^{2+} < Zn^{2+}$
Verl72	5	"	$Fe^{3+} < Mn^{2+} \leqslant Pb^{2+} < Cu^{2+} < Zn^{2+}$
Verl72	6	"	$Mn^{2+} < Fe^{3+} \leqslant Pb^{2+} < Cu^{2+} < Zn^{2+}$
Verl72	7	"	$Mn^{2+} \leqslant Fe^{3+} < Pb^{2+} < Cu^{2+} < Zn^{2+}$
Verl72	8	"	$Mn^{2+} \leqslant Cu^{2+} < Fe^{3+} < Zn^{2+} < Pb^{2+}$
Verl72	9	"	$Mn^{2+} \leqslant Cu^{2+} < Fe^{3+} < Pb^{2+} < Zn^{2+}$
Verl72	10	"	$Mn^{2+} \leqslant Pb^{2+} < Cu^{2+} < Fe^{3+} < Zn^{2+}$
Berg76	—	Soils*5	$Cd^{2+} \simeq Zn^{2+} < Cu^{2+} < Pb^{2+}$
Stev76	—	H.A.*6	$Cd^{2+} < Pb^{2+} \simeq Cu^{2+}$

*1 Fulvic acid originating from a podzol B_h horizon. *2 Peat humic acids. *3 Humic acids originating from one black chernozem and two grey wooded soils of Alberta. *4 Humic acid from a podzol B_h horizon. *5 Soil samples from a moor soil and two podzol soils. *6 Humic acid isolated from Harpster silt loam (humic gley), Ohio peat and a weathered North Dakota lignite.

As to the interaction of heavy metals with clay minerals, many authors point to the occurrence of "retention" phenomena, aside from exchange adsorption. In this context retention comprises adsorption in excess of the CEC as well as irreversible adsorption (i.e. non-extractable with neutral salts or weakly acidic solutions). A number of mechanisms have been proposed to account for such retention phenomena, comprising:

(a) chemisorption, particularly at the edges of clay minerals, involving the hydroxyl groups present because of broken bonds;

(b) specific sorption into the hexagonal holes on the planar side of clay minerals, possibly followed by penetration into the octahedral layer;

(c) adsorption as metal-ion complexes (including OH^- or other ligands, see chapter 6); and

(d) precipitation as hydroxides or insoluble salts.

In Table 5.5 some relevant references are cited, indicating briefly the materials involved and the mechanisms suggested.

The occurrence of the retention mechanisms suggested is undoubtedly influenced by the experimental conditions. Increasing the pH would favor hydrolysis of the metal ion, but the known hydrolysis constants of the ions

TABLE 5.5

"Retention" of heavy metals by clay minerals

Ref.	Ion	Material*1	Suggested retention mechanism
Elga50	Zn	M, K, H, Br, V	lattice penetration*2 + complex adsorption
Menz50	Cu	M, K	complex adsorption
DeMu56	Cu, Zn	M	chemisorption
Hodg60	Co	M	chemisorption + lattice penetration*2
Hodg64a	Co	M	chemisorption
Hodg64b	Co	M, V, Ph, No	complex adsorption
Bing64	Cu, Zn	M	complex adsorption + precipitation
Till68	Cu, Zn, Co, Mn, Ni	H	complex adsorption
DuPl71	Cu	M, I, K	—
McBr74	Cu	M	lattice penetration*2
Redd74	Zn	M, I, K	precipitation + interlattice adsorption
Harm77	Zn, Cd, Cu, Pb	I	chemisorption at edges

*1 Br = brucite, H = hectorite, I = illite, K = kaolinite, M = montmorillonite, No = nontronite, Ph = phlogopite, V = vermiculite. *2 Lattice penetration and imbedding in hexagonal cavities.

involved indicate that in neutral or weakly acidic solutions the degree of hydrolysis is generally insignificant (cf. DeMumbrum and Jackson, 1957; Bingham et al., 1964). Some authors found indications of acetate complexes being adsorbed (Bingham et al., 1964); the very pronounced adsorption of certain metal-(uncharged)ligand complexes is discussed in the chapter following. McBride and Mortland (1974) state that only Cu ions dehydrated by heating can enter the hexagonal holes or become chemically sorbed by edge OH groups. They conclude that, under conditions akin to those in the field, the specific adsorption of Cu by montmorillonite is unlikely to occur.

While in many investigations the role of edge-OH groups of clay minerals in accounting for specific adsorption of heavy metals has been stressed (cf. also Harmsen, 1977), it appears that these groups should perhaps be regarded as a special case of MOH groups abundantly present on the oxide surfaces occurring in soils. It thus seems probable that particularly the (hydr)oxides of Mn, Al, and Fe play a significant role in the retention of heavy metals in soil (McKenzie, 1967, 1972; Grimme, 1968; Taylor, 1968; Forbes et al., 1974; MacNaughton and James, 1974; Bar-Yosef et al., 1975; Kinniburgh et al., 1976).

Finally, in contrast to the situation described with regard to the "mobility" of heavy metals in soil under the influence of organic constituents, the presence of the inorganic ones discussed here leads in general to an immobilization. The inorganic constituents may be considered as immobile themselves, with the exception of suspended solid-phase particles in rivers and the ocean. The overall mobility of heavy metals in soil thus tends to be low, unless there are reasons to believe that low molecular organic constituents play a dominant role in the formation of mobile complexes.

5.4.2. Interlattice fixation of cations by clay minerals

In section 4.5.1 of part A some attention has already been paid to the general mechanism and the practical implications of the fixation of K, NH_4, Rb and Cs in the interlattice spaces of clay minerals. It was mentioned that K-fixation has important consequences for the availability to crops of resident and added soil potassium. The agricultural problems arising from this phenomenon triggered off the research on potassium-fixation early in this century. Schuffelen and Van der Marel (1955) could, even at this date, present an extensive review, citing 180 references, in which they discussed the world-wide occurrence of the problem, the factors affecting it, the methods used to determine the magnitude of potassium-fixation in different soils, and the availability of fixed K for different crops.

The fact that ammonium ions could also be fixed in soils was discovered by McBeth in 1917, but only in the late thirties did it become recognized that interlattice fixation of NH_4 in clays could be responsible for this process (Gruner, 1939, as cited by Schuffelen and Van der Marel, 1955). Fixed NH_4 has, like fixed K, a low availability for crops (Chaminade, as cited by Wikländer, 1954) and is poorly accessible for nitrification (Bower, 1950; Allison et al., 1953).

Interlattice fixation of Cs became of interest after the second world war, when the occurrence of Cs-containing radioactive fall-out became a real possibility (Jacobs, 1963; Klobe and Gast, 1967, 1970), while it was already known that Cs (and Rb) could be fixed in the same way as K and NH_4 (Barshad, 1950; Walker and Milne, 1950; Van der Marel, 1954).

In this section some more information will be given on the nature of the processes involved and the factors affecting it. As an excellent review of some of the more recent findings is presented by Sawhney (1972), this section will be limited to a broad description.

It has for a long time been recognized that the interlattice fixation of cations is due to the fact that they can "sink" into the hexagonal holes (radius 1.40 Å) in the exterior oxygen plane of the tetrahedral layer (Page and Baver, 1940; Stanford, 1948; Wear and White, 1951). In order to fit into these holes the ions have to be completely or partially dehydrated (Hendricks et al., 1940; Barshad, 1948). Whether this can happen or not depends on the balance between the energy needed for dehydration and the gain of potential energy in the electrostatic attraction force field produced by the negative charges of the clay (Kittrick, 1966; Shainberg and Kemper, 1966). This explains not only why cations with a high hydration energy, such as Ca, Mg, Ba, Na and Li, cannot be fixed, but also why Rb and Cs are more easily and strongly fixed than K with its somewhat higher hydration energy (Coleman et al., 1963a,b; Sawhney, 1964; Marshall and McDowell, 1965).

Because of the high electric field strength close to the substitution charges, the polarizability of the ions could also be important (Van der Marel, 1954). As the latter property depends on ionic size in a similar manner as the hydration energy, however, it may be difficult to distinguish between the two properties when comparing the ions of the alkali series.

Wear and White (1951), finally, pointed out that also the stability of the configuration of the fixed cations with the surrounding oxygen ions of the opposite tetrahedral layers plays an important role.

The gain of potential energy upon fixation depends first of all on the magnitude of the electrostatic attraction forces near the entrances to the hexagonal cavities. As tetrahedral substitution charges are seated much closer to the clay surface than octahedral charges, they exert a much higher attraction (Jackson and West, 1930). Wear and White (1951) accept this as the main cause of the differences in fixation behavior between vermiculites and illites (predominantly tetrahedral) and montmorillonites (predominantly octahedral). Other authors consider that within different groups of minerals, for example montmorillonites (Weir, 1965) or vermiculites (Sawhney, 1969), also the total charge density of the minerals is a major influence on the fixation behavior. Finally, it has been observed that dioctahedral minerals fix K ions more tightly than trioctahedral minerals (Jackson and Sherman, 1953; Newman and Brown, 1966), which may be due to a better coordination of the fixed ions with the surrounding oxygen ions (Serratosa and Bradley, 1958) or to a shorter distance of the fixed ions to the negative charges (Radoslovich, 1962; Rich, 1968).

Whatever the main causes of the differences in behavior of clay minerals with respect to interlattice fixation may be, it has been observed that saturation of vermiculite with K or NH_4 ions produces a spontaneous collapse of the basal spacing to 10 Å. This means that the distance between the plates is reduced to practically zero and consequently that the mineral is completely dehydrated (Mackenzie, 1955). In montmorillonite this spacing can only be produced after heating (Jackson and Hellman, 1942) or forced drying (Page and Baver, 1940; Callière and Hénin, 1949). Without heating or drying, K-saturated montmorillonite retains a 12-Å spacing, with corresponds to a monolayer of water in between the plates (Barshad, 1950). Illites show in this respect a behavior more similar to that of the vermiculites.

Release of fixed K from interlattice sites is, in the first instance, a very slow process of film diffusion, in which both the released K ions and the displacing cations participate (Wikländer, 1954; Mortland, 1958; Keay and Wild, 1961; Quirk and Chute, 1968; Von Reichenbach, 1968). The displacement proceeds from the edges of the mineral inward and consequently the rate of release is a falling function of time (Mortland and Ellis, 1959). In principle, however, equilibration between the interlattice K and the soil solution is possible, which is evident from the continuing release of fixed K to solutions in which the K concentration is kept at a low level (Hanway et al., 1957) and from the fact that fixed ^{39}K can be exchanged, though slowly, against an added ^{42}K isotope (Wikländer, 1950; Tendille et al., 1956, as cited by Agarwal, 1960).

Sumner and Bolt (1962), on the other hand, found that Winsum illite samples exchanged some fixed ^{39}K against ^{42}K only after they had been in prolonged contact with a 0.5 M $CaCl_2$ solution. They assumed that the Ca ions had partly opened up the interlattice spaces, presumably at the edges of the mineral. These results were confirmed in experiments on K-release and K-fixation in different neutral salt solutions with the same illite (Bolt et al., 1963; Van Schouwenburg and Schuffelen, 1963). These experiments lead to the identification of a separate K phase on interlayer sites situated near the edges of the minerals, the so-called "edge-interlayer sites". This K was readily exchangeable with the ions in the extraction liquid, even though these sites had a very high preference for K ions. The above authors calculated for these sites an exchange constant K'_N of 300—500 (Table 5.1, no. 68) for the K-Na equilibrium and a Gapon constant of $100\ mol^{-\frac{1}{2}}\ l^{\frac{1}{2}}$ for the K-Mg equilibrium (Table 5.1, no. 121). These values contrast strongly with the Gapon constants determined for the outer planar sites of the minerals, which were of the order of $2\ mol^{-\frac{1}{2}}\ l^{\frac{1}{2}}$ for the K-Ca equilibrium. Jackson (1963b) and Rich (1964) suggested that "frayed edges" of partly weathered micas (illites) produce such highly selective sorption of K ions. The existence of these weathered edges has been confirmed by optical observations (Mortland, 1958; Walker, 1959; Reed and Scott, 1962; Scott and Smith, 1966; Sawhney and Voigt, 1969).

A quite different mechanism for K-fixation and K-defixation has been reported for vermiculite. During fixation a quite sudden collapse, from 15 Å to 10 Å, of alternate basal spacings was observed when the Cs- or K-saturation surpassed a certain level (Rhoades and Coleman, 1967; Sawhney, 1967, 1969, 1972; Klobe and Gast, 1970). Further sorption of K ions produced a collapse of the remaining 15-Å spaces within the interstratified mineral, until the whole sample was converted to 10-Å spacings. Conversely, the replacement of K from collapsed 10-Å layers for Mg or Ca produces expansion in alternate interlattice spaces (Bassett, 1959; Rausell-Colom et al., 1965; Farmer and Wilson, 1970).

Both the model of "frayed edge" collapse in illites or weathered micas and the model of alternate layer collapse in vermiculite agree with observations by several workers, to the effect that the bulk of a fixation process recorded for many days takes place in the first minutes or hours (Chaminade, 1936; Hauser, 1941; Martin et al., 1946; Stanford and Pierre, 1947; Karlsson, 1952, as cited by Wikländer, 1954). It is likely that the subsequent slow fixation takes place by film diffusion after the lattices have closed. Also the observations by Levine and Joffe (1947) are interesting in this respect. They found that, during the early phases of a fixation process, the decrease of the CEC surpasses the amount of K fixed, whereas later on this decrease is smaller than the K-fixation. These two phases could well correspond to the periods before and after the complete collapse of the interlayers or frayed edges.

Finally, it is worthwhile mentioning that the total amount of fixed K that can be released from micas and illites (in meq per 100 g) tends to decrease with a decreasing particle size (Mortland and Lawton, 1961; Von Reichenbach and Rich, 1969; Scott, 1968). Sawhney (1972) suggests that this may be due to a better structural order and consequently a more stable collapse structure of the smaller particles. He mentions that higher ordering of smaller particles has been reported for kaolinite (Ormsby et al., 1962; Wiewiora and Brindley, 1969). Klobe and Gast (1970) found that smaller particles of vermiculite also showed a higher and stronger fixation of Cs than larger particles.

5.4.3. Adsorption of Al^{3+} ions

The strong tendency of Al ions to hydrolyse and polymerize complicates investigations on the adsorption of these ions and is the main cause of the turbulent history of research in this field (Jenny, 1961; Coulter, 1969). The original concept of Daikuhara (1914, cited in Chernov, 1947) and Kappen (1916), i.e. that Al ions are adsorbed by soil components and can be exchanged for the cations of neutral salts, was replaced in Western Europe and in the U.S.A. by the “H-clay hypothesis”. This hypothesis was based on observations regarding the weak acidic nature of acid soils and on the results of electrodialysis studies (Kelly and Brown, 1926; Wilson, 1928, 1929). Even though Paver and Marshall (1934) already proved clearly that Al ions can be adsorbed and exchanged by clay, the so-called H-Al concept became generally accepted after 1950 through the publications of Chernov (1947), Schofield (1949), Coleman and Haward (1953), Haward and Coleman (1954), and Low (1955). This caused an intensification of the research effort, leading to a better understanding of the adsorption process and, more particularly, of the nature of the adsorbed Al species. It also increased the knowledge about the influence of adsorbed Al species on various soil properties.

In the presence of Al ions the CEC of the inorganic soil components becomes pH-dependent. That this should be the case was already suggested in 1954 by Pratt and Holowaychuk. Coleman et al. (1959) introduced the terms "permanent charge" and "pH-dependent charge". Many investigators have proved that the pH-dependent charge of the inorganic components is caused by adsorbed Al polymers, and possibly also Fe polymers, which are not easily exchangeable (Hsu and Rich, 1960; Rich, 1960; Dixon and Jackson, 1962; Shen and Rich, 1962; Coleman and Thomas, 1964; McLean et al., 1964). Also the significance of adsorbed Al for the weak acid character of the inorganic soil components was clear recognized (Heddleson et al., 1960). Suspensions of Al-clay and Fe-clay exhibit a gradual increase of the pH during titration with NaOH (Coleman and Thomas, 1964). This indicates that there is a transformation of monomeric and low-polymeric to high-polymeric forms, during which many forms are present simultaneously. When the suspension is allowed to "age" the number of polymeric forms present tends to decrease, as is shown by the inflection points in the titration curves of aged Al-clay suspensions (Jackson, 1963b; Schwertmann and Jackson, 1964).

Adsorbed Al ions (as well as Fe ions (Blackmore, 1973)) influence the aggregate stability of clay soils. X-ray diffraction patterns of Al-clays show that plate condensation occurs, with a decrease of the interlattice spaces between the clay particles to 4 Å (Hsu and Bates, 1964). Several investigators have found in natural clay samples a so-called 14-Å (basal spacings) mineral, which they described as a transition mineral between montmorillonite or vermiculite and chlorite. The interlattice space of the 14-Å mineral is reduced neither on saturation with potassium ions or drying at 100°C, nor by glycerol treatment. The plate condensation of this mineral is attributed by Rich and Obenshain (1955), Klages and White (1957), and Rich (1960) to the presence of Al polymers between the clay plates. The Al interlayer can be removed with the aid of an Al complexing agent (Dixon and Jackson, 1962), but it is not possible to extract the Al polymers with neutral salt solutions.

Other important effects of adsorbed Al species on soil properties are the influence on the selectivity of the soil for other cations and on the phosphate-fixation behavior of soils (Russell and Low, 1954; Ellis and Truog, 1955; Hemwall, 1957; Coleman et al., 1960; John, 1971; Mokwunge, 1975; see also chapter 8).

In addition to an insight into the influence of adsorbed Al ions on the properties of the soil, the research findings mentioned above have contributed important information about the nature of the adsorbed Al species. At the same time, these results indicated that it is important to understand the processes involved in the adsorption of Al ions in order to gain more knowledge about the nature of the adsorbed Al species and their influence on soil properties. When studying the adsorption of Al ions onto soils and soil components, a number of special phenomena are observed, viz.:

(1) The (excess) adsorption of Al ions by homoionic Al-clay (Ragland and Coleman, 1960).

(2) Partially irreversible adsorption of Al ions (Rich, 1960; Hsu and Rich, 1960; Shen and Rich, 1962; Lin and Coleman, 1960; Kaddah and Coleman, 1967; Kissel et al., 1971; Bruggenwert, 1972; Brown and Newman, 1973).

(3) A decrease of the pH when the salt concentration of an Al-clay suspension is increased (salt effect).

(4) An increase of the pH when an Al salt solution and a clay suspension with equal pH values are mixed (Bruggenwert, 1972).

(5) A higher H/Al ratio in the bulk solution of an Al-clay suspension relative to the H/Al ratio of an Al salt solution with the same Al concentration — this difference decreases with an increasing electrolyte concentration (Kaddah and Coleman, 1967; Bruggenwert, 1972).

These phenomena indicate that the chemical equilibria between the different Al species (degree of hydrolysis) can change under the influence of adsorption reactions. For example, Ragland and Coleman (1960), and Jackson (1963a) attributed the Al adsorption by homoionic Al-clay to an "increased hydrolysis" caused by the clay. Rich (1960), Kaddah and Coleman (1967), and Kissel et al. (1971) were thinking along the same lines to explain the irreversible adsorption of Al ions. They reasoned that the Al-ions will hydrolyse in the adsorption phase on the addition of salts or, alternatively, that the exchanged Al ions will hydrolyse in the bulk solution and then will be adsorbed irreversibly. This mechanism also provided an explanation for the salt effect and for the observation that the H/Al ratio in the bulk solution is always higher than in an Al salt solution with the same Al concentration. Also Coleman and Thomas (1967) suggested that the hydrolysis of Al ions resulting from an increase of the electrolyte concentration is promoted by the clay. This conclusion is, however, not in agreement with the experimental results of Kaddah and Coleman (1967), and Bruggenwert (1972), who found that the H/Al ratio of the bulk solution of an Al-clay suspension decreases with an increase of the electrolyte concentration. Frink (1960), and Frink and Peech (1963) concluded from the adsorption selectivities between Al^{3+}, $Al(OH)^{2+}$ and H^+, and from the H/Al ratio in the bulk solution of a clay suspension, that the degree of hydrolysis is reduced in the presence of clay — "suppressed hydrolysis". Frink and Peech (1963) derived an additional indication for the validity of their hypothesis from the fact that extracted Al ions are always in the Al^{3+} form. The hypothesis of these authors does not, however, explain the irreversible adsorption of Al ions.

Bruggenwert (1972) presented an analysis of the adsorption by clay of Al^{3+}, $Al(OH)^{2+}$ and H^+, in which the (opposing) effects of the accumulation and the selective adsorption on the degree of hydrolysis are distinguished. This "selective accumulation" model, which is based on the existence of Boltzmann accumulation factors, solely accounting for electrostatic interaction, provides a satisfactory explanation for the phenomena described in the literature.

REFERENCES

Agarwal, R.P., 1960. Potassium fixation in soils. Soils Fert., 23: 375—378.

Allison, F.E., Kefauver, M. and Roller, E.M., 1953. Ammonium fixation in soils. Soil Sci. Soc. Am. Proc., 17: 107—110.

Amphlett, C.B. and McDonald, L.A., 1956. Equilibrium studies on natural ion exchange minerals. I. Caesium and strontium. J. Inorg. Nucl. Chem., 2: 403—414.

Amphlett, C.B. and McDonald, L.A., 1958. Equilibrium studies on natural ion exchange minerals: II. Caesium, sodium and ammonium ions. J. Inorg. Nucl. Chem., 6: 145—152.

André, J.B., 1970. Isothermes d'échange ionique sur les sols et réseaux de concentration. Ann. Agron., 21: 703—724.

Ardakani, M.S. and Stevenson, F.J., 1972. A modified ion-exchange technique for the determination of stability constants of metal—soil organic matter complexes. Soil Sci. Soc. Am. Proc., 36: 884—890.

Babcock, K.L. and Schulz, R.K., 1963. Effect of anions on the sodium—calcium exchange in soils. Soil Sci. Soc. Am. Proc., 27: 630—632.

Bache, B.W., 1974. Soluble aluminium and calcium aluminium exchange in relation to the pH of dilute calcium chloride suspensions of acid soils. J. Soil Sci., 25: 320—333.

Bakker, A.C., Emerson, W.W. and Oades, J.M., 1973. The comparative effects of exchangeable calcium, magnesium, and sodium on some physical properties of red-brown earth subsoils: I. Exchange reactions and water content for dispersion of Shepparton soil. Aust. J. Soil Res., 11: 143—150.

Banerjee, S.K. and Mukherjee, S.K., 1972. Physico-chemical studies of the complexes of divalent transitional metal ions with humic and fulvic acids of Assam soil. J. Ind. Soc. Soil Sci., 20(1): 13—18.

Barshad, I., 1948. Vermiculite and its relation to biotite as revealed by base exchange reaction, X-ray analysis, differential thermal curves and water content. Am. Mineral., 33: 655—678.

Barshad, I., 1950. The effect of the interlayer cations on the expansion of the mica type of crystal lattice. Am. Mineral., 35: 225—238.

Bar-Yosef, B., Posner, A.M. and Quirk, J.P., 1975. Zinc adsorption and diffusion in Goethite. J. Soil Sci., 26: 1—21.

Bassett, W.A., 1959. The origin of vermiculite deposit at Libby, Montana. Am. Mineral.,

Beckett, P.H.T., 1964. Potassium—calcium exchange equilibria in soils: specific adsorption sites for potassium. Soil Sci., 97: 376—383.

Benson, H.R., 1966. Zinc retention by soils. Soil Sci., 101: 171—179.

Bergseth, H. and Stuanes, A., 1976. Selektivität von Humusmaterial gegenüber einige Schwermetallionen. Acta Agric. Scand., 26(1): 52—58.

Bingham, F.T., Page, A.L. and Sims, J.K., 1964. Retention of Cu and Zn by H-montmorillonite. Soil Sci. Soc. Am. Proc., 28: 351—354.

Bittel, J.E. and Miller, R.J., 1974. Lead, cadmium and calcium selectivity coefficients on a montmorillonite, illite and kaolinite. J. Environ. Qual., 3: 250—253.

Blackmore, A.V., 1973. Aggregation of clay by the products of iron (III) hydrolysis. Aust. J. Soil Res., 11: 75—82.

Bolt, G.H., 1955. Ion adsorption by clays. Soil Sci., 79: 267—276.

Bolt, G.H., Sumner, M.E. and Kamphorst, A. 1963. A study of the equilibria between three categories of potassium in an illitic soil. Soil Sci. Soc. Am. Proc., 27: 294—299.

Bower, C.A., 1950. Fixation of ammonium in difficulty exchangeable form under moist conditions by some soils of semi-arid regions. Soil Sci., 70: 375—383.

Bower, C.A., 1959. Cation-exchange equilibria in soils affected by sodium salts. Soil Sci., 88: 32—35.

Brown, G. and Newman, A.C.D., 1973. The reactions of soluble aluminium with montmorillonite. J. Soil Sci., 24: 339—355.

Bruggenwert, M.G.M., 1972. Adsorption of Aluminium Ions on the Clay Mineral Montmorillonite. Agric. Res. Rep. No. 768, Wageningen.

Caillère, S. and Hénin, S., 1949. Transformation of minerals of the montmorillonite family into 10 Å micas. Mineral. Mag., 28: 606—611.

Carson, C.D. and Dixon, J.B., 1972. Potassium selectivity in certain montmorillonitic soil clays. Soil Sci. Soc. Am. Proc., 36: 838—843.

Chaminade, R., 1936. La rétrogration du potassium dans les sols. Ann. Agron., 6: 817—830.

Chernov, V.A., 1947. The Nature of Soil Acidity. Academy of Sciences, U.S.S.R.

Clark, J.S. and Turner, R.C., 1965. Extraction of exchangeable cations and distribution constants of ion exchange. Soil Sci. Soc. Am. Proc., 29: 271—274.

Cohen, P. and Gailledreau, C., 1962. Equilibrium parameters of Sr-90 in saclay soil. Soil Sci., 93: 124—138.

Coleman, N.T. and Haward, M.E., 1953. The heats of neutralisation of acid clays and cation exchange resins. J. Am. Chem. Soc., 75: 6045—6046.

Coleman, N.T. and LeRoux, F.H., 1965. Ion exchange displacement of cesium from soil vermiculite. Soil Sci., 99: 243—250.

Coleman, N.T. and Thomas, G.W., 1964. Buffer curves of acid clays as affected by the presence of ferric iron and aluminum. Soil Sci. Soc. Am. Proc., 28: 187—190.

Coleman, N.T. and Thomas, G.W., 1967. The basic chemistry of soil acidity. In: R.W. Pearson and F. Adams (Editors), Soil Acidity and Liming. Am. Soc. Agron., Madison, Wisc.

Coleman, N.T., Weed, S.B. and McCracken, R.J., 1959. Cation exchange capacity and exchangeable cations in Piedmont soils of North Carolina. Soil Sci. Soc. Am. Proc., 23: 146—149.

Coleman, N.T., Thorup, J.T. and Jackson, W.A., 1960. Phosphate sorption reactions that involve exchangeable Al. Soil Sci., 90: 1—7.

Coleman, N.T., Craig, D. and Lewis, R.J., 1963a. Ion exchange reactions of cesium. Soil Sci. Soc. Am. Proc., 27: 287—289.

Coleman, N.T., Lewis, R.J. and Craig, D., 1963b. Sorption of cesium by soils and its displacement by salt solution. Soil Sci. Soc. Am. Proc., 27: 290—294.

Coulter, B.S., 1966. The Exchange of Aluminium in Soils and Clays by Calcium, Potassium and Hydrogen Ions. Ph.D. thesis, Univ. of London.

Coulter, B.S., 1969. The chemistry of hydrogen and aluminium ions in soils, clay minerals and resins. Soils Fert., 32: 215—223.

Coulter, B.S. and Talibudeen, O., 1968. Calcium: aluminium exchange equilibria in clay minerals and acid soils. J. Soil Sci., 19: 237—250.

Cremers, A., 1968. Ionic Movement in a Colloidal Environment; Measurement, Interpretation and Relevance to Plant Nutrition. Ph.D. thesis, Univ. of Louvain.

Davies, R.I., Cheshire, M.V. and Graham-Bryce, I.J., 1969. Retention of low levels of copper by humic acid. J. Soil Sci., 20: 65—71.

Deist, J. and Talibudeen, O., 1967. Ion exchange in soils from the ion pairs K—Ca, K—Rb and K—Na. J. Soil Sci., 18: 125—137.

DeMumbrum, L.E. and Jackson, M.L., 1956. Infra-red adsorption evidence on exchange reaction mechanisms of Cu and Zn with layer silicates and peats. Soil Sci. Soc. Am. Proc., 20: 334—337.

DeMumbrum, L.E. and Jackson, M.L., 1957. Formation of basic cations of copper, zinc, iron and aluminum. Soil Sci. Soc. Am. Proc., 21: 662.

Dixon, J.B. and Jackson, M.L., 1962. Properties of intergradient chlorite-expansible layer silicates of soils. Soil Sci. Soc. Am. Proc., 26: 358—362.

Dolcater, D.L., Lotse, E.G., Syers, J.K. and Jackson, M.L., 1968. Cation exchange selectivity of some clay-sized minerals and soil materials. Soil Sci. Soc. Am. Proc., 32: 795—799.

DuPlessis, S.F. and Burger, R.D.T., 1971. Die spesifiekte adsorpsie van koper deur kleiminerale en grondfraksies. Agrochemophysica, 3: 1—10.

Ehlers, W., Meyer, B. and Scheffer, F., 1967. K Selektivität und Fraktionierung des Austauschkaliums (Beiträge zum K-Austausch des Bodens II). Z. Pflanzenern. Düng. Bodenk., 117: 1—29.

Elgabaly, M.M., 1950. Mechanisms of zinc fixation by colloidal clays and related minerals. Soil Sci., 69: 167—174.

Eliason, J.R., 1966. Montmorillonite exchange equilibria with strontium—sodium—cesium. Am. Mineral., 51: 324—335.

Ellis, R., Jr. and Truog, E., 1955. Phosphate fixation by montmorillonite. Soil Sci. Soc. Am. Proc., 19: 451—454.

El Sayed, M.H., Burau, R.G. and Babcock, K.L., 1970a. A theoretical study of copper (II)—ammonia system in aqueous solutions. Soil Sci., 110: 197—201.

El Sayed, M.H., Burau, R.G. and Babcock, K.L., 1970b. Copper (II)—ammine complexes in clay systems. Soil Sci., 110: 202—207.

El Sayed, M.H., Burau, R.G. and Babcock, K.L., 1970c. Thermodynamics of copper (II)—calcium exchange on bentonite clay. Soil Sci. Soc. Am. Proc., 34: 397—400.

Farmer, V.C. and Wilson, M.J. 1970. Experimental conversion of biotite to hydrobiotite. Nature, 226: 841—842.

Faucher, J.A. Jr. and Thomas, H.C., 1954. Adsorption studies on clay minerals: IV. The system montomorillonite caesium—potassium. J. Chem. Phys., 22: 258—261.

Forbes, E.A., Posner, A.M. and Quirk, J.P., 1974. The specific adsorption of inorganic Hg(II) species and Co(III) complex ions on goethite. J. Colloid Interface Sci., 49: 403—409.

Foscolos, A.E., 1968. Cation exchange equilibrium constants of aluminum saturated montmorillonite and vermiculite clays. Soil Sci. Soc. Am. Proc., 32: 350—354.

Foscolos, A.E. and Barshad, I., 1969. Equilibrium constants between both freshly prepared and aged H-montmorillonites and chloride solutions. Soil Sci. Soc. Am. Proc., 33: 242—247.

Frink, C.R., 1960. Reactions of the Aluminum Ion in Aqueous Solutions and Clay Suspensions. Ph.D. thesis, Cornell Univ.

Frink, C.R. and Peech, M., 1963. Hydrolysis and exchange reactions of the aluminum ion in hectorite and montmorillonite suspensions. Soil Sci. Soc. Am. Proc., 27: 527—530.

Fripiat, J.J., Cloos, P. and Poncelet, A., 1965. Comparaison entre les propriétés d'échange de la montmorillonite et d'une résine vis-à-vis des cations alcalins et alcalino-terreux. I. Réversibilité des processus. Bull. Soc. Chim. Fr., 208—215.

Frysinger, G.R., 1962. Sodium—caesium exchange on clinoptilolite. Nature, 194: 351—353.

Gaines, G.L. and Thomas, H.C., 1955. Adsorption studies on clay minerals. V. Montmorillonite caesium—strontium at several temperatures. J. Chem. Phys., 23: 2322—2326.

Garcia-Miragaya, J. and Page, A.L., 1976. Influence of ionic strength and inorganic complex formation on the sorption of trace amounts of Cd by montmorillonite. Soil Sci. Soc. Am. J., 40: 658—663.

Gast, R.G., 1969a. Standard free energies of exchange for alkali metal cations on Wyoming bentonite. Soil Sci. Soc. Am. Proc., 33: 37—41.

Gast, R.G., 1972. Alkali metal cation exchange on Chambers montmorillonite. Soil Sci. Soc. Am. Proc., 36: 14—19.

Gast, R.G. and Klobe, W.D., 1971. Sodium—lithium exchange equilibria on vermiculite at 25° and 50°C. Clays and Clay minerals, 19: 311—319.

Gast, R.G., Van Bladel, R. and Deshpande, K.B., 1969b. Standard heats and entropies of exchange for alkali metal cations of Wyoming bentonite. Soil Sci. Soc. Am. Proc., 33: 661—664.

Gheyi, H.R. and Van Bladel, R., 1975. Calcium—sodium and calcium—magnesium exchange equilibria on some calcareous soils and a montmorillonite clay. Agrochimica, 19(6): 468—479.

Gheyi, H.R., Frankart, R. and Van Bladel, R., 1974. Chemical and physical properties of some calcareous soils of Nkheila and Gharb region (Morocco). Pédologie, 24: 199—215.

Gilbert, M., 1970. Thermodynamic study of calcium—magnesium exchange on Camp-Berteau montmorillonite. Soil Sci., 109: 23—25.

Gilbert, M. and Laudelout, H., 1965. Exchange properties of hydrogen ions in clays. Soil Sci., 100: 157—162.

Gilbert, M. and Van Bladel, R., 1970. Thermodynamics and thermochemistry of the exchange reaction between NH_4^+ and Mn^{2+} in a montmorillonite clay. J. Soil Sci., 21: 38—49.

Grimme, H., 1968. Die Adsorption von Mn, Co, Cu und Zn durch Goethit aus verdünnten Lösungen. Z. Pflanzenern. Dueng. Bodenkd., 121: 58—65.

Hanway, J.J., Scott, A.D. and Stanford, G., 1957. Replaceability of ammonium fixed in clay minerals as influenced by ammonium or potassium in the extracting solution. Soil Sci. Soc. Am. Proc., 21: 29—34.

Harmsen, K., 1977. Behaviour of Heavy Metals in Soils. Agric. Res. Rep. 866, Wageningen.

Hauser, G.F., 1941. Die nichtaustauschbare Festlegung des Kaliums im Boden. Thesis, Wageningen Univ.

Haward, M.E. and Coleman, N.T., 1954. Some properties of H and Al clays and exchange resins. Soil Sci., 78: 181—188.

Heald, W.R., 1960. Characterization of exchange reactions of strontium or calcium on four clays. Soil Sci. Soc. Am. Proc., 24: 103—106.

Heddleson, H.R., McLean, E.O. and Holowaychuk, N., 1960. Aluminum in soils: IV. The role of aluminum in soil acidity. Soil Sci. Soc. Am. Proc., 24: 91—94.

Hemwall, J.B., 1957. The fixation of phosphorus by soils. Adv. Agron., IX: 95—112.

Hendricks, S.B., Nelson, R.A. and Alexander, L.I., 1940. Hydration mechanism of the clay mineral montmorillonite saturated with various cations. J. Am. Chem. Soc., 62: 1457—1464.

Hodgson, J.F., 1960. Cobalt reactions with montmorillonite. Soil Sci. Soc. Am. Proc., 24: 165—168.

Hodgson, J.F., Geering, H.R. and Fellows, M., 1964. The influence of fluoride, temperature, calcium and alcohol on the reaction of cobalt with montmorillonite. Soil Sci. Soc. Am. Proc., 28: 39—42.

Hodgson, J.F., Tiller, K.G. and Fellows, M., 1964. The role of hydrolysis in the reaction of heavy metals with soil-forming materials. Soil Sci. Soc. Am. Proc., 28: 42—46.

Hodgson, J.F., Lindsay, W.L. and Trierweiler, J.F., 1966. Micronutrient cation complexing in soil solution: II. Complexing of Zn and Copper in displaced solution from calcareous soils. Soil Sci. Soc. Am. Proc., 30: 723—730.

Howery, D.G. and Thomas, H.C., 1965. Ion exchange on the mineral clinoptilolite. J. Phys. Chem., 69: 531—537.

Hsu, P.H. and Rich, C.I., 1960. Aluminum fixation in a synthetic cation exchanger. Soil Sci. Soc. Am. Proc., 24: 21—25.

Hsu, P.H. and Bates, T.F., 1964. Fixation of hydroxy—aluminium polymers by vermiculite. Soil Sci. Soc. Am. Proc., 28: 763—769.

Hutcheon, A.T., 1966. Thermodynamics of cation exchange on clay; Ca—K-montmorillonite. J. Soil Sci., 17: 339—355.

Jackson, M.L., 1963a. Aluminum bonding in soils. A unifying principle in soil science. Soil Sci. Soc. Am. Proc., 27: 1—10.

Jackson, M.L., 1963b. Interlayering of expansible layer silicates in soils by chemical weathering. Clays and Clay Minerals, 11: 29—46.

Jackson, M.L. and Hellman, N.N., 1942. X-ray diffraction procedure for positive differentiation of montmorillonite from hydrous mica. Soil Sci. Soc. Am. Proc., 6: 133—145.

Jackson, M.L. and Sherman, G.D., 1953. Chemical weathering of minerals in soils. Advan. Agron., 5: 219—318.

Jackson, W.W. and West, J., 1930. The crystal structure of muscovite. Z. Krist., 76: 211—227.

Jacobs, D.J., 1963. The effect of collapse-inducing cations on the Cs-sorption properties of hydrobiotite. Int. Clay Conf. 1963. Pergamon Press, New York, N.Y., pp. 239—248.

Jenny, H., 1961. Reflections on the soil acidity merry-go-round. Soil Sci. Soc. Am. Proc., 25: 428—432.

Jensen, H.E., 1973. Potassium—calcium exchange equilibria on a montmorillonite and a kaolinite clay: I. A test on the Argersinger thermodynamic approach. Agrochimica, 17: 181—190.

John, M.K., 1971. Soil properties affecting the retention of phosphorus from effluent. Can. J. Soil Sci., 51: 315—322.

Juo, A.S.R. and Barber, S.A., 1969. An explanation for the variability in Sr—Ca exchange selectivity of soils, clays and humic acid. Soil Sci. Soc. Am. Proc., 33: 364—369.

Kaddah, M.T. and Coleman, N.T., 1967. Salt displacement and titration of $AlCl_3$-treated trioctahedral vermiculite. Soil Sci. Soc. Am. Proc., 31: 328—332.

Kappen, H., 1916. Studies of acidic mineral soils from the neighbourhood of Jena. Die Landw. Vers. Stat., 88.

Keay, J. and Wild, A., 1961. The kinetics of cation exchange in vermiculite. Soil Sci., 92: 54—60.

Kelly, W.P. and Brown, S.M., 1926. Ion exchange in relation to soil acidity. Soil Sci., 21: 289—297.

Khan, S.U., 1969. Interaction between the humic acid fraction of soils and certain metallic cations. Soil Sci. Soc. Am. Proc., 33: 851—854.

Khasawneh, F.E., Juo, A.S.R. and Barber, S.A., 1968. Soil properties influencing differential Ca to Sr adsorption. Soil Sci. Soc. Am. Proc., 32: 209—211.

Kinniburgh, D.G., Jackson, M.L. and Syers, J.K., 1976. Adsorption of alkaline earth, transition and heavy metal cations by hydrous oxide gels of iron and aluminum. Soil Sci. Soc. Am. J., 40: 796—799.

Kishk, F.M. and El-Sheemy, H.M., 1974. Potassium selectivity of clays as affected by the state of oxidation of their crystal structure. Clays Clay Miner., 22: 41—47.

Kissel, D.E., Gentzsch, E.P. and Thomas, G.W., 1971. Hydrolysis of non-exchangeable acidity in soils during salt extractions of exchangeable acidity. Soil Sci., 111: 293—297.

Kittrick, J.A., 1966. Forces involved in ion fixation by vermiculite. Soil Sci. Soc. Am. Proc., 30: 801—803.

Klages, M.G. and White, J.L.A., 1957. A chlorite like mineral in Indiana soils. Soil Sci. Soc. Am. Proc., 21: 16—20.

Klobe, W.D. and Gast, R.G., 1967. Reactions affecting cation exchange kinetics in vermiculite. Soil Sci. Soc. Am. Proc., 31: 744—749.

Klobe, W.D. and Gast, R.G., 1970. Conditions affecting cesium fixation and sodium entrapment in hydrobiotite and vermiculite. Soil Sci. Soc. Am. Proc., 34: 746—750.

Knibbe, W.G.J. and Thomas, G.W., 1972. Potassium—calcium exchange coefficients in clay fractions of some vertisols. Soil Sci. Soc. Am. Proc., 36: 568—572.

Krishnamoorthy, C. and Overstreet, R., 1950. An experimental evaluation of ion-exchange relationships. Soil Sci., 69: 41—53.

Krishnappa, M., Gajanan, G.N., Mithyantha, M.S. and Perur, N.G., 1974. Calcium—magnesium exchange equilibrium in three soils of Karnataka. Mysore J. Agric. Sci., 8(1): 97—102.

Kunishi, H.M. and Heald, W.R., 1968. Rubidium—sodium exchange on kaolinite and bentonite. Soil Sci. Soc. Am. Proc., 32: 201—204.

Kunishi, H.M. and Taylor, A.W., 1968. Adsorption of calcium and strontium by crandallite. Soil Sci. Soc. Am. Proc., 32: 441—443.

Lagerwerff, J.V. and Bolt, G.H., 1959. Theoretical and experimental analysis of Gapon's equation for ion exchange. Soil Sci., 87: 217—222.

Lagerwerff, J.V. and Brower, D.L., 1972. Exchange adsorption of trace quantities of cadmium in soils treated with chlorides of aluminum, calcium and sodium. Soil Sci. Soc. Am. Proc., 36: 734—737.

Lagerwerff, J.V. and Brower, D.L., 1973. Exchange adsorption or precipitation of lead in soils treated with chlorides of aluminum, calcium and sodium. Soil Sci. Soc. Am. Proc., 37: 11—13.

Laudelout, H., 1965. Calorimetric determination of ion selectivity in clay minerals. Soil Sci., 100: 218—219.

Laudelout, H., Van Bladel, R., Bolt, G.H. and Page, A.L., 1968. Thermodynamics of heterovalent cation exchange reactions in a montmorillonite clay. Trans. Faraday Soc., 64: 1477—1488.

Levine, A.K. and Joffe, J.S., 1947. Fixation of potassium in relation to exchange capacity of soils: III. Soil Sci., 63: 329—335.

Levy, R. and Hillel, D., 1968. Thermodynamic equilibrium constants of sodium—calcium exchange in some Israel soils. Soil Sci., 106: 393—398.

Levy, R. and Shainberg, I., 1972a. Calcium—magnesium exchange in montmorillonite and vermiculite. Clays Clay Miner., 20: 37—46.

Levy, R., Shainberg, I., Shalhevet, J. and Alperovitch, N., 1972b. Selectivity coefficients of Ca—Mg exchange for three montmorillonitic soils. Geoderma, 8: 133—138.

Levy, R. and Francis, C.W., 1976. Adsorption and desorption of cadmium by synthetic and natural organo—clay complexes. Geoderma, 15: 361—371.

Levi-Minzi, R., Soldatini, G.F. and Riffaldi, R., 1976. Cadmium adsorption by soils. J. Soil Sci., 27: 10—15.

Lewis, R.J. and Thomas, H.C., 1963. Adsorption studies on clay minerals: VIII. A consistency test of exchange sorption in the system sodium—caesium—barium montmorillonite. J. Phys. Chem., 67: 1781—1783.

Lin, C. and Coleman, N.T., 1960. The measurement of exchangeable aluminum in soils and clays. Soil Sci. Soc. Am. Proc., 24: 444—447.

Loven, A.W. and Thomas, H.C., 1965. Adsorption studies on clay minerals: IX. Ion-exchange properties of natural and thermally altered montmorillonite. Soil Sci. Soc. Am. Proc., 29: 250—254.

Low, P.F., 1955. The role of aluminum in the titration of bentonite. Soil Sci. Soc. Am. Proc., 19: 135—139.

Mackenzie, R.C., 1955. Potassium in clay minerals. Potassium Symp. 1955, Int. Potass. Inst.

MacNaughton, M.G. and James, R.O., 1974. Adsorption of aqueous mercury(II) complexes at the oxide/water interface. J. Colloid Interface Sci., 47: 431—440.

Maes, A., 1973. Ion Exchange of Some Transition Metal Ions in Montmorillonite and Synthetic Faujasites. Ph.D. thesis, Univ. of Louvain.

Mangaroo, A.S., Himes, F.L. and McLean, E.O., 1965. The adsorption of zinc by some soils after various pre-extraction treatments. Soil Sci. Soc. Am. Proc., 29: 242—245.

Marshall, C.E. and McDowell, L.L., 1965. The surface reactivity of micas. Soil Sci., 99: 115—131.

Martin, H. and Laudelout, H., 1963. Thermodynamique de l'échange des cations alcalins dans les argiles. J. Chim. Phys., 60: 1086—1099.

Martin, J.C., Overstreet, R. and Hoagland, D.R., 1946. Potassium fixation in soils in replaceable and non-replaceable forms in relation to chemical reactions in the soil. Soil Sci. Soc. Am. Proc., 10: 94—101.

McBride, M.B. and Mortland, M.M., 1974. Copper(II) interactions with montmorillonite: evidence from physical methods. Soil Sci. Soc. Am. Proc., 38: 408—415.

McKenzie, R.M., 1967. The sorption of cobalt by manganese minerals in soils. Aust. J. Soil Res., 5: 235—246.

McKenzie, R.M., 1972. The sorption of some heavy metals by the lower oxides of manganese. Geoderma, 8: 29—35.

McLaren, R.G. and Crawford, D.V., 1973. Studies on soil copper: I; II. The-fraction of copper in soils. J. Soil Sci., 24: 172—181; 443—452.

McLean, E.O., Hourigan, W.R., Shoemaker, H.E. and Bhumbla, D.R., 1964. Aluminum in soils: V. Form of aluminum as a cause of soil acidity and a complication in its measurement. Soil Sci., 97: 119—126.

McLean, G.W., Pratt, P.F. and Page, A.L., 1966. Nickel—barium exchange selectivity coefficients for montmorillonite. Soil Sci. Soc. Am. Proc., 30: 804—805.

Menzel, R.G. and Jackson, M.L., 1950. Sorption of copper from acid systems by kaolinite and montmorillonite. Trans. 4th Int. Congr. Soil Sci., 1: 125—128.

Merriam, C.N. and Thomas, H.C., 1956. Adsorption studies on clay minerals: VI. Alkali ions on attapulgite. J. Chem. Phys., 24: 993—995.

Mokwunge, U., 1975. The influence of pH on the adsorption of phosphate by soils from Guinea and Sudan savannah zones of Nigeria. Soil Sci. Soc. Am. Proc., 39: 1100—1102.

Mortland, M.M., 1958. Kinetics of potassium release from biotite. Soil Sci. Soc. Am. Proc., 22: 503—508.

Mortland, M.M. and Ellis, B.G., 1959. Release of fixed potassium as a diffusion controlled process. Soil Sci. Soc. Am. Proc., 23: 363—364.

Mortland, M.M. and Lawton, K., 1961. Relationships between particle size and potassium release from biotite and its analogues. Soil Sci. Soc. Am. Proc., 25: 473—476.

Mukherjee, D.C., Basu, A.N. and Adhikari, M., 1973. Exchange reactions of Mn^{2+}, Co^{2+} and Ni^{2+} with Mg^{2+} as a complementary ion in a montmorillonite clay. J. Ind. Soc. Soil Sci., 21: 27—34.

Muller, J., 1960. Échanges des ions cuivriques sur les colloides minéraux: I. Phénomènes d'adsorption. Ann. Agron., I: 75—91.

Naylor, D.V. and Overstreet, R., 1969. Sodium—calcium exchange behavior in organic soils. Soil Sci. Soc. Am. Proc., 33: 848—851.

Newman, A.C.D. and Brown, G., 1966. Chemical changes during the alteration of micas. Clay miner., 6: 297—310.

Nye, P., Craig, D., Coleman, N.T. and Ragland, J.L., 1961. Ion exchange equilibria involving aluminum. Soil Sci. Soc. Am. Proc., 25: 14—17.

Ormsby, W.C., Schartsis, J.M. and Woodside, K.H., 1962. Exchange behavior of kaolins of varying degrees of crystallinity. J. Am. Ceram. Soc., 45: 361—366.

Page, J.B. and Baver, L.B., 1940. Ionic size in relation to fixation of cations by colloidal clay. Soil Sci. Soc. Am. Proc., 4: 150—155.

Paver, H. and Marshall, C.E., 1934. The rôle of aluminium in the reactions of the clays. J. Soc. Chem. Ind., 53: 750—760.

Peterson, F.F., Rhoades, J., Arca, M. and Coleman, N.T., 1965. Selective adsorption of magnesium ions by vermiculite. Soil Sci. Soc. Am. Proc., 29: 327—328.

Pratt, P.F. and Grover, B.L., 1964. Monovalent—divalent cation-exchange equilibria in soils in relation to organic matter and type of clay. Soil Sci. Soc. Am. Proc., 28: 32—35.
Pratt, P.F. and Holowaychuk, 1954. A comparison of ammonium acetate, barium acetate and buffered barium chloride methods of determining cation exchange capacity. Soil Sci. Soc. Am. Proc., 18: 365—368.
Quirk, J.P. and Chute, J.H., 1968. Potassium release from mica-like clay minerals. Trans. 9th Int. Congr. Soil Sci., 1: 709—719.
Radoslovich, E.W., 1962. The cell dimensions and symmetry of layer lattic silicates: II. Regression relations. Am. Mineral., 47: 617—636.
Ragland, J.L. and Coleman, N.T., 1960. The hydrolysis of aluminum salts in clay and soil systems. Soil Sci. Soc. Am. Proc., 24: 457—460.
Randhawa, N.S. and Broadbent, F.E., 1965. Soil organic matter—metal complexes: 6. Stability constants of zinc—humic acid complexes at different pH values. Soil Sci. 99: 362—366.
Rausell—Colom, J.A., Sweatman, T.R., Wells, C.B. and Norrish, K., 1965. Studies in the artificial weathering of mica. In: E.G. Hallsworth and D.V. Crawford (Editors), Experimental Pedology. Butterworths, London, pp. 40—62.
Reddy, M.R. and Perkins, H.F., 1974. Fixation of zinc by clay minerals. Soil Sci. Soc. Am. Proc., 38: 229—231.
Reed, M.G. and Scott, A.D., 1962. Kinetics of potassium release from biotite and muscovite in sodium tetraphenylboron solutions. Soil Sci. Soc. Am. Proc., 26: 437—440.
Rhoades, J.D. and Coleman, N.T., 1967. Interstratification in vermiculite and biotite produced by potassium sorption: I. Evaluation by simple X-ray diffraction pattern inspection. Soil Sci. Soc. Am. Proc., 31: 366—372.
Rich, C.I., 1960. Aluminum in interlayers of vermiculite. Soil Sci. Soc. Am. Proc., 24: 26—32.
Rich, C.I., 1964. Effect of cation size and pH on potassium exchange in Nason soil. Soil Sci., 98: 100—106.
Rich, C.I., 1968. Mineralogy of soil potassium. In: V.J. Kilmer, S.E. Younts and N.C. Brady (Editors), The Role of Potassium in Agriculture. Am. Soc. Agron., Madison, Wisc., pp. 78—108.
Rich, C.I. and Black, W.R., 1964. Potassium exchange as affected by cation size, pH and mineral structure. Soil Sci., 97: 384—390.
Rich, C.I. and Obenshain, S.S., 1955. Chemical and clay mineral properties of a red-yellow podzolic soil derived from muscovite schist. Soil Sci. Soc. Am. Proc., 19: 334—339.
Riffaldi, R. and Levi-Minzi, R., 1975. Adsorption and desorption of Cd on humic acid fraction of soils. Water, Air, Soil Pollut., 5: 179—184.
Riffaldi, R., Levi-Minzi, R. and Soldatini, G.F., 1976. Pb adsorption by soils. II. Specific adsorption. Water, Air, Soil Pollut., 6: 119—128.
Russell, G.C. and Low, P.F., 1954. Reaction of phosphate with kaolinite in dilute solutions. Soil Sci. Soc. Am. Proc., 18: 22—25.
Sawhney, B.L., 1964. Sorption and fixation of micro-quantities of Cs by clay minerals: effect of saturating cations. Soil Sci. Soc. Am. Proc., 28: 183—186.
Sawhney, B.L., 1965. Sorption of cesium from dilute solution. Soil Sci. Soc. Am. Proc., 29: 25—28.
Sawhney, B.L., 1966. Kinetic of cesium sorption by clay minerals. Soil Sci. Soc. Am. Proc., 30: 565—569.
Sawhney, B.L., 1967a. Cesium sorption in relation to lattice spacing and cation exchange capacity of biotite. Soil Sci. Soc. Am. Proc., 31: 181—184.
Sawhney, B.L., 1967b. Interstratification in vermiculite. Clays Clay Miner., 15: 75—84.
Sawhney, B.L., 1969. Regularity of interstratification as affected by charge density in layer silicates. Soil Sci. Soc. Am. Proc., 33: 42—46.

Sawhney, B.L., 1972. Selective sorption and fixation of cations by clay minerals: a review. Clays Clay Miner., 20: 93—100.

Sawhney, B.L. and Voigt, G.K., 1969. Chemical and biological weathering in vermiculite from Transvaal. Soil Sci. Soc. Am. Proc., 33: 625—629.

Schalscha, E.B., Pratt, P.F. and DeAndrade, L., 1975. Potassium—calcium exchange equilibria in volcanic-ash soils. Soil Sci. Soc. Am. Proc., 39: 1069—1072.

Schnitzer, M. and Skinner, S.I.M., 1965. Organo—metallic interaction in soils: 4. Carboxyl and hydroxyl groups in organic matter and metal retention. Soil Sci., 99: 278—284.

Schnitzer, M. and Skinner, S.I.M., 1966. Organo—metallic interactions in soils: 5. Stability constants of Cu^{2+}—, Fe^{2+}—and Zn^{2+}—fulvic acid complexes. Soil Sci., 102: 361—365.

Schnitzer, M. and Skinner, S.I.M., 1967. Organo—metallic interactions in soils: 7. Stability constants of Pb^{2+}—, Ni^{2+}—, Co^{2+}—, Ca^{2+}—and Mg^{2+}—fulvic acid complexes. Soil Sci., 103: 247—252.

Schofield, R.K., 1949. Effect of pH on electrical charges carried by clay particles. J. Soil Sci., 1:1—8.

Schuffelen, A.C. and Van der Marel, H.W., 1955. Potassium fixation in soils. Potass. Symp., 2: 157.

Schwertmann, U., 1961. Der Mineralbestand der Fraktion $< 2\mu$ einiger Böden aus Sedimenten und seine Eigenschaften. Z. Pflanzenernaehr. Dueng. Bodenkd., 95: 209—227.

Schwertmann, U., 1962. Die selektieve Kationensorption der Tonfraktion einiger Böden aus Sedimenten. Z. Pflanzenernaehr. Dueng. Bodenkd., 97: 9—25.

Schwertmann, U. and Jackson, M.L., 1964. Influence of hydroxy aluminum ions on pH titration curves of hydronium—aluminum clays. Soil Sci. Soc. Am. Proc., 28: 179—183.

Scott, A.D., 1968. Effect of particle size on interlayer potassium exchange in micas. Trans. 9th Int. Congr. Soil Sci., 2: 649—659.

Scott, A.D. and Smith, S.J., 1966. Susceptibility of interlayer potassium in micas to exchange with sodium. Clays Clay Miner., 14: 69—81.

Serratosa, J.M. and Bradley, W.F., 1958. Determination of the orientation of OH-bond axes in layer silicates by infrared adsorption. J. Phys. Chem., 62: 1164—1167.

Shainberg, I. and Kemper, W.D., 1966. Hydration status of adsorbed cations. Soil Sci. Soc. Am. Proc., 30: 707—713.

Shainberg, I. and Kemper, W.D., 1967. Ion exchange equilibria on montmorillonite. Soil Sci., 103: 4—9.

Shen, M.J. and Rich, C.I., 1962. Aluminium fixation in montmorillonite. Soil Sci. Soc. Am. Proc., 86: 33—36.

Shuman, L.M., 1975. The effect of soil properties on zinc adsorption by soils. Soil Sci. Soc. Am. Proc., 39: 454—458.

Singh, M.M. and Talibudeen, O., 1971. K—Al exchange in acid soils of Malaya and the use of thermodynamic functions to predict the release of non-exchangeable K in soil to plants. Proc. Int. Symp. on Soil Fert. Eval., New Delhi; Ind. Soc. Soil Sci., I: 85—95.

Singhal, J.P. and Rishi Pal Singh, 1973. Thermodynamic of cobalt(II)—sodium exchange on montmorillonite clay. J. Soil Sci., 24: 271—273.

Soldatini, G.F., Riffaldi, R. and Levi-Minzi, R., 1976. Pb adsorption by soils. I. Adsorption as measured by the Langmuir and Freundlich isotherms. Water, Air, Soil Pollut., 6: 111—118.

Stanford, G., 1948. Fixation of potassium in soils under moist conditions and on drying in relation to type of clay mineral. Soil Sci. Soc. Am. Proc., 12: 167—171.

Stanford, G. and Pierre, W.H., 1947. The relation of potassium fixation to ammonium fixation. Soil Sci. Soc. Am. Proc., 11: 155—160.

Stevenson, F.J., 1976. Stability constants of Cu^{2+}, Pb^{2+} and Cd^{2+} complexes with humic acids. Soil Sci. Soc. Am. J., 40: 665—672.

Stevenson, F.J., Krastanov, S.A. and Ardakai, M.S., 1973. Formation constants of Cu^{2+} complexes with humic and fulvic acid. Geoderma, 9: 129—141.

Sumner, M.E. and Bolt, G.H., 1962. Isotopic exchange of potassium in an illite under equilibrium conditions. Soil Sci. Soc. Am. Proc., 26: 541—544.

Swartzen-Allen, S.L. and Matijević, E., 1974. Surface and Colloid chemistry of clays. Chem. Rev., 74: 385—400.

Szalay, A., 1969. Accumulation of U and other micrometals in coal and organic shales and the role of humic acids in their geo-chemical enrichments. Arkiv Mineralogi Geologi, 5: 23—36.

Szalay, A. and Szilagyi, M., 1967. The association of vanadium with humic acids. Geochim. Cosmochim. Acta, 31: 1—6.

Tabikh, A.A., Barshad, I. and Overstreet, R., 1960. Cation-exchange hysteresis in clay minerals. Soil Sci., 90: 219—226.

Taylor, R.M., 1968. The association of manganese and cobalt in soils — further observations. J. Soil Sci., 19: 77—80.

Tiller, K.G., 1968. Stability of hectorite in weakly acidic solutions. III. Adsorption of heavy metal cations and hectorite solubility. Clay Miner., 7: 409—419.

Tiller, K.G., Honeysett, J.L. and Hallworth, E.G., 1969. The isotopically exchangeable form of native and applied cobalt in soils. Aust. J. Soil Res., 7: 43—56.

Tucker, B.M., 1967. The solubility of potassium from illites: III. Reactivity toward other ions. Aust. J. Soil Res., 5: 173—190.

Tucker, B.M., 1967. The solubility of potassium from illites. IV. Rates of reaction and exchange constants. Aust. J. Soil Res., 5: 191—201.

Udo, E.J., Bohn, H.L. and Tucker, T.C., 1970. Zinc adsorption by calcareous soils. Soil Sci. Soc. Am. Proc., 34: 405—407.

United States Salinity Laboratory Staff, 1969. Saline and Alkali Soils. USDA Handbook 60.

Van Bladel, R., 1966. Thermodynamique et Thermochimie de l'Échange du Potassium par les Cations Alcalins et Alcalino-Terreux dans les Argiles. Ph.D. thesis, Univ. of Louvain.

Van Bladel, R. and Menzel, R., 1969. A thermodynamic study of sodium—strontium exchange on Wyoming bentonite. Proc. Int. Clay Conf., Tokyo, 1: 619—634.

Van Bladel, R., Gaviria, G. and Laudelout, H., 1972. A comparison of the thermodynamic double layer theory and empirical studies of the Na—Ca exchange equilibria in clay water systems. Proc. Int. Clay Congr., Madrid, 385—398.

Van der Marel, H.W., 1954. Potassium fixation in Dutch soils: minerological analysis. Soil Sci., 78: 163—177.

Vansant, E.F. and Uytterhoeven, J.B., 1972. Thermodynamics of the exchange of n-alkylammonium ions on Na-montmorillonite. Clays Clay Miner., 20: 47—54.

Van Schouwenburg, J.Ch. and Schuffelen, A.C., 1963. Potassium exchange behaviour of an illite. Neth. J. Agric. Sci., 11: 13—22.

Vanselow, A.P., 1932. Equilibria of the base exchange reaction of bentonites, permutites, soil colloids and zeolites. Soil Sci., 33: 95—113.

Verloo, M. and Cottenie, A., 1972. Stability and behaviour of complexes of Cu, Zn, Fe, Mn and Pb with humic substances of soils. Pédologie, 22(2): 174—184.

Von Reichenbach, H.G., 1968. Cation exchange in the interlayers of expansible layer silicates. Clay Miner., 7: 331—341.

Von Reichenbach, H.G. and Rich, C.I., 1969. Potassium release from muscovite as influenced by particle size. Clays Clay Miner., 17: 23—29.

Walker, G.F., 1959. Diffusion of exchangeable cations in vermiculite. Nature, 184: 1392—1394.

Walker, G.F.and Milne, A., 1950. Hydration of vermiculite saturated with various cations. Trans. 4th Int. Congr. Soil Sci., 2: 62.

Wear, J.I. and White, J.L., 1951. Potassium fixation in clay minerals as related to crystal structure. Soil Sci., 71: 1—14.

Weir, A.H., 1965. Potassium retention in montmorillonite. Clay Miner., 6: 17—22.

Wiewiora, A. and Brindley, G.W., 1969. Potassium acetate intercalation in kaolinite and its removal; effect of material characteristics. Proc. Int. Clay Conf., Tokyo, Israel Univ. Press, 1: 723—733.

Wikländer, L., 1950. Fixation of potassium by clays saturated with different cations. Soil Sci., 69: 261—268.

Wikländer, L., 1954. Forms of potassium in the soil. Int. Potass. Inst., Potass. Symp., pp. 110—121.

Wild, A. and Keay, J., 1964. Cation exchange equilibria with vermiculite. J. Soil Sci., 15: 135—144.

Wilson, B.D., 1928. Exchangeable cations in soils as determined by means of normal NH_4Cl and electrodialysis. Soil Sci., 26: 407—421.

Wilson, B.D., 1929. Extraction of adsorbed cations from soil by electrodialysis. Soil Sci., 28: 411—421.

Zende, G.K., 1954. The fate of applied cobalt. J. Ind. Soc. Soil Sci., 2: 67—72.

CHAPTER 6

CATION EXCHANGE IN CLAY MINERALS: SOME RECENT DEVELOPMENTS

A. Maes and A. Cremers

The contents of chapters 1—4 show that a number of model theories have been developed in attempting to describe the distribution and exchange of ions between soil constituents and the liquid phase and that, from the experimental point of view, it is a relatively simple task to obtain the thermodynamic state functions for these processes. A common feature to most of these models is that they imply considerations and assumptions which may, or may not, be experimentally verifiable. A typical example in this regard is the recurrent question concerning the extent of (de)hydration of a metal ion upon entering the ion exchanger. Very often, the reasons for the inadequacy of a given model have to do with assumptions whose validity is difficult to assess.

In the case of the exchange of ions of different valence, as treated in chapter 3, it is possible to make a priori calculations, based on purely electrostatic considerations, which lead to a reference value for the free enthalpy effect involved. These calculations lead to a simple linear relationship between the standard free enthalpy and the logarithm of the layer charge density for the electrostatic component in mono—divalent exchange. It is perhaps surprising that this functional relationship with one of the most important characteristics of the ion exchanger remains to be submitted to experimental verification. Usually, the testing is limited to the exchange of a particular pair of ions in some given clay mineral and the experimental values are found to agree fairly well with double layer (DL) predictions. As for the effect of charge density on the exchange of homovalent ions, predictions based on particular models are equally difficult to evaluate for lack of systematic experimental evidence.

The question of whether the charge density of clay minerals has any effect on the extent of ion (de)hydration may appear futile at this point, in view of the seemingly impossible task of experimental checking. However, the question is perhaps less naive than may appear since it is possible to investigate the effect of the ion exchanger on the thermodynamic properties of metal—(uncharged)ligand complexes. The study of the effect of layer charge on these species, which result from the displacement of the solvent molecules in the (adsorbed) hydrated ion by some other uncharged ligands, such as amines, may provide an indirect line of approach to the study of metal—solvent interaction in clay systems.

6.1. ION EXCHANGE EQUILIBRIA AND CHARGE DENSITY EFFECTS

A considerable amount of ion exchange data has been obtained for a variety of organic and inorganic ion exchangers. Judging from the survey of thermodynamic data for clay minerals in chapter 5, significant differences are sometimes found between the state functions referring to a given ion exchange reaction on very similar clay minerals. For example, the enthalpy and free enthalpy data for the Na—Cs equilibrium differ by a factor of nearly 2 in montmorillonite and bentonite. Occasionally, charge density differences have been invoked as a parameter for understanding differences in exchange behavior. This is particularly true in the case of synthetic zeolites which are ideally suited for preparing an isostructural series of varying charge density by changing the Si/Al ratio. In resins, selectivity-dependence on charge density is poorly understood and coherent insight is lacking.

Studies on natural clay minerals and some synthetic clays have focussed attention on the importance of charge density in selectivity phenomena. The so-called fixation of K^+ and Cs^+ is a well-known example in this respect. It should be emphasized, however, that any conclusion regarding the dependence of ion exchange equilibria on charge density should be based on comparisons in isostructural materials of identical charge origin and nature, and preferably of a homogeneous charge density distribution. A rather unambiguous way of elucidating the effect of charge density could therefore be based on a charge reduction, making use of the Hofmann-Klemen effect in a montmorillonite clay, with a charge deficit originating exclusively in the octahedral layer.

Since it is well known that clay minerals and, to a lesser extent, resins, but not zeolites, exhibit varying swelling properties in water, depending on the nature of the saturating cations, the importance of hydration forces will be paramount. In the following, we propose to limit the discussion to dilute aqueous solutions and to materials which show a predominantly strong acid character of the exchange sites.

6.1.1. Reduced charge montmorillonite (RCM): preparation and properties

Hofmann and Klemen (1950) observed that octahedrally substituted clay minerals, containing exchangeable Li^+ ions, exhibit partial loss of their exchange capacity and swelling properties upon heating at 220—240°C for 24 h. Beidellites (tetrahedral substitution) on the contrary retain these properties upon heat treatment of the Li^+ form. The reason for this behavior was explained as a migration of the small Li^+ ion into the vacant octahedral positions of the dioctahedral clay, neutralizing the layer charge in situ. Because of the absence of vacant octahedral positions, trioctahedral clays such as hectorite are not susceptible to this effect.

The charge reduction as measured by the CEC determination in a 1/1 (v/v) ethanol—water mixture is a linear function of the Li-occupancy of the exchange complex before heating. Up to 50% charge reduction, the swelling properties in water can be restored. Beyond 50% reduction, a suitable solvent such as ethanol or a 1/1 ethanol—water mixture is needed to pull the laminae apart. It appears therefore necessary to limit comparisons of ion exchange studies to samples which are less than 50% reduced in charge. The specific surface area of these RCM clays is then comparable with that of the parent clay. Assuming that all interlayers are available for exchange, the specific surface areas calculated from the crystallographic a and b dimensions of the swollen clays (Ravina and Low, 1972) and the weight (from the structural formulas) (Schultz, 1969) are identical within 3% for Camp Berteau, Bayard, and Otay montmorillonites and hectorite. In additon these minerals are completely octahedrally substituted. Their incorporation in the RCM series is therefore justified and is proven in the section 6.1.2 and 6.1.3 (cf. Figs. 6.3 and 6.4).

Since the exchangeable cations remaining after heat treatment can be resolved and exchanged, the Hofmann-Klemen technique can be useful in obtaining the fraction of tetrahedral and/or octahedral substitution and, as proposed by Glaeser et al. (1972), affords a supplementary tool for determining structural formulas. Moreover, RCM prepared by this technique seems a promising material for studying charge density effects on a series of nearly isostructural clay minerals. This procedure partially converts the dioctahedral mineral into a trioctahedral one and complete triotahedral minerals can be integrated in the RCM series; their ion exchange and swelling characteristics do indeed correspond with those of RCM of comparable charge density (Maes and Cremers, 1977, 1978).

The charge density of montmorillonite is not homogeneous throughout the different clay sheets. Some particles are higher, others lower in charge. The charge density distribution is easily obtained through a measurement of the d_{001} spacings of the clay, saturated with n-alkyl-ammonium cations of different chain length. The transition from a monolayer of alkylammonium cations to a double layer depends on the charge density of the mineral and the chain length of the cation. Figure 6.1A shows this transition for the (exclusively octahedrally substituted) Camp Berteau (CB) montmorillonite (Stul and Mortier 1974). In between the integral values of 13.6 Å (monolayer) and 17.6 Å (double layer), a series of non-integral reflections is observed from which the interlayer cation density distribution, shown in Fig. 6.1B can be calculated. A series of RCM samples was prepared by the Li treatment corresponding to 0.95 RCM (original Na-CB, heated at 220°C for 24 h, has a CEC decrease of 5%), 0.74 RCM and 0.59 RCM (notation refers to the fraction of the original CEC). X-ray analysis of the alkylammonium data for these samples showed a shift towards lower charge densities, corresponding with observed CEC decrease, with minor variations in charge density distribution, i.e. a charge reduction which proceeds rather homogeneously throughout the various density classes.

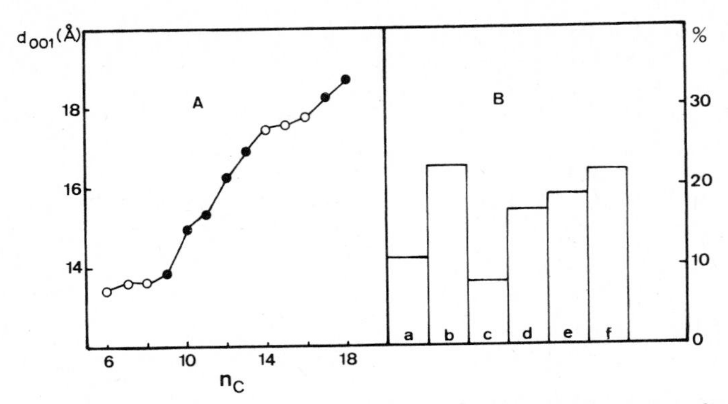

Fig. 6.1.A. Basal spacing of the n-alkylammonium complexes of the original Camp Berteau as a function of the number of carbon atoms, n_c, in the alkyl chain. Open and filled circles respectively refer to integral and non-integral reflections. (From Stul and Mortier, 1974)
B. Interlayer cation density distribution in different classes a to f. The mean density in electrons $(SiAl)_4O_{10}$ for a particle radius of 150 Å is a, 0.409; b, 0.378; c, 0.351; d, 0.328; e, 0.308; f, 0.291. (From Stul and Mortier, 1974)

6.1.2. Heterovalent exchange equilibria

Application of double layer equations for non-interacting double layers in the case of heterovalent exchange has hitherto met with variable success. On illite clays, predicted and observed values are in good agreement, while, in montmorillonite, reasonable values of the midway potential have to be introduced to make the data fit.

The influence of charge density on heterovalent exchange reactions is easily deduced from Eriksson's equation (cf. (3.58) and (3.59)) and predicts an increasing selectivity for the bivalent ion at higher charge density. A series of thermodynamic data, taken from the literature, for the mono—divalent ion exchange reactions is shown in Table 6.1. The free enthalpy of the Na—Ca exchange on Chambers montmorillonite, as calculated from the combined data of Hutcheon (1966) and Gast (1972), is exergonic for 0.175 kcal/eq and decreased to 0.100 kcal/eq for CB (Van Bladel et al., 1972) and 0.017 kcal/eq for Wyoming Bentonite (WB) (Dufey, 1974), the respective capacities being 120, 100, and 85 meq/100 g. Almost no difference is observed for the Na—Sr equilibrium on Chambers (−0.165 kcal/eq) and WB(−0.123) while CB (0.01) with an intermediate CEC does not fit in at all. Barrer and Jones (1971) find even the reverse order on synthetic fluorhectorites of greatly differing capacity (150 and 90 meq/100 g). Of course, a comparison of literature data on these ion exchange reactions is not conclusive in view of the rather small free energy changes involved and the varying fraction of tetrahedral charge.

A series of isostructural montmorillonites with varying charge, prepared

TABLE 6.1

Standard free enthalpy, enthalpy (kcal/eq) and entropy (e.u./eq) of heterovalent exchange in Chambers, Camp Berteau, and Wyoming Bentonite

	Na → Ca			Na → Sr		
	Chambers[a]	CB[b]	WB[c]	Chambers[d]	CB[e]	WB[f]
CEC (meq per 100 g)	120	100	85	120	100	85
ΔG°	−0.175	−0.10	−0.017	−0.165	≃0	−0.123
ΔH°	+1.35	+1.00	nd	nd	+1.00	+0.30
ΔS°	+6	+3.50	nd	nd	+3	+1.50

[a] Hutcheon (1966); Gast (1972).
[b] Van Bladel et al. (1972).
[c] Dufey (1974).
[d] Eliason (1966).
[e] Laudelout et al. (1968); Martin and Laudelout (1963).
[f] Van Bladel and Menzel (1969).

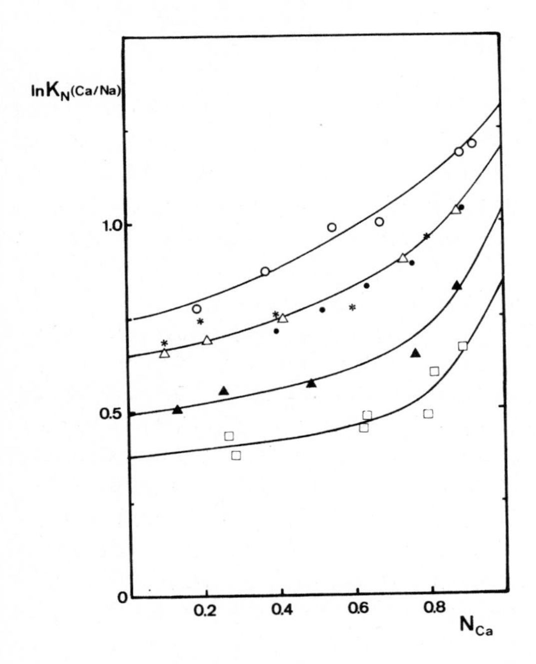

Fig. 6.2. Dependency of $\ln K_N$ on the bivalent ion occupancy, N_{Ca}, for Otay (circles), 0.95 RCM (open triangles), 0.74 RCM (filled triangles), and 0.59 RCM (squares). (From Maes and Cremers, 1977.) Data calculated from Van Bladel et al. (1972) (asterisks) and Dufey (1974) (dots) on the original Camp Berteau are given for comparison.

TABLE 6.2

Exchange capacities, charge densities (calculated from Na-CEC, taking $S = 750\,m^2/g$), d_{001} spacings, and thermodynamic data (ΔG°, ΔH°, ΔS°) for the Na-Ca exchange on montmorillonite of varying charge density (Maes and Cremers, 1978)

	Otay	CB	0.95 RCM	0.74 RCM	0.59 RCM
Na-CEC (meq/100 g)	121	100	95	74	59
Ca-CEC	130	110	108	86	68
$\Gamma \times 10^9$ (keq/m^2)	1.60	1.33	1.27	0.99	0.78
d_{001} (Å)	18.6	18.6—18.8	—	—	20.1—20.5
$\Delta G^\circ_{25^\circ}$ (kcal/eq)	−0.267	−0.188	−0.188	−0.075	+0.019
ΔH° (kcal/eq)	+0.410	+0.337	—	+0.240	+0.256
ΔS° (e.u./eq)	2.27	1.76	—	1.06	0.80

by in situ neutralization of charge through the Hofmann-Klemen effect, offers a better opportunity and satisfies the requirements for an experimental verification. Figure 6.2 demonstrates unequivocally that the selectivity decreases with charge density. The experimentally observed selectivity coefficients K_N, as defined in (2.14), are plotted as a function of the fractional bivalent cation content, N_{Ca}, of the 0.95, 0.74, and 0.59 RCM samples (Maes and Cremers, 1977). Data observed on Otay montmorillonite with a higher CEC than CB (see Table 6.2) are also shown. The selectivity data on the original CB after Van Bladel et al. (1972) and the CB heated to 220°C (0.95 RCM) coincide as expected. All the exchange reactions were conducted at 10^{-2} total normality (Na + Ca) and at 25°C. The constancy of the CEC over the whole distribution range is verified.

The experimental free enthalpy change on reducing the charge by 25% is about 0.12 kcal/eq. Although a good agreement between theoretical and experimental data can be obtained in the case of the 0.95 and 0.74 RCM using a midway potential of 85 mV in Eriksson's equation, the data on the 0.59 RCM and Otay fail to do so. Application of Eriksson's equation at different total normalities and extrapolation to zero ionic strength, as proposed by Bolt and Winkelmolen (1968), or carrying out the hypothetical charging—discharging procedure (see chapter 3) leads to the purely electrostatic component for the free enthalpy of exchange for the monovalent—divalent ion exchange. Equation (3.44), which predicts a linear dependency of the free enthalpy of exchange with charge density can be rewritten in the form:

$$\ln K^\circ_{ex} = \left(\ln \frac{\sqrt{\beta}}{2} - 1\right) + \ln \Gamma \simeq \langle \ln K_N \rangle - 1/2 \qquad (6.1)$$

Figure 6.3 shows that, for the Na—Ca exchange on a series of montmorillonite samples with identical charge localization, $\langle \ln K_N \rangle$ is indeed linearly related to $\ln \Gamma$. All the relevant data (CEC, Γ, ΔG°, ΔH°, ΔS°, X-ray spacings) corresponding to the systems in Fig. 6.3 are summarized in Table 6.2. The

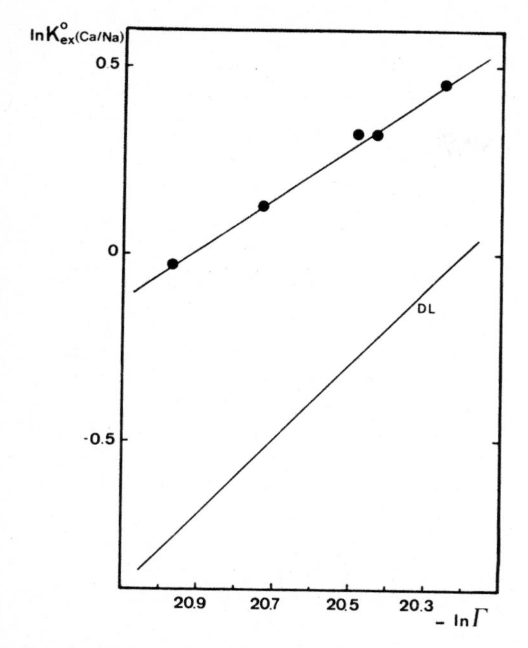

Fig. 6.3. Dependency of $\langle \ln K_N \rangle - 1/2 \simeq \ln K^{\circ}_{ex}$ on the natural logarithm of the charge density. Both the experimentally obtained (upper curve) and the theoretically calculated (DL) relationship are shown. (From Maes and Cremers, 1977)

slope of $\langle \ln K_N \rangle$ as a function of $\ln \Gamma$ equals 0.7 and, remembering that $\ln K$ data were obtained at 10^{-2} total normality, would be expected to increase by about 0.05, applying the expected shift at infinite dilution (see Table 3.3). Figure 6.3 shows that the experimental free enthalpy effect is more exergonic by a value of 0.5—0.65 RT per equivalent. A similar difference was found for the Na—Mg equilibrium on CB montmorillonite by Laudelout et al. (1968).

It is tempting to ascribe these small differences to non-purely electrostatic effects, particularly so in view of the linear correlations observed between ΔH° and ΔG° terms and polarizability differences, leading rather prematurely to a nearly exact agreement between experiment and DL predictions. However, Fig. 6.3 reveals the rather disturbing fact that the predictive value of DL theory improves with increasing charge density! (Exact agreement would occur at charge densities of about $6—7 \cdot 10^{-9}$ keq/m^2.) If anything, one would expect the dependency on charge density to be more pronounced, and not less than DL predictions which are based on purely electrostatic considerations.

The seemingly good agreement between DL theory and experimental data may very well be coincidental, as indicated by the thermodynamic data in Tables 6.1 and 6.2. The enthalpy effect, calculated from (3.44), making due corrections for the temperature-dependence of the dielectric constant, is slightly endothermic by some 0.025 kcal/eq. The experimental data show that the displacement of Na by Ca is indeed accompanied by an enthalpy gain which, however, is not only far greater but becomes increasingly important with higher charge density. It is seen that, in all cases, the enthalpy effect is nearly exactly compensated by an entropy gain which increases similarly with charge density. Perhaps it is worthwhile to mention that a similar compensatory effect occurs in vermiculite (Wild and Keay, 1964), the absolute values being, as expected, significantly larger (5.56—6 kcal/eq) than in montmorillonite.

Evidently, the increasing values of enthalpy and entropy with increasing charge density point towards effects other than purely electrostatic ones, such as (changes in) ion solvent interactions. If the enthalpy change is postulated to be due to the partial shedding of water molecules from the exchangeable bivalent ions, the deviation from the theoretical ΔH° value which is almost zero and relates to point charges in a water continuum (no structuring of water molecules), is in the expected direction. Indeed, at decreasing charge, bivalent ions can more freely keep their water of hydration, which is reflected in a concommitant decrease in endothermicity of the reaction, as observed from the data in Tables 6.2 and 6.2. The lowering of the entropy with decreasing charge density agrees with a smaller number of water molecules being expelled from the interlammelar regions. Evidence is found in the d_{001} spacings of the Ca-clay samples, which indeed increase with decreasing charge density.

6.1.3. Homovalent exchange equilibria

Charge density differences have occasionally been invoked for explaining qualitatively the difference in selectivity between ions of the same valence. Tabikh et al. (1960) report that the K-affinity increases with CEC: the equilibrium constant K_N (K/Na) changes from 1.75 for a Colony Wyoming bentonite (95 meq/100 g) to 2.23 for Belle Fourche (108 meq/100 g) to 3.09 for Plymouth Utah (124 meq/100 g). Foscolos (1968) points out that the increased K-selectivity with increasing charge results in K-fixation in Jeffersite vermiculite (175 meq/100 g).

Table 6.3 shows some representative thermodynamic data for the exchange of alkali metal ions on different clay minerals. Comparison of the CEC data in Chambers and WB bentonite (both containing about 25% tetrahedral charge) shows that free energy effects increase with CEC. A straightforward analysis of charge density effects upon selectivity should preferably be made on materials with the same charge origin, all other parameters remaining constant.

Table 6.3 indicates that the most marked variations are to be expected for the Na—Cs equilibrium. The effect of CEC on $-\Delta H^\circ$ and $-\Delta G^\circ \simeq$

TABLE 6.3

Thermodynamic data at 25°C in kcal/eq (ΔG°, ΔH°) and e.u./eq (ΔS°) for alkali metal cation exchange on Chambers, Camp Berteau, and Wyoming Bentonite

	Chambers[a]			CB[b]			WB[c]		
CEC (meq/100 g):	120			100			85		
	ΔG°	ΔH°	ΔS°	ΔG°	ΔH°	ΔS°	ΔG°	ΔH°	ΔS°
Reaction									
Li → Na	− 0.080	− 0.111	− 0.1	− 0.14	+ 0.14	+ 1	− 0.048	− 0.150	− 0.5
Na → K	− 0.725	− 1.16	− 1.5	− 0.83	− 1.08	− 0.8	− 0.306	− 0.605	− 1.0
Na → Rb	− 1.347	− 1.92	− 1.9	− 1.94	− 2.29	− 1.2	− 0.634	− 1.505	− 2.9
Na → Cs	− 1.886	− 2.648	− 2.6	− 2.48	− 3.84	− 4.7	− 1.081	− 2.557	− 4.9
				− 2.07[d]	− 4.36[d]	− 7.7[d]			

[a]Gast (1972).
[b]Martin and Laudelout (1963).
[c]Gast et al. (1969).
[d]Cremers and Thomas (1968).

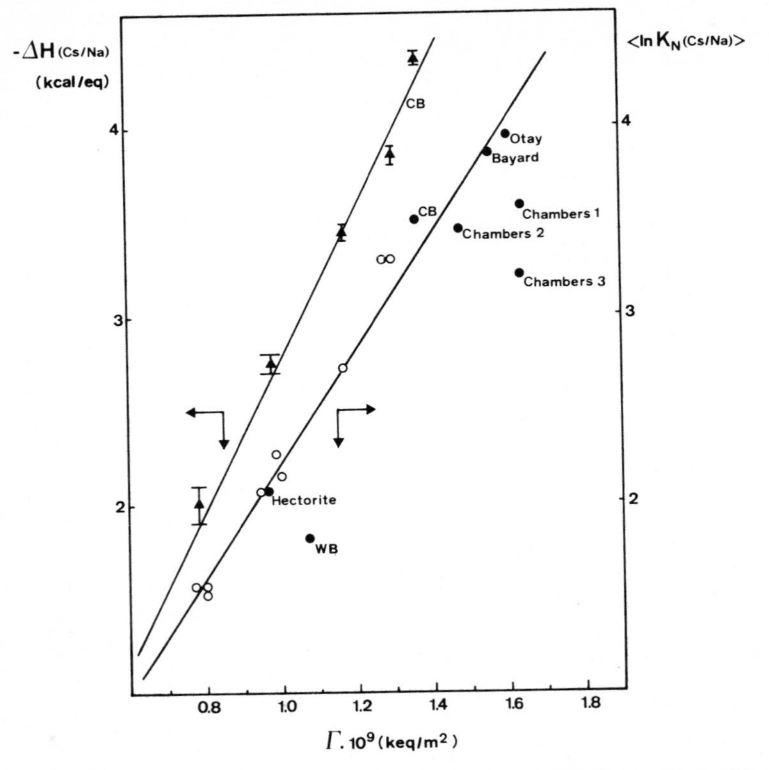

Fig. 6.4. ln K_N as a function of Γ for the Na—Cs exchange, for different species of RCM (circles) and clays as indicated, viz. Otay, hectorite (from Maes and Cremers, 1978); original Camp Berteau montmorillonite (from Cremers and Thomas, 1968); Baynard montmorillonite (from Eliason, 1966); Wyoming bentonite (from Gast, 1969); and Chambers montmorillonite (1, from Lewis and Thomas, 1973; 2, from Eliason, 1966; 3, from Gast, 1972). The enthalpy of the Na—Cs exchange reaction on the RCM samples and the original Camp Berteau (from Cremers and Thomas, 1968) are also shown. The numerical values of $-\Delta H$ and $\langle \ln K_N \text{(Cs/Na)} \rangle$ are arbitrarily put on the same scale.

RT$\langle \ln K_n \rangle$ for the Na → Cs exchange, as carried out on the RCM samples is shown in Fig. 6.4 and compared with some literature data. Enthalpy data were obtained calorimetrically (RCM) or from temperature effects on equilibrium constants. It is seen that both free enthalpy and enthalpy decrease linearly with decreasing charge and tend to vanish at zero charge. Obviously, so does the entropy. The free enthalpy effect is seen to decrease by about 1 kcal/eq on reducing the charge to one-half of its original value. Bayard and Otay montmorillonites are known to be completely octahedrally substituted montmorillonites and fit the linear relation. Wyoming bentonite and Chambers bentonite, which have about 25% tetrahedral charge, show a somewhat lower free enthalpy effect than would be expected from their CEC.

The exchange of a non-preferred ion for a preferred is accompanied by a

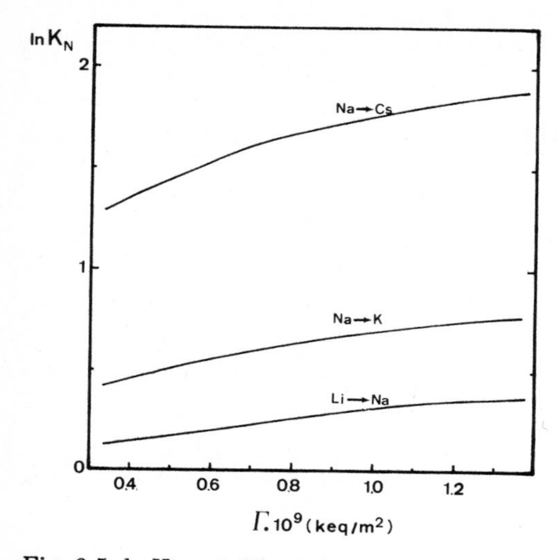

Fig. 6.5. ln K_N at $N = 0.5$ as calculated according to Shainberg and Kemper (1966a, 1966b, 1967), and Bolt et al. (1967). The potential energy difference between hydrated and unhydrated ions is taken as $-9.5 \cdot 10^{-21}$ (Li), $-1.7 \cdot 10^{-21}$ (Na), $+5.5 \cdot 10^{-21}$ (K) and $13.6 \cdot 10^{-21}$ (Cs) J/ion; total concentration is 0.01 *N*.

decrease in energy and entropy of the system. The absolute values of ΔH° and ΔS° increase with increasing selectivity or ΔG°. Confining attention to the Na—Cs exchange, the increase of ΔG° with charge logically results in higher $-\Delta H^\circ$, and $-\Delta S^\circ$ values. The parallelism between charge density effects on Na—Cs and the alkali metal exchange is also noticeable in the hydration pattern. Admittedly, the hydration of the ions on the surface is lower than in equilibrium solution. If entropy changes are only due to changes in hydration, the relation:

$$\Delta S = (\Delta S_1 - \Delta S_2)_{surface} - (\Delta S_1 - \Delta S_2)_{solution} \tag{6.2}$$

shows that the hydration status of the adsorbed ions is decreased when ΔS is negative, the more so at increasing negative values of ΔS. The change in hydration is thus most pronounced in the Na—Cs system when comparing the alkali metals, and this increases with charge density when focusing on the Na—Cs exchange, for example. In other words, the lower the charge density the less becomes the impact of charge on the hydration of the interlayer cations, which tend more towards their behavior in equilibrium solutions, resulting in vanishing selectivity.

Introduction of a pair formation constant (Shainberg and Kemper, 1966a, 1966b, 1967; Bolt et al., 1967) into the amended double layer equation renders the possibility of calculating ion exchange selectivities. In Fig. 6.5 the free enthalpy change as estimated from K_N at 50% occupancy of both exchanging ions is plotted against charge density for the Na—Cs, Na—K and Li—Na equilibrium. A gradual decrease in free enthalpy of exchange of the preferred ion is observed. Comparison with the data in Table 6.2 shows good agreement of Na—K and Li—Na with the data at 10^{-9} keq/m^2 which correspond to the WB clay, while Na—Cs does not. The prediction of the dependency on charge density of ΔG° is too small although Na—K and Na—Li data agree better in this respect than the data for Na—Cs. It is important, however, that double layer theory corrected for hydration predicts an almost linear decrease of ΔG° up to 50% charge reduction in a semi-quantitative way.

By comparing the selectivity data in Fig. 6.4 the charge-dependency is seen to be more pronounced as the potential energy of the two exchanging ions differs. It follows that the hydration status of the adsorbed ions is an important parameter in selectivity; or, apart from specific interactions, the action of "forces" that tend to dehydrate the interlamellar region (like charge density or external electrolyte concentration) enhances the selectivity for the preferred species. It should be emphasized, however, that this selectivity enhancement of the preferred ion on increasing the charge density exceeds that produced on increasing the electrolyte concentration. As regards the Na—Cs equilibrium, a two-fold increase in charge density results in a selectivity rise by a factor of 5, while the increase in electrolyte concentration from 0.02 to 0.2 results in only a 10% rise in selectivity (Laudelout et al., 1972).

It is mentioned in passing that the d_{001} spacing of the Ca form of 0.59 RCM was found to exceed that of the original clay (cf. Table 6.2). Without going into details of the mechanism of crystalline swelling of montmorillonite, the following explanation appears plausible. If the first stage of swelling may be regarded primarily as a result of the build-up of a somewhat ordered water layer under the influence of the forces of hydration of the clay mineral itself, then the presence of cations close to the surface would oppose such an ordering. A decreasing number of cations in such a Stern layer accompanying a reduction of the surface charge would thus favor crystalline swelling. The above reasoning is also in line with the observed lyotropic series in crystalline swelling of clays (Norrish, 1972). The increased number of cations with a smaller hydrated radius in a Stern layer, as will be discussed in Chapter 7 in connection with anion exclusions, leads to a smaller value of d_{001}.

6.1.4. Comparison with other exchangers

Charge density variations in synthetic zeolites are easily obtained through a different Si/Al ratio, i.e. changing the number of isomorphic substitutions of Si for Al and concomitantly the charge density. Free swelling in zeolites is prevented because of their rigid three-dimensional channel system, an example of which is shown in Fig. 6.6. Although the well-known lyotropic series is found as in most other exchangers, the change of selectivity for a certain ion with charge density is generally the opposite of what is observed in exchangers which can take up considerable amounts of water, as clay minerals and resins are able to do. The selectivity of the preferred ion decreases as the charge density increases or, referring to Table 6.4, the ion with the highest radius is preferred by the zeolite of lower charge density.

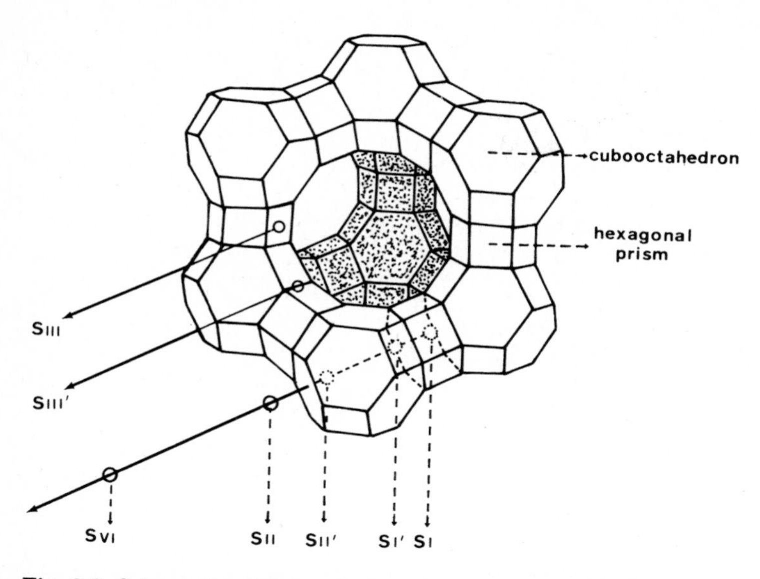

Fig. 6.6. Schematic picture of the zeolite X and Y framework. Each line represents the diameter of an oxygen ion. The apexes are Si or Al atoms. The possible ion exchange sites are indicated.

The pore-system characteristics of each type of zeolite determines its ion exchange behaviour. For example, in the synthetic zeolites X and Y (Breck, 1973) only the big cavities (see Fig. 6.6) present the opportunity for the exchangeable cations to fully hydrate. The cubooctahedra can accommodate a few water molecules, while the hexagonal prims can only contain bare ions. Localization of the cations by X-ray diffraction techniques reveals a characteristic neutralization pattern for each ion. This is made possible through

TABLE 6.4

Standard free enthalpy of exchange (kcal/eq) on two types of zeolite: faujasite (X and Y) and chabazite

	Faujasite type		Chabazite	
	Y[a]	X[a]	natural[b]	synthetic[c]
Si/Al ratio	2.51	1.22	4.90	2.14
CEC (meq/100 g)	243	625	349	616
Reaction				
	−2.7	−1.60	−1.72	−1.12
	−0.19	+0.14	−1.60	−0.55

[a]Sherry (1966).
[b]Barrer et al. (1969).
[c]Barrer and Klinowski (1972).

TABLE 6.5

Standard free enthalpies, enthalpies (kcal/eq) and entropies (e.u./eq) at 298.2 K for the exchange of the alkali metal cations with sodium ion in cross-linked polystyrene sulphonates (Boyd, 1970)

Nominal cross-linking (%DVB)	$Na^+ \rightarrow H^+$			$Na^+ \rightarrow Li^+$			$K^+ \rightarrow Na^+$			$Cs^+ \rightarrow Na^+$		
	$-\Delta G^\circ$	$-\Delta H^\circ$	$-\Delta S^\circ$	$-\Delta G^\circ$	$-\Delta H^\circ$	$-\Delta S^\circ$	$-\Delta G^\circ$	$-\Delta H^\circ$	$-\Delta S^\circ$	ΔG°	ΔH°	ΔS°
0.5	0.005	0.26	0.9	0.06	0.33	0.9	0.12	0.23	0.3	0.23	0.36	0.4
2	0.05	0.46	1.4	0.07	0.65	1.9	0.17	0.29	0.4	0.25	0.56	1.0
4	0.14	0.86	2.4	0.23	1.08	2.8	0.20	0.41	0.7	0.29	0.62	1.1
8	0.23	1.18	3.0	0.36	1.46	3.6	0.26	0.55	1.0	0.33	0.77	1.5
16	0.28	1.25	3.3	0.46	1.70	4.0	0.36	0.88	1.8	0.37	1.19	2.7
24	0.48	1.40	3.1	0.54	1.83	4.2	0.41	1.17	2.6	0.45	1.48	3.5

the available excess of crystallographic sites over the isomorphic negative charges.

In sulphonic acid resins a selectivity enhancement for the less hydrated ion is observed with increasing cross-linking. Examination of the thermodynamic data on the effect of increasing cross-linking in Table 6.5 shows that these data are completely analogous to those on montmorillonite of varying charge density. The heat evolved in the reactions is counter-balanced by an almost equal entropy loss. The magnitude of the thermodynamic values increase with the percentage of divinylbenzene suggesting here again that the ion—solvent interactions in the aqueous solutions and the exchanger differ and become more important as the hydration inside the exchanger is reduced. The solvent content indeed decreases as the cross-linking increases.

On linear polystyrene sulfonic acid resin Hutschneker and Deuel (1956) found an increase for K_N (NH_4/Na) and K_N (NH_4/Li) with rising specific capacity. At increasing specific capacity and constant cross-linking the selectivity for the systems Cs/Na, NH_4/Na, Cs/NH_4, and Ag/Na, however, decreases on polystyrene sulfonic acid resins. The K/Na selectivity, on the contrary, decreases with cation exchange capacity increase at low degrees of cross-linking while the reverse is observed at high divinylbenzene content (Reichenberg, 1966). Both the capacity and cross-linking affect the swelling of the exchanger and consequently the amount of water available per exchange grouping. In general, the magnitude of the selectivity decreases as the degree of swelling increases; or, as the counterions are more diluted inside the exchanger the selectivity too is "diluted" (Reichenberg, 1966), which is analogous to the observations on the RCM samples.

On zeolites, however, swelling is prohibited and electrostatic forces compete more and more with hydration forces with increasing charge. Since the ion with the smallest radius would be preferred if electrostatic forces only were involved, a decrease in selectivity is predicted as the charge increases and hydration is counteracted; the more so, since in zeolites, swelling is prevented and consequently the hydration of the counterions even decreases at increasing capacity.

6.2. ION EXCHANGE OF METAL—(UNCHARGED)LIGAND COMPLEXES IN CLAYS

The adsorption behavior of organic molecules and heavy metal ions in clay minerals are two areas of research which have received considerable attention in recent years and the advance made in these fields have been reviewed in the recent past. Such studies have a considerable practical importance from the environmental point of view in that they are concerned with the fate and elution behavior in the soil of toxic heavy metal ions and pesticides. Very often, these studies are focussing attention on the adsorption behavior of either pesticides or metal ions without much regard to the possible reciprocal effects of organics on metal ions or vice versa. Of course, it is well known that the adsorption properties of clays towards organic molecules such as amines are drastically improved when the clay is saturated with transition metal ions, and the phenomenon is readily explained in terms of complex formation in the clay interlayers. Occasionally, strong adsorption affinities of clays for metal complexes have been reported in the literature and suggestions were made regarding their usefulness as index cations for the determination of the cation exchange capacity of clays. However, insofar as the characterization of these adsorbed species is concerned, most of these studies are more of a qualitative and diagnostic nature in that they are related to questions such as structure and orientation and the possibility of accommodating such complexes within the clay interlayers.

In this section, the adsorption by clays of metal ions bonded to non-ionic ligands will be discussed. Such complexes, when formed or adsorbed on clays, can be characterized in terms of thermodynamic state functions which may differ quite significantly from the values for the bulk solution analogs. The point of view which is here taken is that the clay—solution interface is to be considered as a reaction medium with unique properties which may profoundly alter the thermodynamic stability of simple complexes, without any apparent effect on their structure or composition. Admittedly, this idea is not new and has been advanced in the past. We merely take these ideas one step further by demonstrating quantitatively the nature of these effects. The scope of the present approach is limited to completely "wet" systems and is not concerned with dry or semi-dry systems (Mortland, 1970).

6.2.1. Adsorption affinity and stability of adsorbed complexes

a. Adsorption affinity

A large number of literature data exist which show that cationic metal—(uncharged) ligand complexes can be exchanged against hydrated metal cations adsorbed by clays (Bodenheimer et al., 1962, 1963a, b, 1973; Fripiat and Helsen, 1966; Heller and Yariv, 1969; Laura and Cloos, 1970; Das Kanongo and Chakravarti, 1973; Theng, 1974; Swartzen Allen and Matijevic, 1975). Some of these findings indicate that the exchange adsorption of such complexes by montmorillonite is thermodynamically much more favorable than that found for the aqueous ions (Mantin, 1969; El Sayed et al., 1971; Pleysier and Cremers, 1975; Peigneur, 1976; Pleysier, 1976). This phenomenon is entirely similar to what is found in the case of some metal—amine complexes in sulfonic acid resins (Stokes and Walton, 1954; Cockerell and Walton, 1962; Suryaraman and Walton, 1962). Remembering that such metal complexes are formed by the displacement of the water ligands by simple molecules such as ammonia, diamines, thiourea, etc., leaving the cationic charge unchanged, this phenomenon is rather surprising, particularly in view of the considerable magnitude of the effect.

The extent of the synergistic effect of complex formation upon metal adsorption has rarely been expressed in thermodynamically quantitative terms; in some cases the reports were more of a qualitative nature, whereas in other cases difficulties were encountered which relate to the fact that the uptake of the metal complex occurred non-stoichiometrically. The exchange adsorption of the aqueous metal ion may be compared with the one for the metal complex by studying their ion exchange selectivity with respect to a given non-complex-forming reference cation of the same valence (Maes et al., 1977). Representing the complex formed upon complete coordination of a metal ion, M, with n uncharged ligand molecules, L, as L_nM, this amounts to studying the following exchange equilibria:

$$\bar{B} + L_nM \rightleftharpoons \overline{L_nM} + B \tag{6.3}$$

$$\bar{B} + M \rightleftharpoons \bar{M} + B \tag{6.4}$$

where the bar denotes the adsorbed species, while the reference ion B has the same valence, z_+, as both M and L_nM. The systems used to study the adsorption of L_nM must obviously contain a free ligand concentration consistent with the formation of the "saturated" complex L_nM.

The thermodynamic equilibrium constants of the above reactions involving cations with valence z_+, if specified per equivalent of exchanger (see chapter 2) are then given by:

$$K_L^\circ = \left[\frac{(\overline{L_nM})(B)}{(\bar{B})(L_nM)}\right]^{1/z_+} \tag{6.5}$$

$$K_{ex}^\circ = \left[\frac{(\bar{M})(B)}{(\bar{B})(M)}\right]^{1/z_+} \tag{6.6}$$

where parentheses indicate activities. Standard states for the ion exchanger are taken as the homoionic forms of the exchanger, while the usual convention is used for the equilibrium solution.

Referring to section 2.2 of chapter 2, the thermodynamic constants may be obtained via integration of $\ln K_N$ over the full range of the exchanger composition, where the selectivity coefficient K_N is defined in terms of equivalent fractions (cf. (2.14)). Furthermore, as in the present case all cations involved are of the same valence, it is often acceptable to replace the activity ratio of the ions in solution by their concentration ratio.

If the metal complex has the same nature in the two phases (which should be verified experimentally) and if the free enthalpy function of the reference ion is unaffected by the presence of the ligand (a reasonable assumption in the case where the reference cation does not form a complex), then the equilibrium constant for the hypothetical exchange of one equivalent of the aqueous metal ion against its complex is given by the ratio (K_L°/K_{ex}°).

b. Adsorption affinity and thermodynamic stability of the metal complex in, or on, the exchanger: quantitative relationships

The overall thermodynamic stability of a metal complex according to the reaction:

$$M^{z+} + nL \rightleftharpoons L_nM^{z+} \tag{6.7}$$

is usually expressed in terms of the "gross" constant, $\mathscr{B}_n$, defined as

$$\mathscr{B}_n = \frac{(L_nM)}{(M)(L)^n} \tag{6.8}$$

Referring to Sillén and Martell (1964) for details, the above gross constant is the product of the n stability constants, $K_1, K_2 \ldots K_n$, governing the stepwise addition of ligand molecules to the metal cation, with $K_n = (L_nM)/(L_{n-1}M)(L)$.

Similarly, the formation of the complex *on* or *in* the exchanger could be characterized by:

$$\bar{\mathscr{B}}_n = \frac{(\overline{L_nM})}{(\bar{M})(\bar{L})^n} \tag{6.9}$$

where the standard state of the (uncharged) ligand *on* the exchanger has been taken as similar to its standard state in solution (i.e. at the same concentration) while those of the adsorbed ions again refer to the homoionic exchanger.

Combining (6.8) and (6.9) leads to the equilibrium constant for the reaction:

$$\bar{M} + \overline{nL} + L_nM \rightleftharpoons \overline{L_nM} + M + nL \tag{6.10}$$

according to:

$$K^\circ_{syn} = \frac{\bar{\mathscr{B}}_n}{\mathscr{B}_n} = \frac{(\overline{L_nM})(M)(L)^n}{(L_nM)(\bar{M})(\bar{L})^n} \tag{6.11}$$

where K°_{syn} expresses the synergistic effect of the adsorption on complex formation.

Alternatively, one may express K°_{syn} in terms of a differential free enthalpy of complex formation according to:

$$-RT\ln K^\circ_{syn} = [\bar{G}^\circ_{\overline{L_nM}} - \bar{G}^\circ_{\bar{M}} - n\bar{G}^\circ_{\bar{L}}] - [\bar{G}^\circ_{L_nM} - \bar{G}^\circ_M - n\bar{G}^\circ_L] \tag{6.12}$$

Substitution of (6.8) and (6.9) into (6.5) and (6.6) now gives:

$$(K^\circ_L/K^\circ_{ex})^{z+} = K^\circ_{syn}(\bar{L})^n/(L)^n \tag{6.13}$$

where the ratio $(\bar{L})/(L)$ could be regarded as a partition coefficient of the uncharged ligand between the exchanger phase (e.g. clay interlayer space) and the solution. This partition coefficient cannot be evaluated experimentally and, accordingly, the stability constant in the exchanger phase as derived from the experimentally accessible value of $(K_L^\circ / K_{ex}^\circ)$ remains subject to an arbitrary assumption with regard to the value of this partition coefficient.

The direct evaluation of the stability constant of the adsorbed complex obviously leads to the same situation. Experimentally, the ligand number of the adsorbed ions is measured as a function of the free ligand concentration *in solution*, yielding:

$$\bar{\mathscr{B}}_n(\text{exp}) = \frac{(\overline{L_n M})}{(\bar{M})(L)^n} = \bar{\mathscr{B}}_n \frac{(\bar{L})^n}{(L)^n} \tag{6.14}$$

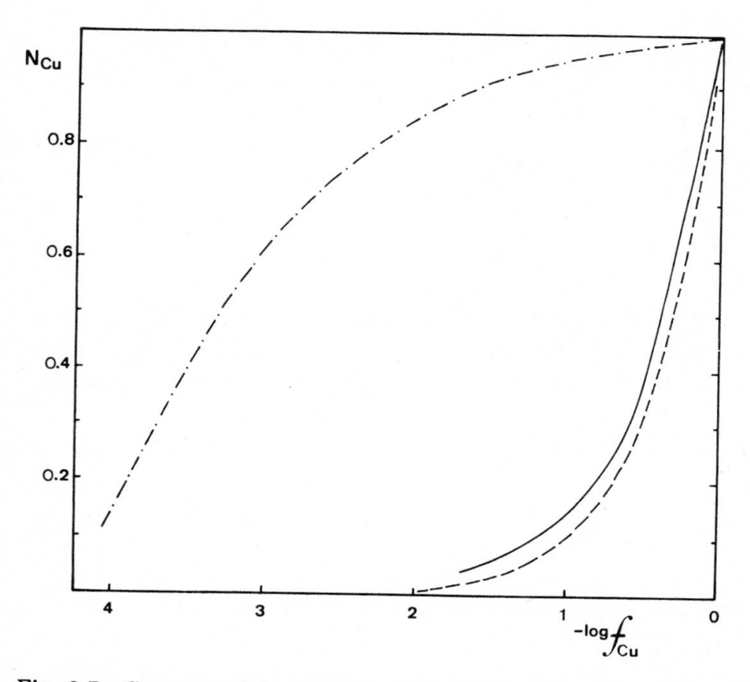

Fig. 6.7. Copper—calcium ion exchange isotherms on CB montmorillonite in the absence (full line) and presence (broken line) of ethylenediamine. Dashed line indicates unit selectivity coefficient; N and f refer to equivalent fraction in the clay and the equilibrium solution respectively. (From Maes et al., 1978)

6.2.2. Some representative cases

a. Metal—ethylenediamine complexes

Metal ions such as copper, nickel, cadmium, zinc, mercury, and silver form rather stable complexes with ethylenediamine, all of which are strongly stabilized in clays (Peigneur, 1976) such as montmorillonite and illite. The effect amounts to roughly three orders of magnitude on the overall stability constant. The most representative case, for which a fair amount of data has been accumulated, is the copper—ethylenediamine complex (Maes et al., 1977) which is ideally suited to illustrate the nature of the stability effect.

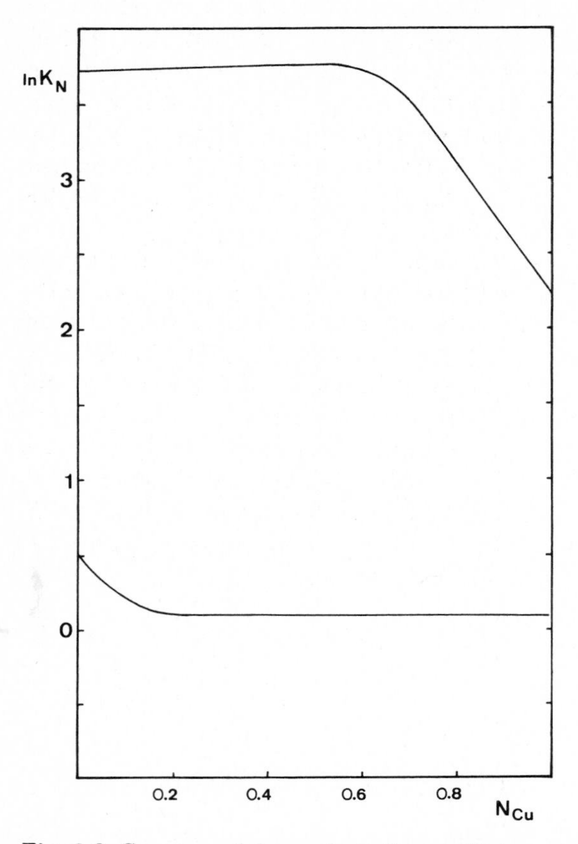

Fig. 6.8. Copper—calcium selectivity coefficients versus ion exchanger composition in the absence (lower curve) and presence of ethylenediamine in CB montmorillonite. (From Maes et al., 1978)

The $Cu(en)_2^{2+}$ complex is a rather stable one ($\log K_1 = 10.7$; $\log K_2 = 9.3$). Figure 6.7 shows a comparison of the exchange isotherms of the aqueous copper ion and the $Cu(en)_2^{2+}$ complex in calcium—CB montmorillonite. The copper species involved was identified as the two-complex and the equilibrium was shown to be both reversible and stoichiometric. Figure 6.7 shows clearly that complex formation strongly enhances the adsorption affinity: it is seen that, at a given composition of the clay, the concentration of copper in the equilibrium solution is shifted to lower values of some three orders of magnitude. The ion exchange data are summarized in Fig. 6.8 showing the logarithm of the selectivity coefficient as a function of clay composition. The standard free enthalpies, as obtained from graphical integration (Gaines and Thomas, 1953) (cf. chapter 2), are -0.08 kcal/eq (Cu/Ca) and -2.08 kcal/eq ($Cu(en)_2$/Ca), corresponding to a free enthalpy effect of -2 kcal/eq for the displacement of the aqueous copper ion by the complex. On the assumption that the partition coefficient of the uncharged ligand can be taken as unity, this corresponds to a synergistic effect given by $K^{\circ}_{syn} \simeq 10^3$. On the basis of X-ray (rational spacing of 12.6 Å) and spectroscopic evidence, the structure of the complex in the interlayer space is identified as a square planar complex, similar to the one in bulk solution, differing only in the sense that, in the clay, the copper ion is axially coordinated to lattice oxygen instead of water molecules (a similar situation occurs in the case of the aqueous copper ion; Clementz et al., 1974).

The stabilization effect of the layer charge on the copper complex can be "dissected" into two terms by following the build-up of the complex in the clay as a function of free ligand concentration in solution (calculated from pH, copper content, en content, pK values of en and stability constants of the complex). This is illustrated in Fig. 6.8 showing a comparison of the ligand number versus the log of the free ligand concentration in bulk solution and montmorillonite. It is clearly seen that complex formation in the clay is initiated at a much earlier stage and is nearly completed at ligand concentrations at which complex formation in solution is barely starting. The adsorbed complex can be characterized in terms of the stepwise formation constants within the clay interlayers which are nearly identical: $\log K_1$ =

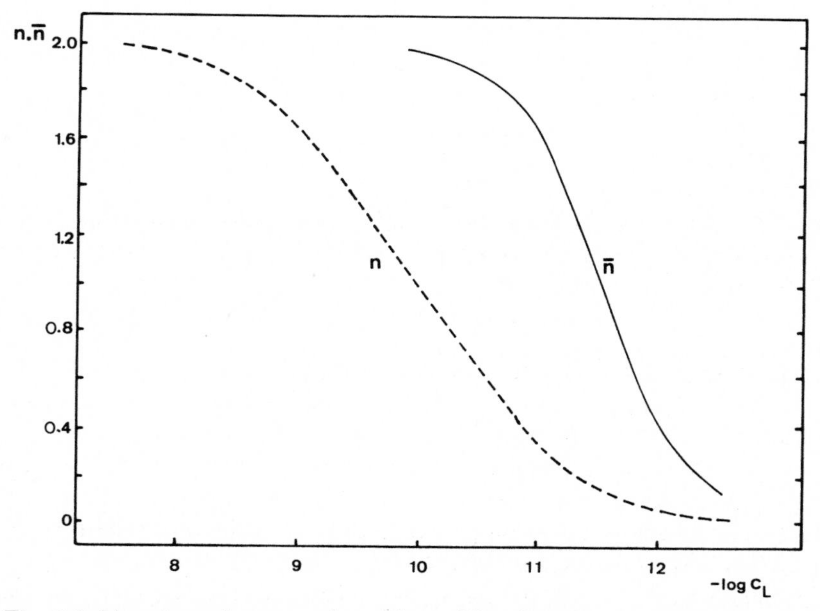

Fig. 6.9. Ligand number versus logarithm of free ligand concentration, c_L, for the copper—ethylenediamine complex in solution (n) and for CB montmorillonite ($\bar{n}$). (From Maes et al., 1978)

11.6, and $\log K_2 = 11.5$. Comparison of the bulk solution constants show that the overall stability constant in the ion exchanger is higher by three orders of magnitude, a result which agrees remarkably well with that obtained by ion exchange.

The effect of charge density upon the properties of the $Cu(en)_2^{2+}$ complex was studied by carrying out similar distribution measurements on reduced charge montmorillonite. The excess stability of the complex was found to increase linearly with charge density and, as expected, extrapolated to zero at vanishing charge density.

The stabilization of the ethylenediamine complexes with nickel, zinc, cadmium, and mercury is of the same order of magnitude and varies between -1.7 to -2.7 kcal/eq. Moreover, a comparison of ion exchange data between the ethylenediamine complexes of these ions showed that these adsorption processes are thermodynamically reversible (Peigneur, 1976).

b. The silver—thiourea complex

A second rather typical case of complex stabilization, which may be put to use in soil chemistry, is that of silver—thiourea (AgTU) in clays. Silver forms a series of very stable complexes with thiourea, as exemplified in the stepwise stability constants:

$$\log K_1 = 7.04, \quad \log K_2 = 3.57, \quad \log K_3 = 2.23, \quad \log K_4 = 0.78.$$

The ion exchange affinity of the AgTU complex for clay minerals is extremely high, as illustrated in Fig. 6.10 which shows the ion exchange isotherms at 0.01 total normality in Ca- and Al-montmorillonite. It appears that, at equivalent fractions of 1% of the silver complex in the solution, the clay is nearly saturated with AgTU. The free enthalpy of exchange for the reversible displacement of sodium, calcium, and aluminum ions is near -4 kcal/eq (Pleysier and Cremers, 1975) and can be described in terms of a drastic stabilization of the one-complex in the clay ($\log K_1 = 10.4$) and a slight destabilization of the three-complex ($\log K_3 \simeq 1.3$).

Stabilization of the silver—thiourea complex is not limited to montmorillonite but occurs in other clay minerals as well, such as illite and kaolinite, the overall effect being of the same order of magnitude (-3.5 to -4 kcal/eq). This is illustrated in Fig. 6.11 for Na-kaolinite in which the ion exchange adsorption of the silver complex is nearly quantitative up to complete saturation, which is reached at a 0.01 equivalent fraction of silver in the solution. Surprisingly enough, the same phenomenon occurs in vermiculite in which the complex penetrates into the interlamellar space. This is illustrated in Fig. 6.12, which shows a comparison of the ion exchange adsorption of the silver complexes with thiourea and ethylamine. Apparently the exchange of the silver—ethylamine complex (with shows a stabilization effect of about 10^3 in montmorillonite; Pleysier, 1976) is restricted to the external surface.

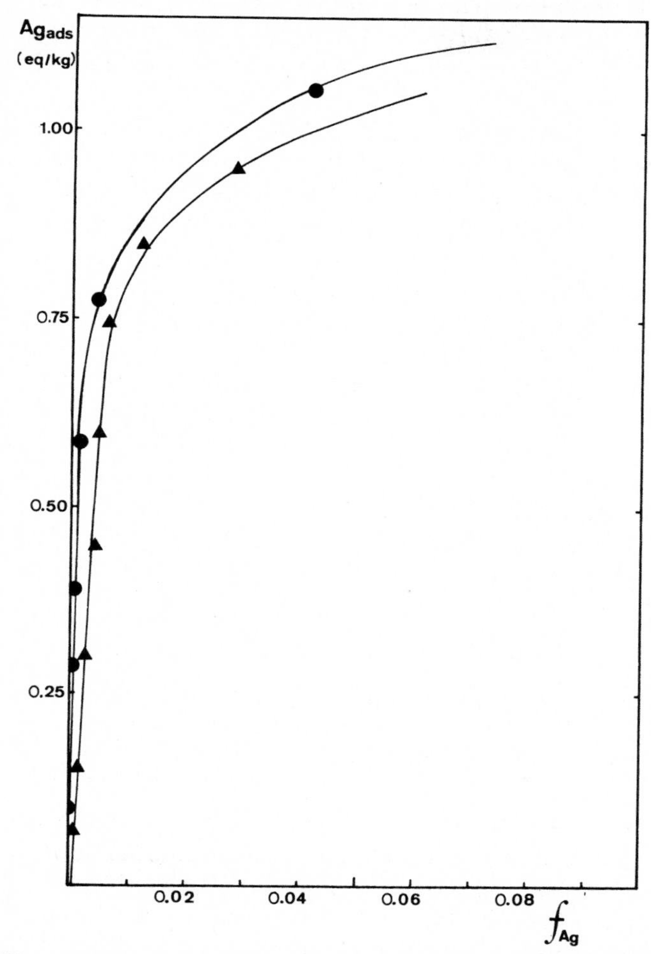

Fig. 6.10. Silver—thiourea exchange isotherms in calcium (dots) and aluminum montmorillonite (triangles); f_{Ag} refers to the equivalent fraction of AgTU in the equilibrium solution (0.01 total normality). (From Pleysier and Cremers, 1975)

Evidently, the very high affinity of this complex for clays may be useful in determining cation exchange capacities and exchangeable cations in soils. The possibilities of relying on the silver—thiourea complex as an index cation in these measurements have been tested and a new method has been proposed (Chabra et al., 1976) which agrees very closely with the currently used ammonium acetate method. The one-step method is based upon a single treatment of the soil sample with a silver—thiourea solution, buffered at pH = 7 with 0.1 *M* ammonium acetate. From the analysis of the extract for silver, calcium, magnesium, sodium, and potassium, the exchange capacity and the exchangeable ions can be obtained. The agreement with other

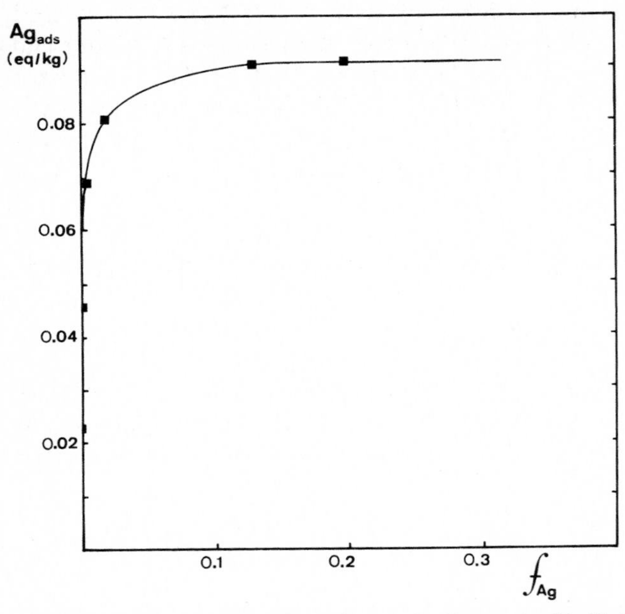

Fig. 6.11. Silver—thiourea exchange isotherm in Na-Zettlitz kaolinite (0.01 total normality). (From Pleysier, 1976)

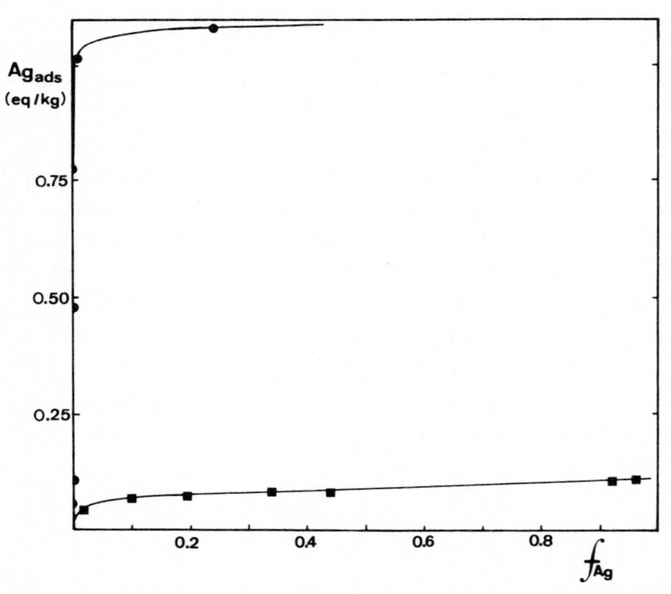

Fig. 6.12. Adsorption isotherm of the silver—thiourea (dots) and silver—ethylamine (squares) complexes in vermiculite. (Pleysier, 1976)

methods was very good for a wide variety of soils, including calcareous and saline and alkali soils, and for soils containing high amounts of organic matter. The nature of the agreement is perhaps best illustrated for the data on CEC and base-saturation percentage obtained for 30 podzolic and anthropogenic soils. The correlation for the CEC data (in meq per 100 g) between the AgTU and NH_4OAc methods is given by the relation:

$$\text{CEC(AgTU)} = -0.42 + 1.006\ \text{CEC}(NH_4OAc)$$

with r = 0.987. The corresponding relation for the base-saturation percentage (BS) is:

$$\text{BS(AgTU)} = -0.19 + 1.052\ \text{BS}(NH_4OAc)$$

with r = 0.983. The use of a high affinity index cation such as AgTU may be particularly useful in calcareous soils containing high amounts of $CaCO_3$, in which the continuous solubilization of calcium ions prevents complete saturation of the exchange complex.

6.3. CONCLUDING COMMENTS

The foregoing shows that the displacement of solvent molecules from the metal ion coordination sphere by other uncharged ligands occurs much more favorably in the (hydrated) clay than in the bulk solution, and that these effects are promoted by increasing charge density. The fact that the clay charge induces such drastic changes in thermodynamic stability without any apparent effect on the composition or conformation of these complexes is rather significant and it is tempting to ascribe these effects, at least in part, to possible changes in hydration of adsorbed metal ions. Of course, the observed synergistic effects refer to the overall water—ligand displacement reaction and may be due to interaction changes in either metal—solvent or metal-ligand, or both. In any case, the finding that the dependence on charge density of the stabilization of the metal complex is of a similar nature to what is found in homovalent exchange reactions, such as sodium—cesium, deserves some serious thought. In conclusion, it should be emphasized that small differences in ion exchange affinity, as found among transition metal complexes and alkaline earth metal ions, do not necessarily imply that the properties of these species are not altered upon entering the clay phase.

REFERENCES

Barrer, R.M., Davies, J.A. and Rees, L.V.C., 1969. Thermodynamics and thermochemistry of cation exchange in chabazite. J. Inorg. Nucl. Chem., 31: 219—232.

Barrer, R.M. and Jones, D.L., 1971. Ion exchange and ion fixation in fluorhectorites. J. Chem. Soc. A., 3: 503—508.

Barrer, R.M. and Klinowski, J., 1972. Influence of framework charge density on ion exchange properties of zeolites. J. Chem. Soc. Faraday, Trans. I, 68: 1950—1963.

Bodenheimer, W., Heller, L., Kirson, B. and Yariv, S., 1962. Organo-metallic clay complexes part II. Clay Miner. Bull., 5: 145—154.

Bodenheimer, W., Heller, L., Kirson, B. and Yariv, S., 1963. Organo-metallic clay complexes IV. Nickel and mercury aliphatic polyamines. Isr. J. Chem., 1: 391—403.

Bodhenheimer, W., Kirson, B. and Yariv, S., 1963b. Organo-metallic clay complexes, Part I. Isr. J. Chem., 1: 69—78.

Bodenheimer, W., Heller, L., Kirson, B. and Yariv, S., 1973. Organo-metallic clay complexes. Part III. Copper-polyamine-clay complexes. Proc. Int. Clay Conf., 1: 351—363.

Bolt, G.H. and Winkelmolen, C.J.G., 1968. Calculation of standard free energy of cation exchange in clay systems. Isr. J. Chem., 6: 175—187.

Bolt, G.H., Shainberg, I. and Kemper, W.D., 1967. Discussion of the paper by Shainberg and Kemper in Soil Sci., 103: 4—9; Soil Sci., 104: 444—453.

Boyd, C.E., 1970. Thermal effects in ion-exchange reactions with organic exchangers: enthalpy and heat capacity changes. In: Ion Exchange in the Process Industries. Soc. Chem. Ind., London, pp. 261—269.

Breck, D.W., 1973. Zeolite Molecular Sieves. Wiley, New York, N.Y., 784 pp.

Chabra, R., Pleysier, J. and Cremers, A., 1976. The measurement of the cation exchange capacity and exchangeable cations in soils: a new method. Proc. Int. Clay Conf., 1: 439—449.

Clementz, D.M., Mortland, M.M. and Pinnavaia, T.J., 1974. Properties of reduced charge montmorillonite: hydrated Cu(II) ions as a spectroscopic probe. Clays Clay Miner., 22: 49—57.

Cloos, P., Fripiat, J.J., Poncelet, G. and Poncelet, A., 1965. Comparison entre les propriétés d'échange de la montmorillonite et d'une résine vis-à-vis des cations alcalins et alcalino-terreux. I. Réversibilité des processus. II. Phénoméne de sélectivité. Bull. Soc. Chim. Fr., pp. 208—219.

Cockerell, L. and Walton, H.F., 1962. Metal-amine complexes in ion exchange. II. 2-Aminoethanol and ethylenediamine complexes. J. Phys. Chem., 66: 75—78.

Cremers, A. and Pleysier, J., 1973. Adsorption of the silver-thiourea complex in montmorillonite. Nature, Phys. Sci., 243: 86—87.

Cremers, A. and Thomas, H.C., 1968. The thermodynamics of sodium-cesium exchange on Camp Berteau montmorillonite: an almost ideal case. Isr. J. Chem., 6: 949—957.

Das Kanongo, J. L. and Chakravarti, S.K., 1973. Equilibrium behavior of exchange of $Co(en)_3^{3+}$ on vermiculite. Kolloid Z. Z. Polym. 251: 154—158.

Dufey, J., 1974. Ion Mobilities in Clay Suspensions. Thesis, Univ. Louvain.

Eliason, J.R., 1966. Montmorillonite exchange equilibria with strontium-sodium, cesium. Am. Mineral., 51: 324—335.

El Sayed, M.H., Burau, R.G. and Babcock, K.L., 1971. Reaction of copper tetrammine with bentonite clay. Soil Sci. Soc. Am. Proc., 35: 571—574.

Foscolos, A.E., 1968. Cation exchange equilibrium constants of aluminum saturated montmorillonite and vermiculite clay. Soil Sci. Soc. Am. Proc., 32: 350—354.

Fripiat, J.J. and Helsen, J., 1966. Use of cobalt hexamine in the cation exchange capacity determination of clays. Clays Clay Miner., 14: 163—169.

Gaines, G.L. and Thomas, H.C., 1953. Adsorption studies on clay minerals. II. A formulation of the thermodynamics of exchange adsorption. J. Chem. Phys., 21: 714—718.

Gast, R.G., 1969. Standard free energies of exchange for alkali metal cations on Wyoming bentonite. Soil Sci. Soc. Am. Proc., 33: 37—41.

Gast, R.G., 1972. Alkali metal cation exchange on Chambers Montmorillonite. Soil Sci. Soc. Am. Proc., 36: 14—19.

Gast, R.G., Van Bladel, R. and Deshpande, K.B., 1969. Standard heats and entropies of exchange for alkali metal cations on Wyoming bentonite. Soil Sci. Soc. Am. Proc., 33: 661—664.

Glaeser, R., Beguinot, S. and Mering, M.J., 1972. Détection et dénobbrement des charges à localisation tétraédriques dans les smectites dioctaédriques. C.R. Acad. Sci., Paris, 274: 1—4.

Glaeser, R. and Mering, M.J., 1967. Effet de chauffage sur les montmorillonites saturées de cations de petit rayon. C.R. Acad. Sci., Paris, 265: 833—835.

Glaeser, R. and Mering, M.J., 1971. Migration des cations Li dans les smectites di-octaédriques (effet Hofmann-Klemen). C.R. Acad. Sci., Paris, 173: 2399—2402.

Heller, L. and Yariv, S., 1969. Sorption of some anilines by Mn, Cs, Ni, Cu, Zn and Cd-montmorillonite. Proc. Int. Clay Conf., 1: 741—755.

Hofmann, U. and Klemen, R., 1950. Verlust der Austauschfähigkeit von Lithium-ionen an Bentoniet durch Erhitzung. Z. Anorg. Allg. Chem., 262: 95—99.

Hutcheon, A.T., 1966. Thermodynamics of cation exchange on clay. Ca-K montmorillonite. J. Soil Sci., 17: 339—355.

Hutschneker, K. and Deuel, H., 1956. Ionengleichgewichte an Kationenaustauschern verschiedener Austauschkapazität. Helv. Chim. Acta, 39: 1038—1045.

Laudelout, H., Van Bladel, R., Bolt, G.H. and Page, A.L., 1968. Thermodynamics of heterovalent cation exchange reactions in a montmorillonite clay. Trans. Faraday Soc., 64: 1477—1488.

Laudelout, H., Van Bladel, R. and Robeyns, J., 1972. Hydration of cations adsorbed on a clay surface from the effect of cation activity on ion exchange selectivity. Clays Clay Miner., 36: 30—34.

Laura, R.D. and Cloos, P., 1970. Adsorption of ethylenediamine on montmorillonite saturated with different cations. Reun. Hispano-Belga de Minerales de la Arcilla, Madrid, pp. 76—86.

Lewis, R.J. and Thomas, H.C., 1963. Adsorption studies on clay minerals VIII. A consistency test of exchange sorption in the systems sodium-cesium, barium montmorillonite. J. Phys. Chem., 67: 1781—1783.

Maes, A. and Cremers, A., 1977. Charge density effects in ion exchange. Part 1. Heterovalent exchange equilibria. J. Chem. Soc. Faraday Trans. I, 73: 1807—1814.

Maes, A. and Cremers, A., 1978. Charge density effects in ion exchange. Part 2. Homovalent exchange equilibria. J. Chem. Soc. Faraday Trans. I, 74: 1234—1241.

Maes, A., Marynen, P. and Cremers, A., 1977. Stability of metal—(uncharged) ligand complexes in ion exchangers. Part i. Quantitative characterization and thermodynamic basis. J. Chem. Soc. Faraday Trans. I, 73: 1297—1301.

Maes, A., Peigneur, P. and Cremers, A., 1978. Stability of metal—(uncharged) ligand complexes in ion exchangers. Part 2. The copper-ethylene-diamine complex in montmorillonite and sulphonic acid resin. J. Chem. Soc. Faraday Trans. I, 74: 182—189.

Mantin, I., 1969. Mesure de la capacité d' échange des minéraux argileux par l'éthylénediamine et les ions complexes de l'éthylénediamine. C.R. Acad. Sci. Paris, 269: 815—818.

Martin, H. and Laudelout, H., 1963. Thermodynamique de l'échange des cations alcalins dans les argiles. J. Chim. Phys., 60: 1086—1099.

Mortland, M.M., 1970. Clay organic complexes and interactions. Adv. Agron., 22: 75—117

Norrish, K., 1972. Forces between clay particles. Proc. 15th Int. Clay Conf., 375—389.

Peigneur, P., 1976. Stability and Adsorption Affinity of Some Transition Metal-Amine Complexes in Aluminosilicates. Thesis, Univ. Louvain.

Pleysier, J., 1976. Silver Uncharged Ligand Complexes in Aluminosilicates: Adsorption and Stability. Thesis, Univ. Louvain.

Pleysier, J. and Cremers, A., 1975. Stability of silver-thiourea complexes in montmorillonite clay. Chem. J. Soc. Faraday Trans. I, 71: 256—264.

Ravina, I. and Low, P.F., 1972. Relation between swelling, water properties and b-dimension in montmorillonite-water systems. Clays Clay Miner., 20: 109—123.

Reichenberg, D., 1966. Ion exchange selectivity. In: J.A. Marinsky (Editor), Ion Exchange. Marcel Dekker, New York, N.Y., pp. 227—276.

Schultz, L.G., 1969. Lithium and potassium adsorption, dehydroxylation temperature and structural water content of aluminous smectites. Clays Clay Miner., 17: 115—149.

Shainberg, I. and Kemper, W.D., 1966a. Hydration status of adsorbed cations. Soil Sci. Soc. Am. Proc., 30: 707—713.

Shainberg, I. and Kemper, W.D., 1966b. Electrostatic forces between clay and cations as calculated and inferred from electrical conductivity. Clays Clay Miner., Proc. 14th Natl. Conf., pp. 117—132.

Shainberg, I. and Kemper, W.D., 1967. Ion exchange equilibria on montmorillonite. Soil Sci., 103: 4—9.

Sherry, H.S., 1966. The ion exchange properties of zeolites I. Univalent ion exchange in synthetic faujasite. J. Phys. Chem., 70: 1158—1168.

Sillén, L.G. and Martell, A.E., 1964. Stability Constants of Metal-Ion Complexes. The Chemical Society, London.

Stokes, R.H. and Walton, H.F., 1954. Metal-amine complexes in ion exchange. J. Am. Chem. Soc., 76: 3327—3331.

Stul, M.S. and Mortier, W.J., 1974. The heterogeneity of the charge density in montmorillonite. Clays Clay Miner., 22: 391—396.

Suryaraman, M.G. and Walton, H.F., 1962. Metal-amine complexes in ion exchange. III. Diamine complexes of silver(I) and nickel (II). J. Phys. Chem., 66: 78—81.

Swartzen Allen, S.L. and Matijevic, E., 1975. Colloid and surface properties of clay suspensions. II. Electrophoresis and cation adsorption of montmorillonite. J. Colloid Interface Sci., 50: 143—153.

Tabikh, A., Barshad, I. and Overstreet, R., 1960. Cation exchange hysteresis in clay minerals. Soil Sci., 90: 219—226.

Theng, B.V.G., 1974. The Chemistry of Clay Organic Reactions. Adam Hilger, London.

Van Bladel, R. and Menzel, R., 1969. A thermodynamic study of sodium-strontium exchange on Wyoming bentonite. Proc. Int. Clay Conf., 1: 619—634.

Van Bladel, R., Gaviria, G. and Laudelout, H., 1972. A comparison of the thermodynamic double layer theory and empirical studies of the Na-Ca exchange equilibria in clay water systems. Proc. 15th Int. Clay Conf., pp. 385—398.

Wild, A. and Keay, J., 1964. Cation exchange equilibria with vermiculite. J. Soil Sci., 15: 135—144.

CHAPTER 7

ANION EXCLUSION IN SOIL

G.H. Bolt and F.A.M. de Haan

As was pointed out in chapter 1 the exclusion of anions from the liquid zone adjacent to negatively charged surfaces offers good perspectives for quantitative application of the theory of the diffuse double layer. Summarizing the arguments in favor of this conclusion one may state that although the co-ionic concentration very near to the surface is subject to the same uncertainties as that of the counterions, the *deficit* of these ions, $(c_{0,-} - c_{s,-})$, remains insensitive to those as long as $c_{0,-} \gg c_{s,-}$. With respect to the influence of the type of anion on its exclusion one may thus conclude that the exclusion will deviate considerably from predicted theoretical values only if a specific binding energy between surface and co-ion exists in excess of several times RT.

The influence of the accompanying counterion on the co-ion exclusion likewise remains small, unless specific adsorption of these counterions in a first layer reduces grossly the extent of the double layer. With reference to section 7.4 for details it is mentioned here that halving the density of charge of clay surfaces by tightly bonded counterions causes a reduction in the extent of the double layer of only a few Å (cf. (1.37) which relates the cut-off distance δ to the charge density).

In case of truncated double layers the double layer theory appears amply sufficient for estimates as in this case one soon arrives at almost complete exclusion from the entire layer. At the same time predictions then become dominated by the often existing uncertainty of the thickness of truncated layers.

The applications of the calculations of the anion exclusion in soil systems are manyfold. In chapter 3 these were already used to estimate the change of the selectivity coefficient for cations, K_N, as it is influenced by the total electrolyte concentration. Other applications comprise the determination of the surface area of charged surfaces present in soil, as used by Schofield (1947, 1948), Bolt and Warkentin (1956, 1958), Edwards et al. (1965), De Haan (1965a, 1965b), the determination of positive adsorption of particular anions from the observed net adsorption corrected with the calculated anion exclusion, as given by De Haan (1964, 1965b) and the effect on the concentration of the pressure filtrate of soil suspensions, as treated by Bolt (1961). In section 7.4 some illustrative examples of these applications will be given.

7.1. BASIC EQUATIONS

The deficit of an anion a, with valence z_a (taken negative), in a diffuse layer formed on a negatively charged planar surface is found as:

$$\Gamma_a = f_a C_0 \int_{\delta}^{x_t} (1 - 1/u^{|z_a|}) dx \tag{7.1}$$

where $f_a C_0$ is the concentration of the anion in the equilibrium solution with total concentration C_0, u signifies again the Boltzmann accumulation factor for monovalent cations, δ and x_t are the cut-off values on the distance axis as introduced in chapter 1. Referring to Fig. 7.1, giving the trajectory of

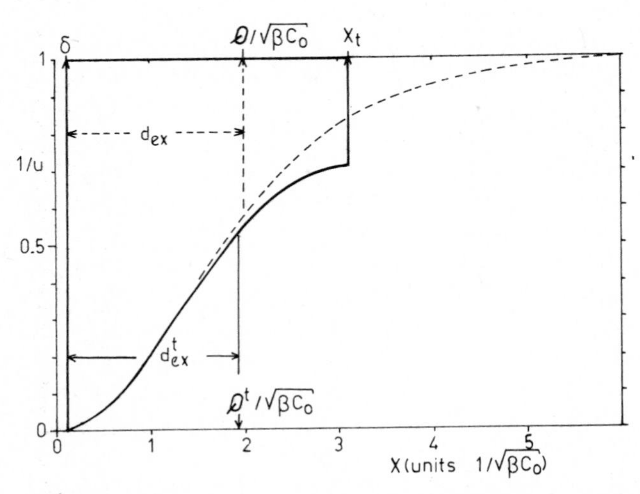

Fig. 7.1. Distribution of co-ions in a diffuse double layer. Solid line: concentration of co-ions relative to the equilibrium concentration, $c/c_0 \equiv 1/u$, in a truncated layer with thickness $(x_t - \delta)$.
Dashed line: same in freely extended layer.
Area bounded by heavy lines: relative amount of co-ions excluded per unit surface area, $\Gamma\text{-}/C_0$, which corresponds to an equivalent distance of exclusion, d^t_{ex}.
Area bounded by dashed line: same in freely extended double layer, corresponding to $d_{ex} > d^t_{ex}$.
The distance δ corresponds to a surface with charge density $\Gamma = 10^{-9}$ keq/m^2 (≈ 0.1 Coul/m^2) in equilibrium with 0.01 *N* mono-monovalent salt.

1/u, one may identify the exclusion of a monovalent anion a with the product of the surface area bordered by heavy lines as shown and the equilibrium concentration $f_a C_0$. This surface area, corresponding to the integral of (7.1), has the dimension of length and will be referred to as the equivalent distance of exclusion of the anion, d_{ex}: the exclusion is equivalent to an anion-free zone of thickness d_{ex}. For the present truncated double layer with a liquid film of thickness $d_1 = (x_t - \delta)$ one may thus write:

$$d^t_{ex,a} \equiv \Gamma_a/f_a C_0 = \int_\delta^{x_t} (1 - 1/u^{|z_a|})dx \tag{7.2}$$

Making use of (1.9) one finds for the exclusion distance of anion a present in a mixture containing a number of ion species k with valence z_k and concentration f_k:

$$\sqrt{\beta C_0}\, d^t_{ex,a} = -\int_{u_s}^{u_t} \frac{(1 - 1/u^{|z_a|})du}{u\sqrt{\Sigma_k\{f_k u^{z_k}/|z_k| - f_k u_t^{z_k}/|z_k|\}}}$$

where the subscripts s and t refer to the surface and plane of truncation. Although (7.3) may be solved in its present form for a wide variety of mixture compositions (either numerically or analytically), such a solution requires first the evaluation, by means of the procedures suggested in chapter 1, of u_s and u_t for given composition, surface charge density and thickness of the liquid layer. For most practical purposes this involved solution may be avoided, as $\sqrt{\beta C_0}\, d^t_{ex}$ is rather insensitive towards the precise values of u_s (if sufficiently large) and u_t (if not too far from unity). This is demonstrated in Fig. 7.1, where the course of 1/u for the freely extended double layer (formed on a surface with $\Gamma \rightarrow \infty$ for the same composition of the equilibrium solution) shows that the anion exclusion for such a system is only slightly larger than in the previous system (cf. the surface area above the dashed curve ranging from $1/u = 0$ to $1/u = 1$). This maximum value of the anion exclusion for a given composition of the equilibrium solution then corresponds to the integral of (7.3) with u_t in the numerator replaced by unity, and the limits ranging from $u = \infty$ to $u = 1$. This integral then delivers a number Q_a which depends solely on the mixture composition as characterized by all f_k. It specifies the exclusion distance for the freely extended double layer on surfaces with infinite charge density, according to:

$$\sqrt{\beta C_0}\, d_{ex,a} \equiv Q_a = \int_1^\infty M(u)du \tag{7.4}$$

where M(u) is the integrand of (7.3) with $u_t = 1$. The parameter Q_a thus constitutes the limiting (maximum) value of d^t_{ex} in units of $1/\sqrt{\beta C_0}$. In the case pictured in Fig. 7.1 one may observe that $Q_a = 2$ as against $d^t_{ex,a} = 1.8$ units of $1/\sqrt{\beta C_0}$.

Recalling that the extent of a double layer for a given value of C_0 depends particularly on the valence of the dominant cation, while the exclusion of a particular anion for a given extent of the double layer depends on its own valence, one may easily establish in turn the limits of Q_a. Thus a maximum is found in a system containing essentially a mono-monovalent salt (i.e. maximum extent), for the exclusion distance pertaining to a trace of a polyvalent anion. As will be shown in the following section Q_a then reaches a value of about three (for trivalent anions), implying an exclusion distance of 3 units of $1/\sqrt{\beta C_0}$. Conversely Q_a reaches a minimum value of about 0.6 for a trace of monovalent anions in a tri-trivalent salt.

Once Q_a has been found, one may apply corrections allowing for the effects of truncation and of a finite charge density of the surface.

7.2. ANION EXCLUSION IN THE FREELY EXTENDED DOUBLE LAYER

Following De Haan (1965b), (7.4) may be written for the common soil system containing mono- and divalent cations and anions. Characterizing the equilibrium solution for such systems with:

$$c_{0,2+} = f_c C_0 \,;\, c_{0,+} = (1-f_c)C_0 \,;\, c_{0,2-} = f_a C_0 \,;\, c_{0,-} = (1-f_a)C_0$$

one finds:

$$Q_- = \int_1^{\infty} \frac{(1-1/u)du}{\sqrt{f_c u^4/2 + (1-f_c)u^3 - (4-f_c-f_a)u^2/2 + (1-f_a)u + f_a/2}} \qquad (7.5)$$

$$= \int_1^{\infty} \frac{\sqrt{2}du}{u\sqrt{f_c u^2 + 2u + f_a}} \qquad (7.6a)$$

For the divalent anion an expression is found similar to (7.5) except that the factor in the numerator now equals $(1-1/u^2)$. This gives upon simplification:

$$Q_{2-} = \int_1^{\infty} \frac{\sqrt{2}(1+1/u)du}{u\sqrt{f_c u^2 + 2u + f_a}} \qquad (7.6b)$$

Solution of these integrals gives:

$$Q_- = \sqrt{2/f_a}\left[\text{arg sinh}\sqrt{\frac{f_a(f_c+2+f_a)}{1-f_c f_a}} - \text{arg sinh}\sqrt{\frac{f_a f_c}{1-f_c f_a}}\right] \qquad (7.7a)$$

and:

$$Q_{2-} = \sqrt{2/f_a}\left[\sqrt{\frac{f_c+2+f_a}{f_a}} - \sqrt{\frac{f_c}{f_a}}\right] - \frac{1-f_a}{f_a}Q_- \qquad (7.7b)$$

In the case of single salt systems the values of f_c and f_a are zero or unity and the inverse hyperbolics may be replaced by their arguments. This gives the expressions used by Schofield (1947), i.e.:

$Q_- = 2$ for a mono-monovalent salt,

$Q_- = (\sqrt{6} - \sqrt{2})$ for a di-monovalent salt,

$Q_{2-} = \sqrt{6}$ for a mono-divalent salt,

$Q_{2-} = \sqrt{2}$ for a di-divalent salt.

TABLE 7.1

Values of Q for mixed systems containing mono- and divalent cations and anions

f_{2+}		f_{2-}					
		0.0	0.2	0.4	0.6	0.8	1.0
	Q_-						
0.0		2.000	1.968	1.938	1.912	1.886	1.862
0.2		1.465	1.441	1.419	1.398	1.379	1.361
0.4		1.296	1.277	1.259	1.242	1.226	1.209
0.6		1.171	1.168	1.153	1.139	1.125	1.109
0.8		1.102	1.088	1.074	1.062	1.050	1.037
1.0		1.035	1.023	1.008	1.000	0.989	0.980
	Q_{2-}						
0.0		2.667	2.615	2.572	2.526	2.486	2.449
0.2		2.066	2.028	1.991	1.958	1.920	1.897
0.4		1.854	1.822	1.792	1.764	1.738	1.713
0.6		1.708	1.681	1.655	1.631	1.609	1.588
0.8		1.596	1.573	1.551	1.530	1.510	1.492
1.0		1.507	1.486	1.471	1.448	1.431	1.414

In Table 7.1, taken from De Haan (1965b), the values of Q are given for mixed systems. It may be noted that the column with $f_a = 1$ for Q_- corresponds to systems containing a trace of monovalent anions, while Q_{2-} in the column $f_a = 0$ pertains to trace amounts of divalent anions.

In the absence of divalent cations the equilibrium solution may also contain high concentrations of trivalent anions, e.g. phosphate ions. As was shown by De Haan (1965b) the relevant form of (7.4) may then be solved in terms of elliptic integrals. Specifying the solution composition in this case as:

$$c_{0,+} = C_0\,;\, c_{0,-} = (1 - f_{2-} - f_{3-})C_0\,;\, c_{0,2-} = f_{2-}C_0\,;\, c_{0,3-} = f_{3-}C_0$$

one finds the corresponding values for Q as given in Table 7.2, taken from De Haan (1965b). It is noted that in this case a maximum value of Q_{3-} is found for $f_{2-} = f_{3-} = 0$, corresponding to a trace amount of e.g. PO_4^{3-} ions in a system containing otherwise only a mono-monovalent salt.

In the presence of divalent cations it is usually satisfactory to treat the trivalent anions as being present in negligible amounts compared to the other ions in the system. In that case (7.6a) and (7.6b) remain valid and may be supplemented with a third equation specifying the exclusion distance for trivalent anions present in trace amounts in terms of:

$$Q_{(3-)} = \int_1^\infty \frac{\sqrt{2}\,(1 + 1/u + 1/u^2)\,du}{u\sqrt{f_c u^2 + 2u + f_a}} \qquad (7.6c)$$

TABLE 7.2

Values of Q for systems containing mono-, di- and trivalent anions and monovalent cations

f_{2-}		f_{3-}					
		0.0	0.2	0.4	0.6	0.8	1.0
	Q_-						
0.0		2.000	1.948	1.905	1.866	1.829	1.793
0.2		1.968	1.919	1.876	1.838	1.806	—
0.4		1.939	1.892	1.854	1.819	—	—
0.6		1.912	1.870	1.832	—	—	—
0.8		1.886	1.848	—	—	—	—
1.0		1.863	—	—	—	—	—
	Q_{2-}						
0.0		2.667	2.578	2.510	2.449	2.391	2.334
0.2		2.615	2.534	2.469	2.410	2.356	—
0.4		2.572	2.492	2.432	2.378	—	—
0.6		2.526	2.457	2.400	—	—	—
0.8		2.487	2.424	—	—	—	—
1.0		2.450	—	—	—	—	—
	Q_{3-}						
0.0		3.067	2.952	2.857	2.778	2.708	2.648
0.2		3.002	2.906	2.816	2.739	2.674	—
0.4		2.945	2.854	2.773	2.701	—	—
0.6		2.890	2.807	2.729	—	—	—
0.8		2.840	2.760	—	—	—	—
1.0		2.796	—	—	—	—	—

TABLE 7.3

$Q_{(3-)}$ in systems containing a trace of trivalent anions in a mixture of mono-and divalent cations and anions

f_{2+}		f_{2-}					
		0.0	0.2	0.4	0.6	0.8	1.0
	$Q_{(3-)}$						
0.0		3.067	3.009	2.946	2.890	2.840	2.794
0.2		2.442	2.384	2.341	2.298	2.264	2.222
0.4		2.211	2.157	2.122	2.086	2.053	2.019
0.6		2.049	1.999	1.969	1.939	1.910	1.883
0.8		1.929	1.883	1.850	1.825	1.799	1.773
1.0		1.824	1.782	1.755	1.732	1.706	1.687

The solution of this integral may be written as:

$$Q_{(3-)} = \left[\sqrt{\frac{f_c + 2 + f_a}{2f_a^2}} + \frac{2f_a - f_c}{2f_a} Q_- + \frac{2f_a - 3}{2f_a}(Q_{2-} - Q_-) \right] \tag{7.7c}$$

and is tabulated in Table 7.3. As was shown by De Haan (1965b) the values given in Table 7.3 may safely be used for values up to $f_{3-} = 0.4$, at which value the overestimate of $Q_{(3-)}$ amounts to less than about 6%.

In order to find the exclusion distance pertaining to surfaces with *finite* charge density the integral given in (7.3) should also be evaluated between the limits $u = \infty$ and $u = u_s$. For high values of u_s, however, the term $1/u^{|z_A|}$ in the numerator remains insignificant throughout the range of integration. The integral then yields $\sqrt{\beta C_0}\,\delta$ in accordance with (1.10) with $x \equiv \delta$ at $u = u_s$. This is also demonstrated in Fig. 7.1 where the amount of anions situated between $x = 0$ and $x = \delta$ is insignificant. This then yields the convenient expression valid for freely extended double layers:

$$d_{ex,a} \approx \frac{Q_a}{\sqrt{\beta C_0}} - \delta \tag{7.8}$$

The applicability of (7.8) is indeed satisfactory for most systems of conern in soil chemistry, that is for surfaces with a charge density around 10^{-9} keq/m^2.

In that case δ is found to be about 2—4 Å (cf. (1.37)), while at 0.1 molar total electrolyte u_s equals about 5 for divalent counterions and 25 for monovalent counterions (cf. (1.39)). Neglect of $1/u$ in (7.3) may then cause an overestimate of about 10% for the distance to be subtracted from $Q/\sqrt{\beta C_0}$, i.e. no more than about 0.2 Å for the system with divalent cations. As the exclusion distance may never be evaluated experimentally within 1 Å this is of no concern. Only in cases of surfaces with a much lower charge density would it be worthwhile to evaluate (7.3) more precisely (cf. the relevant discussion in 7.4).

In conclusion it may be stated that for the freely extended double layer the exclusion distance amounts to 1—3 units of $1/\sqrt{\beta C_0}$ as read from the Tables 7.1—7.3, diminishing with δ as given by (1.37) and (1.38) (and usually not exceeding a few Å units).

7.3. THE EFFECT OF DOUBLE LAYER TRUNCATION ON ANION EXCLUSION

The depression of the anion exclusion due to double layer truncation is partly offset by the accompanying increase (in absolute value) of the electric potentials in the truncated layer. As in the case of extreme compression of the double layer, anions become virtually completely excluded, the exclusion distance will eventually approach the thickness of the liquid layer itself. Accordingly $d_{ex,a}$ will show a smooth transition between d_1 and $(Q_a/\sqrt{\beta C_0} - \delta)$.

In view of the rather cumbersome procedure involved in solving (7.3) for a wide variety of system compositions in conjunction with a series of values for d_1 it appears useful to attempt to summarize the situation in terms of one graph of general applicability as used by De Haan (1964). Introducing for this purpose $Q_a^t(x_t)$ — i.e. the value of Q_a as applicable in cases of truncation of the double layer at a distance from the charged surface equal to $d_1 = (x_t - \delta)$ — one finds in analogy with (7.8) the distance of exclusion as:

$$d_{ex,a}^t = Q_a^t/\sqrt{\beta C_0} - \delta \tag{7.9}$$

As $Q_a^t(x_t)$ approaches asymptotically to $x_t\sqrt{\beta C_0}$ for small values of x_t and to the constant Q_a for large values of x_t as compared to $(d_{ex} + \delta)$, the generalized graph sought is found by plotting Q_a^t/Q_a as a function of $x_t/(d_{ex} + \delta) = x_t\sqrt{\beta C_0}/Q_a$ for a variety of system compositions. As will be shown below (cf. Fig. 7.2) all systems then fall within a fairly narrow, curved wedge because of the common asymptotes for all systems. The maximum deviation of Q^t/Q from its enveloping asymptotes must then fall at unit value of $x_t\sqrt{\beta C_0}/Q$.

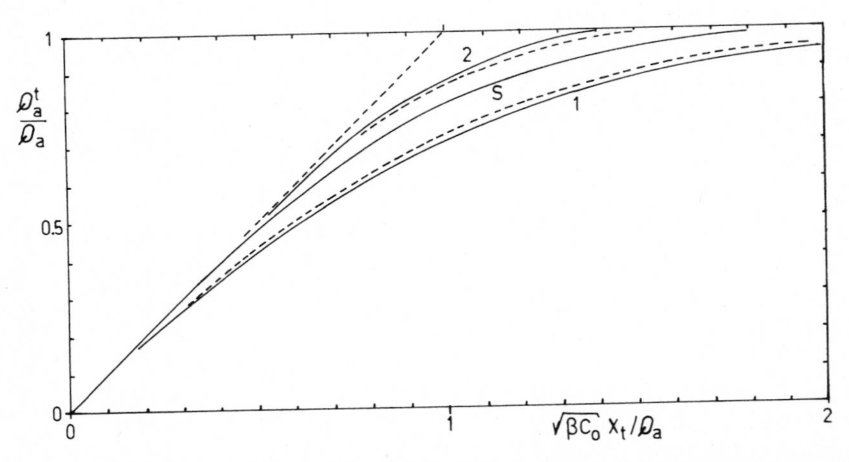

Fig. 7.2. Effect of double layer truncation on anion exclusion. Relative value of Q_a as a function of the thickness of the liquid layer, expressed as $(d_1 + \delta) \equiv x_t$ relative to the distance of exclusion in the freely extended double layer, $(d_{ex} + \delta) = Q_a/\sqrt{\beta C_0}$. Different lines refer to salt systems with valence compositon: s, single symmetric salt; 2, solid line (+ 1, − 2), dashed line (+ 1, − 1, − 2); 1, solid line (+ 2, − 1), dashed line (+ 2, − 2, − 1).

A "heart" line of the above wedge is now found for systems containing a single symmetric salt. The anion exclusion in such systems was calculated by Bolt and Warkentin (1958). Using (7.3) for a system containing a salt with $z_+ = -z_- = z$, at concentration C_0, gives:

$$Q^t_{z-} = \sqrt{\beta C_0}\,(d_{ex} + \delta) = \int_{u_t}^{\infty} \frac{(1 - 1/u^z)du}{u\sqrt{\{(u + u^{-z}) - (u_t^z + u_t^{-z})\}/z}} \tag{7.10}$$

$$= \frac{1}{\sqrt{z}} \int_{u_t}^{\infty} \frac{\{(1 - u_t^z) + (u_t^z - u^{-z})\}du^z}{\sqrt{u^z(u^z - u_t^z)(u^z - u_t^{-z})}} \tag{7.11}$$

Putting $(u_t/u)^z \equiv \sin^2\phi$ and $u_t^{-z} = \sin\alpha$ the above integral is readily expressed in terms of complete elliptic integrals of the first and second kind according to:

$$Q^t_{z-} = \frac{2}{\sqrt{z}}[\sqrt{u_t^z}\,E(\alpha, \pi/2 - (\sqrt{u_t^z} - \sqrt{u_t^{-z}})F(\alpha, \pi/2)] \tag{7.12}$$

With α going to zero for high values of u_t^z, $E(\alpha, \pi/2)$ and $F(\alpha, \pi/2)$ may be expanded to $(\pi/2)(1 - 1/4u_t^{2z})$ and $(\pi/2)(1 + 1/4u_t^{2z})$, respectively. In that case the E-term cancels part of the F-term, leaving only the term:

$$\frac{2}{\sqrt{zu_t^z}}\,F(\alpha, \pi/2) = x_t\sqrt{\beta C_0}$$

as may be verified with the help of (1.14). Conversely $u_t = 1$ causes cancellation of the F-term, leaving $E(\pi/2, \pi/2) = 1$. This gives $Q_- = 2$ and $Q_{2-} = \sqrt{2}$ which checks with Table 7.1.

In order to establish the width of the wedge in Fig. 7.2 one may solve (7.3) for single asymmetric salts of the mono-divalent (+ 1, − 2) and di-monovalent (+ 2, − 1) types. In the first case this gives:

$$Q^t_{2-} = \int_{u_t}^{\infty} \frac{(1 - 1/u^2)du}{\sqrt{(u - u_t)(u^2 - u/2u_t^2 - 1/2u_t)}} \tag{7.13}$$

$$= 2\sqrt{u_t - r_3}\,E(\alpha, \pi/2) - \frac{2(u_t - 1)}{\sqrt{u_t - r_3}}\,F(\alpha, \pi/2) \tag{7.14}$$

in which:

$$r_3 = (1/2u_t)^2 - \sqrt{(1/2u_t)^4 + 1/2u_t}$$

$$r_2 = (1/2u_t)^2 + \sqrt{(1/2u_t)^4 + 1/2u_t}$$

$$\alpha = \arcsin\sqrt{(r_2 - r_3)/(u_t - r_3)}$$

The corresponding curve of Q^t_{2-}/Q_{2-} as a function of $x_t\sqrt{\beta C_0}/Q_{2-}$ is shown in Fig. 7.2 and indicated by 2 as it refers to divalent anions. Similarly the di-monovalent system gives rise to:

$$Q^t_{-} = \int_{u_t}^{\infty} \frac{(1 - 1/u)du}{\sqrt{(u - u_t)u(u^2/2 + uu_t/2 - 1/u_t)}} \tag{7.15}$$

This equation leads to an expression for Q^t_- in terms of incomplete elliptic integrals of first and second kind, which has been given in De Haan (1965b) but is too lengthy to reproduce here. The corresponding curve 1 is also given in Fig. 7.2. Together the three curves show the entire influence of the cationic valence on the truncation effect, as the central curve s corresponds to both mono- and divalent anions in symmetric systems. The difference between curves s and 1 then gives the change of the truncation effect for monovalent anions if the cationic composition is changed from only monovalent cations to only divalent ones. Similarly the difference between 2 and s gives the effect on the divalent anions. As was shown by De Haan (1965b) the mixed cationic compositions are situated in between these extremes at a position roughly proportional to f_+.

In addition to the above effect of the cationic composition, which appears rather small for the chosen way of plotting, there is also a barely detectable influence of the overall *an*ionic composition in a mixture, on the truncation effect for a particular anion. As is shown by the dotted lines in Fig. 7.2 this effect amounts to two percent at most.

This effect of the overall anionic composition is conveniently assessed by contrasting the above single asymmetric salt systems with a *symmetric* one containing the same cation but only a trace of the anion a. Thus the value of Q^t_- in the above $(+2, -1)$ system may be compared with Q^t_- in the symmetric divalent system containing a trace of monovalent anions, here indicated as $\{+2, -2, (-1)\}$. Using equations (7.10) and (7.11) for this case gives:

$$Q^t_- = \frac{1}{\sqrt{2}} \int_{u_t}^{\infty} \frac{(1-1/u)\,du^2}{\sqrt{u^2(u^2-u_t^2)(u^2-u_t^{-2})}} \tag{7.16}$$

as the presence of a trace of the monovalent anion leaves the polynomial in the denominator unchanged. The first term in the numerator then produces again the elliptic integral of the first kind, while the second term reverts to a simpler integral in terms of u^2, according to:

$$Q^t_- = \sqrt{\frac{2}{u_t^2}}\,F(\alpha, \pi/2) - \frac{1}{\sqrt{2}} \int_{u_t}^{\infty} \frac{du^2}{u^2\sqrt{(u^2-u_t^2)(u^2-u_t^{-2})}}$$

$$= \sqrt{\frac{2}{u_t^2}}\,F(\alpha, \pi/2) - \ln\frac{u_t^2-1}{u_t^2+1} \tag{7.17}$$

with again $\alpha = \arcsin(u_t^{-1})$. The corresponding curve is dotted in Fig. 7.2 next to the previous curve 1.

Similarly the exclusion distance for a trace of divalent anions in a mono-monovalent system, to be indicated as $\{+1, -1, (-2)\}$ gives:

$$Q^t_{2-} = \int_{u_t}^{\infty} \frac{(1-1/u^2)\,du}{\sqrt{(u-u_t)(u^2-u/u_t)}} \tag{7.18}$$

$$= \frac{4\sqrt{u_t}}{3}\left[(u_t+1/u_t)E(\alpha, \pi/2) - (u_t-1/u_t)F(\alpha, \pi/2)\right] \tag{7.19}$$

with $\alpha = \arcsin(1/u_t)$, and shown as the dotted line adjacent to curve 2 in Fig. 7.2.

Equation (7.18) may also be amended to cover the system $\{+1, -1, (-3)\}$, i.e. trace amounts of trivalent anions in a symmetric monovalent salt. In that case the multiplier of the numerator reads $(1 - 1/u^3)$ and the ensuing equation may be solved by means of recurrence equations for elliptic integrals (cf. Bateman, 1953). The resulting curve practically coincides with curve 2 as shown in Fig. 7.2.

The two-step procedure to be used to estimate the anion exclusion in a mixed ionic system with truncated double layers (i.e. for a given thickness of the liquid layer, d_l) thus amounts simply to:
(a) The evaluation of the appropriate value of Q_a from Tables 7.1—7.3;
(b) Reading the correction factor Q_a^t/Q_a pertaining to the value of $(d_l + \delta)\sqrt{\beta C_0}/Q_a$, using linear interpolation between the curves for mixed cationic compositions.
The anion exclusion then follows from:

$$\Gamma_{-a} \equiv d_{ex,a}^t \, c_{0,a}$$
$$= c_{0,a} \left\{ \frac{Q_a^t}{Q_a} \frac{Q_a}{\sqrt{\beta C_0}} - \delta \right\} \qquad (7.20)$$

where δ is estimated by (1.37).

Using as an example a system containing 0.005 N each of Na^+, Ca^{2+}, Cl^-, SO_4^{2-} ions plus a trace level of PO_4^{3-} ions, one finds from Tables 7.1 and 7.3, respectively: $Q_- = 1.18$, $Q_{2-} = 1.72$, $Q_{(3-)} = 2.04$. At the chosen total electrolyte level of 0.01 N, $1/\sqrt{\beta C_0}$ equals about 31 Å units. Using $\delta = 2$ Å for a surface with a charge density of 10^{-9} keq/m^2 containing predominantly adsorbed cations of the divalent type one finds for a system with a liquid layer thickness $d_l = 40$ Å the appropriate value of $x_t\sqrt{\beta C_0}$ as 1.35. Reading Fig. 7.2 then gives for the monovalent anions at $x_t\sqrt{\beta C_0}/Q_- = 1.15$ the value of Q_-^t/Q_- as 0.82 (halfway between curves 1 and s).

Similarly Q_{2-}^t/Q_{2-} is read at $x_t\sqrt{\beta C_0}/Q_{2-} = 0.79$ midway between curves 2 and s to give the value 0.72. Finally $Q_{(3-)}^t/Q_{(3-)}$ is read at 0.66 on the distance axis and gives a value between 0.64 and 0.60, most probably 0.63. This then gives d_{ex}^t for the three anions as 27 Å(Cl^-), 36 Å(SO_4^{2-}) and 38 Å (PO_4^{3-}), respectively, implying that the PO_4^{3-} ions are almost completely excluded from the entire liquid layer, while the Cl^- content of the system is about one third of its maximum value equal to the product of concentration and liquid content of the system.

7.4. SOME EXPERIMENTAL INFORMATION ON ANION EXCLUSION PHENOMENA IN SOIL SYSTEMS AND CLAY SUSPENSIONS

Perhaps the most important aspect of the calculations given in the previous sections is its application to systems exhibiting simultaneously positive adsorption of anions on certain sites and anion exclusion because of the presence of negatively charged surfaces (cf. part A, section A5.2). With positive adsorption of anions (notably phosphates) by soil constituents which is often a case of highly specific adsorption or chemisorption, one finds steep adsorption isotherms. Thus the capacity for such adsorption becomes filled at low concentrations. In contrast, the exclusion of the same anion at negatively charged sites is non-specific and increases only proportionally to

$d_{ex} \, c_{0,a} \sim \sqrt{c_{0,a}}$ for systems containing only the particular anion. Accordingly, the net adsorption of such an anion, if plotted against its equilibrium concentration, may go through a maximum value. An example of such a system is given in Fig. 7.3, taken from De Haan (1965b). Addition to these net adsorption data of the anion exclusion values as calculated according to the previous sections gave a corrected isotherm for the positive adsorption of a shape to be expected. This indicates the need for scrutinizing all net adsorption data obtained with anions in soil systems for the possibly significant correction necessary because of anion exclusion. Estimates of the latter are easily obtained with the help of the Tables in the previous sections, provided an estimate of the surface area involved is available (see below).

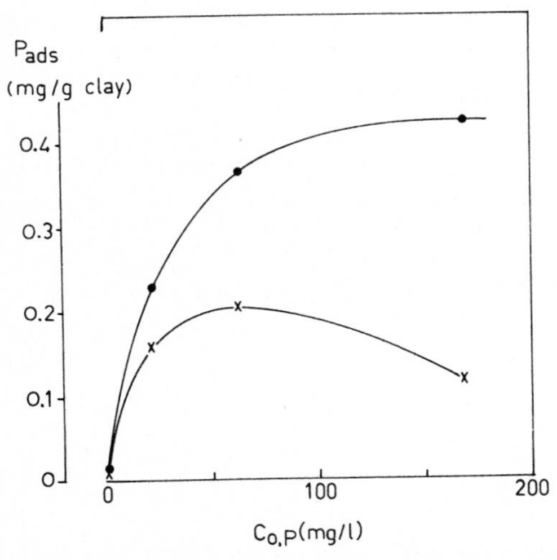

Fig. 7.3. Phosphate adsorption by Winsum illite, fraction $< 2\,\mu$ (after De Haan, 1965b). Crosses refer to net adsorption, dots are the values obtained after addition of the calculated exclusion of phosphate on the negatively charged surface.

The high specificity in the case of positive adsorption of anions may be used to advantage to separate the sites responsible for positive and negative adsorption (i.e. exclusion) of anions. Thus any positive adsorption of e.g. Cl^- ions (at low pH values) by edges of clay plates and oxide surfaces is effectively suppressed by maintaining a low level of, for example, phosphate ions in a mixture containing excess Cl^- ions. In such a case the anion exclusion is dominated by the chloride ions, while the positive adsorption is taken care of by the phosphate ions. This is of particular importance if anion exclusion data are to be used for the determination of the surface area of negatively charged soil constituents.

7.4.1. Determination of the specific surface area of soil colloids

The surface area determination by anion exclusion was introduced by Schofield (1947) and applied by several authors. For this purpose (7.2) is written as:

$$\frac{\bar{\gamma}_a}{f_a C_0} = \frac{S F_a}{f_a C_0} = S d_{ex,a} \tag{7.21}$$

in which S indicates the specific surface area in m^2/kg, while $\bar{\gamma}_a$ is the anion exclusion in keq per kg. In this equation the LHS connotes a volume of exclusion of the anions, V_{ex}.

The magnitude of V_{ex} is conveniently and accurately measured by adding a small amount of radio-active tracer of the anion studied to a known volume of a soil or clay suspension. The evaluation of the activity concentration of the added tracer in a small dialysis bag included with the suspension volume then gives the volume accessible to the tracer. In turn the latter volume is subtracted from the total volume to yield V_{ex}.

It is noted in passing that for dilute suspensions V_{ex} is often a small fraction of the total liquid volume, so usually a high degree of accuracy is needed for the determination of the concentration in the equilibrium solution (preferably 0.1% if an extended concentration range is to be covered). Although a potentiometric chloride titration may attain such accuracy (cf. Bolt, 1961) generally radioactive tracer methods will be preferable as the counting time may be adjusted, provided the total liquid volume is determined by weighing.

Making use of (7.8) gives for sufficiently dilute suspensions:

$$V_{ex,a} = S\left(\frac{Q_a}{\sqrt{\beta C_0}} - \delta\right) \tag{7.22}$$

The experimental quantity $V_{ex,a}$ is thus plotted against $Q_a/\sqrt{\beta C_0}$ for a range of concentrations, to yield S as the value of the slope of this line. In fact it would suffice to determine V_{ex} at one particular concentration, provided the corresponding value of d_{ex} exceeds grossly the a priori expected value of δ. Thus in the case of Na-clays δ was seen to range between 2 and 4 Å, which is certainly small compared to $d_{ex,-}$ in systems with the "convenient" concentration level of 0.01 *N*. The determination of V_{ex} at a range of concentrations then serves as a check on the applicability of the underlying theory.

Examples of the dependability of the determination of the specific surface area of soils and clays when brought in the Na-form have been given by De Haan and Bolt (1963) and De Haan (1965a, b), from which part of the data are reproduced in Fig. 7.4. The three soils (fraction $< 50\,\mu$) show reasonably linear behavior for the chosen way of plotting, where it may be noted that the lines drawn correspond to the surface areas as determined from a separate series of 5 determinations around the convenient concentration level of 10^{-2} N (cf. De Haan, 1965a). The credibility of the surface areas found in these cases was enhanced by the corresponding values of the surface density of charge (as calculated from the known cation exchange capacity) which

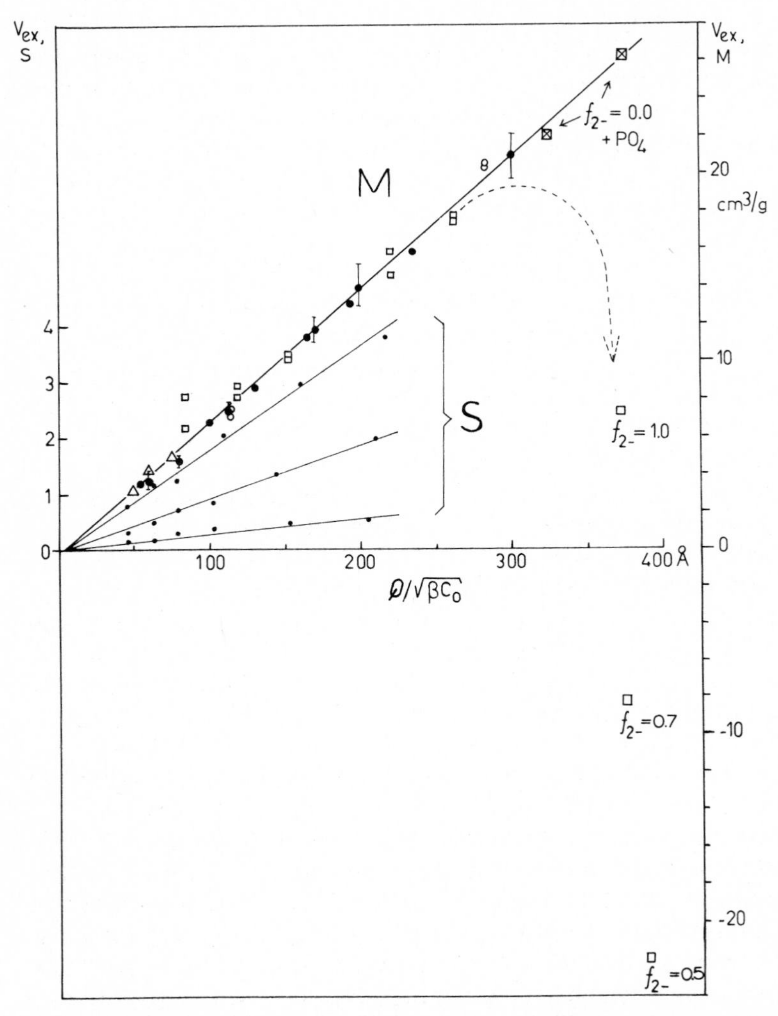

Fig. 7.4. Exclusion of anions by soils and a montmorillonite suspension, plotted as $V_{ex}(Q/\sqrt{\beta C_0})$ (compiled from De Haan, 1965b).
S: three soils: River basin clay soil with a surface area of 182 m^2/g, Randwijk soil with 93 m^2/g and a loess soil with 28 m^2/g.
M: fraction $< 2\,\mu$, of Osage bentonite. Dots: Cl-exclusion in NaCl systems (several series, showing range of variation within quadruplet sets).
Open circles: Cl-exclusion in NaCl-Na_2SO_4 systems.
Open squares: Sulfate-exclusion in NaCl—Na_2SO_4 systems.
Crossed squares: same after addition of about 10^{-4} N Na_3PO_4.
Triangles: HPO_4^{2-} exclusion in mixed sodium system after pretreatment with Na-silicate.

were all situated around $2.5 \cdot 10^{-9}$ keq/m^2, a reasonable value for soils containing illitic clay.

All other data shown in Fig. 7.4 refer a fraction $< 2\mu$ of a Wyoming bentonite (Osage) and refer to numerous series of measurements (De Haan, 1965b). The surace area of $700\ m^2/g$ corresponding to the line drawn was chosen as the nearest round number. Taking into account that these data correspond to a wide range of anionic compositions including chloride, sulfate and phosphate ions they appear to give an impressive support for the validity of the previous calculations of exclusion of anions of different valence as applied to sodium systems. Of particular interest are the severe deviations noted for sulfate ions at concentration levels below about $10^{-3}\ N$, i.e. the points marked $f_{2-} = 1.0$ at $4 \cdot 10^{-4}$; $f_{2-} = 0.7$ at $3 \cdot 10^{-4}$ and $f_{2-} = 0.5$ at $2 \cdot 10^{-4}\ N$ sulfate. These data indicate the presence of a positive adsorption of sulfate ions of the order of $1\text{—}5 \cdot 10^{-3}$ meq per gram of clay. The fact that the addition of comparable amounts of Na_3PO_4 suppressed completely the observed deviations of $V_{ex,2-}$ from the straight line corroborates this conclusion. Similarly the three points giving the exclusion of HPO_4^{2-} ions refer to systems which had been pretreated with Na-silicate solutions to suppress the positive adsorption of phosphate. In non-treated systems of comparable composition the net adsorption of phosphate was positive as expected.

Experimental information about anion exclusion in systems containing cations other than Na gives evidence of complications. Bolt and Warkentin (1956) produced data covering Cl-exclusion of Na- and Ca-montmorillonite in the range of $Q/\sqrt{\beta C_0}$ from 20 to 100 Å which yielded the same value of S around $700\ m^2/g$, although the spread of the data was around $\pm 100\ m^2/g$ and the values for Ca-clay tended to be slightly lower than the values for Na-clay. Edwards and Quirk (1962), however, came to quite different results, indicating a severe suppression of the exclusion of Cl ions in a Ca-montmorillonite as compared to the same in Na-montmorillonite. In Fig. 7.5 the data given in separate graphs at different scales by the above authors are reproduced together. Whereas the data for the Na-clay follow a straight line passing at the expected distance δ from the origin and with a slope corresponding to $560\ m^2/g$, the data for Ca-clay not only follow a line with a very different slope but the extension of this line does not intercept the distance axis at the required positive value of δ. Authors presumably correctly interpreted their data to indicate the presence of plate condensates in the Ca-system, leaving only a limited external area of $85\ m^2/g$ as derived from the slope of the line through these data. As the anion exclusion on the external surface should, however, follow a line passing through $\delta = 2$ Å, the observed exclusion exceeds the predicted value from the external surface area with a roughly constant amount of $0.3\ cm^3/g$, which for an internal surface area (i.e. within the plate condensates) of about 5—$600\ m^2/g$ would produce an internal exclusion distance of around 5 Å corresponding neatly to an anion-free interplate distance of about 10 Å, i.e. a basal spacing close to the 19 Å reported in the literature for Ca-montmorillonite.

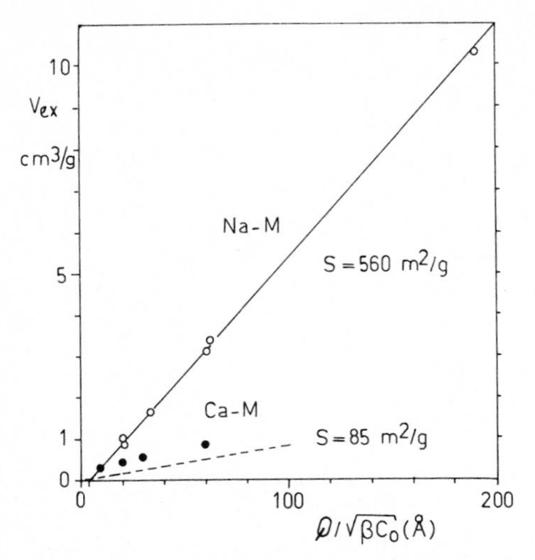

Fig. 7.5. Exclusion of Cl-ions in Na-montmorillonite and Ca-montmorillonite. (After Edwards et al., 1965)

Again in contrast with Edwards and Quirk (1962), De Haan (1965b) produced data on Na-, Ca- and Na/Ca-montmorillonite which all followed a line corresponding to 700 m^2 /g over the full range from $Q/\sqrt{\beta C_0}$ = 25 to 250 Å, within an experimental error of less than 3% at the dilute end. In conclusion it may be stated that Ca-clays will not necessarily exhibit anion exclusion commensurate with the formation of diffuse ionic layers on the entire surface area, as plate condensates may be formed presumably depending on the method of preparation. In view of the rather perfect behavior often observed for Na-clays the determination of the total area of charged surface should be done with Na-clays and soils, while the relative magnitude of the anion exclusion of the same in Ca-form could then supply a measure of the relative amounts of internal and external surfaces in such materials. In a later paper Edwards et al. (1965) produced data on the exclusion of Cl-ions by a series of homoionic clays saturated with different cations. These data, covering a range of $Q/\sqrt{\beta C_0}$ from 20 to about 120 Å indicated again a linear relationship between V_{ex} and $Q/\sqrt{\beta C_0}$, passing roughly through the origin of the axes but exhibiting very different slopes. The corresponding values of the specific surface area are tabulated below. Offhand these data indicate again the presence of plate condensates. Thus for a two-third reduction of S (K-M:Li-M) it would suffice if one-third of the plates occurred as singlets and the remainder as duplets. Taking into account that the internal volume would probably exclude anions completely one would thus find the following relationship for V_{ex}:

$$V_{ex} = \frac{2}{3} S \left(\frac{2}{\sqrt{\beta C_0}} - \delta \right) + \frac{1}{3} Sd \qquad (7.23)$$

TABLE 7.4

Specific surface area derived from $V_{ex}/(Q/\sqrt{\beta C_0})$ for homoionic alkali-montmorillonite (M) and -illite (I) (from Edwards et al., 1965)

	M (m^2/g)	I (m^2/g)
Li	650	80
Na	560	70
K	436	35
NH_4	256	22
Rb	—	10
Cs	156	0

in which 2d is the thickness of the waterlayer within a duplet. With $\delta = 4$ Å this would imply that for $d < 8$ Å, V_{ex} would give a straight line of slope (2/3) S, which upon extrapolation intersects the axis representing $Q/\sqrt{\beta C_0}$ somewhere between 0 and 4 Å, i.e. within the limits of detection afforded by the plot given in Edwards et al. (1965). As $d < 8$ Å corresponds to a basal spacing of less than 26 Å this seems completely within reason. Also the more severe reductions of S observed in Table 7.4 could be interpreted in similar fashion, taking into account that then higher order condensates with more severe restrictions on d are necessary to prevent the lines from intersecting the V_{ex}-axis at considerable positive values (cf. the Ca-montmorillonite discussed above). Thus, in order to account for the Cs-M surface one would need quadruplets, giving:

$$V_{ex} = \frac{1}{4} S \left(\frac{2}{\sqrt{\beta C_0}} - \delta \right) + \frac{3}{4} Sd \tag{7.24}$$

indicating that the basal spacing of such units should be less than about 13 Å if the line corresponding to V_{ex} still passes through the origin.

Edwards et al. (1965), in discussing their data, tacitly reject the explanation based on plate condensation because they were unable to find evidence for changes in crystal size from X-ray diffraction and BET analysis. Instead they advanced a hypothesis based on the presence of single plates with a reduced charge density, resulting from the formation of tightly bonded surface ions ("ion-pairs"). Although such ion pairs are likely to be present (cf. chapter 3), the ensuing trajectory for a surface with reduced (but constant) charge density shows no likeness with the data produced. In fact, as was discussed in section 7.2, the effect of charge reduction on d_{ex} is negligible as long as Γ remains of the order of 10^{-9} keq/m^2. If, however, the density of "unpaired" charges is brought down one or two orders of magnitude the effect becomes quite noticeable, as is shown in Fig. 7.6. In this Figure $d_{ex} = V_{ex}/S$ is plotted against $2/\sqrt{\beta C_0}$, as applies to the present mono-monovalent systems, for different values of Γ. Using for this purpose the actual value of the integral in (7.3) between the limits $u = u_s$ and $u = \infty$, one finds (cf. De Haan, 1965b):

$$\sqrt{\beta C_0}\,\delta_{ex} \equiv \int_{u_s}^{\infty} \frac{(1-1/u)du}{u\sqrt{(u+1/u-2)}} = 2/\sqrt{u_s} \qquad (7.25)$$

Making use of (1.39), u_s may be expressed in terms of Γ, according to:

$$\sqrt{u_s} = (\bar{\Gamma}/4 + \sqrt{\bar{\Gamma}^2/16+1}) \qquad (7.26)$$

with $\bar{\Gamma} \equiv \Gamma\sqrt{\beta}/\sqrt{C_0}$ as before.
Combination of these equations and introduction into (7.8) then gives:

$$d_{ex} = \frac{2}{\sqrt{\beta C_0}} - \frac{4}{\beta\Gamma}\,\frac{2/\sqrt{\beta C_0}}{1/\sqrt{\beta C_0} + \sqrt{1/\beta C_0 + (4/\beta\Gamma)^2}} \qquad (7.27)$$

Over the range of $2/\sqrt{\beta C_0}$ between 20 and 120 Å explored by Edwards et al. (1965) only the curve representing $\Gamma = 10^{-11}$ keq/m^2, i.e. 1% free charge for the montmorillonite considered, could faintly resemble a straight line, leading in that case to an apparent value of S equal to about 12% of the full value applicable to $\Gamma = 10^{-9}$ keq/m^2. In order to interpret, nevertheless, the bundle of straight lines found for the clays saturated with different cations in terms of charge reduction by ion-pairing, Edwards et al. then forwarded a hypothesis based on the formation of individual exclusion zones present around the widely spaced remaining free charges. Without going into the necessity of assuming such individual exclusion zones, the present authors object to the further assumption needed and made by Edwards et al. (1965) in order to obtain a linear relation between V_{ex} and $Q/\sqrt{\beta C_0}$, viz. a steep *increase* in the percentage of paired surface charges upon a *decrease* in concentration of the equilibrium solution. Moreover, it appears that the absence of X-ray evidence cited by Edwards et al. (1965) as pleading against plate condensates is to be expected in cases where such condensates are simply duplets. Some further discussion on this matter is given in Bolt and de Haan (1965).

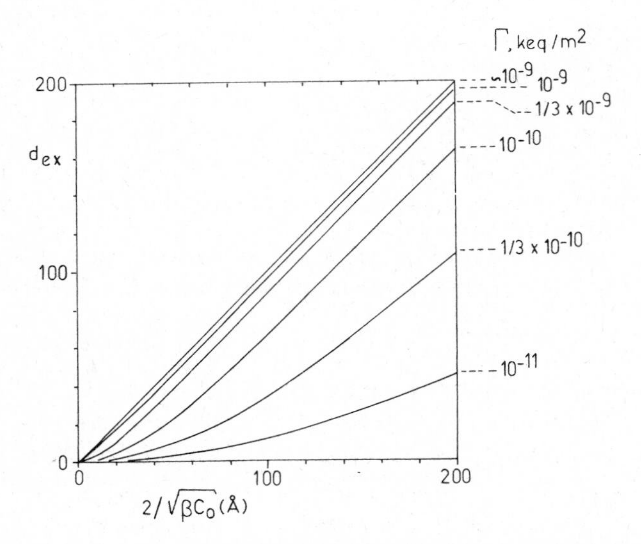

Fig. 7.6. Exclusion distance for monovalent anions in a symmetric salt system, as a function of the surface charge density, Γ.

7.4.2. Dilution and pressure filtration of the soil solution

Anion exclusion may be of concern when it is attempted to determine the salt concentration in the soil solution under field conditions. In principle its influence upon the relation between salt concentration and water content of the soil may be estimated with the help of the definition equation (A4.5) used in part A:

$$T_{an} = Wc_{0,an} - \bar{\gamma}_{an} \tag{7.28}$$

where the total amount T, the water content W and $\bar{\gamma}$ must obviously all be related to the same amount of soil. As $\bar{\gamma}_{an}$ is a function of $c_{0,an}$ it follows immediately that $c_{0,an}$ is not inversely proportional to W as is the case in aqueous solution. Assuming for the time being that all anion species are of the same valence and do not exhibit positive adsorption, the above equation may be used as well to describe the behavior of the total salt concentration. In the absence of solid phase salts one then finds in accordance with the previous sections:

$$\frac{T}{S_{ex}} = \frac{WC_0}{S_{ex}} - \left(\frac{Q}{\sqrt{\beta C_0}} - \delta\right) C_0 \tag{7.29}$$

where S_{ex} is the surface area exhibiting anion exclusion as measured or estimated, again specified for the same amount of soil as used for T and W. Since T/S_{ex} is a system constant, i.e. the total amount of dissolved salts per unit area of surface exhibiting anion exclusion, while Q is a known function of the liquid film thickness $d_l = W/S_{ex}$ and C_0, (7.29) prescribes in principle the relation between C_0 and W/S_{ex}. In practice the precise calculation of $C_0(W)$ will soon require computer assistance, because Q depends also on the mixing ratio of cationic species, which in turn depends on C_0. Only in cases of a single salt system might one venture to work out this equation by hand. Doing so for relatively high liquid contents (e.g. $d_l > 3Q/2\sqrt{\beta C_0}$) both the truncation effect and the correction distance δ may be neglected and (7.29) reverts to a simple quadratic equation in $\sqrt{C_0}$. With $Q = 2$ for the mono-monovalent salt this gives:

$$C_0 = \left[\frac{1}{\sqrt{\beta}\, W/S_{ex}} + \sqrt{\frac{1}{\beta (W/S_{ex})^2} + \frac{T/S_{ex}}{W/S_{ex}}}\right]^2 \tag{7.30}$$

Using as an example a soil with $S_{ex} = 2 \times 10^8$ cm^2/kg and containing a total amount of monovalent salt of 2 meq/kg, the salt concentration is found as 0.0186 normal at $W = 200$ cm^3/kg (20% by weight) as against 0.0078 normal at $W = 400$ cm^3/kg.

In view of the many simplifying conditions used in constructing (7.30) it should only be regarded as an illustration of the effect of anion exclusion on dilution, thus serving as a warning if concentrations measured at a particular moisture content must be back translated to another one. At the same time too many factors are usually unknown to warrant a more precise calculation with computer aid, taking into account possible shifts in the cationic composition as a result of changes in moisture content. In practice an iterative guess on the expected magnitude of the anion exclusion with the Tables and graph given in the previous section should suffice to estimate the change in total concentration upon dilution.

Related to the above phenomenon is the effect of anion exclusion on the concentration of salts in a pressure filtrate of a soil or clay paste. Here again the interest arises because of the technical difficulty in analyzing the soil solution in situ, i.e. at field moisture content. Instead of diluting to, for example, liquid limit and subsequent back calculation to field moisture content, one may attempt to isolate the equilibrium solution by pressure filtration at the original moisture content.

Again a "complete" calculation for a soil solution of mixed composition in equilibrium with an exchange complex would require computer assistance because any change in salt level would give rise to changes in the mixing ratio of cations and anions. Lack of precise information on the magnitude of the surface area contributing to anion exclusion, changes in this area following compression, etc. render the value of such calculations doubtful. Thus it will only be attempted to demonstrate the principle behind observed changes in salt concentration upon progressing filtration by applying the theory to a system containing only one mono-monovalent salt (cf. Bolt, 1961). As in the previous example one may describe the system considered in terms of the parameters T_{an}/S_{ex} (total anion content per unit area of surface excluding anions) and $W/S_{ex} = d_l$ (liquid content per unit area). If such a system is subjected to excess gas pressure in a standard pressure membrane apparatus the instantaneous concentration of the pressure filtrate, C_p, is found as:

$$C_p = \frac{d(T_{an}/S_{ex})}{d(W/S_{ex})} \tag{7.31}$$

Starting with a very wet suspension (i.e. $d_l \gg d_{ex}$), C_p will simply equal the initial equilibrium concentration of the system, $C_{0,i}$, until the remaining liquid has decreased to the point where $d_l \approx 1.5\, d_{ex}$. Then truncation of double layers will arise and the amount of salt removed from the system per unit amount of filtrate must become less than $C_{0,i}$.

As was pointed out by Bolt (1961) one may then distinguish two extreme conditions with regard to the filtration process, depending on whether or not the filtrate drops leaving the system remain in equilibrium with the suspension from which they are derived. If so, the decreasing concentration of the drops will induce a build up of the anion-repelling electric field in the double layers. In turn this leads to an increasing anion exclusion which, as will be shown below, induces a preeemptive effect with regard to the removal of salt from the system: the salt will have disappeared before all of the liquid phase has been removed. This condition obviously requires a filtration process that is slow enough to allow continuous equilibration between filtrate drops as they are formed and the remaining suspension. Such a process will be referred to as *slow* filtration.

The other extreme is a filtration process *fast* enough to prevent any equilibration (by diffusion through the membrane) between remaining suspensions and filtrate drops formed, the suspension remaining in equilibrium with its original solution. Reality should then be situated between these extremes.

Using again (7.29) to relate T/S_{ex} with W/S_{ex}, while using (7.12) and (7.12a) to account for the presence of truncated double layers one finds:

$$\frac{W}{S_{ex}} = x_t - \delta = \frac{2}{\sqrt{\beta C_0 u_t}} F(\alpha, \pi/2) - \delta \tag{7.32}$$

and:

$$\frac{T}{S_{ex}} = \frac{2\sqrt{C_0 u_t}}{\sqrt{\beta}} [F(\alpha, \pi/2) - E(\alpha, \pi/2)] \tag{7.33}$$

where $\alpha = \arcsin(1/u_t)$, while δ may safely be regarded as a constant. This gives in cases of *slow* filtration:

$$C_p = \left(\frac{\partial T}{\partial W}\right)_{C_0 = C_p} \tag{7.34}$$

$$= \frac{\sqrt{C_p}\, d\sqrt{u_t}(F - E) + \sqrt{u_t}(F - E) d\sqrt{C_p}}{(1/\sqrt{C_p}) d(F/\sqrt{u_t}) + (F/\sqrt{u_t}) d1/\sqrt{C_p}} \tag{7.35}$$

where F and E are the complete elliptic integrals with modulus $1/u_t$ as defined above. Separating the variables C_p and u_t leads to:

$$\frac{d\sqrt{C_p}}{\sqrt{C_p}} = \frac{d(F/\sqrt{u_t}) - d\sqrt{u_t}(F - E)}{\sqrt{u_t}(F - E) + F/\sqrt{u_t}} \tag{7.36}$$

Differentiating the complete elliptic integrals with respect to the modulus $(1/u_t)$ gives the relation between C_p and $1/u_t$ as:

$$\frac{d\sqrt{C_p}}{\sqrt{C_p}} = \frac{d(1/u_t)}{(2/u_t)} - \frac{d(1/u_t)}{(1 + 1/u_t)} \tag{7.37}$$

Integration from the boundary value $C_p \equiv C_0 = C_{0,i}$ at $u_t = 1$ (or $W \to \infty$) then gives:

$$\sqrt{C_p/C_{0,i}} = \frac{2/\sqrt{u_t}}{(1 + 1/u_t)} \tag{7.38}$$

Substitution of (7.38) into (7.32), with $C_0 \equiv C_p$, allows one to express W/S_{ex} in terms of u_t:

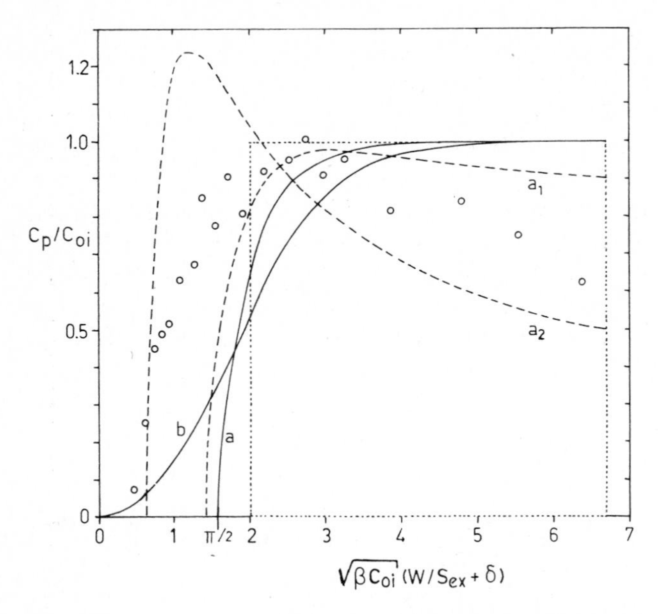

Fig. 7.7. Relative concentration of the pressure filtrate of a colloid paste containing one symmetric salt, as a function of the liquid content expressed as the thickness of the liquid layer, W/S_{ex}, in units $1/\sqrt{\beta C_0}$. Curve a, *slow* filtration; b *fast* filtration; a_1, *slow* filtration with a filter reflection coefficient equal to 0.1; a_2, *slow* filtration with a filter reflection coefficient equal to 0.5. Initial liquid content immaterial for a and b, for a_1 and a_2: $(W_i/S_{ex} + \delta)\sqrt{\beta C_0} = 6.7$.
Area bounded by dotted line: $(T_{an}/S_{ex})(\sqrt{\beta C_0}/C_0)$.
Circles: experimental data as found for a montmorillonite in 0.0028 N NaCl with S_{ex} = 650 m^2/g and W_i = 25.1 cm^3/g, corresponding to 6.7 on the horizontal axis. (From Bolt, 1961).

$$\sqrt{\beta C_{0,i}}\left(\frac{W}{S_{ex}} + \delta\right) = (1 + 1/u_t)F(\alpha, \pi/2) \tag{7.39}$$

Equations (7.38) and (7.39) then serve as a set of parameter equations determining the relation between $C_p/C_{0,i}$ and W/S_{ex}, which is plotted in Fig. 7.7.

In cases of *fast* filtration C_0 appearing in (7.32) and (7.33) remains constant and equal to $C_{0,i}$. Differentiating both equations with respect to $1/u_t$ then gives:

$$C_p/C_{0,i} = \frac{(u_t + 1/u_t)E(\alpha, \pi/2) - (u_t - 1/u_t)F(\alpha, \pi/2)}{2E(\alpha, \pi/2) - (1 - 1/u_t^2)F(\alpha, \pi/2)} \tag{7.40}$$

Together with (7.32), using $C_0 = C_{0,i}$, this allows one to plot the relation between $C_p/C_{0,i}$ and W/S_{ex}, as is shown also in Fig. 7.7. Obviously the area underneath both types of filtration curves must be equal as it represents T/S_{ex} (in units of $1/\sqrt{\beta C_{0,i}}$) multiplied by $C_{0,i}$. The area above these curves bounded by $C_p/C_{0,i} = 1$ then represents the anion exclusion of the original system (per unit area S_{ex}) in the same units, i.e. the exclusion distance in units of $1/\sqrt{\beta C_{0,i}}$. In the present mono-monovalent system this corresponds with $Q = 2$. It may also be noted that the curve representing *slow* filtration gives $C_p/C_{0,i} = 0$ (with $u_t \to \infty$) at a *finite* value for W/S_{ex}, viz. at $\pi/2$ units of $1/\sqrt{\beta C_{0,i}}$.

Experimental information about the concentration of pressure filtrates from Na-clay suspensions may be found in Bolt (1961), from which one set of data has been reproduced in Fig. 7.7. Although indicating a rather steep decline in the filtrate concentration, roughly at the expected values of W/S_{ex}, these and the other data given in the mentioned publication show convincingly that the present model is still far too simple. Thus the ascent of C_p following an initial value considerably below $C_{0,i}$ as determined by dialysis indicates that partial retention of the salt by filter plus clay mat formed is operative (cf. Kemper (1960) and Chapter 11.5).

Following Bolt (1961) one may illustrate the effect of a "reflection coefficient", σ (of clay mat plus filter) upon C_p by introducing a membrane passage factor $m \equiv (1 - \sigma)$ into (7.35). Taking this factor as a constant one finds

$$C_p^* = mC_0 = \frac{\sqrt{C_0}\,d\sqrt{u_t}\,(F-E) + \sqrt{u_t}\,(F-E)\,d\sqrt{C_0}}{(1/\sqrt{C_0})\,d(F/\sqrt{u_t}) + F/\sqrt{u_t}\,d\,1/\sqrt{C_0}} \qquad (7.35m)$$

Separating the variables C_0 and u_t now leads to:

$$\frac{d\sqrt{C_p^*}}{\sqrt{C_p^*}} \equiv \frac{d\sqrt{C_0}}{\sqrt{C_0}} = \frac{m\,d(F/\sqrt{u_t}) - d\sqrt{u_t}\,(F-E)}{\sqrt{u_t}\,(F-E) + mF/\sqrt{u_t}} \qquad (7.36m)$$

which gives in turn:

$$\frac{d\sqrt{C_p^*}}{\sqrt{C_p^*}} = \frac{d\,1/u_t}{(2/u_t)} - \frac{m(1+1/u_t)F - (m - 1/u_t)E/(1-1/u_t)}{(m/u_t + 1)F - E}\,\frac{d\,1/u_t}{(1+1/u_t)} \qquad (7.37m)$$

Subtracting (7.37) allows one to relate C_p^* to C_p (with $m = 1$), according to:

$$\ln\sqrt{C_p^*/C_p} = \int_{u_{t,i}} \frac{1 - M(1/u_t)}{(1+1/u_t)}\,d(1/u_t) \qquad (7.41)$$

where $M(1/u_t)$ indicates the function of m and $1/u_t$ appearing in front of the second term of (7.37m). The outcome of the above integral has been tabulated in a generalized form in Bolt (1961), for $m = 0.5$ and $m = 0.9$, allowing one to calculate the trajectory of $C_p^*/C_{0,i}$ for any chosen starting point of the pressure filtration as characterized by $u_{t,i}$. The above was applied to construct the curves marked a_1 and a_2 in Fig. 7.7.

In view of the uncertainties sketched above, attempts to execute calculations based on the present model for more involved ion mixtures appear futile. The important conclusion remains, however, that the isolation of the soil solution at field moisture content by means of pressure filtration — for the purpose of determining its composition — is in general not a dependable method. Possibly the removal in this manner of only a very small amount of solution may give more reliable results. In this context the isolation of an equilibrium solution by means of filter paper tablets described by Schuffelen et al. (1964) appears promising. The tablets are embedded in a fairly large soil sample which is then compressed for a short time period (less than one hour). The liquid phase taken up by the tablets represents a very small amount of pressure filtrate, while the short exposure time prevents diffusion — which would lead to an adsorption equilibrium with the filter paper — to take effect. Results obtained in this manner were satisfactory for an artificial porous medium (carborundum powder) at fairly low moisture content. Even if this method proved to be satisfactory for soils, it should be remembered that once the moisture content sinks considerably below "double layer truncation value" (e.g. $2S_{ex}/\sqrt{\beta C_0}$), in principle the equilibrium solution does no longer exist as such in soil. In other words, in truncated diffuse layers electroneutrality of the ionic composition is not preserved, implying that the composition of an electrically neutral equilibrium solution is found only in a "pressure dialysate" of such a system. In practice such an equilibrium dialysate is hardly obtainable from a soil system. In view of the variability of the moisture contents in the field it thus seems most practical to dilute the sample to, for example, liquid limit, determine the composition of a limited amount of filtrate and attempt to estimate iteratively the composition at lower moisture contents taking into account both cation exchange and anion exclusion.

As a side line to the above, the possibility of the existence of semi-occluded pore space in clay aggregates is mentioned. As was shown by Blackmore (1976), sufficiently narrow voids connecting such space with e.g. inter-aggregate pores may become closed off for the passage of anions upon a decrease of the electrolyte level. Prolonged soaking (or leaching) of the soil with, for example, destilled water will then lead to removal of salt from the semi-occluded pores until a certain threshold value has been reached. This value then corresponds to the composition of the interior solution at which near complete exclusion of anions in the narrow connecting pores occurs, thus effectively suppressing further diffusional equilibration with the leaching solution.

REFERENCES

Bateman, H., 1953. Higher Transendental Functions, Vol. II. Mc-Graw-Hill, New York, N.Y.

Blackmore, A.V., 1976. Salt sieving within clay soil aggregates. Aust. J. Soil Res., 14: 149—158.

Bolt, G.H., 1961. The pressure filtrate of colloidal suspensions, I and II. Kolloid. Z., 175: 33—39, 144—150.

Bolt, G.H. and De Haan, F.A.M., 1965. Interactions between anions and soil constitutes. IAEA, Vienna, Techn. Rep. Ser., 48: 94—110.

Bolt, G.H. and Warkentin, B.P., 1956. Influence of the method of sample preparation on the negative adsorption of anions in montmorillonite suspensions. 6th Int. Congr. Soil Sci., Paris, Trans., B: 33—40.

Bolt, G.H. and Warkentin, B.P., 1958. The negative adsorption of anions by clay suspensions. Kolloid. Z., 156: 41—46.

De Haan, F.A.M., 1964. The negative adsorption of anions (anion exclusion) in systems with interacting double layers. J. Phys. Chem., 68: 2970—2976.
De Haan, F.A.M., 1965a. Determination of the specific surface area of soils on the basis of anion-exclusion measurements. Soil Sci., 99: 379—386.
De Haan, F.A.M., 1965b. The interaction of certain inorganic anions with clay and soils. Agric. Res. Rep. (Wageningen), 655: 1—168.
De Haan, F.A.M. and Bolt, G.H., 1963. Determination of anion adsorption by clays. SSSA Proc., 27: 636—640.
Edwards, D.G. and Quirk, J., 1962. Repulsion of chloride by montmorillonite. J. Colloid Sci., 17: 872—882.
Edwards, D.G., Posner, A.M. and Quirk, J., 1965. Repulsion of chloride ions by negatively charged clay surfaces, I, II and III. Trans. Faraday Soc., 61: 2808—2823.
Kemper, W.D., 1960. Water and ion movement in thin filters as influenced by the electrostatic charge and diffuse layer of cations associated with clay mineral surfaces. SSSA Proc., 24: 10—16.
Schofield, R.K., 1947. Calculation of Surface Areas from measurements of negative adsorption. Nature, 160: 408—410.
Schofield, R.K. and Talibuddin, O., 1948. Measurement of the internal surface by negative adsorption. Discuss. Faraday Soc., 3: 51—56.
Schuffelen, A.C., Koenigs, F.F.R. and Bolt, G.H., 1964. The isolation of the soil solution with the aid of filter paper tablets. 8th Int. Congr. Soil Sci., Bucharest, III: 519—528.

INTERACTION OF ORTHOPHOSPHATE IONS WITH SOIL

J. Beek and W.H. van Riemsdijk

The chemical properties of the (partly dissociated) orthophosphate ion cause it to be a species that strongly reacts with various soil components (e.g. aluminum and iron oxides*, clay minerals, solid carbonates, or the corresponding metal ions in soluble form). The most commonly occurring reaction mechanisms involved are chemisorption processes and the formation of solid phosphates. In view of the complexity of these processes detailed models, as have been used in the previous chapters for physical (e.g. Coulombic) adsorption processes of (hydrated) cations and some anions, like e.g. Cl^-, NO_3^-, and SO_4^{2-} with adsorbents of mainly the constant charge type, are not available. Instead one has to take recourse for the time being to a more qualitative description of the presumed transition processes between the soluble (and thus mobile forms of phosphate) and the immobile forms. The latter may include "stable" phosphate minerals, adsorbed phosphate and various organic phosphate compounds. All these immobilized phosphate fractions can in principle be mobilized, the inorganic forms by lowering the solution phosphate concentration and the organic forms by mineralization of organic matter.

The chemical processes of interaction of orthophosphate ions with soil components are thus important from the point of view of plant nutrition. Furthermore, these processes have important implications with respect to the prevention of the pollution of natural waters due to losses of phosphate (e.g. by leaching or runoff) from agricultural soils. The same holds for the situation where soil is used for the disposal of waste waters or of excessive amounts of liquid manure in areas with intensive animal production (Beek et al., 1973, 1977a; Hook et al., 1973; Fordham and Schwertmann, 1977). In this chapter the attention has been focused primarily on the chemical processes of interaction of orthophosphate ions with soil or its single minerals. The interaction of condensed** inorganic phosphate species or organic phosphate compounds with the soil have not been treated in detail; the same holds for the microbial immobilization and mineralization processes of soil

---

* The term oxides is used as a comprehensive term in this chapter for oxides, hydrous oxides and hydroxides.

**Condensed phosphates are defined as multiple units of PO_4 tetrahedra joined by shared oxygen atoms between the tetrahedra; the linear chain polymers are known as polyphosphates.

phosphate. Finally, it is stressed that no formal literature review is given in this chapter. Rather the literature is referred to where relevant in the authors' opinion.

8.1. FORMS AND OCCURRENCE OF SOIL PHOSPHATES

The soil phosphate compounds are predominantly derived from orthophosphate, the stable form of phosphorus in nature. The presence of condensed phosphate is possible temporarily, because living cells may synthesize these in situ (Ghonsikar and Miller, 1973; Anderson and Malcolm, 1974). In cultivated soils these compounds may further originate from additions of condensed phosphates serving as fertilizer or from disposal of waste waters. Reduced forms of phosphorus, like e.g. phosphite (H_3PO_3), hypophosphite (H_2PO_2) or phosphine (PH_3), the last compound being a gas, cannot be excluded entirely, though the available evidence with regard to the formation of such compounds in reduced (water-logged) soils is scant and often ambiguous (Tsubota, 1959; Burford and Bremner, 1972). The total soil phosphate content of many virgin soils is in the range of 0.04—0.1 wt.% P* of the dry soil (Black, 1968; Larsen, 1967). Usually higher values are found in the topsoil than in the subsoil, very likely as a consequence of the upward cycling of phosphate by plants. The soil phosphate compounds are derived mainly from apatite minerals and phosphorus present as a substitution for silicon in silicate rocks of the earth's crust (McKelvey, 1973). Apart from the commonly occurring phosphate bearing mineral apatite about 200 other phosphate minerals have been identified (Van Wazer, 1958; Fisher, 1973). In the weathering zones of deposits of apatite or guano a large number of secondary minerals occur. During weathering and soil development processes transformation and translocation of the original phosphate compounds takes place.

The application of direct methods (viz. petrographic microscopy, X-ray and electron diffraction, differential thermal analysis) for the identification of inorganic soil phosphates is generally not successful. From this lack of success, however, it is premature to conclude a priori that the inorganic phosphate compounds are of amorphous nature only, as is often claimed. An alternative explanation may be that in view of the low phosphate content of most soils and the mixing of the inorganic phosphate compounds with the soil minerals present in the fine fraction, these methods are bound to fail. Besides, it is hardly possible to separate the inorganic phosphates from the other minerals by physical methods (Black, 1968) or chemical pretreatment of the bulk sample (Campbell et al., 1972). According to Norrish (1968), Campbell et al. (1972), Adams et al. (1973) and Black (1968)

---

* In order to avoid terminological confusion the elemental form of expression has been used throughout this chapter. To convert P to P_2O_5 as is common practice in mineralogy and agriculture, multiply percentage P by 2.29.

— the last author summarizing earlier reported evidence — the following crystalline phosphates have been identified: apatites, minerals of the plumbo-gummite group (hydrated aluminophosphates, identified members are gorceixite, florencite and crandallite), wavellite and vivianite.

Apatite, in particular fluorapatite, may be present in soils for several reasons:
(1) it is the main phosphate mineral in the earth's crust,
(2) it is highly stable in calcareous soils,
(3) it may occur as large crystals of sand or silt size,
(4) it can be present as an occlusion in other soil minerals (Syers et al., 1967; Cescas et al., 1970) as a result of which it is less readily transformed.
The occurrence of secondary minerals as wavellite and members of the plumbo-gummite group confirms the suggestion of Altschuler (1973) to the effect that crandallite and wavellite are the stable secondary minerals upon weathering of deposits of apatite and guano in the presence of silicate minerals. In view of these results the often expressed opinion that variscite or otherwise strengite are the stable aluminum or iron phosphate phases in soils may need to be reconsidered. Vivianite, a ferrous phosphate, is stable only under reducing conditions (Nriagu, 1972) and it is found only in poorly drained or waterlogged soils.

A different approach to the characterization of the soil inorganic phosphate fraction is based on the difference in solubility in various extractants. The fractionation scheme of Chang and Jackson (1957) in its original or modified form (Petersen and Corey, 1966; Williams et al., 1967, 1971a, b; Kurmies, 1972) has been used frequently. Although the selectivity of the extractants for different phosphate compounds is limited (Bromfield, 1967a, b; Vahtras and Wiklander, 1970) and also changes in the original phosphate fraction may occur during extraction (Bromfield, 1970; Rajendran and Sutton, 1970), this method provides information concerning general trends of phosphate transformation reactions. Thus it is found that with increasing soil pH the relative proportion of aluminum- and iron-bound phosphate is decreasing in favor of presumably calcium-bound phosphate. Furthermore, as a result of progressive soil development the calcium-bound phosphate fraction is decreasing and replaced by aluminum- and iron-bound forms. Also the organic-bound phosphate fraction increases with soil development (Wells and Saunders, 1960; Walker and Syers, 1976).

The fraction of total soil P that is present in organic combination varies widely, and ranges roughly from 20 to 90% (Mattingly and Talibudeen, 1967). These large differences are explained, at least partly, by the strong dependence of the largely biological processes of formation and breakdown (mineralization) of organic phosphates upon local conditions of rainfall and temperature, various chemical and physical properties of the soil as well as plants and organisms present. For instance the rate of mineralization of organic phosphates (which is a prerequisite for its uptake by plants) is depressed in general if the organic phosphate compounds or the enzymes (phosphatases) catalyzing the dephosphorylation become adsorbed onto the soil minerals (Goring and Bartholomew, 1952; Mortland and Gieseking, 1952; Cosgrove, 1967; Ivanov and Sauerbeck, 1971). In particular, inositolhexa-phosphates exhibit adsorption and precipitation characteristics that correspond with the behavior of orthophosphate ions (Jackman and Black, 1951; Anderson, 1963, Anderson et al., 1974). On account of this behavior, inositol(hexa) phosphates may be less readily mineralized than other organic phosphates. This may offer an explanation for its relative abundance (up to 50%) in the soil organic phosphate fraction. Other identified compounds, present in smaller proportions, belong to or are derived from phospholipids and nucleic acids. Trace amounts of other organic phosphates have been isolated from the soil (Anderson, 1967; Halstead and Anderson, 1970; Anderson and Malcolm, 1974; Fares et al., 1974). This indicates that approximately 30—50% of the soil organic phosphate compounds can be partly or completely identified.

Due to the strong interactions of soluble phosphate species with some soil constituents low solution concentrations result. These vary generally from 10^{-7} to 10^{-5} M (0.003—0.3 mg P/l). Part of the inorganic and organic phosphate is present in colloidal form. This fraction may be suspended under suitable conditions (low ionic strength) and thus will be a part of the "soil solution". Indications exist that the suspended forms are leached more easily from the soil than the dissolved species, as the latter may interact more strongly with the soil minerals than the former ones (Hannapel et al., 1964; Hilal, 1969). The dissolved phosphate species may consist of dissociated or complexed orthophosphate, condensed phosphates, and organic phosphates. The distribution of all these species depends on the stability constants, pH, and other relevant solution concentrations.

8.2. THE REACTION OF PHOSPHATE WITH SOIL MINERALS

The introduction has shown that phosphate reactions in a soil system are of a complex nature. Adsorption may for instance take place on different minerals in competition with, or enhanced by, other (specifically) adsorbing species present in the soil system. Mineralization and microbial immobilization, precipitation and dissolution reactions may all occur concurrently with adsorption-desorption processes.

If one wishes to study the mechanisms of the chemical interaction of phosphate with soil minerals one has to limit oneself to the study of pure systems which are in itself still quite complex. The insight gained from these studies may be of use for the interpretation of results from soil-phosphate studies. Tentative reaction schemes derived from soil-phosphate studies can also be tested with pure systems. The reaction with single soil minerals (e.g.

metal oxides, clay minerals and calciumcarbonate) in pure systems will first be discussed, followed by a discussion of some literature concerning the phosphate reaction with the "whole soil".

8.2.1. The reaction with metal oxides

The reaction of phosphate with metal oxides may in principle be divided into an adsorption process and a process during which the corresponding metal phosphate may be formed. The latter is usually a slow process (days—weeks). Adsorption, although often considered to be fast, may still be quite slow (minutes—days) compared with e.g. cation adsorption on surfaces with a negative charge. The stability diagram of the three-phase system [$M(OH)_3$(s) —MPO_4(s)—solution] has been shown in chapter 6 of part A, Fig. A6.11. The phosphate equilibrium concentration calculated for such a system is strongly dependent on the pH and on the solubility products used for the different minerals. The formation of $AlPO_4.2H_2O$ in the presence of $Al(OH)_3$ at pH 5 is, for the chosen stability constants of Fig. A6.11, thus feasible above a phosphate concentration of $10^{-4.5}$ *M*. Classical thermodynamics, however, does not predict the rate of such a reaction nor its mechanism. Kinetic studies show that the reaction rate depends on the pH and on the phosphate concentration, in other words it depends on the supersaturation of the system. At low phosphate concentration and for short reaction periods the reaction will thus be limited practically to adsorption at the oxide—water interface. The possibilities for the study of adsorption reactions in these cases are limited, because uncertainty remains whether the adsorption is at equilibrium, both with respect to solution concentration and with respect to the arrangement on the surface, at the chosen reaction time. The restriction of low concentration further leads to problems with the determination of the maximum adsorption capacity. These difficulties lead Boehm (1971) to the conclusion that: "Even phosphate adsorption on alumina from Na_2HPO_4 solution cannot be measured because insoluble aluminium phosphates are formed." Adsorption experiments with crystalline iron oxides, however, seem to give satisfactory results.

Experimental evidence from different authors working with different techniques indicates that adsorption, which is considered to be chemisorption, occurs by exchange with singly coordinated surface hydroxyl or water groups (Atkinson et al., 1972; Russell et al., 1974; Parfitt et al., 1975; Yates and Healy, 1975; Parfitt et al., 1976). This reaction may be visualized for H_2PO_4 displacing a hydroxyl group as:

$$\left[\begin{array}{l} -M(OH_2)(OH) \\ \quad | \\ \quad HO \\ \quad | \\ -M(OH_2)(OH_2) \end{array}\right] + H_2PO_4^- \rightarrow \left[\begin{array}{l} -M(OH_2)(OPO_3H_2) \\ \quad | \\ \quad HO \\ \quad | \\ -M(OH_2)(OH_2) \end{array}\right] + OH^- \qquad (8.1)$$

Corresponding reactions can be given for the displacement of water or hydroxyl groups by the other (dissociated) orthophosphate ions.

The adsorbed phosphate may react further with another surface OH or H_2O group either coordinated to the same metal ion or to a neighboring metal ion. In the latter case the adsorbed phosphate is usually referred to as a binuclear surface complex. The maximum amount of phosphate which may be bound in this manner is thus determined by the amount of singly coordinated surface OH(H) groups and by the stoichiometry of the adsorption reaction. The surface of the oxide can further be characterized by the point of zero charge (pzc) which is defined as the pH value at which the total charge on the solid, arising from all sources, is zero and indicated as pH_0. The pzc in the absence of specific adsorption for the metal oxides of iron, aluminum, and titanium falls in the range pH 5—9. The surface will be positive for $pH < pH_0$ and negative for $pH > pH_0$. Phosphate adsorption will in general lower the (positive) surface charge of the oxide and even charge reversal may occur. The increase in negative charge opposes further adsorption of phosphate, which is reflected in the decreasing amounts of phosphate adsorbed with increasing pH (Hingston et al., 1967, 1968, 1972; Chen et al., 1973a; Huang, 1975). The addition of negative charge to the surface by phosphate adsorption depends on the pH, i.e. on the average charge of the phosphate ions in solution, and on the amount of hydroxyl ions produced during adsorption. The production of hydroxyls can be measured, provided the proper experimental conditions have been chosen, and can thus be related to the amount of phosphate adsorbed at constant pH.

Schematically the reaction may then be presented for a unit surface area containing *s* singly coordinated surface M—O groups carrying on an average (1.5 + x) H ions and reacting with *a* phosphate molecules carrying y H ions, according to:

$$\underset{sx}{\text{surface}} + \underset{a(y-3)}{aPO_4H_y} \rightarrow \underset{sx+a(y-3)+r}{\text{surface} \sim P} + bH_2O + \underset{-r}{rOH} \qquad (8.2)$$

charge: (given beneath each term)

where surface ~ P indicates the phosphated surface. The parameters x and y are determined by the dissociation equilibrium of surface-OH_2 and orthophosphate ions, respectively. Obviously $x \geqslant 0$ for $pH \leqslant pH_0$. The parameter y will be about two at pH 5, 1.5 at pH 7.2 and about unity at pH 9. Equation (8.2) is valid irrespective of the form of the surface complex; whether phosphate occurs singly coordinated, in a binuclear complex, as PO_4^{3-} ion, or whatever may be assumed.

If the equation is expressed per unit phosphate group reacting with *s/a* surface groups one finds the OH^- produced per phosphate group added (*r/a*). This ratio, $\bar{r}$, was introduced by Breeuwsma and Lyklema (1973) and is often measured (Breeuwsma and Lyklema, 1973; Rajan et al., 1974; Rajan,

1975a, 1976). It is tempting to suggest reaction mechanisms from the found values of $\bar{r}$. The proposed mechanisms are, however, always speculative because it is only an overall ratio. Particularly it should be recognized that the OH^- produced per unit surface area adsorbing *a* molecules of phosphate reflects the net effect of proton adsorption—desorption of the phosphated-surface as compared to the reactants and the hydroxyl released by exchange reactions. No information is gained about the position of the retained H ions with respect to the PO_4 group and the neighboring MOH groups.

For monodentate ligand exchange, (8.2) can be rewritten per molecule of PO_4 with $s/a \equiv \bar{s}$, as:

$$\bar{s}MOH_{(3/2+x)} + PO_4H_y \rightarrow MOPO_3H_{y'} + (\bar{s}-1)MOH_{(3/2+x')} + \bar{r}OH + (1-\bar{r})H_2O \quad (8.3)$$

One may now construct a proton balance according to:

$$\bar{s}(1\tfrac{1}{2} + x) + y = y' + (\bar{s}-1)(1\tfrac{1}{2} + x') + 2 - \bar{r}$$

This gives with $(y' - y) \equiv \Delta y$ and $(x' - x) \equiv \Delta x$

$$\bar{r} = \tfrac{1}{2} + \Delta y + (\bar{s}-1)\Delta x - x \quad (8.4)$$

If it is assumed that Δx and Δy are about zero (i.e. no change of dissociation upon adsorption) then $\bar{r} = \frac{1}{2} - x$. In other words, $\bar{r}$ is now simply determined by x, ranging from zero at $x = \frac{1}{2}$ (all the surface groups being positively charged as $M{-}OH_2^{1/2+}$) to unity at $x = -\frac{1}{2}$ (all surface groups present as $MOH^{1/2-}$). It follows that $\bar{r}$ will thus increase with increasing pH.

The inference sometimes made that the phosphate would react preferentially with MOH_2 groups if only a limited number of these is present seems paper chemistry: once $MOPO_3H_y$ has been formed the binding energy will be the same whether or not this group contained originally one or two protons. In short, $MOH_2^{1/2+}$ groups as such are not recognizable on an incompletely protonated surface, which must instead be described as a dense packing of O atoms with an amount of associated protons lying between one and two per O atom.

The assumption of Δx and Δy being zero also implies a $\bar{r}$-value which is independent of the amount adsorbed. The experimental evidence that $\bar{r}$ sometimes varies with surface coverage (Rajan et al., 1974; Rajan, 1976) might thus imply that $[(\bar{s}-1)\Delta x + \Delta y]$ is not negligible and changes with surface coverage. At the extreme, viz. x approaching $-\frac{1}{2}$ and with a positive value for $[(\bar{s}-1)\Delta x + \Delta y]$, $\bar{r}$ in (8.4) may exceed one, which is experimentally found at pH 8.5 (Rajan, 1975a). If, however, a binuclear surface complex is assumed, $\bar{r}$ may exceed one for less extreme conditions. The reaction equation for the formation of a binuclear complex may be given by:

$$\bar{s}MOH_{(3/2+x)} + PO_4H_y \rightarrow MO_2PO_2H_{y'} + (\bar{s}-2)MOH_{(3/2+x')} + \bar{r}OH + (2-\bar{r})H_2O \quad (8.5)$$

The *r* value inferred from the proton balance for this reaction is then given by:

$$\bar{r} = 1 + (\bar{s} - 2)\Delta x + \Delta y - 2x \qquad (8.6)$$

For $[(\bar{s} - 2)\Delta x + \Delta y] \approx 0$ equation (8.6) reduces to $\bar{r} = 1 - 2x$, which then ranges from zero (for $x = \frac{1}{2}$) to two (for $x = -\frac{1}{2}$). The recorded $\bar{r}$ value of 1.44 at pH 8.5, which is near the pH_0 of the adsorbent, indicates that $[(\bar{s} - 2)\Delta x + \Delta y]$ has to be positive for the assumed reaction mechanism.

The $\bar{r}$ value which can be used to derive the change in surface charge upon adsorption (cf. (8.2)), is often found to depend on the amount adsorbed. This led Rajan et al. (1974) to introduce a differential $\bar{r}$ value, d(OH)/d(P), which is a function of the amount adsorbed. The change of the surface charge on adsorption may also be inferred from the change in concentration of the indifferent electrolyte ions following phosphate adsorption at low ionic strength (Ryden and Syers, 1975a; Rajan, 1976; Ryden et al., 1977b).

Phosphate adsorption data are often expressed in the form of adsorption-isotherms which relate the amount of P adsorbed per gram (m^2) to the solution concentration at equilibrium. Different theoretical relationships exist, derived for uncharged systems, each implying different presupposed conditions of the adsorption system. Most authors try to fit their data to one of these well known isotherms (Hsu and Rennie, 1962; Muljadi et al., 1966; Chen et al., 1973a; Kuo and Lotse, 1974b; Rajan et al., 1974; Rajan, 1975a; Ryden et al., 1977b). The Langmuir isotherm seems favorite in this respect. In practice, however, it is often possible only to fit phosphate adsorption data covering a wide concentration range to an isotherm of the Langmuir type, if the existence of more than one type of adsorption sites is invoked, each type having its own adsorption maximum and binding constant. Two or three Langmuir terms are usually necessary to describe satisfactorily the adsorption data (Muljadi et al., 1966; Rajan 1975a; Ryden et al., 1977b). The Langmuir adsorption maximum obtained by this procedure, which then corresponds to the sum of the adsorption maxima of the different regions, depends on the pH at which the adsorption experiments were done (Hsu and Rennie, 1962; Muljadi et al., 1966; Chen et al., 1973a; Rajan et al., 1974; Rajan, 1975a; Ryden et al., 1977b). This indicates that this fitting procedure (to Langmuir isotherms) produces no physically relevant parameters, because the maximum possible adsorption is determined by the number of reactive surface groups only (which is a pH-independent constant for a given adsorbent), if ligand exchange is assumed.

A model for adsorption on variable charge surfaces, which is not based on the mechanism of ligand exchange, has been developed by Bowden (1973) and Bowden et al. (1973, 1974).This model places the adsorbed phosphate in a "Stern layer" with potential (ψ_i) differing from the surface potential (ψ_s) and presumably often opposite in sign. Balancing the total charge then requires that $\sigma_s + \sigma_i + \sigma_d = 0$, with σ_d indicating the charge present in a diffuse layer. It involves only one phosphate adsorption maximum and two phosphate binding constants, one for the $H_2PO_4^-$ ion and the other for the

HPO_4^{2-} ion. Phosphate adsorption data on goethite for different pH values have been fitted with the help of this model. Competition with, for example, silicate and selenite and enhancement of adsorption by coadsorption of cations in the same region of potential can also be described.

When desorption of phosphate is measured in solutions of the same pH and ionic strength it is found that desorption does not follow the adsorption isotherm (Hingston et al., 1974). It seems that a part of the adsorbed phosphate has been adsorbed irreversibly. Experiments with ^{32}P, however, show that all of the adsorbed phosphate can be exchanged, although in part very slowly (Atkinson et al., 1972). The slowness of exchange is then explained by assuming that part or all of the adsorbed phosphate exists with two coordinate bonds to the surface.

So far adsorption was considered not to exceed the available singly coordinated hydroxyl or water groups. On goethite it is found (at pH 3.8) that the maximum adsorption equals just half of the available singly coordinated OH surface groups, which is in agreement with the supposed binuclear surface complex (Parfitt et al., 1976). The measured "adsorption" (Muljadi et al., 1966; Hingston et al., 1974; Helyar et al., 1976) on gibbsite, however, may easily exceed the capacity of the singly coordinated surface groups located on the edge faces (111 and 011) of the gibbsite crystal (Pariftt et al., 1977). The question then arises whether adsorption is possible beyond the capacity of the singly coordinated surface groups. Ryden et al. (1977b) suggest that phosphate adsorbs, in addition to ligand exchange, physically on amorphous iron hydroxide. Furthermore Rajan et al. (1974) suggest that phosphate may react with doubly coordinated surface OH groups on amorphous $Al(OH)_3$ at high phosphate concentrations. The fact that in supersaturated systems the reaction between phosphate ions and metal oxide proceeds beyond the capacity of ligand exchange is hardly surprising, however. In this context it is of interest to note that isotopic exchange experiments using ^{32}P with phosphate adsorbed on gibbsite could be interpreted satisfactorily only if it were assumed that a fraction of the adsorbed phosphate did not take part in this exchange (Kyle et al., 1975). This fraction increased with increasing supersaturation of the system.

With longterm experiments (or in short time at high supersaturation) it is found that the amount of phosphate removed from solution in alumina-phosphate systems may exceed the amount removed initially (1 day) several times. This is logically associated with the formation of a discrete aluminum-phosphate phase (Haseman et al., 1950; Kittrick and Jackson, 1955, 1956; Chen et al., 1973b; Van Riemsdijk et al., 1975, 1977). The exact mechanism of the longterm reaction is not known.

The amount of phosphate removed, measured after different reaction times and with various initial P concentrations, may thus give a picture as shown in Fig. 8.1.

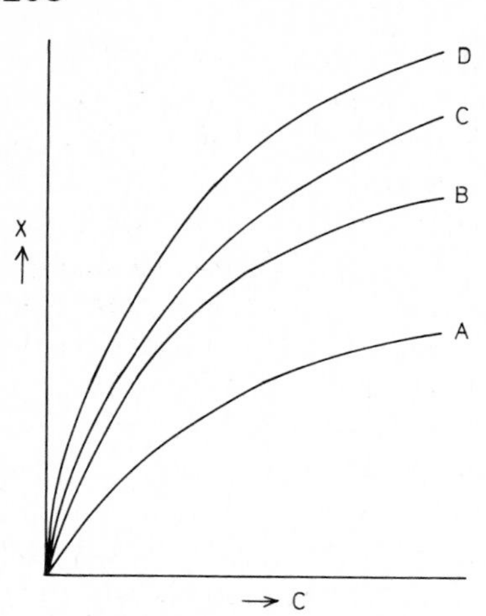

Fig. 8.1. The amount of phosphate removed per unit amount adsorbent (X) as a function of the solution concentration (C) after various reaction times; A, 1 day; B, 1 week; C, 2 weeks; D, 1 month.

It will be clear from the foregoing discussion that curve D is not the best estimate of the "equilibrium" adsorption isotherm. The real adsorption isotherm for these systems has to be estimated, either by trying to separate the longterm reaction from the adsorption or with the help of exchange experiments with ^{32}P.

In this context the question is raised whether the adsorption sites, initially occupied with phosphate, remain blocked during the longterm process, or that desorption of phosphate occurs. Despite the progress that has been made more work is needed, however, before a theoretically sound model can be developed on the basis of experimental facts which describes all the features of the reaction of phosphate with metal oxides (adsorption, desorption and longterm reaction).

8.2.2. The reaction with clay minerals and solid carbonate

The nature and mechanism of phosphate adsorption onto clay minerals is assumed to be a ligand exchange reaction with the OH(H) groups coordinated to the exposed Al atoms on the edge face of the clay crystal, in analogy with the presumed adsorption mechanism with metal oxides. The planar faces of the clay minerals, however, exhibit negative adsorption of anions due to the constant negative charge arising from isomorphous substitution (cf. part A of this text). Thus when the adsorption of phosphate by clay minerals is calculated from the observed decrease of the phosphate concentration in solution, a correction for negative adsorption may be necessary in order to avoid an underestimation of the amount adsorbed (cf. Fig. 7.3, or chapter 7 for a discussion of negative adsorption). Whether the adsorption of phosphate by clay minerals, at least at low equilibrium concentrations, is restricted to singly coordinated hydroxyls — as was found with infra-red studies of phosphated iron and aluminum oxides — has not been proven so far. Likewise, there is no direct evidence available that a binuclear surface phosphate complex is formed. The formation of such a complex has been proposed by Kafkafi et al. (1967) on the basis of results of isotopic exchange studies with ^{32}P of phosphated kaolinite. The adsorption capacity of clay minerals for phosphate will thus depend on, for example, the proportion of the total surface area occupied by the edge faces (which depends on the dimensions of the clay platelets), the number of reactive sites per unit area edge face and the stoichiometry of the adsorption. When dealing with kaolinite, a 1:1 type of clay mineral, one of the planar surface sides also consists of AlOH groups. The hydroxyls, however, are coordinated to two Al-atoms and for that reason probably unreactive (Parfitt et al., 1977), at least at low concentrations of the adsorbate. Obviously the estimate of the phosphate adsorption capacity of clays obtained in the above manner implies that the clay minerals should be free from substances either adsorbed or deposited on the clay surface that may influence the phosphate adsorption.

The effect of pH on phosphate adsorption by clays usually reveals a maximum within the pH range of 4 to 7, either in the form of a sharp peak (generally around pH 4) or as a more or less smooth curve. The actual shape of the curve seems to depend on, for example the type of clay mineral, its pretreatment prior to the adsorption, solution phosphorus concentration and reaction time (cf. e.g. Black, 1942; De Haan, 1965; Muljadi et al., 1966; Chen et al., 1973a; Edzwald et al., 1976). Because clay minerals always exhibit exclusion of anions at the constant-charge planar sides the net adsorption of phosphate by clays should be influenced by the addition of an indifferent electrolyte (at a given pH) for two reasons, viz. the decrease of the non-specific exclusion at the planar side in addition to a change of the specific adsorption at the edges. Results presented by De Haan (1965) and Bowden (1973) show that the phosphate adsorption by montmorillonite increases with increasing ionic strength (at neutral and acid pH values, respectively). The possible reaction mechanisms that have been mentioned in the preceding section concerning the adsorption beyond the available singly coordinated hydroxyls on oxide surfaces may be applicable to clay minerals in the same manner. In addition replacement of silica from the silicate framework at higher solution phosphate concentrations could be a possible mechanism (Low and Black, 1950; Rajan, 1975b; Rajan and Perrott, 1975). In super-saturated solutions formation of an aluminum phosphate phase may occur and be responsible for a slow but continuous removal of phosphate from the solution (Black, 1942; Coleman, 1944; Chen et al., 1973b). These reactions may lead to degradation of the original clay structure, particularly at high phosphate concentrations (Low and Black, 1948; Haseman et al., 1950; Kittrick and Jackson, 1955, 1956).

In view of the prevailing conditions in soils, solid calcium carbonate is usually present only at pH values around neutral or higher (cf. chapter 6 of part A); thus indicating the pH values of practical interest for the reaction of orthophosphate with the above solid.

The reactive sites for adsorption on calcite are assumed to be exposed surface Ca^{2+} ions of which the vacant coordinate positions may be occupied with water molecules, bicarbonate ions or hydroxyl ions in an aqueous suspension of the solid. Phosphate ions may replace these molecules or ions, the adsorption process being chemical in nature (Kuo and Lotse, 1972). In supersaturated solutions the adsorption reaction is followed by the formation of a calcium phosphate.

The solubility of calcite is relatively high, in contrast to the solubilities of the discussed metal oxides (e.g. the solubility at pH 8.25 is about $4.5 \cdot 10^{-4}$ M). This implies that direct precipitation is possible already at moderate phosphate additions, because both calcium and phosphate ions are then present in reasonably high concentrations. In pure solutions with low supersaturation, nucleation is, however, hardly possible; thus limiting precipitation (Griffin and Jurinak, 1974). Nuclei are formed fairly fast, however, upon adsorption on calcite at the same supersaturation. These nuclei, initially amorphous, are then slowly transformed into crystalline apatite followed by crystal growth (Stumm and Leckie, 1971). Results of Cole et al. (1953), Clark and Turner (1955), and Griffin and Jurinak (1973) support the supposition that the type of heteronuclei formed initially depends on the Ca/P ratio. At higher Ca/P ratios the formation of apatite is favored while at lower ratios (i.e. at higher phosphate concentrations) the formation of octocalcium phosphate nuclei or possibly dicalcium phosphate is encouraged. The presumed dependence of the type of nuclei formed on the Ca/P ratio is derived from a plot of the solution activities of Ca^{2+}, $H_2PO_4^-$ and of pH in the form of a solubility diagram of the solid carbonate (usually calcite) and the different crystalline calcium phosphates.

The kinetics of the reactions discussed above indicates that the surface adsorption is a very rapid endothermic reaction which is nearly or totally completed in the order of minutes (Griffin and Jurinak, 1974). The subsequent formation of nuclei and crystalline solids is much slower and the rate constants depend on, for example, the temperature and the degree of supersaturation of the system. It has also been found that the rates of the transformation into apatite and the crystal growth increases with the introduction of fluoride ions in the system and upon lowering the CO_3/PO_4 ratio, probably because the carbonate ions compete with phosphate for adsorption on crystal growth sites. Organic substances and Mg^{2+} ions may have a depressing effect on apatite formation (Stumm and Leckie, 1971; Griffin and Jurinak, 1973).

The term "retention" (cf. Wild, 1949) is used to emphasize the point that when dealing with soils, the removal of phosphate from solution represents the net effect of physical, chemical and microbial processes involved in the transformation of soluble and solid phosphate. Though this implies that the separate contributions of the different processes cannot be determined in general, it is sometimes possible — under laboratory conditions — to take measures that favor the removal by a single mechanism.

If soluble orthophosphate is added to the soil the solution phosphate concentration decreases rapidly initially, followed by a slow decline that may last for months (Barrow and Shaw, 1975a; Munns and Fox, 1976), implying that the amount retained increases with time (Black, 1942; Enfield, 1974; Sawhney, 1977). The slow reaction may represent the formation of a solid phosphate phase (cf. section 8.2), though diffusion into the soil crumbs (Vaidyanathan and Talibudeen, 1968), or gel-like adsorbents (McLaughlin et al., 1977) and/or uptake by microorganisms (White, 1964) may also play a role. The rapid decline is then ascribed to adsorption processes. Conclusive evidence, however, that the formation of a solid phase has occurred, is difficult to provide either because these products are amorphous or when crystalline not accessible to direct methods of identification in the bulk of the soil sample (cf. section 8.1). If the degree of supersaturation is high, these processes may proceed faster and (metastable) crystalline solids may be found in detectable amounts. Such supersaturated conditions may exist around dissolving granules of chemical fertilizers where the phosphate concentration may reach values of a few moles per liter and where the pH may attain either very low or very high values depending on the fertilizer (Lindsay et al., 1962). Besides direct precipitation of less soluble solids from the solution it may further induce dissolution and degradation of the soil minerals and subsequent formation of solid phosphates (Lindsay and Stephenson, 1959a, b, c; Lindsay et al., 1959, 1962; Moreno et al., 1960; Tamini et al., 1964, 1968). Part of these solids are metastable according to the conditions prevailing in the bulk of the soil, thus leading to subsequent dissolution and formation of less soluble solids.

Returning to conditions that are less extreme it has been found that during the long term reaction the retained phosphate becomes less readily replaceable by other specifically adsorbing anions, like e.g. arsenate, molybdate and hydroxide (Barrow, 1973, 1974a). Furthermore, the fraction of the added phosphate exchangeable with radioactive phosphate and the fraction which may be desorbed decreases likewise (Barrow and Shaw, 1975 b, c). All these changes are greater at higher temperatures. Owing to these changes occurring at longer reaction times the effectiveness of the added phosphate for plant growth decreases (Barrow, 1974b). If the above effects are formulated in terms of labile phosphate (Mattingly, 1975;

Holford and Mattingly, 1976; Ryden and Syers, 1977) it means a decrease in the labile soil phosphate fraction at longer reaction times (Talibudeen, 1958; Larsen et al., 1965; Larsen and Widdowson, 1971).

The retention of phosphate by means of adsorption reactions is in view of the preceding discussion best estimated at relative low solution phosphate concentrations and after short reaction times. The number of adsorption sites will vary for different soils and depend mainly on, for example, the amounts present of the compounds mentioned in the preceding section and on other forms of occurrence of reactive aluminum, iron or calcium (Coleman et al., 1960; Conesa, 1969; Perrott et al., 1974). Also the surface area and surface structure of these compounds are important parameters (cf. section 8.2, Holford and Mattingly, 1975 a, b).

The total retention capacity of soils for phosphate will also depend on the different compounds referred to above as these should deliver the metal ions necessary to form the solid phosphate (Williams et al., 1958; Fassbender, 1969; Harter, 1969; John, 1971; Lopez-Hernandez and Burnham, 1973, 1974a; Ballard and Fiskell, 1974; Vyayachandran and Harter, 1975; Beek et al., 1978).

The fraction of the available adsorption sites occupied with phosphate ions depends on the solution phosphate concentration and on factors, for example, like pH, ionic strength, and the presence of substances competing for the same sites as the phosphate ions. The effect of pH on phosphate adsorption is of practical significance in those cases that (very) acid soils are supplied with lime in order to raise the pH to a value favorable for plant growth. Since in acid soils the adsorption sites are located mainly on aluminum and iron compounds it is to be expected (cf. section 8.2) that raising the pH decreases the amount adsorbed (Lopez-Hernandez and Burnham, 1974b). In addition, mineralization of organic phosphate is favored at higher pH values (Cosgrove, 1967).

Exceptions to the general rule have been noticed, however. These were ascribed to the hydrolysis and polymerization of exchangeable Al^{3+} ions upon pH increase, as these polymerized compounds are more effective with respect to phosphate retention than Al^{3+} ions (Coleman et al., 1960; Mokwunge, 1975).

On the other hand higher salt concentrations usually increase the amount adsorbed, in particular if a calcium salt is used to increase the concentration (Lehr and Wesemael, 1952; Pissarides et al., 1968; Barrow, 1972; Volkweiss et al., 1973; Ryden and Syers, 1975b; Ryden et al., 1977 a, b). Liming thus implies an increase in the pH as well as in soluble calcium, which have opposing effects with respect to phosphate adsorption (Stoop, 1974; Amarasiri and Olsen, 1973; Lopez-Hernandez and Burnham, 1974b). Inorganic anions that may compete with orthophosphate ions are, for example, silicate (Deb and Datta, 1967; Obihara and Russell, 1972; Rajan, 1975b; Rajan and Fox, 1975), molybdate (Barrow, 1970, 1973), arsenate (Barrow, 1974a; Fassbender, 1974; Wauchope, 1975), selenite (Gebhardt and Coleman, 1974;

Rajan and Watkinson, 1976), bicarbonate (Nagarajah et al., 1968; Barrow and Shaw, 1976) and fluoride (Kuo and Lotse, 1974a). Nitrate, chloride and sulfate ions are actually no competitors for phosphate (Barrow, 1970; Kinjo and Pratt, 1971; Bailey, 1974). Organic substances that may compete, are, e.g. inositol hexaphosphate (Anderson et al., 1974), simple organic anions (EDTA, citrate, oxalate, phtalate, benzoate), fulvic and humic acids and other complex organic substances (Deb and Datta, 1967, Nagarajah et al., 1968, 1970; Kuo and Lotse, 1974a; Mossi et al., 1974; Appelt et al., 1975). Another phenomenon accompanying the adsorption of phosphate ions by soil minerals may be the co-adsorption of cations, thus leading to an increase in the cation exchange capacity of these soils (Schalscha et al., 1972, 1974; Sawhney, 1974; Ryden and Syers, 1976). At the completion of the adsorption processes the relationships between amount adsorbed and solution phosphate concentration are described in general by a Freundlich or Langmuir equation. Prior to the construction of the isotherm it is usually necessary to account for a certain portion of the native soil phosphate occupying adsorption sites at the start of the experiment (Holford et al., 1974; Fitter and Sutton, 1975). The Langmuir isotherm usually gives a reasonable fit of the experimental data over narrow concentration ranges (1—20 mg P/l), though it may still be necessary to introduce 2 or 3 different sets of adsorption sites (Syers et al., 1973; Holford et al., 1974; Ryden et al., 1977 a, b). The adsorption data of Holford and Mattingly (1975c) for calcareous soils strongly suggest that the sites with the higher bonding energy are located on (Fe) hydrous oxides and the low energy sites on the carbonate particles.

Phosphate applied in the form of chemical fertilizer (usually superphosphate) accumulates in general in the top layers of agricultural soils and once retained is hardly subject to leaching (Roscoe, 1960; Wiechmann, 1966, 1971; Van Diest, 1968; Henkens, 1972; Kolenbrander, 1972; Hanway and Laflen, 1974). Obviously, additional factors, like, for example, method and size of application, soil type, rainfall, slope and soil cover may bring about (slight) changes in this general pattern (Kao and Blanchar, 1973; Munn et al., 1973; Rauschkolb et al., 1976). Exceptions to this rule are found if it concerns soils with a low phosphate retention capacity, like, for example, organic soils and some sandy soils (Fox and Kamprath, 1971; Humphreys and Pritchett, 1971) or with excessive additions of fertilizers (Fiskell and Spencer, 1964; Logan and McLean, 1973 a, b).

In the case of liquid manure or sewage water additions, convective transport of orthophosphate species (as well as inorganic condensed phosphate species and the organic phosphate compounds present in these additions) to deeper layers is possible. This is in contrast to the situation where chemical fertilizer is applied to soil, in which case diffusion processes are predominant (Bouldin and Black, 1954; Olsen et al., 1962, 1965). The removal of phosphate from waste water and its distribution in the soil profile thus depends on the ratio of the carrier flux density to the rate constants of the processes by which the phosphate species are retained by the solid phase. These problems have stimulated the development of models aimed at a quantitative description of the displacement of phosphate in soil subject to a given water flow regime in combination with the phosphate retention processes. Foregoing details of the modelling of water movement in the soil, conceptually different models have been proposed for the phosphate retention processes, which vary from a description of the retention of added phosphate in the form of a single process to a distinction in concurrently occurring processes, mentioning here adsorption, precipitation and other forms of chemical immobilization — like, for example, occlusion, where the adsorbed phosphate becomes surrounded by matrices of Fe and Al components (Enfield and Shew, 1975; Novak et al., 1975; Shah et al., 1975; Harter and Foster, 1976; Overman et al., 1976; Mansell et al., 1977 a, b). In these models the adsorption is treated as a reversible process, i.e. the adsorbed phosphate may desorb, though with a slower rate than during the adsorption step. Chemical immobilization and precipitation are usually considered as irreversible processes, generally with a much slower rate than adsorption. Kinetic expressions for these processes may vary from instantaneous equilibrium for adsorption to simple or complex rate equations (Enfield, 1974; Novak and Adriano, 1975; Enfield et al., 1976). For lack of precise information chemical immobilization and precipitation are usually expressed in terms of first-order rate equations (Overman et al., 1976; Mansell et al., 1977 a, b). The rate expressions and the numerical values of the corresponding parameters are then derived from batch experiments, whereas results of percolation experiments on soil columns are used to test the complete model. The predictive value of the models must then be tested for long term additions. For practical reasons the results obtained from (existing) field experiments (Erickson et al., 1971; Beek et al., 1973, 1977a, b; Adriano et al., 1975; Kardos and Hook, 1976) would have to be used for such a purpose. A convincing check on the long term applicability of phosphate displacement models has not been reported so far. Whether inorganic condensed phosphates and organic phosphate compounds present in waste water or liquid manure give different distribution patterns will depend on the retention characteristics of these compounds and the rate constants of the hydrolysis or mineralization process by which these compounds are transformed into orthophosphate (Bowman et al., 1967; Hashimoto and Lehr, 1973; Khasawneh et al., 1974; Rolston et al., 1975). Obviously the presence of substances in the added waste water or liquid manure that compete for the same sites as the phosphate compounds also influence the distribution pattern of the latter (Campbell and Raez, 1975).

REFERENCES

Adams, J.A., Howarth, D.T. and Campbell, A.S., 1973. Plumbogummite minerals in a strongly weathered New Zealand soil. J. Soil Sci., 24: 224—231.
Adriano, D.C., Novak, L.T., Erickson, A.E., Wolcott, A.R. and Ellis, B.G., 1975. Effect of long term land disposal by spray irrigation of food processing wastes on some chemical properties of the soil and subsurface water. J. Environ. Qual., 4: 242—248.
Altschuler, Z.S., 1973. The weathering of phosphate deposits — geochemical and environmental aspects. In: E.J. Griffith, A. Beeton, J.M. Spencer, and D.T. Mitchell (Editors). Environmental Phosphorus Handbook. Wiley, New York, N.Y., pp. 33—96.
Amarasiri, S.L. and Olsen, S.R., 1973. Liming as related to solubility of P and plant growth in an acid tropical soil. Soil Sci. Soc. Am. Proc., 37: 716—721.
Anderson, G., 1963. Effect of iron/phosphorus ratio and acid concentration on the precipitation of ferric inositol hexaphosphate. J. Sci. Food Agric., 14: 352—359.
Anderson, G., 1967. Nucleic acids, derivatives and organic phosphates. In: A.D. McLaren, and G.H. Peterson (Editors), Soil Biochemistry. Marcel Dekker, New York, N.Y., pp. 67—90.
Anderson, G. and Malcolm, R.E., 1974. The nature of alkali-soluble soil organic phosphates. J. Soil Sci., 25: 282—297.
Anderson, G., Williams, E.G. and Moir, J.O., 1974. A comparison of the sorption of inorganic orthophosphate and inositol hexaphosphate by six acid soils. J. Soil Sci., 25: 51—62.
Appelt, H., Coleman, N.T. and Pratt, P.F., 1975. Interactions between organic compounds, minerals and ions in volcanic ash derived soils: II. Effects of organic compounds on the adsorption of phosphate. Soil Sci. Soc. Am. Proc., 39: 628—630.
Atkinson, R.J., Posner, A.M. and Quirk, J.P., 1972. Kinetics of heterogeneous isotopic exchange reactions. Exchange of phosphate at the α-FeOOH aqueous solution interface. J. Inorg. Nucl. Chem., 34: 2201—2211.
Bailey, J.M., 1974. Changes in the adsorbed sulphate status of a yellow-brown earth after phosphate fertilisation and legume growth. N.Z.J. Agric. Res., 17: 257—265.
Ballard, R. and Fiskell, J.G.A., 1974. Phosphorus retention in Coastal Plain Forest soils: I. Relationship to soil properties. Soil Sci. Soc. Am. Proc., 38: 250—255.
Barrow, N.J., 1970. Comparison of the adsorption of molybdate, sulfate and phosphate by soils. Soil Sci., 109: 282—288.
Barrow, N.J., 1972. Influence of solution concentration of calcium on the adsorption of phosphate, sulfate, and molybdate by soils. Soil Sci., 113: 175—180.
Barrow, N.J., 1973. On the displacement of adsorbed anions from soil: 1. Displacement of molybdate by phosphate and by hydroxide. Soil Sci., 116: 423—431.
Barrow, N.J., 1974a. On the displacement of adsorbed anions from soil: 2. Displacement of phosphate by arsenate. Soil Sci., 117: 28—33.
Barrow, N.J., 1974b. The slow reactions between soil and anions: 1. Effects of time, temperature, and water content of a soil on the decrease in effectiveness of phosphate for plant growth. Soil Sci., 118: 380—386.
Barrow, N.J. and Shaw, T.C., 1975a. The slow reactions between soil and anions: 2. Effect of time and temperature on the decrease in phosphate concentration in the soil solution. Soil Sci., 119: 167—177.
Barrow, N.J. and Shaw, T.C., 1975b. The slow reactions between soil and anions: 3. The effects of time and temperature on the decrease in isotopically exchangeable phosphate. Soil Sci., 119: 190—197.

Barrow, N.J. and Shaw, T.C., 1975c. The slow reactions between soil and anions: 5. Effects of period of prior contact on the desorption of phosphate from soils. Soil Sci., 119: 311—320.

Barrow, N.J. and Shaw, T.C., 1976. Sodium bicarbonate as an extactant for soil phosphate: I. Separation of the factors affecting the amount of P displaced from soil from those affecting secondary adsorption. Geoderma, 16: 91—107.

Beek, J. and De Haan, F.A.M., 1973. Phosphate removal by soil in relation to waste disposal. In: J. Tomlinson (Editor), Proc. of the Int. Conf. on Land for Waste Management. The Agric. Inst. of Canada, Ottawa, pp. 77—86.

Beek, J., De Haan, F.A.M., and Van Riemsdijk, W.H., 1977a. Phosphates in soils treated with sewage water: 1. General information on sewage farm, soil, and treatment results. J. Environ. Qual., 6: 4—7.

Beek, J., De Haan, F.A.M., and Van Riemsdijk, W.H., 1977b. Phosphates in soils treated with sewage water: II. Fractionation of accumulated phosphates. J. Environ. Qual., 6: 7—12.

Beek, J., Van Riemsdijk, W.H. and Koenders, K., 1978. Aluminum and iron fractions affecting phosphate bonding in a sandy soil treated with sewage water. In: A. Banin (Editor), Proc. Congr. Agrochemicals in Soils. Hebrew Univ., Rehovot, Israel., in press.

Black, C.A., 1942. Phosphate fixation by kaolinite and other clays as affected by pH, phosphate concentration and time of contact. Soil Sci. Soc. Am. Proc., 7: 123—133.

Black, C.A., 1968. Soil-Plant Relationships. Wiley, New York, N.Y., pp. 558—653.

Boehm, H.P., 1971. Acidic and basic properties of hydroxylated metal oxide surfaces. In: Surface chemistry of oxides. Disc. Faraday Soc., 52: 264—275.

Bouldin, D.R. and Black, C.H. 1954. Phosphorus diffusion in soils. Soil Sci. Soc. Am. Proc., 18: 255—259.

Bowden, J.W., 1973. Models for Ion Adsorption on Mineral Surfaces. Ph.D. Thesis, Univ. of Western Australia.

Bowden, J.W., Bolland, M.D.A., Posner, A.M. and Quirk, J.P., 1973. Generalized model for anion and cation adsorption at oxide surfaces. Nature London, Phys. Sci., 245: 81—83.

Bowden, J.W., Posner, A.M. and Quirk, J.P., 1974. A model for ion adsorption on variable charge surfaces. Trans. 10th Int. Congr. Soil Sci., II: 29—36.

Bowman, B.T., Thomas, R.L. and Elrick, D.E., 1967. The movement of phytic acid in soil cores. Soil Sci. Soc. Am. Proc., 31: 477—481.

Breeuwsma, A. and Lyklema, J., 1973. Physical and chemical adsorption of ions in the electrical double layer on hematite (α—Fe_2O_3). J. Colloid Interface Sci., 43: 437—448.

Bromfield, S.M., 1967a. Phosphate sorbing sites in acid soils. An examination of ammonium fluoride as a selective extractant for aluminum-bound phosphate in phosphated soils. Aust. J. Soil Res., 5: 93—102.

Bromfield S.M., 1967b. An examination of the use of ammonium fluoride as a selective extractant for aluminum-bound phosphate in partially phosphated systems. Aust. J. Soil Res., 5: 225—234.

Bromfield, S.M., 1970. The inadequacy of corrections for resorption of phosphate during the extraction of aluminum-bound soil phosphate. Soil Sci., 109: 388—390.

Burford, J.R. and Bremner, J.M., 1972. Is phosphate reduced to phosphine in waterlogged soils. Soil Biol. Biochem., 4: 489—495.

Campbell, A.S., Adams, J.A. and Howarth, D.T., 1972. Some problems encountered in the identification of plumbogummite minerals in soils. Clay Min., 9: 415—423.

Campbell, L.B. and Raez, G.J., 1975. Organic and inorganic P content, movement and mineralization of P in soil beneath a feedlot. Can. J. Soil Sci., 55: 457—466.

Cescas, M.P., Tyner, E.H. and Syers, J.K., 1970. Distribution of apatite and other mineral inclusions in a rhyolitic pumice ash and beach sands from New Zealand: an electron microprobe study. J. Soil Sci., 21: 78—84.
Chang, S.C. and Jackson, M.L., 1957. Fractionation of soil phosphorus. Soil Sci., 84: 133—144.
Chen, Y.S.R., Butler, J.N. and Stumm, W., 1973a. Adsorption of phosphate on alumina and kaolinite from dilute aqueous solutions. J. Colloid Interface Sci., 43: 421—436.
Chen, Y.S.R., Butler, J.N. and Stumm, W. 1973b. Kinetic study of phosphate reaction with aluminum oxide and kaolinite. Environ. Sci. Technol., 7: 327—332.
Clark, J.S. and Turner, R.C., 1955. Reactions between solid calcium carbonate and orthophosphate solutions. Can. J. Chem., 33: 665—761.
Cole, C.V., Olsen, S.R. and Scott, C.V., 1953. The nature of phosphate sorption by calcium carbonate. Soil Sci. Soc. Am. Proc., 17: 352—356.
Coleman, N.T., Thorup, J.T. and Jackson, W.A., 1960. Phosphate sorption reactions that involve exchangeable Al. Soil Sci., 90: 1—7.
Coleman, R., 1944. The mechanism of phosphate fixation by montmorillonitic and kaolinitic clays. Soil Sci. Soc. Am. Proc., 9: 72—78.
Conesa, A.P., 1969. Quelques aspects de la distribution du phosphore en sol calcaire. Ann. Agron., 20: 225—244.
Cosgrove, D.J., 1967. Metabolism of organic phosphates in soil. In: A.D. McLaren and G.H. Peterson (Editors), Soil Biochemistry. Marcel Dekker, New York, N.Y., pp. 216—228.
Deb, D.L. and Datta, N.P., 1967. Effect of associating anions on phosphorus retention soil. I, Under variable phosphorus concentration; II, Under variable anion concentration. Plant and Soil, 26: 303—316, 432—444.
De Haan, F.A.M., 1965. The Interaction of certain inorganic anions with clays and soils Agr. Res. Rep., Wageningen, 655, 167 pp.
Edzwald, J.K., Toensing, D.C. and Leung, M.C., 1976. Phosphate adsorption reactions with clay minerals. Environ. Sci. Technol., 10: 485—490.
Enfield, C.G., 1974. Rate of phosphorus sorption by five Oklahoma soils. Soil Sci. Soc. Am. Proc., 38: 404—407.
Enfield, C.G. and Shew, D.C., 1975. Comparison of two predictive nonequilibrium one-dimensional models for phosphorus sorption and movement through homogeneous soils. J. Environ. Qual., 4: 198—202.
Enfield, C.G., Harlin, C.C. and Bledsoe, B.E., 1976. Comparison of five kinetic models for orthophosphate reactions in mineral soils. Soil Sci. Soc. Am. Proc., 40: 243—249.
Erickson, A.E., Tiedje, J.M., Ellis, B.G. and Hansen C.M., 1971. A barried landscape water renovation system for removing phosphate and nitrogen from liquid feedlot waste. In: Proc. Int. Symp. on Livestock Wastes. Am. Soc. Agric. Eng., St. Joseph, Mich., pp. 232—234.
Fares, F., Fardeau, J.C. and Jacquin, F., 1974. Quantitative survey of organic phosphorus in different soil types. Phosphorus Agric., 63: 25—40.
Fassbender, H.W., 1969. Phosphorus fixation in tropical soils. Agric. Digest, 18: 20—28.
Fassbender, H.W., 1974. Gehalt, Formen und Fixierung von Arsenat im Vergleich zu Phosphat in Waldböden, Z. Pflanzenernaehr. Bodenkd., 137: 188—203.
Fisher, J.D., 1973. Geochemistry of minerals containing phosphorus. In: E.J. Griffith, A. Beeton, J.M. Spencer and D.T. Mitchell (Editors), Environmental Phosphorus Handbook. Wiley New York, N.Y., pp. 141—152, 153—168.
Fiskell, J.G.A. and Spencer, W.F., 1964. Forms of phosphate in lakeland fine sand after six years of heavy phosphate and lime applications. Soil Sci., 97: 320—327.

Fitter, A.H. and Sutton, C.D., 1975. The use of the Freundlich isotherm for soil phosphate sorption data. J. Soil Sci., 26: 241—246.

Fordham, A.W. and Schwertmann, U., 1977. Composition and reactions of liquid manure (Gülle), with particular reference to phosphate: 1. Analytical composition and reaction with poorly crystalline iron oxide (ferrihydrite). J. Environ. Qual., 6: 133—136.

Fox, R.L., and Kamprath. E.J., 1971. Adsorption and leaching of P in acid organic soils and high organic matter sand. Soil Sci. Soc. Am. Proc., 35: 154—156.

Gebhardt, H. and Coleman, N.T., 1974. Anion adsorption by allophanic tropical soils. III, Phosphate adsorption. Soil Sci. Soc. Am. Proc., 38: 263—266.

Ghonsikar, C.P. and Miller, R.H., 1973. Soil inorganic polyphosphates of microbial origin Plant and Soil, 38: 651—655.

Goring, C.A.I. and Bartholomew, W.V., 1952. Adsorption of monoculeotides, nucleic acids, and nucleoproteins by clays. Soil Sci., 74: 149—164.

Griffin, R.A. and Jurinak, J.J., 1973. The interaction of phosphate with calcite. Soil Sci. Soc. Am. Proc., 37: 847—850.

Griffin, R.A. and Jurinak, J.J., 1974. Kinetics of the phosphate interaction with calcite. Soil Sci. Soc. Am. Proc., 38: 75—79.

Halstead, R.L. and Anderson, G., 1970. Chromatographic fractionation of organic phosphates from alkali, acid, and aqueous acetylacetone extracts of soil. Can. J. Soil Sci Sci., 50: 111—119.

Hannapel, R.J., Fuller, W.H. and Fox, R.H., 1964. Phosphorus movement in a calcareous soil: II, Soil microbial activity and organic phosphorus movement. Soil Sci., 97: 421—427.

Hanway, J.J. and Laflen, J.M., 1974. Plant nutrient losses from tile-outlet terraces. J. Environ. Qual., 3: 351—356.

Harter, R.D., 1969. Phosphorus adsorption sites in soils. Soil Sci. Soc. Am. Proc., 33: 630—632.

Harter, R.D. and Foster, B.B., 1976. Computer simulation of phosphorus movement through soils. Soil Sci. Soc. Am. J., 40: 239—242.

Haseman, J.F., Brown, E.H. and Whitt, C.D., 1950. Some reactions of phosphate with clays and hydrous oxides of iron and aluminum. Soil Sci., 70: 257—271.

Hashimoto, I. and Lehr, J.R., 1973. Mobility of polyphosphates in soil. Soil Sci. Soc. Am. Proc., 37: 36—41.

Helyar, K.R., Munns, D.N. and Burau, R.G., 1976. Adsorption of phosphate by Gibbsite II Formation of a surface complex involving divalent cations. J. Soil. Sci., 27: 315—323.

Henkens, Ch.H., 1972. Fertilizer and the quality of surface water. Stikstof (Dutch Nitrogenous Fertilizer Rev.), 15: 28—39.

Hilal, M.H.M., 1969. Characterization of mobile forms of P in a calcareous soil. J. Sci. Food Agric., Abstr., 20: i 241.

Hingston, F.J., Atkinson, R.J., Posner, A.M. and Quirk, J.P., 1967. Specific adsorption of anions. Nature, 215: 1459—1461.

Hingston, F.J., Atkinson, R.J., Posner, A.M. and Quirk, J.P., 1968. Specific adsorption of anions on goethite. Trans. 9th Int. Congr. Soil Sci., 1: 669—678.

Hingston, F.J., Posner, A.M. and Quirk, J.P., 1972. Anion adsorption by goethite and gibbsite, I. The role of the proton in determining adsorption envelopes. J. Soil Sci., 23: 177—192.

Hingston, F.J., Posner, A.M. and Quirk, J.P., 1974. Anion adsorption by goethite and gibbsite, II. Desorption of anions from hydrous oxide surfaces. J. Soil Sci., 25: 16—26.

Holford, I.C.R. and Mattingly, G.E.G., 1975a. Surface areas of calcium carbonate in soils. Geoderma, 13: 247—255.

Holford, I.C.R. and Mattingly, G.E.G., 1975b. Phosphate sorption by Jurassic oolitic limestones. Geoderma, 13: 257—264.

Holford, I.C.R. and Mattingly, G.E.G., 1975c. The high- and low-energy phosphate adsorbing surfaces in calcareous soils. J. Soil Sci., 26: 407—417.

Holford, I.C.R. and Mattingly, G.E.G., 1976. A model for the behaviour of labile phosphate in soil. Plant and Soil, 44: 219—229.

Holford, I.C.R., Wedderburn, R.W.M. and Mattingly, G.E.G., 1974. A Langmuir two-surface equation as a model for phosphate adsorption by soils. J. Soil Sci., 25: 242—255.

Hook, J.E., Kardos, L.T., and Sopper, W.E. 1973. Effects of land disposal of wastewaters on soil phosphorus relations. In: W.E. Sopper and L.T. Kardos (Editors), Recycling Treated Municipal Wastewater and Sludge through Forest and Cropland. Proc. Symp. The Pennsylvania State University Press, pp. 200—219.

Hsu, P.H. and Rennie, D.A., 1962. Reactions of phosphate in aluminum systems, I. Adsorption of phosphate by X-ray amorphous "aluminum hydroxide". Can. J. Soil Sci., 42: 197—221.

Huang, C.P., 1975. Adsorption of phosphate at the hydrous γ-Al_2O_3-electrolyte interface. J. Colloid Interface Sci., 53: 178—186.

Humphreys, F.R. and Pritchett, W.L., 1971. Phosphorus adsorption and movement in some sandy forest soils. Soil Sci. Soc. Am. Proc., 35: 495—500.

Ivanov, P. and Sauerbeck, D., 1971. Festlegung, Umwandlung und Aufnehmbarkeit von Phytin-Phosphor im Boden. Z. Pflanzenernaehr. Bodenkd., 129: 113—125.

Jackman, R.H. and Black, C.A., 1951. Solubility of iron, aluminum, calcium and magnesium inositol phosphates at different pH-values. Soil Sci., 72: 179—186.

John, M.K., 1971. Soil properties affecting the retention of phosphorus from effluent. Can. J. Soil Sci., 51: 315—322.

Kafkafi, U., Posner, A.M. and Quirk, J.P., 1967. Desorption of phosphate from kaolinite. Soil Sci. Soc. Am. Proc., 31: 348—353.

Kao, C.W., and Blanchar, R.W., 1973. Distribution and chemistry of phosphorus in an Albaqualf soil after 82 years of phosphate fertilization. J. Environ. Qual., 2: 237—240.

Kardos, L.T. and Hook, J.E., 1976. Phosphorus balance in sewage effluent treated soils. J. Environ. Qual., 5: 87—90.

Khasawneh, F.E., Sample, E.C. and Hashimoto, I., 1974. Reactions of ammonium ortho- and poly- phosphate fertilizers in soil: I, Mobility of phosphorus. Soil Sci. Soc. Am. Proc., 38: 446—451.

Kinjo, T. and Pratt, P.F., 1971. Nitrate adsorption: II, In competition with chloride, sulfate, and phosphate. Soil Sci. Soc. Am. Proc., 35: 725—728.

Kittrick, J.A. and Jackson, M.L., 1955. Rate of phosphate reaction with soil minerals and electron microscope observations on the reaction mechanism. Soil Sci. Soc. Am. Proc., 19: 292—295.

Kittrick, J.A. and Jackson, M.L., 1956. Electron-microscope observations of the reaction of phosphate with minerals, leading to a unified theory of phosphate fixation in soils. J. Soil Sci., 7: 81—89.

Kolenbrander, G.J., 1972. The eutrophication of surface water by agriculture and the urban population. Stikstof (Dutch Nitrogenous Fertilizer Rev.), 15: 56—67.

Kuo, S. and Lotse, E.G., 1972. Kinetics of phosphate adsorption by calcium carbonate and Ca- kaolinite. Soil Sci. Soc. Am. Proc., 36: 725—729.

Kuo, S. and Lotse, E.G., 1974a. Kinetics of phosphate adsorption and desorption by lade sediments. Soil Sci. Soc. Am. Proc., 38: 50—54.

Kuo, S. and Lotse, E.G., 1974b. Kinetics of phosphate adsorption and desorption by hematite and gibbsite. Soil Sci., 116: 400—406.

Kurmies, B., 1972. Zur Fraktionierung der Bodenphosphaten (with English summary). Die Phosphorsäure, 29: 118—151.

Kyle, J.H., Posner, A.M. and Quirk, J.P., 1975. Kinetics of isotopic exchange of phosphate adsorbed on gibbsite. J. Soil Sci., 26: 32—43.

Larsen, S., 1967. Soil phosphorus. In: A.G. Norman, (Editor). Advances in Agronomy, 19, Academic Press, New York, N.Y., pp. 151—210.

Larsen, S. and Widdowson, A.E., 1971. Ageing of phosphate added to soil. J. Soil Sci., 22: 5—7.

Larsen, S., Gunary, D. and Sutton, D.C., 1965. The rate of immobilization of applied phosphate in relation to soil properties. J. Soil Sci., 16: 141—148.

Lehr, J.J. and Van Wesemael, J.Ch., 1952. The influence of neutral salts on the solubility of soil phosphate. J. Soil Sci., 3: 125—135.

Lindsay, W.L. and Stephenson, H.F., 1959a. Nature of the reactions of monocalcium-phosphate monohydrate in soils: I. The solution that reacts with the soil. Soil Sci. Soc. Am. Proc., 23: 12—18.

Lindsay, W.L. and Stephenson, H.F., 1959b. Nature of the reactions of monocalcium-phosphate monohydrate in soils: II. Dissolution and precipitation reactions involving iron, aluminium, manganese and calcium. Soil Sci. Soc. Am. Proc., 23: 18—22.

Lindsay, W.L. and Stephenson, H.F., 1959c. Nature of the reactions of monocalcium-phosphate monohydrate in soils: IV. Repeated reactions with metastable triple-point solution. Soil Sci. Soc. Am. Proc., 23: 440—445.

Lindsay, W.L., Lehr, J.R. and Stephenson, H.F., 1959. Nature of the reactions of monocalciumphosphate monodydrate in soils: III. Studies with metastable triple-point solution. Soil Sci. Soc. Am. Proc., 23: 342—345.

Lindsay, W.L., Frazier, A.W. and Stephenson, H.F., 1962. Identification of reaction products from phosphate fertilizers in soils. Soil Sci. Soc. Am. Proc., 26: 446—452.

Logan, T.J. and McLean, E.O., 1973a. Effects of phosphorus application rate, soil properties and leaching mode on ^{32}P movement in soil columns. Soil Sci. Soc. Am. Proc., 37: 371—374.

Logan, T.J. and McLean, E.O., 1973b. Nature of phosphorus retention and adsorption with depth in soil columns. Soil Sci. Soc. Am. Proc., 37: 351—355.

Lopez-Hernandez, D. and Burnham, C.P., 1973. Extraction methods for aluminum and iron in relation to phosphate adsorption. Commun. Soil Sci. Plant Anal., 4: 9—16.

Lopez-Hernandez, D. and Burnham, C.P., 1974a. The covariance of phosphate sorption with other soil properties in some British and tropical soils. J. Soil Sci., 25: 196—206.

Lopez-Hernandez, D. and Burnham, C.P., 1974b. The effect of pH on phosphate adsorption in soils. J. Soil Sci., 25: 207—216.

Low, P.F. and Black, C.A., 1948. Phosphate-induced decomposition of kaolinite. Soil Sci. Soc. Am. Proc., 12: 180—184.

Low, P.F. and Black, C.A., 1950. Reactions of phosphate with kaolinite. Soil Sci., 70: 273—290.

Mansell, R.S., Selim, H.M., Kanchanasut, P., Davidson, J.M. and Fiskell, J.G.A., 1977a. Experimental and simulated transport of phosphorus through sandy soils. Water Resour. Res., 13: 189—194.

Mansell, R.S., Selim, H.M. and Fiskell, J.G.A., 1977b. Simulated transformations and transport of phosphorus in soils. Soil Sci., 124: 102—109.

Mattingly, G.E.G., 1975. Labile phosphate in soils. Soil Sci., 119: 369—375.

Mattingly, G.E.G. and Talibudeen, O., 1967. Progress in the chemistry of fertilizer and soil phosphorus. In: M. Grayson, and J. Griffith (Editors), Topics in Phosphorus Chemistry, 4. Interscience, New York, N.Y., pp. 157—290.

McKelvey, V.E., 1973. Abundance and distribution of phosphorus in the lithosphere. In: E.J. Griffith, A. Beeton, J.M. Spencer, and D.T. Mitchell (Editors), Environmental Phosphorus Handbook. Wiley, New York, N.Y., pp. 13—31.

McLaughlin, J.R., Ryden J.C., and Syers, J.K., 1977. Development and evaluation of a kinetic model to describe phosphate sorption by hydrous ferric oxide gel. Geoderma, 18: 295—307.

Mokwunge, U., 1975. The influence of pH on the adsorption of phosphate by soils from the Guinea and Sudan savannah zones of Nigeria. Soil Sci. Soc. Am. Proc., 39: 1100—1102.

Moreno, E.C., Lindsay, W.L. and Osborn, G., 1960. Reactions of dicalcium phosphate dihydrate in soils. Soil Sci., 90: 58—68.

Mortland, M.M. and Gieseking, J.S., 1952. The influence of clay minerals on the enzymatic hydrolysis of organic phosphorus compounds. Soil Sci. Soc. Am. Proc., 16: 10—13.

Mossi, A.D., Wild A. and Greenland, D.J., 1974. Effect of organic matter on the charge and phosphate adsorption characteristics of Kikuyu red clay from Kenya. Geoderma, 11: 275—285.

Muljadi, D., Posner, A.M. and Quirk, J.P., 1966. The mechanism of phosphate adsorption by kaolinite, gibbsite, and pseudoboehmite. Part 1. The isotherms and the effect of pH on adsorption. J. Soil Sci., 17: 212—229.

Munn, D.A., McLean, E.O., Ramirez, A. and Logan, T.J., 1973. Effect of soil, cover, slope, and rainfall factors on soil and phosphorus movement under simulated rainfall conditions. Soil Sci. Soc. Am. Proc., 37: 428—431.

Munns, D.N. and Fox, R.L., 1976. The slow reaction which continues after phosphate adsorption: Kinetics and equilibrium in some tropical soils. Soil Sci. Soc. Am. Proc. 40: 46—51.

Nagarajah, S., Posner, A.M. and Quirk, J.P., 1968. Desorption of phosphate from kaolinite by citrate and bicarbonate. Soil Sci. Soc. Am. Proc., 32: 507—510.

Nagarajah, S., Posner, A.M. and Quirk, J.P., 1970. Competitive adsorption of phosphate with polygalacturonate and other organic anions on kaolinite and oxide surfaces. Nature, 228: 83—85.

Norrish, K., 1968. Some phosphate minerals in soils. Trans. 9th Int. Congr. Soil Sci., 2: 713—723.

Novak, L.T. and Adriano, D.C., 1975. Phosphorus movement in soils. Soil-orthophosphate reaction kinetics. J. Environ. Qual., 4: 261—266.

Novak, L.T., Adriano, D.C., Coulman, G.A. and Shah, D.B., 1975. Phosphorus movement in soils: Theoretical aspects. J. Environ. Qual., 4: 93—99.

Nriagu, J.O., 1972. Stability of vivianite and ion-pair formation in the system $Fe_3(PO_4)_2$-H_3PO_4-H_2O. Geochim. Cosmochim. Acta, 36: 459—470.

Obihara, C.H., and Russell, E.W., 1972. Specific adsorption of silicate and phosphate by soils. J. Soil Sci., 23: 105—111.

Olsen, S.R., Kemper, W.D. and Jackson, R.D., 1962. Phosphate diffusion to plant roots. Soil Sci. Soc. Am. Proc., 26: 222—227.

Olsen, S.R., Kemper, W.D. and Van Schaik, J.C., 1965. Self-diffusion coefficient of phosphorus in soil measured by transient and steady state methods. Soil Sci. Soc. Am. Proc., 29: 154—158.

Overman, A.R., Chu, R. and Lesemann, W.G., 1976. Phosphorus transport in a packed bed reactor. J. Water Pollut. Control Fed., 48: 880—888.

Parfitt, R.L., Atkinson, R.J. and Smart, R.St.C., 1975. The mechanism of phosphate fixation by iron oxides. Soil Sci. Soc. Am. Proc., 39: 837—841.

Parfitt, R.L., Russell, J.D. and Farmer, V.C., 1976. Confirmation of the surface structure of goethite and phosphated goethite. J. Chem. Soc. Faraday Trans. I, 72: 1082—1087.

Parfitt, R.L., Fraser, A.R., Russell, J.D. and Farmer, V.C., 1977. Adsorption on hydrous oxides. II, Oxalate, benzoate and phosphate on gibbsite. J. Soil Sci., 28: 40—47.

Perrott, K.W., Langdon, A.G. and Wilson, A.T., 1974. Sorption of anions by the cation exchange surface of muscovite. J. Colloid Interface Sci., 48: 10—19.

Petersen, G.W. and Corey, R.B., 1966. A modified Chang and Jackson procedure for routine fractionation of inorganic soil phosphates. Soil Sci. Soc. Am. Proc., 30: 563—565.

Pissardes, A., Stewart, J.W.B. and Rennie, D.A., 1968. Influence of cation saturation on phosphorus adsorption by selected clay minerals. Can. J. Soil Sci., 48: 151—157.

Rajan, S.S.S., 1975a. Adsorption of divalent phosphate on hydrous aluminum oxide. Nature, 253: 434—436.

Rajan, S.S.S., 1975b. Phosphate adsorption and the displacement of structural silicon in an allophane clay. J. Soil Sci., 26: 250—256.

Rajan, S.S.S., 1976. Changes in net surface charge of hydrous alumina with phosphate adsorption. Nature, 262: 45—46.

Rajan, S.S.S. and Fox, R.L., 1975. Phosphate adsorption by soils: II, Reaction in tropical acid soils. Soil Sci. Am. Proc., 39: 846—851.

Rajan, S.S.S. and Perrott, K.W. 1975. Phosphate adsorption by synthetic amorphous aluminosilicates. J. Soil Sci., 26: 257—266.

Rajan, S.S.S. and Watkinson, J.H., 1976. Adsorption of selenite and phosphate on an allophane clay. Soil Sci. Soc. Am. J., 40: 51—54.

Rajan, S.S.S., Perrott, K.W. and Saunders, W.M.H., 1974. Identification of phosphate-reactive sites of hydrous alumina from proton consumption during phosphate adsorption at constant pH values. J. Soil Sci., 25: 438—447.

Rajendran, N. and Sutton, D.C., 1970. Re-sorption of soil phosphate during fractionation. J. Soil Sci., 21: 199—202.

Rauschkolb, R.S., Rolston, D.E., Miller, R.J., Carlton, A.B. and Bureau, R.G., 1976. Phosphorus fertilization with drip irrigation. Soil Sci. Soc. Am. J., 40: 68—72.

Rolston, D.E., Rauschkolb, R.S. and Hoffman, D.L., 1975. Infiltration of organic phosphate compounds in soils. Soil Sci. Soc. Am. Proc., 39: 1089—1094.

Roscoe, B., 1960. The distribution and condition of soil phosphate under old permanent pasture. Plant and Soil, 12: 17—19.

Russell, J.D., Parfitt, R.L., Fraser, A.R. and Farmer, V.C., 1974. Surface structures of gibbsite, goethite and phosphated goethite. Nature, 428: 220—221.

Ryden, J.C. and Syers, J.K., 1975a. Charge relationships of phosphate sorption. Nature, 255: 51—53.

Ryden, J.C. and Syers, J.K., 1975b. Rationalization of ionic strength and cation effects on phosphate sorption by soils. J. Soil Sci., 26: 395—406.

Ryden, J.C. and Syers, J.K., 1976. Calcium retention in response to phosphate sorption by soils. Soil Sci. Soc. Am. J., 40: 845—846.

Ryden, J.C. and Syers, J.K., 1977. Origin of the labile phosphate pool in soils. Soil Sci., 123: 353—361.

Ryden, J.C., Syers, J.K. and McLaughlin, J.R., 1977a. Effects of ionic strength on chemisorption and potential-determining sorption of phosphate by soils. J. Soil Sci., 28: 62—71.

Ryden, J.C., McLaughlin, J.R., and Syers, J.K., 1977b. Mechanism of phosphate sorption by soils and hydrous ferric oxide gel. J. Soil Sci., 28: 72—92.

Sawhney, B.L., 1974. Charge characteristics of soils as affected by phosphate sorption. Soil Sci. Soc. Am. Proc., 38: 159—160.

Sawhney, B.L., 1977. Predicting phosphate movement through soil columns. J. Environ. Qual., 6: 86—89.

Schalscha, E.B., Pratt, P.F., Kinjo, T. and Amar, A.J., 1972. Effect of phosphate salts as saturating solutions in cation-exchange capacity determinations. Soil Sci. Soc. Am. Proc., 36: 912—914.

Schalscha, E.B., Pratt, P.F. and Soto, D., 1974. Effects of P-adsorption on the cation exchange capacity of volcanic ash soils. Soil Sci. Soc. Am. Proc., 38: 539—540.

Shah, D.B., Coulman, G.A., Novak, L.T. and Ellis, B.G., 1975. A mathematical model for phosphorus movement in soils. J. Environ. Qual., 4: 87—92.

Stoop, W.A., 1974. Interactions between Phosphate Adsorption and Cation Adsorption by Soils and implications for Plant Nutrition. Ph.D. Thesis, Univ., of Hawaii.

Stumm, W., and Leckie, J.O., 1971. Phosphate exchange with sediments; its role in the productivity of surface waters. In: S.H. Jenkins (Editor), Advances in Water Pollution Research, 2. Pergamon Press, Oxford, III. 26/1—16.

Syers, J.K., Williams, J.D.H., Campbell, A.S. and Walker, T.W., 1967. The significance of apatite inclusions in soil phosphorus studies. Soil Sci. Soc. Am. Proc., 31: 752—756.

Syers, J.K., Browman, M.G., Smillie, G.W. and Corey, R.B., 1973. Phosphate sorption by soils evaluated by the Langmuir adsorption equation. Soil Sci. Soc. Am. Proc., 37: 358—36.

Talibudeen, O., 1958. Isotopically exchangeable phosphorus in soils: III. The fractionation of soil phosphorus. J. Soil Sci., 9: 120—129.

Tamini, Y.N., Kanchiro, Y. and Sherman, G.D., 1964. Reactions of ammonium phosphate with gibbsite and with montmorillonite and kaolinitic soils. Soil Sci., 98: 249—255.

Tamini, Y.N., Kanchiro, Y. and Sherman, G.D., 1968. Effect of time and concentration on the reactions of ammonium phosphate with a humic latosol. Soil Sci., 105: 434—439.

Tsubota, G., 1959. Phosphate reduction in the paddy field, I. Soil Plant Food Tokyo, 5: 10—15.

Vahtras, K. and Wiklander, L., 1970. Phosphate studies in soils: With special reference to Chang and Jackson's fractionation procedure. Lantbruks hoegsk. Ann., 36: 115—134.

Vaidyanathan, L.V. and Talibudeen, O., 1968. Rate-controlling processes in the release of soil phosphate. J. Soil Sci., 19: 342—353.

Van Diest, A., 1968. Biological immobilization of fertilizer phosphorus: I. Accumulation of soil organic phosphorus in coastal plain soils of New Jersey. Plant and Soil, 29: 241—247.

Van Riemsdijk, W.H., Weststrate, F.A. and Bolt, G.H., 1975. Evidence for a new aluminum phosphate phase from reaction rate of phosphate with aluminum hydroxide. Nature, 257: 473—474.

Van Riemsdijk, W.H., Weststrate, F.A. and Beek, J., 1977. Phosphates in soils treated with sewage water: III. Kinetic studies on the reaction of phosphate with aluminum compounds. J. Environ. Qual., 6: 26—29.

Van Wazer, J.R., 1958. Phosphorus and Its Compounds, 1. Interscience, New York, N.Y.,

Volkweiss, S.J., Robarge, W.P. and Corey, R.B., 1973. Effect of associated cations on phosphate sorption by minerals and soils. Agron. Abstr., p. 87.

Vyayachandran, P.K. and Harter, R.D., 1975. Evaluation of phosphorus adsorption by a cross section of soil types. Soil Sci., 119: 119—126.

Walker, T.W. and Syers, J.K., 1976. The fate of phosphorus during pedogenesis. Geoderma, 15: 1—19.

Wauchope, R.D., 1975. Fixation of arsenical herbicides, phosphate and arsenate in alluvial soils. J. Environ. Qual., 4: 355—358.

Wells, N. and Saunders, W.M.H., 1960. Soil studies using sweet vernal to assess element availability. IV, Phosphorus. N.Z.J. Agric. Res., 3: 279—299.

White, R.E., 1964. Studies on the phosphate potentials of soils: II. Microbial effects. Plant and Soil, 20: 184—193.

Wiechmann, H., 1966. Phosphor-Verteilung in Böden bei Düngung mit Superphosphat und Thomasphosphat. Z. Pflanzenernaehr. Dueng. Bodenkd., 115: 114—123.

Wiechmann, H., 1971. Uberlegungen zum Phosphataustrag aus Böden. Die Phosphorsäure, 29: 67—84.

Wild, A., 1949. The retention of phosphate by soil. A review. J. Soil Sci., 1: 221—238.
Williams, E.G., Scott, N.M. and McDonald, M.J., 1958. Soil properties and phosphate sorption. J. Sci. Food Agric., 9: 551—559.
Williams, J.D.H., Syers, J.K. and Walker, T.W., 1967. Fractionation of soil inorganic phosphate by a modification of Chang and Jackson's procedure. Soil Sci. Soc. Am. Proc., 31: 736—739.
Williams, J.D.H., Syers, J.K., Harris, R.F. and Armstrong, D.E., 1971a. Characterization of inorganic phosphate in noncalcareous lake sediments. Soil Sci. Soc. Am. Proc., 35: 556—561.
Williams, J.D.H., Syers, J.K., Harris, R.F. and Armstrong, D.E., 1971b. Fractionation of inorganic phosphate in calcareous lake sediments. Soil Sci. Soc. Am. Proc., 35: 250—255.
Yates, D.E. and Healy, T.W., 1975. Mechanism of anion adsorption at the ferric and chromic oxide/water interface. J. Colloid Interface Sci., 52: 222—228.

MOVEMENT OF SOLUTES IN SOIL: PRINCIPLES OF ADSORPTION/EXCHANGE CHROMATOGRAPHY

G.H. Bolt

9.1. INTRODUCTORY REMARKS

The recognition of the problem of predicting the motion of dissolved components through soil stems from the time that fertilizers came into use in agriculture. Actually, the implications of solute displacement have been manifest ever since it was noted that agricultural land may become spoiled following the regular application of water for irrigation purposes.

Attempts towards a quantitative description of the motion of solutes through soil date from the period following the second world war and may have been prompted in part by the world-wide fear of the effects of radioactive fall-out products when reaching the soil (cf. Thornthwaite et al., 1960). In that period the synthetic ion exchangers had made their entry into the chemical world and considerable attention was paid in the chemical engineering literature to models describing the performance of ion exchange columns (De Vault, 1943; Thomas, 1944; Glueckauf and Coates, 1947; Klinkenberg, 1948; Lapidus and Amundson, 1952; Hiester and Vermeulen, 1952).

The penetration of this information from the chemical literature into soil science became evident in Rible and Davis (1955), applying De Vault's model, Van der Molen (1957), who followed particularly the Glueckauf model, and Bower et al. (1957), using the Thomas/Hiester and Vermeulen approach.

The influence of dispersion phenomena in modifying the shape of a solute front passing through a soil column was studied in a series of papers by Nielsen and Biggar (Nielsen and Biggar, 1961; Biggar and Nielsen, 1962; Nielsen and Biggar, 1962; Biggar and Nielsen, 1963). Reviews of the application of chromatographic theory in soil systems were given by Ritchie (1966) and by Frissel and Poelstra (1967). A more detailed discussion of the assumptions underlying the various theories and their presumed applicability is given in Reiniger and Bolt (1972).

In the last decade interest in solute displacement has increased substantially, as it forms one of the basic issues in soil pollution phenomena, and so has relevant literature available. With the development of computer facilities and simulation language it has become possible to calculate the expected distribution of soil solutes under any set of circumstances, provided these may be formulated in terms of dependable parameters. A thorough analysis of the results that may be obtained in this manner is given in Van Genuchten et al. (1974, 1977) and Van Genuchten and Wierenga (1976, 1977).

Recognizing that in practice the relevant parameters which determine solute displacement in soil are often not known with accuracy, the following discussion is directed particularly towards finding a simplified procedure for making a first guess about the expected behavior. Such a procedure should allow one to assess quickly whether, for example, the penetration of a particular solute into soil should approach a critical depth in a chosen time period. In that case one might be able to single out the parameters whose values are likely to be of decisive importance. These should then be measured with the required accuracy. Finally a more accurate prediction could be made by means of a computer simulation procedure.

With the above purpose in view, the different aspects of the displacement process will be investigated in some detail. From this analysis it follows how one could obtain a first guess about the magnitude of the relevant parameters, even if only scant experimental information is available, and also under what conditions this magnitude is likely to be of critical importance. The conclusions of this chapter were used in part as the basic ingredients for chapter 7 of part A of this text.

9.2. THE CONSERVATION EQUATION

The displacement of an (ionic) solute k in soil is subject to a conservation equation of the type:

$$\frac{\partial A_k}{\partial t} = -\operatorname{div} j_k + P_k \tag{9.1}$$

in which A_k specifies the amount of the ion present per unit volume of the system under consideration, j_k is its flux, and P_k signifies a production term covering the rate of local appearance (or disappearance) of the ion from the specified system. It is important to recognize that particularly the first and last terms of (9.1) are complementary rate terms, the dividing line between the two being determined by the boundaries of the system as specified. Thus if this system comprises "whole" soil, including for example plant roots, P_k would cover only the rate of disappearance of the ion into those parts of the plant outside the volume under consideration. Actually one might in this case omit P_k and include this rate of disappearance into the divergence of the flux, which should then comprise the ionic fluxes both outside and inside any living parts of plants. Another extreme choice would be to limit the system under consideration to the mobile liquid phase present in a unit volume of soil. In that case P_k must cover the rate of (dis)appearance attributable to adsorption, diffusion into immobile liquid, etc.

In practice it seems wise to select the system boundaries for convenience with regard to experimental observations and available theoretical models. In the present discussion P_k will be used to cover all those processes which could be termed chemical or biochemical reactions, including uptake by living organisms and also precipitation and/or dissolution processes. Accordingly the chosen system boundaries include only the liquid phase (mobile or immobile) and the adsorbed phase; so:

$$A_k \equiv q_k + \theta c_k \tag{9.2}$$

in which q_k is the amount adsorbed (reversibly) per unit volume of soil and θc_k signifies the amount present in the solution phase. If the solution phase or the adsorbed phase are inhomogeneous, the above terms must be split into the relevant subphases (cf. the introduction of a stagnant liquid phase in sections 9.4 and 9.6).

With the exception of certain volatile compounds it seems warranted to consider only the flux in the liquid phase. This solution flux is then composed of a convective part and an autonomous flux of the solute with respect to the carrier liquid owing to diffusion and/or dispersion phenomena, according to:

$$j_k = J^V c_k + j_k^D \tag{9.3}$$

in which J^V equals the flux (density) of the solution in m^3/sec per m^2 of soil column, taken as a positive number throughout this chapter, and j_k^D signifies the mentioned autonomous flux.

9.3. EQUILIBRIUM CHROMATOGRAPHY

As will be discussed in sections which follow, the situation in soil columns leached under gravity at about field capacity is often not too far removed from (local) exchange or adsorption equilibrium. Accordingly, equilibrium chromatography will be considered here as the basic backbone of the process to be discussed. In that situation it may safely be assumed that any immobile part of the liquid phase has also attained equilibrium with the mobile liquid phase and the entire liquid phase may then be treated as homogeneous in composition. Writing out the equation for steady state leaching* in a column (i.e. $\partial\theta/\partial t = 0$) one thus finds in the absence of significant transpiration losses in the column (i.e. $\partial J^V/\partial x = 0$):

---

* The present choice for steady state leaching may seem a severe limitation for application of the theory to field situations. Granted that such applications may require that due allowance be made for transient aspects of the flow of the carrier liquid, the incidental nature of these precludes their inclusion in a tentative *general* theory describing solute movement in soil. Once the actual flow regime is known for a particular time period the latter may be taken into account, usually without difficulty, by solving the complete equation employing mathematical simulation procedures (see also the closing remark of section 9.8).

$$\frac{\partial(q_k + \theta c_k)}{\partial t} + J^V \frac{\partial c_k}{\partial x} + \frac{\partial j_k^D}{\partial x} = 0 \qquad (9.4)$$

In order to solve (9.4) for the appropriate boundary conditions an additional relation between q_k and c_k must be found. It is precisely at this point that *exchange* chromatography presents difficulties, as in general q_k depends on the concentration of all competing ions. At best this leads to a set of differential equations of the type (9.4), in which the first term of each contains a function of all ionic concentrations. Such a set of equations could in principle be solved simultaneously with the help of a computer, provided complete information is available about the composition of the exchanger for all different solution compositions (cf. e.g. Frissel and Reiniger, 1974). As this is rarely the case, this generalized situation will not be considered here.

Fortunately, it is often only necessary to follow a column exchange process which involves two strongly dominant ions. A typical example is the sodication (alkalization) process mentioned in chapters 8 and 9 of part A, where Na-ions accumulate in competition only with Ca-ions (or the rather similarly behaving pair of Ca- and Mg-ions). In that case q_k may be treated as a function of c_k and C, the total electrolyte concentration in the column. Aside from precipitation reactions, however, the total electrolyte level does not change during exchange of counterions, so the distribution of C in the column follows the equation:

$$\theta \frac{\partial C}{\partial t} + J^V \frac{\partial C}{\partial x} + \frac{\partial j^D}{\partial x} = 0 \qquad (9.5)$$

A "step feed" of a solution with total concentration C then forms a "concentration front" in the neighborhood of the penetration depth of the feed solution situated at $J^V t/\theta \equiv V/\theta$ (cf. chapter 7 of part A). Under those conditions C may safely be treated as a constant in the region where the exchange front of the counterions is situated. Accordingly (9.4) will now be developed further subject to the condition that q_k does not depend on concentrations other than c_k. In this respect the following treatment, though making use largely of examples based on the exchange between a given pair of cations, applies also to adsorption reactions.

Next a preliminary remark is made about the diffusion/dispersion term of (9.4) and (9.5). As will be covered in section 9.4, the flux of concern is proportional to $\partial c/\partial x$. Integrating its divergence over a "complete" front, where $\partial c/\partial x$ changes from zero via a finite value back to zero, the net flux across the front vanishes. The diffusion/dispersion term then induces solely a spreading of the passing concentration front. As its magnitude under the stated field conditions is often rather limited, this term will be considered for the time being as a perturbation term in comparison to the effect of the first two terms of (9.4).

In order to assess the main features of the concentration front for the case of equilibrium exchange or adsorption chromatography, one should first solve (9.4) in the absence of its third term. In that situation it is convenient to introduce the input volume into a unit area of the column, V, according to $dV \equiv J^V dt$, which yields:

$$\frac{\partial(q + \theta c)}{\partial V} + \frac{\partial c}{\partial x} = 0 \qquad (9.6)$$

in which the subscript k has been omitted. As long as $(\partial c/\partial x)$ remains finite, a general solution of this equation is found straight away. Introducing the differential capacity of the exchanger, according to:

$$q'(c) \equiv dq/dc \qquad (9.7)$$

(9.6) may then be written as:

$$\left(\frac{\partial c}{\partial x}\right)\left[1 - \{q'(c) + \theta\}\left(\frac{\partial x}{\partial V}\right)_c\right] = 0 \qquad (9.8)$$

This implies that the rate of propagation of any particular concentration through the column, $(\partial x/\partial V)_c$, is found from:

$$\left(\frac{\partial x}{\partial V}\right)_c = \frac{1}{q'(c) + \theta} \qquad (9.9)$$

Upon integration this gives for a *homogeneous* column (i.e. q' and θ independent of x) the position of a particular concentration, x_c, at:

$$x_c = \frac{V}{q'(c) + \theta} + \text{constant} \qquad (9.10)$$

in which the constant is a function of c. The latter may be expressed in terms of an inverse "feed function". Thus if $c_f(V)$ signifies the concentration of the feed solution as a function of the volume fed into the column, the inverse of this function, $V_0(c)$, specifies the volume that has been fed into the column at the moment that the concentration at the entrance boundary reaches the value c. One then finds $V = V_0(c)$ for $x_c = 0$, and (9.10) becomes:

$$x_c = \frac{V - V_0(c)}{q'(c) + \theta} \qquad (9.11)$$

In the case of a step increase of the concentration of the feed solution, $V_0(c)$ is obviously a constant for the concentration range covered in the step. If only one single step is involved ranging from c_i to c_f, it is convenient to define V as the volume of solution with concentration c_f that has entered the column. Then $V_0(c) \equiv 0$ and the location of the penetrating front is found from the corresponding form of (9.11), which is also the basis of (A7.4) as used in part A of this text.

Eq. (9.11) may further be used to define the penetration depth of the combined front in solution and adsorbed phase for a step feed according to:

$$x_p \equiv \int_{c_i}^{c_f} x_c(q' + \theta)dc \Big/ \int_{c_i}^{c_f} (q' + \theta)dc$$

$$= \frac{V}{\theta + (q_f - q_i)/(c_f - c_i)} \qquad (9.12)$$

An important limitation of the validity of (9.11) follows from the earlier condition stated in conjunction with the differential equation (9.6), i.e. $(\partial c/\partial x)$ must remain finite during passage of the front through the column. If anywhere in the column, at any moment, the concentration front exhibits a jump in concentration, the conservation equation for the passage of such a jump must be written as:

$$(\Delta q + \theta \Delta c)dx = \Delta c\, dV \qquad (9.6a)$$

giving:

$$\left(\frac{dx}{dV}\right)_{\Delta c} = \frac{1}{\theta + \Delta q/\Delta c} \qquad (9.9a)$$

Together (9.9) and (9.9a) allow one to describe the rate of propagation of the front in all situations, for the present simplified case. For this purpose one should distinguish between favorable exchange and unfavorable exchange between the resident cation of the column and the cation fed into the column. For the time being favorable exchange will be defined as the situation in which the exchange isotherm of the *incoming* cation is convex upward, i.e. $q''(c) \equiv d^2q/dc^2 < 0$. An example of such an isotherm was presented in Fig. A7.3 of part A of this text, both in its actual form (as $q(c)$) and in its normalized form (i.e. as $N(f)$). Obviously the exchange isotherm of the competing cation is then concave and may be read from the same graph by inverting the axes, as was shown in Fig. A7.5. Finally the transition between favorable and unfavorable exchange will be termed linear exchange as it corresponds to a linear exchange isotherm.

Since for the incoming cation — or any adsorbate — $\partial c/\partial V$ must be positive at all V and x, the slope of the concentration front of this cation must always be negative, i.e. $\partial c/\partial x < 0$. Differentiating (9.10) with respect to c at constant V (taking the constant at zero for the chosen step feed) one finds, however:

$$\left(\frac{\partial x}{\partial c}\right)_V = -\frac{V}{(q'(c) + \theta)^2}\, q''(c) \qquad (9.13)$$

Accordingly the validity of (9.9) and (9.10) is limited to unfavorable exchange. Then $q''(c)$ is positive and the front has a negative slope as required. Inverting (9.13) shows furthermore that $(\partial c/\partial x)_V$ is inversely proportional to V, indicating that the front "flattens" upon passage through the column.

In the transition case of linear exchange $q''(c) = 0$, so $(\partial x/\partial c)_V$ becomes (minus) zero. This corresponds to a concentration jump which implies that a step feed of a cation with a linear exchange isotherm passes unchanged through the column, though at a reduced velocity in comparison to the velocity of the carrier, J^V/θ. As then $q'(c) = \Delta q/\Delta c$, (9.9) and (9.9a) become identical.

In the case of favorable exchange, (9.9) and (9.10) become invalid as they imply that the higher concentrations of the incoming cation would run ahead of the lower concentrations. Instead all concentrations are propagated at the same velocity and the concentration jump of the feed solution passes unchanged through the column, akin to the situation for linear exchange. Accordingly, favorable exchange following a step feed is described by (9.6a) and (9.9a) instead of (9.6) and (9.9).

While the above information suffices for construction of the concentration front following a single step feed into a column containing an adsorber exhibiting an isotherm without inflection points, more information is needed in other cases. Considering, because of the simplicity of notation, a system in which the concentration of the incoming ion covers the full range from $c = 0$ to $c = C$, one may define the mean depth of penetration for the fractional concentration range from 0 to c as:

$$\langle x\rangle_c \equiv \int_0^c x_c(q' + \theta)dc \Big/ \int_0^c (q' + \theta)dc = \frac{V}{\theta + q/c} \tag{9.14}$$

Insofar as the concentration front — as determined by the shape of x_c — could not exhibit a positive slope, the same condition must obviously apply to $\langle x\rangle_c$: the mean depth of penetration over a given range from 0 to c_2 cannot exceed the same for the range from 0 to c_1 (with $c_2 > c_1$). Accordingly:

$$\left[\frac{\partial\langle x\rangle_c}{\partial c}\right]_V \leqslant 0 \tag{9.15}$$

Differentiating (9.14) one thus finds:

$$\left[\frac{\partial\langle x\rangle_c}{\partial c}\right]_V = -\frac{V}{(\theta + q/c)^2}\,\frac{(q' - q/c)}{c} \leqslant 0 \tag{9.16}$$

Accordingly the condition to be fulfilled in constructing $\langle x\rangle_c$ reads simply:

$$q/c < q' \tag{9.17}$$

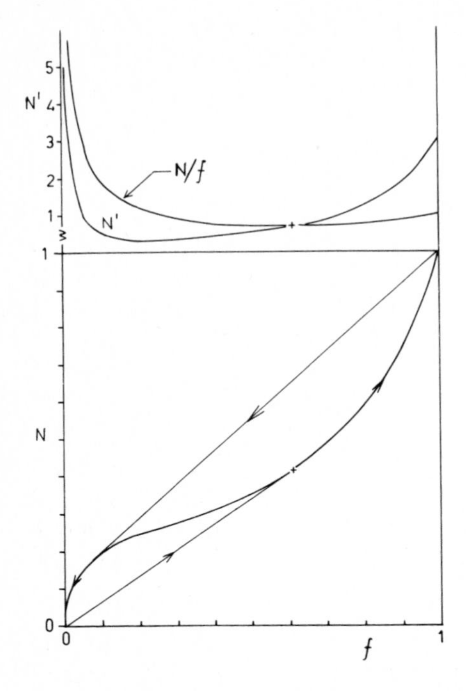

Fig. 9.1. Normalized exchange isotherm for a system with two types of sites: 20% of the exchange capacity favoring strongly the incoming cation, 80% favoring mildly the resident cation. Also shown are the chord envelopes for forward exchange (i.e. from $N = 0$ to $N = 1$) and backward exchange. The upper section gives the slope of the chord from 0 to N and the slope of the isotherm, both as a function of the equivalent fraction in solution, *f*.

Now q/c indicates the slope of the chord of the exchange isotherm over the range from 0 to c, while q' signifies the slope of the isotherm at c. If then the trajectory of $\langle x \rangle_c$ is constructed, using (9.14) and beginning with $\langle x \rangle_C \equiv x_p$ (penetration depth of the complete front), one may follow the exchange isotherm down to the point where q/c equals q', i.e. the point where the chord to $c = 0$ is a tangent to the isotherm. Having reached that point, $\langle x \rangle_c$ becomes constant with decreasing c down to $c = 0$. In other words, the construction of $\langle x \rangle_c$ must follow the trajectory of the lower chord envelope of the exchange isotherm. As, however, $\langle x \rangle_c$ determines fully the trajectory of x_c (cf. 9.10), the latter must also be based on this chord envelope.

The above is demonstrated in Figs. 9.1 and 9.2. Figure 9.1 shows a hypothetical convex—concave (normalized) isotherm corresponding to an exchanger with two different types of sites, i.e. 20% of the total exchange capacity strongly favoring the incoming cation, while the remaining 80% favors mildly the resident cation. The lower "chord envelope" is then the

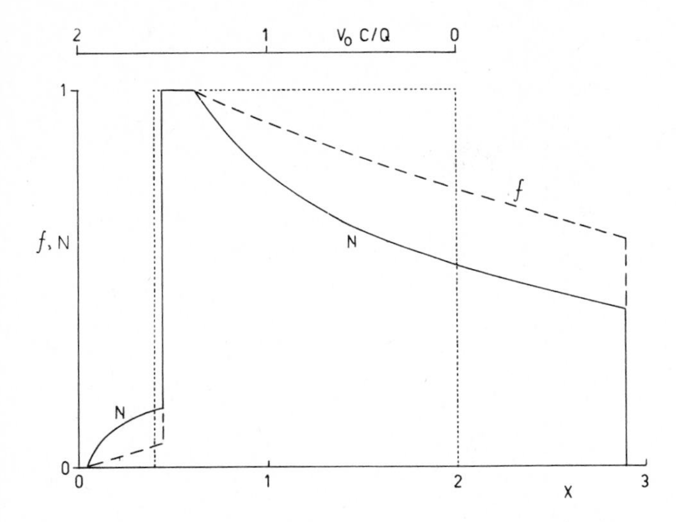

Fig. 9.2. Fronts in solution and adsorbed phase resulting from a block feed (cf. dotted line) into a column following the exchange isotherm of Fig. 9.1. Diffusion and dispersion effects are neglected.

basis for the construction of the concentration front developing in the column as is presented in Fig. 9.2. Remembering, furthermore, that reversal of incoming and resident cation necessitates reading the exchange isotherm backwards, one may construct the front shape for the latter case from the upper chord envelope. In fact both fronts will develop consecutively when passing a block feed through the column, as is shown also in Fig. 9.2. Aside from diffusion/dispersion effects, which will rapidly remove any "sharp corners" of the concentration front, full deployment of the entire double front depends critically on the distance to which it has moved, relative to the width of the block feed, into the column. Thus the "fast" portion of the backward front will eventually catch up with the slow part of the forward front, leading to a decrease of the maximum concentration in the column. In practice the above concentration fronts are measured most easily by following the break-through of the block feed from a column of appropriate length. In the next chapter the above effects are demonstrated with distribution curves computed numerically, for a block feed entering a column, taking into account both diffusion/dispersion and non-linear isotherms.

Finally it should be pointed out that, for isotherms with more than one inflection point, the above reasoning remains valid, i.e. the concentration front follows the lower chord envelope, even if this comprises several concave portions of the isotherm connected by different chords, tangent to these portions.

The above simplified approach (i.e. assuming local exchange equilibrium and negligible influence of diffusion/dispersion effects) is basic for the method of high-pressure chromatography in the determination of the adsorption isotherm of a given adsorbent—adsorbate system. Constructing small columns of very finely ground adsorbent, the adsorbate (dissolved in a suitable carrier) is forced through the narrow pores under high pressure gradients. It is assumed that the extremely small particle size of the adsorbent ensures rapid adsorption equilibrium while the comparatively high rate of flow prevents longitudinal diffusion playing a significant role. By passing a pulse of the adsorbate one may study the concentration fronts formed upon increase and subsequent decrease of the adsorbate in the feed. If the adsorption is reversible, one should thus find in principle a spreading front on either one side (or both sides) of the propagated pulse. Using Ca—Na-exchange as an example, one should thus expect that a Ca-pulse fed into a Na-column would exhibit a concentration jump at the forward end of the pulse, and an extended front at the backward end. According to (9.11) this extended second front would then contain the necessary information to construct the exchange isotherm by an integration procedure.

Some further remarks may elucidate the application of the above. While in Fig. 9.1 a complete (normalized) exchange isotherm was depicted, giving $N(0 \rightarrow 1)$ as a function of $f(0 \rightarrow 1)$, any portion of such an isotherm may be used in a similar fashion to construct the shape of the corresponding exchange front. It should be remembered, however, that the shape and the mean location of a front covering a concentration step from, for example, f_i to f_f, will often differ from that of the same range in a complete front formed when the step feed goes from zero to unity, even if the same amount of feed solution is applied. This is seen immediately for the case of favorable exchange, following a convex isotherm. Then the entire concentration front from $f = 0$ to $f = 1$ is formed as a concentration jump located at a "penetration depth", x_P, equal to $V/(\theta + Q/C)$. A step feed from f_i to f_f, however, leads to a corresponding concentration jump in the column located at $x_p = V/\{\theta + (q_f - q_i)/(c_f - c_i)\}$. As for the convex isotherm the slope of the chord of a particular concentration range in general does not equal the slope of the chord of the complete isotherm, Q/C, $x_P \neq x_p$. In contrast, unfavorable exchange — if based on a simple concave curve — leads to the same front for a certain concentration range regardless whether the step feed covers only this range, or complete exchange. In short the situation may be summarized as follows. As was shown above, the exchange front is determined by the shape of the lower chord envelope of the exchange isotherm of the incoming cation. If this chord envelope changes upon a change of f_i and/or f_f, the shape and location of the exchange front will change accordingly. An alternative to the above is to construct a normalized exchange isotherm in terms of the variables

$$\bar{c} \equiv (c - c_i)/(c_f - c_i) = (f - f_i)/(f_f - f_i)$$

and

$$\bar{q} \equiv (q - q_i)/(q_f - q_i) = (N - N_i)/(N_f - N_i).$$

This enables one to employ the equations and procedures used before and hereafter for the variables f and N, as now the range is again from zero to unity. On the other hand this implies the construction of a separate isotherm for each range of c.

In some cases a graphical construction of the concentration front may be convenient. Referring to Fig. 9.3 as an example, one may design such a construction by rewriting

(9.10) in terms of $N' = dN/df$, according to:

$$x_f = \frac{VC}{Q}\frac{df}{d(N + \theta Cf/Q)}$$
$$= \frac{V}{\theta \bar{R}_D}\frac{df}{d(N + f/\bar{R}_D)} \tag{9.18}$$

in which $\bar{R}_D \equiv Q/\theta C$ signifies the overall distribution ratio of the system as introduced in chapter 7 of part A of this text. Similarly:

$$\langle x\rangle_f = \frac{V}{\theta \bar{R}_D}\frac{f}{N + f/\bar{R}_D} \tag{9.19}$$

and thus:

$$\langle x\rangle_C \equiv x_p = \frac{V}{\theta \bar{R}_D}\frac{1}{1 + 1/\bar{R}_D} \tag{9.20}$$

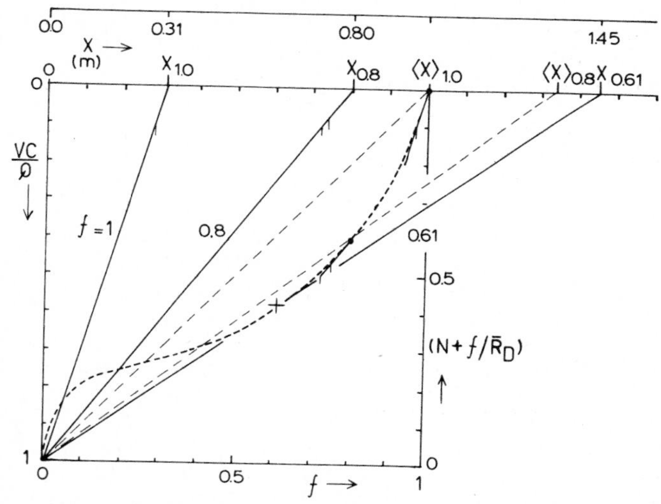

Fig. 9.3. Construction of the forward front pictured in Fig. 9.2. Note that the column storage curve (dashed line) has been arbitrarily taken equal to the exchange isotherm of Fig. 9.1, implying $\bar{R}_D \gg 1$. The bundle of solid lines parallel to the slope of the storage curves are marked with the corresponding value of f. Also shown are some chords marking the mean position of the front, $\langle x\rangle_f$. The chord tangent to the storage curve at $f = 0.61$ determines the position of the step front covering the range $f = 0$ to $f = 0.61$.

As is obvious from (9.18)—(9.20) the relation between x or $\langle x\rangle$ and the feed volume in units of $\theta\bar{R}_D$ is now fully determined by the trajectory of $(N + f/\bar{R}_D)$ as a function of f. This curve will be designated the (normalized) storage curve of the column. It is easily constructed by "addition" of the straight line representing $f/\bar{R}_D$ to the normalized isotherm $N(f)$. For large values of $\bar{R}_D$, as occur in soils at low electrolyte level, the storage curve lies very close to the isotherm. As the actual shape of the storage curve is immaterial for the construction procedure to be discussed, the following example will be based on the exchange isotherm as given in Fig. 9.1. This curve then serves as a storage curve for very large values of $\bar{R}_D$. As now $x_f(V/\theta\bar{R}_D)$ depends on the local slope of the storage curve, its graphical construction is facilitated by superimposing two new axes upon the graph of the storage curve given in Fig. 9.1. Introducing an x-axis, originating at $f = 0$ and parallel to the f-axis, and a V-axis, originating at $N = 1$ and anti-parallel to the N-axis, one may first establish from Fig. 9.3 (applying equation 9.20) that x_p may be read on the x-axis at the intersection with a line run parallel to the chord of the storage curve, originating on the V-axis at the appropriate value of V in units of $\theta \bar{R}_D$ (or Q/C). For any chosen value on the V-axis the course of $\langle x \rangle_f$ may then be determined from the intersection points of a bundle of lines parallel to the chords of the storage curve corresponding to the chosen values of f. In the case of a single step feed it is most convenient to use the chords themselves for this purpose, which then deliver the values of $\langle x \rangle_f$ (in m) if $V = \theta \bar{R}_D$. For any other value of V the readings on the x-axis are adjusted proportionally. Recalling the conclusion from (9.17) it may then also be ascertained that the maximum value of $\langle x \rangle_f$ is found at 1.45 for the chord tangent to the storage curve — corresponding to $f = 0.61$. From thereon down to $f = 0$, $\langle x \rangle_f$ becomes a constant. The construction of the actual concentration front then consists of running a bundle of lines, parallel to the tangents to the storage curve and originating again at $V = \theta \bar{R}_D$. Again in this case x_f becomes constant and equal to 1.45 for $f < 0.61$. The curve as found from the intersection points with the x-axis corresponds to the one shown in Fig. 9.2 (forward front). As actually only the course of x_f is of interest, the construction of $\langle x \rangle_f$ may be omitted except for the determination of the chord tangent to the storage curve which determines the limiting value of f. A similar construction may be applied to the backward front formed if a block feed is employed.

In principle the above procedure may also be applied in the case of a gradual increase of concentration of the feed solution according to an arbitrary "feed function", $V_0(f)$. Taking the total feed volume at unit length of the V-axis one may then construct the curve representing $V_0(f)/V$ as a function of f (cf. Fig. 9.4). To find x_f one then constructs again the bundle of lines parallel to the tangents of the storage curve at chosen values of f, but now each line of the bundle must originate on the V-axis at the corresponding value of $\{V - V_0(f)\}/V$ in accordance with (9.11). Because a monotonically increasing concentration of the feed solution leads to a smaller input volume for a higher concentration, the concentration front in the column will obviously be flatter than for a

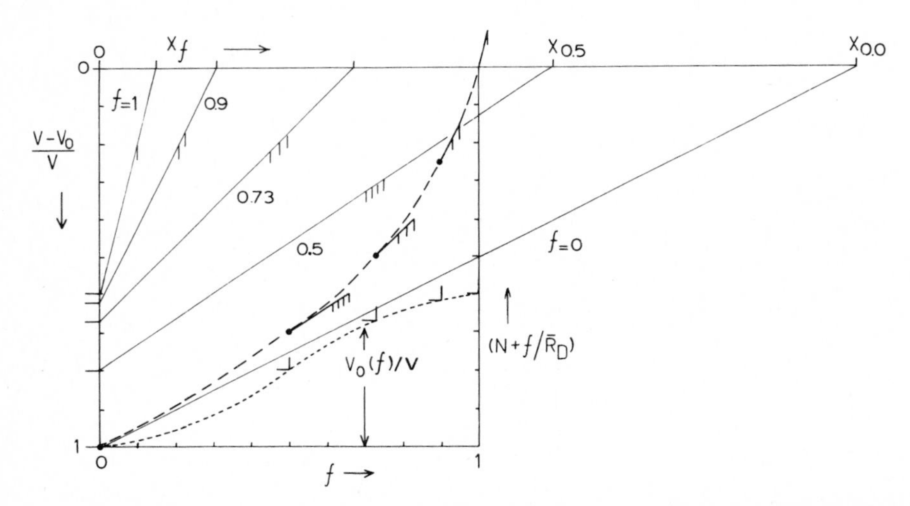

Fig. 9.4. Construction of the concentration front in a column following the dashed storage curve, in the case of a gradually increasing concentration of the feed solution (cf. dotted line representing the feed function $V_0(f)/V$).

step feed. Applying the above to a convex storage curve, the concentration front will be a step front either over the full range or only over a part of the range, depending on the particular shape of the feed curve. Obviously, doubling V while maintaining $V_0(f)$ causes the feed curve to "shrink" to one-half of the values pictured in Fig. 9.4, thus increasing the range of f over which the concentration front has a step shape. The particular value $f \equiv f^*$ below which the concentration front exhibits a step shape may be found in the following manner. Rewriting (9.18) and (9.19) for $V_0(f) \neq 0$ gives:

$$x_f = \frac{V - V_0(f)}{\theta \bar{R}_D} \frac{df}{d(N + f/\bar{R}_D)} \tag{9.18a}$$

$$\langle x \rangle_f = \frac{V - (1/f) \int V_0(f) df}{\theta \bar{R}_D} \frac{f}{N + f/\bar{R}_D} \tag{9.19a}$$

As $x_f = \langle x \rangle_f$ in a step front ranging from $f = 0$ to, say, $f = f^*$, one may find f^* by determining graphically the intersection point of the above two functions of f. Whereas this procedure is rather cumbersome in the case of an arbitrary feed function in combination with an involved isotherm, it lends itself very well to the construction of the concentration front arising for a two-step feed, even in the case of the convex—concave isotherm used in Fig. 9.1.

As an example the front is constructed for a two-step increase, firstly from $f = 0$ to $f = 0.25$ (the inflection point of the isotherm, corresponding to $N = 0.265$) with $V_0 = 0$, and secondly from $f = 0.25$ to $f = 1$ with $V_0 = a$. As the first step covers the fully convex part of the isotherm, it will create a step front in the column, its position being found from the appropriate chord (with slope $0.265/0.25 = 1.06$) originating at $V = 1$ (cf. Fig. 9.5, where again the isotherm has been used as the storage curve). As the second step covers the concave range one constructs again the bundle of lines parallel to the slope

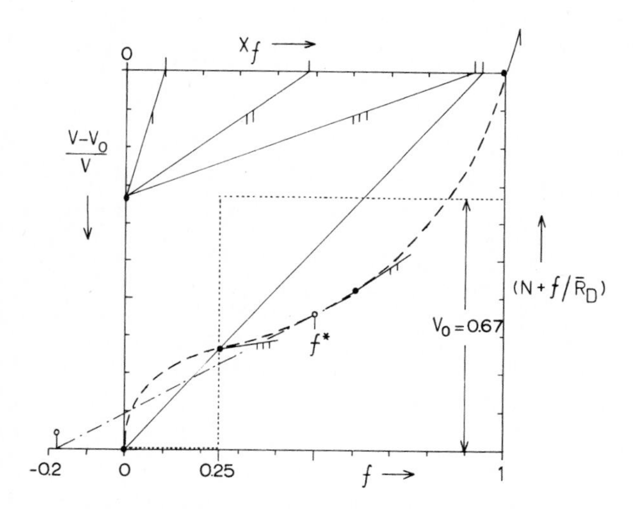

Fig. 9.5. Construction of the concentration front following the dashed storage curve (same as in Fig. 9.3) in the case of a two-step feed function (dotted line). The toe-end of the extended front resulting from the second step has not yet merged with the step front formed by the first step (cf. also Fig. 9.6). Also shown is the construction of the point of merger, f^*, arising if $V_0/V = 0.72/1.72$ (cf. text).

of the storage curve, but originating from the point $(V-a)/V$ on the V-axis. The line corresponding to $f = 0.25$ (with a slope equal to 0.36) determines the position of the toe of this extended front which will fall short of the step front covering the first step as long as $\{(V-a)/V\} < 0.36/1.06 = 0.34$. In Fig. 9.6 the entire front is shown, using VC/Q equal to 3 units of length, with $aC/Q = 2$. In that case $(V-a)/V = 0.33$ and the toe-end of the extended front falls just behind the step front. Introducing also a two step decrease of f in the feed concentration (with aC/Q at 2.7 and 2.9, respectively), the above procedure may be repeated for the construction of the backward fronts as shown in Fig. 9.6.

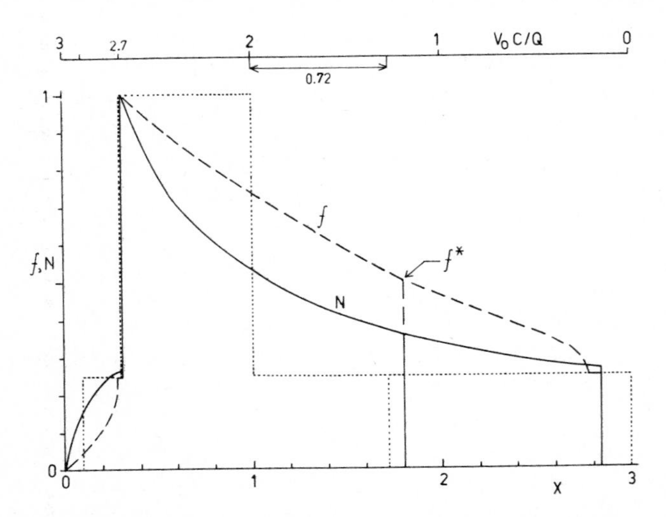

Fig. 9.6. Fronts in solution and adsorbed phase corresponding to a two-step increase followed by a two-step decrease of the feed concentration (outer dotted line), the forward front corresponding to the construction shown in Fig. 9.5. The inner dotted line represents a decrease of the "head start" of the first step, leading to partial merger of the two fronts at f^* (cf. text).

Obviously a further increase of V will cause $(V-a)/V$ to increase above 0.34 and the toe-end of the extended front will merge with the step front, the merger extending to f^* as determined by (9.18a) and (9.19a). Similarly this occurs if parameter 'a' is decreased by the same amount as V (thus diminishing the "head start" of the first step). Using for this purpose $VC/Q = 1.72$ with $aC/Q = 0.72$ (which leaves the backward front intact), one finds the partially merged front ranging from 0 to f^*, as also shown in Fig. 9.6. Since in the present case V_0 is a constant for both ranges of f, one may determine f^* from the relation:

$$(1.72-0.72)\frac{df}{d(N+f/\bar{R}_D)} = \frac{1.72f^* - 0.72(f^*-0.25)}{f^*}\;\frac{f^*}{N+f^*/\bar{R}_D}$$

or:

$$\frac{df}{d(N+f/\bar{R}_D)} = \frac{f^*+0.18}{N+f^*/\bar{R}_D}$$

Constructing now the tangent to the storage curve originating from the point located to the left of the origin of the f-axis at the position -0.18, then gives $f^* = 0.50$ as the value where the slope of the curve just matches that of the tangent as constructed, thus fulfilling the condition above.

From this construction it may again be concluded that merger occurs if $1 < V/(V-a) < 1.06/0.36$, where the lower limit, with $a = 0$, corresponds to maximal merger as obtained for a single step feed (cf. Fig. 9.2.).

9.4. DIFFUSION AND DISPERSION EFFECTS

As was indicated in the previous section, diffusion (and dispersion) are in principle operative in any moving concentration front. As concentration gradients occur (at least) in the length direction of the column, longitudinal diffusion will take place following Fick's law, according to:

$$j_{dif} = -\theta D \frac{\partial c}{\partial x} \tag{9.21}$$

in which D is found from the diffusion coefficient in free solution, D_0, appropriately corrected for the pore tortuosity, λ, i.e. $D = D_0/\lambda$. When assessing the effect of the diffusion flux on the rate of propagation of a concentration front through an exchanger column, it should first be ascertained that its net value across any region bounded on both sides by zero concentration gradients vanishes. Thus, if a rapidly stirred solution with concentration c_f is fed into a column, the contribution of longitudinal diffusion to the total flux entering the column disappears as soon as the developing concentration front in the column has advanced to such a depth that the concentration gradient at the entrance side has approached sufficiently close to zero. As will be shown below, this depth depends on the value of a "diffusion length" parameter of the system, $L_{dif} = \theta D/J^V$.

Although, in theory, the situation of a zero concentration gradient at the entrance is approached asymptotically with increasing time, in practice it appears preferable to ignore this aspect. Thus solutions fed into soil columns in the field are usually not stirred, while on the other hand the total amount of a given cation fed into the column remains fixed. Then initial diffusion across the entrance is accompanied by a decrease in the feed concentration in the initial stages, which offsets the diffusion contribution to the influx. As the precise behavior is hard to describe in general terms, it is preferable to treat the system from the beginning as one with a constant influx equal to $J^V c$.

Assuming that in practice the total influx in a column is a given quantity, the mean depth of penetration of a concentration front, x_p, is found from the integral conservation condition, given in (9.12) and here restated as:

$$J^V t \Delta c = x_p(\theta \Delta c + \Delta q) \tag{9.22}$$

The actual concentration front will be spread around x_p, and diffusion then affects only this spreading, in addition to any spreading caused by non-linear exchange isotherms as treated in the previous section. The rate of spreading of a given front due to longitudinal diffusion may be demonstrated by substituting (9.21) into (9.4), according to:

$$(q' + \theta)\frac{\partial c}{\partial t} + J^V \frac{\partial c}{\partial x} - \frac{\partial \theta D(\partial c/\partial x)}{\partial x} = 0 \qquad (9.23)$$

Rewriting the above in terms of $(\partial x/\partial t)_c$, while making use of the definition of L_{dif}, then gives for a constant value of θD in the column:

$$\left(\frac{\partial x}{\partial t}\right)_c = \frac{J^V}{(q' + \theta)}\left\{1 - L_{dif}\frac{\partial(\partial c/\partial x)}{\partial c}\right\} \qquad (9.24)$$

The diffusion contribution to the rate of propagation of any point of a concentration front ascending in the backward direction is then positive for the concave regions, negative for the convex regions (and zero for any linear region). For a front with concentrations descending in the backward direction the reverse holds, implying a spreading of both types of fronts. Upon integration of (9.24) over a complete sigmoid front, the diffusion term vanishes as indicated above. Thus one finds for the range from $c = c_i$, where $\partial c/\partial x = 0$, to $c = c_f = c_i + \Delta c$, where $\partial c/\partial x = 0$:

$$\int_{c_i}^{c_f} (q' + \theta)\left(\frac{\partial x}{\partial t}\right)_c dc = J^V \Delta c \qquad (9.25)$$

where the left-hand side constitutes the rate of advance of x_p, multiplied by $(\Delta q + \theta\Delta c)$. Once a front has spread to a sufficient extent (depending on the value of L_{dif}), the diffusion term becomes negligible everywhere in the concentration front and (9.24), with $J^V dt \equiv dV$, reverts to (9.9), as used in the previous section.

For soil columns in the field, J^V ranges from about 10^{-8} m/sec for the year-average downward flux in temperate climates to, for example, 10^{-5} m/sec for saturated leaching in the gravity field. The value of θD in soil should vary from about $5 \cdot 10^{-10}$ m^2/sec near saturation to less then 10^{-10} m^2/sec in rather dry soils. The parameter L_{dif} may thus be taken at a value of 2—5 cm when considering the yearly downward flux. For faster leaching processes its value will be smaller by one or more orders of magnitude.

In addition to (molecular) diffusion, which may be traced back to the velocity distribution of individual molecules, one must also consider the effect of the distribution of the convective velocity in a porous medium on the propagation of concentration fronts. Such a velocity distribution is inherent to flow through capillary pores and ranges from an intracapillary distribution of the Poiseuille type, to intercapillary distributions attributable to a variation of pore sizes and random connections via pore "windows". Considerable attention has been paid to dispersion phenomena in the literature. While referring to existing texts (e.g. Bear et al., 1968; Bear, 1972) for a thorough treatment, in the present section an attempt will be made to render plausible the equations commonly used in describing dispersion in soil columns under field conditions.

Foregoing any specific models, while following the reasoning employed by Feynman et al. (1963) in discussing diffusion phenomena, one may derive a general expression for the (barycentric) dispersion flux of a solute across a plane in which the linear carrier velocity varies around its average value $\langle v \rangle$, such that the average value of the excess velocity, $\Delta v \equiv v - \langle v \rangle$, equals zero, i.e. $\langle \Delta v \rangle = 0$. If in this plane the solute concentration gradient equals dc/dx, and the local wet cross-section equals θ, the excess flux of solute in forward direction is found (for $dc/dx < 0$) as:

$$\Delta^+ j_s = \theta \left\langle \Delta^+ v \left(c - l \frac{dc}{dx} \right) \right\rangle \tag{9.26a}$$

where the averaging takes place over those regions which exhibit a velocity in excess of $\langle v \rangle$, while $-l(dc/dx) \equiv \Delta c$ specifies the excess concentration arising in the fast flowing regions owing to the fact that in these regions the concentration front advances with respect to a plane moving with the average carrier velocity, $\langle v \rangle$. In similar fashion one finds a net backward flux in the "slow" regions (relative to $\langle v \rangle$), as:

$$\Delta^- j_s = -\theta \left\langle \Delta^- v \left(c + l \frac{dc}{dx} \right) \right\rangle \tag{9.26b}$$

Addition of these fluxes gives the dispersion flux, for either sign of dc/dx, as:

$$j_{dis} = 0 - \theta \langle l \Delta v \rangle \frac{dc}{dx} \tag{9.27}$$

The validity of (9.27) is clearly limited by the condition that j_{dis}, if opposing the convective flux, θvc, may not exceed the latter (cf. small print section on top p. 346).

In analogy with Fick's diffusion equation one may thus introduce a dispersion coefficient according to:

$$D_{dis} = \langle l \Delta v \rangle \tag{9.28}$$

The evaluation of this generalized expression requires further specification of the expected correlation between l and Δv. A system composed of parallel tubes with different, but internally homogeneous, flow velocities may serve as an example. Ignoring longitudinal diffusion, one may consider the situation as it develops with progressing time if, at a particular time t_0 all tubes exhibit an identical (linear) concentration front of a solute in the carrier fluid. Obviously the average solute flux through the complete set of tubes at time t_0 equals the product $\langle v \rangle c_x$ at any location, x, within the front region; accordingly j_{dis} and l are zero at $t = t_0$. At $t = t_0 + \Delta t$, however, the concentration fronts in the different tubes have moved over a distance $v\Delta t$. The concentration in a particular tube in the plane located at $x + \langle v \rangle \Delta t$ (i.e. the plane moving with the mean velocity of the fluid starting from x at $t = t_0$)

is then found as $c = c_x + \Delta v \Delta t(dc/dx)$, identifying the distance l introduced in (9.26a, b) as the product of the local excess velocity and the time elapsed since the concentrations in this plane moving with the average carrier velocity were identical. Thus, in the early stages of the present case, the dispersion coefficient is found to be proportional to Δt, according to $D_{dis} = (\Delta^2 v)\Delta t$. As for a given set of tubes $\Delta v \sim \langle v \rangle$, D_{dis} is conveniently related to the relative spread of the velocities, $(\sigma')^2 \equiv (\Delta^2 v)/\langle v \rangle^2$, which should be a system constant. This gives for the present extreme situation:

$$D_{dis} = (\sigma')^2 \langle v \rangle^2 \Delta t \tag{9.29}$$

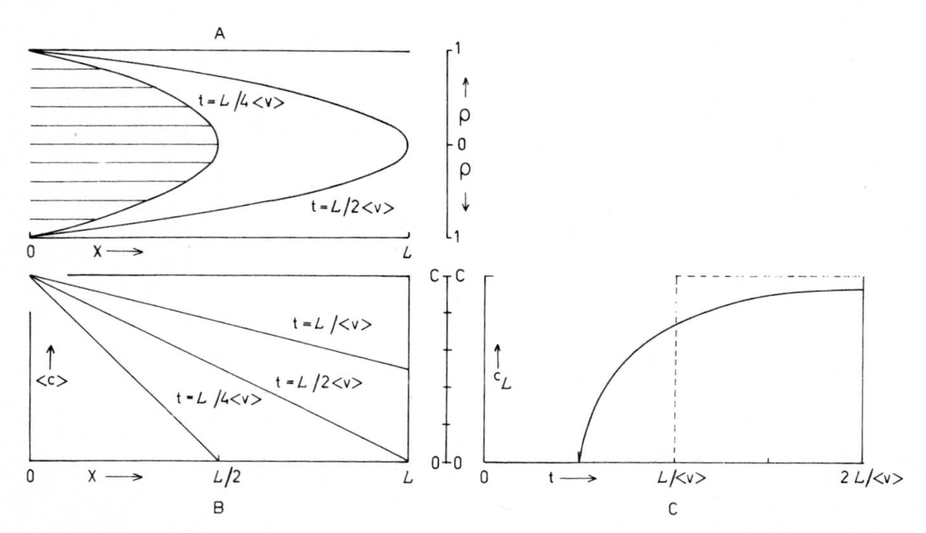

Fig. 9.7. Penetration of a solute front in a Poiseuille tube in the absence of diffusion.
A: Hatched area represents the paraboloid-shaped intrusion of a solution with concentration C entering at t = 0.
B: Volume-averaged concentration in the tube, $\langle c \rangle$, at three different values of t.
C: Break-through curve at L (solid line) as compared to that pertaining to complete radial mixing in the tube (dashed line), indicating an advance break-through of 25% of the incoming solution at a feed volume equal to the tube volume.

Although the simplified derivation above appears to be limited to regions with linear concentration fronts, the validity of (9.29) extends beyond the chosen system. This may be demonstrated with the capillary tube system analyzed by Taylor (1953). Ignoring the effect of diffusion, a step feed of a solute with concentration C introduced at the entrance of a capillary tube becomes spread inside the tube as the carrier velocity varies with position according to:

$$v = 2\langle v \rangle(1 - \rho^2) \tag{9.30}$$

in which ρ indicates the fractional value of the radial coordinate, varying from 0 at the centre to 1 at the circumference of the tube, and $\langle v \rangle$ is the average velocity inside the tube.

The step front at the entrance thus gives rise to a paraboloid-shaped intrusion of the solution at a concentration C, surrounded by the displaced solution here assumed to be free of solute. Indicating the value of ρ at which the feed solution borders the displaced solution with ρ^*, one finds (see Fig. 9.7):

$$\rho^*(t, x) = (1 - x/2\langle v\rangle t)^{1/2} \tag{9.31}$$

Accordingly the average concentration in any cross-section of the tube equals:

$$\langle c\rangle = C(\rho^*)^2 = C(1 - x/2\langle v\rangle t) \tag{9.32}$$

giving a linear "front" extending from $x = 0$, where $\langle c\rangle = C$, to $x = 2\langle v\rangle t$, where $\langle c\rangle = 0$, as is shown in Fig. 9.7. The gradient of the average concentration is found as:

$$\frac{d\langle c\rangle}{dx} = -\frac{C}{2\langle v\rangle t} \tag{9.33}$$

The dispersion flux, as defined above, may now be calculated by integrating the solute flux and subtracting the convection flux equal to $\langle v\rangle\langle c\rangle = \langle v\rangle C(\rho^*)^2$. This gives per unit cross-section:

$$j_{dis} = C\int_0^{\rho^*} (v - \langle v\rangle)\,d\rho^2 = \langle v\rangle C\,\{(\rho^*)^2 - (\rho^*)^4\}$$

which specifies the dispersion flux as a function of x and t with the help of (9.31). Making use of (9.32) and (9.33), the dispersion flux is conveniently expressed in terms of $\langle c\rangle/C$, as:

$$j_{dis} = -\langle v\rangle^2 t \cdot 2(1 - \langle c\rangle/C)(\langle c\rangle/C)\frac{d\langle c\rangle}{dx} \tag{9.34}$$

in which the factors in front of the gradient of the average concentration then constitute D_{dis}. This dispersion coefficient thus depends on the local value of $\langle c\rangle/C$ and its mean value over the entire front is found as:

$$\langle D_{dis}\rangle = \langle v\rangle^2 t \cdot 2\int_0^1 (1 - \langle c\rangle/C)(\langle c\rangle/C)\,d(\langle c\rangle/C) = \tfrac{1}{3}\langle v\rangle^2 t \tag{9.35}$$

For the chosen parabolic velocity pattern the factor $\frac{1}{3}$ corresponds precisely to $(\sigma')^2 \equiv \Delta^2 v/\langle v\rangle^2$, confirming (9.29), for the present case. More important, though, seems the implication that the effective dispersion coefficient in systems resembling a set of flow channels — without cross connections via diffusion or mixing — is proportional to $\langle v\rangle^2 t$, as the parameter l equals $\Delta v \cdot t$.

Characterizing the situation discussed above as "non-limited convective dispersion", one may now consider the more realistic situation arising if Δt is limited by certain counter-effects. Thus the set of tubes discussed in the previous case could be interspersed at regular intervals with "mixing compartments", allowing for complete (transversal) mixing at the points $x = L$, $x = 2L$, etc. In that case the dispersion effect of one section with length L is obtained again from (9.29), but Δt will now correspond to a mean residence time in this section, or $\Delta t = L/\langle v\rangle$. Following a concentration front as it passes through consecutive sections of the set of tubes, one may thus expect that the effective dispersion coefficient will grow from zero at the entrance to a maximum value corresponding to $\Delta t = L/\langle v\rangle$ at the end of the first section. As a result of the mixing at that point it then falls back to zero, to grow again to its maximum value in the next section, and so on. Accordingly, the dispersion coefficient of a set of tubes interspersed with mixing compartments should be given by:

$$D_{dis} = \sigma'^2 \langle v\rangle L/2 \tag{9.36}$$

where L signifies the average distance between mixing points. Comparing (9.36) and (9.29) it is further noted that actually D_{dis}, as given by (9.29), is a pseudo dispersion coefficient in the sense that the effect caused by the velocity distribution in that case is still, in principle, reversible. Thus when the carrier velocity is reversed at any given moment, the observed spread of $\langle c\rangle$ will shrink back and the concentration "front" will steepen up to a step front once it reaches the entrance of the tubes again. An example of partial reversibility of dispersion phenomena in a packed fluorite column was discussed by Heller (1960, 1972); another example pertaining to gas flow is found in Scotter and Raats (1968). In contrast, the dispersion effect of the sectioned set of tubes is irreversible owing to the mixing process involved. Reversing the carrier velocity then has no influence and the concentration front will keep on spreading irrespective of its direction of motion.

Again, the above generalization may be compared with the results obtained for the specific model consisting of a tube (of length L) in which a Poiseuille velocity distribution exists. The concentration in the mixing compartment at $x = L$ then follows the break-through curve of the concentration, according to $c(t)/C = 1 - (L/2\langle v\rangle t)^2$ for $t > L/2\langle v\rangle$ (cf. Fig. 9.7C). Compared with piston displacement, which produces a step increase of c at $t = L/\langle v\rangle$, one finds (at $t = L/\langle v\rangle$) that the relative amount of solute that has come out ahead in the mixing compartment, owing to the existing velocity distribution, equals $\int c(t)dt/(CL/\langle v\rangle) = 1/4$, where the integration is carried out from $t = L/2\langle v\rangle$ to $L/\langle v\rangle$. If expressed in terms of a constant "diffusion" coefficient, such a spread in a break-through curve requires that $\sqrt{Dt_{res}}/\sqrt{\pi} = L/4$. Using (9.35) for the value of D in the present system, one may estimate the necessary residence time as:

$$t_{res} = \frac{L}{\langle v\rangle}\sqrt{\frac{3\pi}{16}} \approx 0.8\,\frac{L}{\langle v\rangle} \tag{9.37}$$

which checks satisfactorily with the general estimate used in (9.36).

Summarizing the situation for beds of granular material, it may be concluded that the frequent connections between pores of different diameter together with the random orientation of pore directions should provide ample opportunity for mixing at short intervals. Accordingly one should expect that in such systems dispersion will occur, characterized by a coefficient of the form:

$$D_{dis} = L_{dis}\langle v\rangle \tag{9.38}$$

in which L_{dis} is a dispersion length parameter related to the relative spread of the velocity distribution, σ', and the effective distance between mixing points, L. Although, at first sight, one would perhaps be inclined to equate L with the "grain size" of the bed, one should actually expect that mixing is not complete at a particular junction. Thus L should then be seen as the distance that must be travelled (on average) before a concentration difference arising locally in adjacent pores is effectively annulled as a result of repeated partial mixing. Thinking in terms of a multiple of the grain diameter for L, it is not unreasonable to expect that L_{dis}, as the product of L and $\sigma'^2/2$, should be of the same order of magnitude as the grain diameter.

In soils, which usually exhibit a composite structural arrangement, the appropriate value of L_{dis} should be determined empirically. Some data available in the literature, e.g. Passioura and Rose (1971), Frissel et al. (1970), indicate that in homogeneous soils (i.e. finely aggregated, without cracks) L_{dis} is often less than 1 cm, though occasionally somewhat larger.

In addition to turbulent mixing at connections between pores, one should also take into account transverse diffusion as an effect countering the convective dispersion proper. Actually this effect will not be of concern if the system considered consists of an assembly of pores of different size, connected solely at certain junctions. The transverse diffusion through these junctions would simply be one component of the assumed mixing in these junctions and (9.38) should still apply. If, however, attention is directed towards intracapillary dispersion arising from a Poiseuille velocity pattern inside a single tube, transverse diffusion will become increasingly effective with increasing length of tube in checking the value of the transverse concentration gradient arising as a result of this velocity distribution. A superficial inspection of the situation then shows that if the convective dispersion proper is expressed in terms of $D_{dis} = \langle\Delta^2 v\rangle\Delta t$, the effective value of Δt must be proportional to the diffusion time needed to traverse the width of the zone over which the velocity distribution is spread, R. Accordingly $\sqrt{D\,\Delta t_{dif}} \sim R$, and the net effect of convective dispersion proper plus transverse diffusion should be governed by:

$$D_{dis} = \sigma'^2 \langle v\rangle^2 g R^2/D \qquad (9.39)$$

in which g is a coefficient which depends on the geometry of the system. The corresponding value of L_{dis} then reads:

$$L_{dis} = g\sigma'^2 R^2 \langle v\rangle/D \qquad (9.40)$$

For the cylindrical tube discussed before, Taylor (1953) derived an expression for D_{dis} valid in the case where $L/\langle v\rangle \gg R^2/D$. Then the time needed to traverse the capillary is much larger than the effective diffusion time in the radial direction and a pseudo steady state arises in which the radial distribution of the solute assumes a roughly constant shape depending solely on the local value of $\partial\langle c\rangle/\partial x$ (cf. the discussion below about the situation arising when a stagnant phase is present). This then leads to:

$$D_{dis} = \frac{R^2}{48D} \langle v \rangle^2 \tag{9.41}$$

indicating that in the cylindrical tube $g \approx \frac{1}{16}$.

Applying the above first to a single tube with length L and radius R, one may ascertain whether the dispersion effect of such a tube is controlled by radial diffusion or by the mixing at the end. As in this case *both* mechanisms work simultaneously, the one yielding the *smallest* value of D_{dis} will dominate. Taking D as 10^{-9} m^2/sec and $L/2gR$ at 10 (corresponding to about isodiametrical pores between grains) one finds from (9.36) and (9.39) that transverse diffusion will control the dispersion caused by a single pore if $\langle v \rangle < 10^{-8}/R$ m/sec. For soil systems, where R should not exceed 1 mm (aside from cracks, see below), this condition will always be fulfilled under field conditions. This indicates that *intra*capillary dispersion is diffusion-controlled. In contrast, *inter*capillary dispersion — where transversal diffusion is not operative because pores are separated by a solid phase — is controlled by (turbulent) mixing at certain intervals and often much larger than the intracapillary case. In conclusion it appears that the dispersion coefficient in soils under field conditions should at least contain a term as specified by (9.38), which covers the effect of the intercapillary velocity distribution.

Admitting, however, that, in a composite structural arrangement as usually prevails in soils, the "grains" are actually aggregates containing a part of the pore space, the above description is still incomplete. Depending on the moisture content at which flow takes place, one should expect that, near saturation, the flow velocity inside the aggregates is negligible in comparison to that in the interaggregate pores. Treating in that case the moisture content of the aggregates, θ_a, as a stagnant phase, the similarity with the diffusion-controlled intracapillary dispersion treated above becomes apparent. At the same time the criteria used before to dismiss this effect in comparison to intercapillary dispersion are now not applicable. In the first place the size of these macropores may easily exceed 1 mm in length, so the effect could be much more severe. Perhaps more important, however, is the fact that any turbulent mixing at the joints between macropores will not include the stagnant phase, so the mixing between stagnant and mobile phase (θ_m) should be entirely dependent on transverse diffusion. Accordingly the dispersion attributable to θ_a should follow an equation such as (9.39), without the restriction that D_{dis} must remain smaller than the value following from (9.38). Actually the validity of (9.39) must be checked itself, as it is based on the simplifying condition that the effective diffusion time, $\Delta t_{dif} = R^2/D$, is small in comparison to the time needed for the front to pass by (cf. the condition given in conjunction with (9.41) where the latter time was taken as $L/\langle v \rangle$). Deferring a discussion of the situation arising if this condition is not fulfilled to the following section on rate processes, it will here be assumed that this condition is met. Moreover, it may be ascertained that dispersion, whatever equation it follows initially, always leads to front spreading and thus diminishes longitudinal concentration gradients. Thus the condition specified is always met at some depth and some time, after which (9.39) will apply.

Following the extensive discussion of Passioura (1971) one finds that a stagnant reservoir, if exposed to a linearly increasing concentration in a mobile phase at its boundary, will fill up after some time according to (cf. Crank, 1964, e.g. equations (5.23) and (6.29)):

$$\langle c_a \rangle = k(t - gR_a^2/D) \qquad (9.42)$$

In this equation $\langle c_a \rangle$ is the volume average concentration in the stagnant phase θ_a with radius R_a; D refers to the diffusion per unit area of the liquid phase inside the aggregate, so $D = D_0/\lambda$; $k = \partial c_m/\partial t$, with c_m referring to the mobile phase. The geometry factor, g, varies between $\frac{1}{8}$ for infinitely extended cylinders to $\frac{1}{15}$ for a sphere, while t must exceed about $0.3\, R_a^2/D$ for (9.42) to apply within a few percent. If the latter condition is met, $\langle c_a \rangle$ will lag behind c_m by an amount directly proportional to k, according to:

$$c_m - \langle c_a \rangle = kgR_a^2/D \qquad (9.43)$$

Reconstructing now the solute flux through a unit cross-section of the entire liquid phase one finds:

$$\begin{aligned} j_s' \equiv \langle v \rangle c_m &= \langle v \rangle \langle c \rangle + (c_m - \langle c_a \rangle)\langle v \rangle \theta_a/\theta \\ &= \langle v \rangle \langle c \rangle + \frac{g\theta_a R_a^2 \langle v \rangle}{\theta D}\, k \end{aligned} \qquad (9.44)$$

The second term of the RHS then constitutes the dispersion flux under the stipulated conditions of a pseudo steady state. In that case k may be replaced by $\partial\langle c \rangle/\partial t$, which in turn may be approximated by $-\langle v \rangle \partial \langle c \rangle/\partial x$. The corresponding dispersion coefficient then equals:

$$D_{dis} = \frac{g\theta_a R_a^2 \langle v \rangle^2}{\theta D} \qquad (9.45)$$

in which g should be about $\frac{1}{15}$ for isodiametrical aggregates. The corresponding value of the effective dispersion length resulting from the finite rate of equilibrium between mobile and stagnant phase, L_r, is then found as:

$$L_r \equiv \frac{D_{dis}}{\langle v \rangle} = \frac{g\theta_a R_a^2}{\theta D}\langle v \rangle \qquad (9.46)$$

Comparing this expression with the value of L_{dis} attributable to intercapillary dispersion, specified before as having the same order of magnitude as the grain diameter, one finds for $\theta_a/\theta = \frac{1}{2}$, $D = 10^{-9}$ m²/sec and $R_a = 10^{-3}$ m (i.e. 2-mm aggregates) the value of L_r (in cm) as about $30\langle v \rangle$ (in cm/sec). If $\langle v \rangle$ is smaller than the previously assumed "field" limit of 10^{-5} m/sec (i.e. 10^{-3}

cm/sec) it is clear that the effect of the stagnant phase remains negligible in comparison to the intercapillary dispersion. For larger size aggregates or clods separated by cracks, however, the situation may be entirely different, as will be commented on below.

Finally a last complication must be mentioned in connection with a stagnant phase. Where the latter is presumed to be situated mainly inside aggregates surrounded by macropores, it becomes obvious that the lagging in solution composition (as compared to the mobile solution in the inter-aggregate pores) implies that the composition of the exchanger sites inside the aggregate cannot be in equilibrium with the composition of the mobile solution. Although non-equilibrium chromatography will be treated more extensively in a following section, it is mentioned here that, for a sufficiently slow rate of increase in concentration in the mobile solution, the situation becomes analogous to the one treated above. In fact the derivations used here were already given much earlier by Glueckauf (1955a, b) for precisely the same situation of exchange sites residing in a (spherical) stagnant phase surrounded by a mobile phase with changing concentration. Thus (9.45) remains valid (at large times) provided D in the denominator is corrected for the presence of a distributed sink in the form of exchange sites. As will be commented on in the next section, this may be taken care of by multiplying D with $\theta_a/(q' + \theta_a)$, in which q' signifies the differential capacity of the exchanger, as used before. Although this will greatly increase the effective diffusion time into the aggregate, the rate of propagation of a concentration front through the column — and thus the value of $\partial c_m/\partial t$ — tends to be reduced by a factor of similar magnitude, leaving (9.46) roughly intact. Referring to a following section for details, the effect of the presence of exchange sites in the aggregate will be acknowledged here by the addition of factor A, signifying a ratio of the distribution ratio inside the aggregate to that of the entire soil. It is noted here that this factor equals unity in the absence of adsorption and approaches θ/θ_a for a very high concentration of adsorption sites.

Combining the different effects discussed above it appears not unreasonable to sum them since they are acting more or less independently. Specifying now the autonomous flux of the solute arising from longitudinal diffusion and convective dispersion (including the presence of a stagnant phase θ_a) per unit area of the total column, one finds:

$$j_s^D = -\left[\theta D + \theta L_{dis}\langle v\rangle + A^2\theta_a R_a^2\langle v\rangle^2/15D\right]\frac{\partial\langle c\rangle}{\partial x} \tag{9.47}$$

In this equation it has been assumed that longitudinal diffusion is operative over the entire liquid phase, while $D \equiv D_0/\lambda$ inside the aggregate has been taken equal to D in the column, which might be a slight overestimate. Replacing now $\theta\langle v\rangle$ by J^V, one may introduce the length parameter L_D covering the combined effect of diffusion and dispersion according to:

$$L_D = L_{dif} + L_{dis} + L_r$$
$$= \theta D/J^V + L_{dis} + A^2(\theta_a/\theta)R_a^2 J^V/15D\theta \qquad (9.48)$$

such that:

$$j_s^D = -J^V L_D(\partial c/\partial x) \qquad (9.49)$$

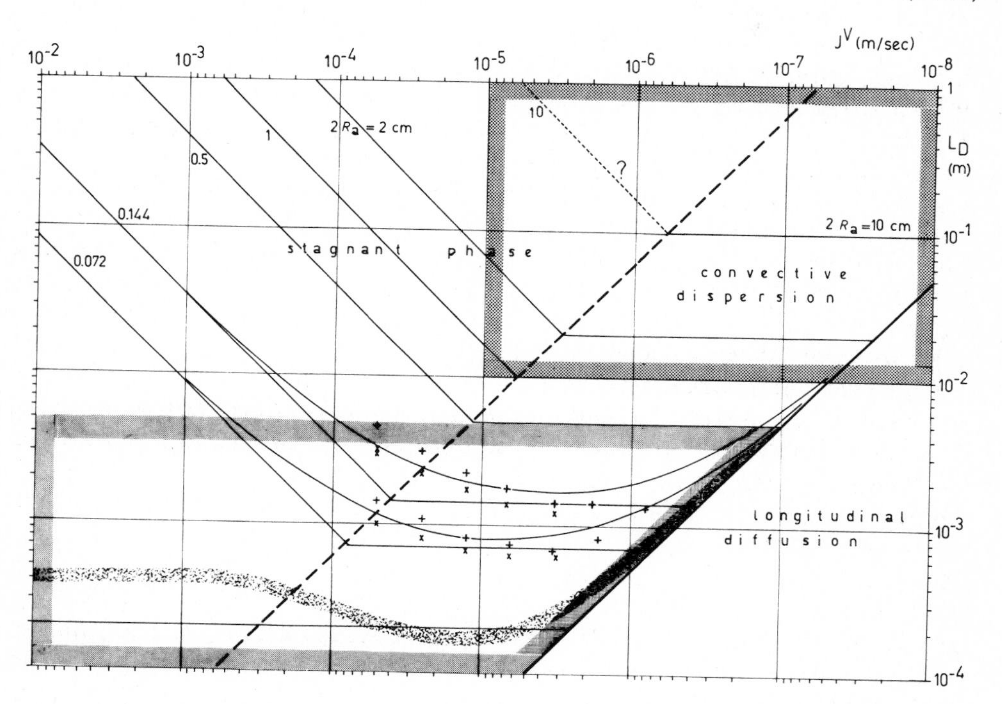

Fig. 9.8. Estimate of the combined diffusion/dispersion length parameter as a function of the volume flux J^V, in m^3 per m^2 column area. The three components resulting from, respectively, longitudinal diffusion (L_{dif}), convective dispersion due to mixing (L_{dis}), and diffusion into a stagnant phase (L_r) may be read off from the appropriate lines, of which the latter two depend on the radius of aggregates, R_a. Summing up gives curves as shown for two aggregate sizes, together with experimental data given by Passioura and Rose (1971). See also text.

The situation is summarized in the composite graph shown in Fig. 9.8, giving the values of the different parameters L_D plotted on double log scale against J^V. In ascending direction of J^V one finds first $L_{dif} = \theta D/J^V$, which is in principle independent of the grain size. Thinking in terms of soils at fairly high moisture contents, θD was taken as $5 \cdot 10^{-10}$ m^2/sec (i.e. corresponding to $\theta = 0.5$, $\lambda = 2$). Next L_{dis} is given as a horizontal line, its magnitude depending on the size of grains or aggregates and in this case taken equal to $2R_a$*. Finally L_r is plotted according to (9.48), for $\theta_a/\theta = 1/2$, corresponding to low suction values and $A = 1$, i.e. in the absence of adsorption. This gives

* See small print section on top p. 346.

a family of ascending lines for different aggregates sizes as indicated. As the sum of the three L parameters is assumed to be operative, one finds that the actual course of L_D corresponds to a smoothed curve as shown for aggregates of 0.144- and 0.072-cm diameter. For these two sizes the experimental data obtained by Passioura and Rose (1971) are shown, indicating a rather remarkable coincidence in view of the fairly rough guesses made. Also the experimental curve given by Pfannkuch (1963) is arbitrarily plotted for $2R_a = 0.02$ cm, the original curve giving the relation between the dimensionless parameters $L_D J^V/\theta D$ and $2J^V R_a/\theta D$, the Péclet number for molecular diffusion. This original curve contains experimental data of many authors and covers hundreds of points corresponding to *non-porous* grains of sizes ranging from 0.01 cm to 0.5 cm. The present re-plot thus serves only as a check on the validity of the curves representing L_{dif} and L_{dis} for the entire area surrounded by the shaded band. This curve appears to indicate that L_{dis} actually approaches $4R_a$ at very high velocities, while the extended "dip" is explained by Pfannkuch as an attenuating interference between longitudinal diffusion and convective dispersion in the transition region.

The Péclet number, Pé, is a dimensionless parameter of the form vL/D. In the present case it characterizes the effect of convective dispersion relative to that of molecular diffusion, as it represents the ratio of the second term of (9.48) to its first term (if L_{dis} is taken equal to $2R_a$). Expressing the complete equation in terms of Pé one finds:

$$L_D/2R_a = 1/\text{Pé} + 1 + \text{Pé}(A^2\theta_a/60\theta) \qquad (9.48a)$$

For the value of $A^2\theta_a/60\theta = 1/120$ as used above, this gives L_{dif} as the dominant term of L_D for $\text{Pé} < 1$ and L_r as the dominant one for $\text{Pé} > 120$. Expressed in terms of the spreading of a concentration front caused by diffusion/dispersion, an increase of Pé — as defined here — beyond about 100 will result in increased spreading.

A parameter similar to the one above, but containing the length of a particular column, L, and the effective value of D, reflecting the combined action of diffusion and dispersion, has also been named the Péclet number (of a column, cf. chapter 10, p.364). This number then characterizes the importance of convective *transport* relative to that of diffusion/dispersion, as it is reflected in the break-through curve at $x = L$. Obviously a high value of Pé (column) now leads to a relatively sharp break-through (aside from front spreading resulting from non-linear adsorption isotherms). Against this background it seems wise to refer to the latter column parameter as the Bodenstein number rather than the Péclet number, as was proposed in the C.E.P. tables of dimensionless numbers, Boucher and Alves (1959, 1963) (cf. also Rose and Passioura (1971), Rose (1977) who proposed to use the term Brenner number for this column parameter, again limiting the use of Pé to the material parameter defined previously).

Several remarks should be made about Fig. 9.8. In the first place the accuracy of predictions based on it is limited to ±50% of L_D, which, however, is often good enough. In the second place experimental verification, though quite good, has been limited to aggregate sizes smaller than several mm. Its application for the present purpose is, on the other hand, of particular interest in the upper right-hand quadrangle, i.e. $10^{-8} < J^V < 10^{-5}$ m/sec and $10^{-2} < L_D < 1$ m, as the front spreading is hardly significant if

$L_D < 1$ cm (see the next section). In this region the part to the right of the dashed line corresponding to $J^V 2R_a = 6 \cdot 10^{-8}$ m²/sec (actually corresponding to a Péclet number less than about 100) seems rather safe for predictive purposes, diffusion and intercapillary convective dispersion dominating the situation. To the left of this line the stagnant phase effect should enter, and lines parallel to the ones drawn for different values of R_a at an appropriate distance (proportional to R_a^2) are a first guess.

Actually, the present lines were constructed for $A^2\theta_a/\theta = 0.5$ and could be replaced by bands covering other values of θ_a/θ and A. Nevertheless it seems that the applicability to systems with clods of, for example, 10 cm or more at filter fluxes exceeding 10^{-6} m/sec (10 cm/day) is highly speculative. Other models than the present one would seem more appropriate. Obviously the mobile phase could in such a case be either located in the cracks (being held up by non-cracked layers underneath) or actually flow through the clods while the cracks are empty because of the build-up of a small suction. In short, the present theory largely serves to support the common experience that, in fairly homogeneous soil, L_D is of the order of a few cm, while it is hardly useful to predict the leaching efficiency for saturated leaching through cracked soils.

9.5. SOLUTION OF THE CONSERVATION EQUATION CONTAINING A SIGNIFICANT DIFFUSION/DISPERSION TERM

Substituting (9.49) into (9.4), the equation to be solved reads:

$$\{q'(c) + \theta\}\frac{\partial c}{\partial t} + J^V\frac{\partial c}{\partial x} - J^V\frac{\partial L_D(\partial c/\partial x)}{\partial x} = 0 \tag{9.50}$$

where thus the first coefficient is a function of c only, while the last term contains L_D which is in principle a function of x in a layered soil — aside from its dependence on J^V according to Fig. 9.8. In general, numerical methods are necessary to obtain a solution even for simple boundary conditions. Nevertheless there are several situations where analytical solutions are applicable. In support of the effort to obtain such solutions it is pointed out that L_D is only approximately known, while it was shown by Frissel et al. (1970) that replacing L_D by a length-averaged value for the entire column gives good results if the situation is not too extreme. As J^V was taken to be independent of x (zero divergence of the carrier flux), the last term then consists of a second-order differential of c multiplied by the constant $-J^V L_D$. If, in addition, the coefficient of $\partial c/\partial t$ is a constant (inferring a linear exchange isotherm over the concentration range of concern), (9.50) may be rewritten as:

$$\frac{\partial c}{\partial t} + v^*\frac{\partial c}{\partial x} - D^*\frac{\partial^2 c}{\partial x^2} = 0 \tag{9.51}$$

in which $v^* \equiv J^V/(q' + \theta)$ and $D^* \equiv J^V L_D/(q' + \theta) = L_D v^*$, with $q' = \Delta q/\Delta c$, the integral capacity of the exchanger over the appropriate range. Referring to (9.24) and (9.25), v^* is recognized as the mean rate of propagation of the concentration front through the column.

Equation (9.51) is known as the Fokker-Planck equation and several solutions pertaining to different boundary conditions may be found in Carslaw and Jaeger (1959, reprinted in 1967). For the present situation three types of boundary conditions are of particular interest (cf. Gershon and Nir, 1969), viz.:

(O) the infinite system stretching from $x \to -\infty$ to $x \to \infty$,

with $c = c_i$ for $x > 0$ at $t \leqslant 0$, and for $t > 0$ at $x \to \infty$

$c = c_f$ for $x < 0$ at $t \leqslant 0$, and for $t > 0$ at $x \to -\infty$;

(A) the semi-infinite system stretching from $x = 0$ to $x \to \infty$,

with $c = c_i$ for $x > 0$ at $t \leqslant 0$, and for $t > 0$ at $x \to \infty$

$c = c_f$ for $x = 0$ at $t > 0$;

(B) the semi-infinite system as above,

with $c = c_i$ for $x > 0$ at $t \leqslant 0$, and for $t > 0$ at $x \to \infty$

$$v^* c - D^* \frac{\partial c}{\partial x} = v^* c_f \text{ for } t > 0 \text{ at } x = 0.$$

Case (O) may serve as a guideline for (A) and (B) as the latter approach (O) for large values of t. Introducing a reduced barycentric coordinate in the column (i.e. following the mean motion of the concentration front) according to $\xi \equiv (x - v^* t)/2\sqrt{D^* t}$, (9.51) may be written as the ordinary differential equation:

$$2\xi \frac{dc}{d\xi} + \frac{d^2 c}{d\xi^2} = 0 \qquad (9.52)$$

subject to $c = c_i$ for $\xi \to \infty$

$c = c_f$ for $\xi \to -\infty$

and valid for $t > 0$. Solution gives the well-known expression:

$$\bar{c} \equiv (c - c_i)/(c_f - c_i) = \tfrac{1}{2} \operatorname{erfc} \xi = \tfrac{1}{2} \operatorname{erfc} \{(x - v^* t)/2\sqrt{D^* t}\} \qquad (9.53)$$

such that $\bar{c} = \frac{1}{2}$ at $\xi = 0$, or at $x = v^* t$.

The front thus develops as a regular diffusion front around its mean depth of penetration $x_p \equiv v^* t$. Obviously the shortcoming of this simple solution for the chosen problem is the neglect of the effect of the "moving" boundary at $\xi = -v^* t/2\sqrt{D^* t}$ (i.e. at the entrance where $x = 0$). Using the boundary conditions as specified under (A), this boundary is treated as a "reflecting" one, as no diffusion/dispersion is allowed in the region $\xi < -v^*t/2\sqrt{D^*t}$, while the concentration remains fixed at $\bar{c} = 1$ for $\xi = -v^*t/2\sqrt{D^*t}$. Especially in the early stages of the process this condition induces an extra solute flux at the boundary proportional to $-D^*(\partial c/\partial x)_0$. The solution of (9.51) subject to (A) is found with the help of Laplace transforms (cf. Crank, 1964) as:

$$\bar{c} = \tfrac{1}{2}\,[\text{erfc}\,\xi + \exp(-\xi^2)\exp(\xi+\alpha)^2\,\text{erfc}\,(\xi+\alpha)] \tag{9.54}$$

in which $\alpha \equiv v^*t/\sqrt{D^*t} = \sqrt{(v^*t)/L_D} = \sqrt{x_p/L_D}$. The contribution of the second term at $x = x_p$, or $\xi = 0$, is found as $\tfrac{1}{2}\exp(\alpha^2)\,\text{erfc}\,\alpha$, which decays rather slowly with increasing α from 0.5 at $x_p = 0$ at the outset to 0.09 at $\alpha = 3$, or $x_p = 9L_D$. For $(\xi + \alpha) > 2$ this second term may be approximated quite well by:

$$\Delta\bar{c} \approx \frac{1 - 0.41/(\xi+\alpha)^2}{2\sqrt{\pi}(\xi+\alpha)}\exp(-\xi^2) \tag{9.55}$$

which gives indeed $\Delta\bar{c} = 0.09$ at $\alpha = 3$, $\xi = 0$. Actually a further approximation according to:

$$\Delta\bar{c} \approx \exp(-\xi^2)/2(\xi+\alpha)\sqrt{\pi} \tag{9.55a}$$

seems good enough once $(\xi + \alpha)$ exceeds the value of 3.

Whereas condition (A) gave a simple solution which differs considerably from (9.53) at a moderate value of x_p/L_D, condition (B) gives a fairly involved solution which approaches (9.53) much faster, except at very early stages. Using the Laplace transform as given by Carslaw and Jaeger (1959, section 14.2 III), Gershon and Nir (1969) worked out this solution which may be given in the above symbols as:

$$\bar{c} = \tfrac{1}{2}\,\text{erfc}\,\xi - \tfrac{1}{2}\{\exp(-\xi^2)\} \cdot [\{1 + 2\alpha(\xi+\alpha)\}\exp(\xi+\alpha)^2\,\text{erfc}\,(\xi+\alpha) - 2\alpha/\sqrt{\pi}] \tag{9.56}$$

The correction to be applied to (9.53) then equals:

$$\Delta\bar{c} = -\alpha\{\exp(-\xi^2)\} \cdot [\{1 + 1/2\alpha(\xi+\alpha)\}(\xi+\alpha)\exp(\xi+\alpha)^2\,\text{erfc}\,(\xi+\alpha) - 1/\sqrt{\pi}] \tag{9.57}$$

which for $(\xi + \alpha) > 2$ becomes:

$$\Delta\bar{c} \approx -\frac{\alpha\exp(-\xi^2)}{\sqrt{\pi}}[\{1 + 1/2\alpha(\xi+\alpha)\}\{1 - 0.41/(\xi+\alpha)^2\} - 1]$$

$$\approx -\frac{\exp(-\xi^2)}{\sqrt{\pi}}\left[\frac{1}{2(\xi+\alpha)} - \frac{0.41\alpha}{(\xi+\alpha)^2} - \frac{0.41}{2(\xi+\alpha)^3}\right] \tag{9.57a}$$

At $x = x_p$, or $\xi = 0$, this gives for $\alpha = 3$ a correction equal to -0.012, i.e. much smaller than for condition (A). At the column entrance $\xi = -\alpha/2$; so, at that point, (9.56) is more easily handled as:

$$\bar{c}_{x=0} = \tfrac{1}{2}\operatorname{erfc}(-\alpha/2) - \tfrac{1}{2}(\alpha^2 + 1)\operatorname{erfc}(\alpha/2) + \frac{\alpha}{\sqrt{\pi}}\exp(-\alpha/2)^2 \qquad (9.58)$$

which gives $\bar{c} = 0$ for $\alpha = 0$ (i.e. at $t = 0$).

In Fig. 9.9 a normalized plot of $\bar{c}$ against ξ is given, allowing one to assess on sight the shape of the concentration front to be expected for given values of L_D and $v^*t \equiv x_p$, for the boundary conditions specified above as (O), (A) and (B).

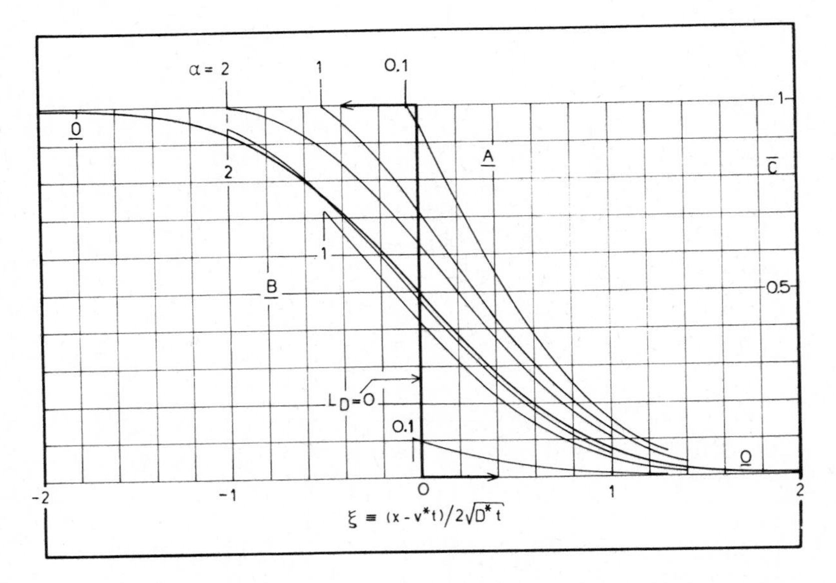

Fig. 9.9. Fully scaled plot of diffusion/dispersion fronts formed in columns following a linear exchange isotherm, given as $\bar{c} \equiv (c - c_i)/(c_f - c_i)$ as a function of $\xi = (x - v^*t)/2\sqrt{D^*t}$, for three different boundary conditions, viz. the infinite system, O; the semi-infinite system with fixed concentration at the entrance, A; the semi-infinite system with a constant solute flux at the entrance, B. The parameter α specifies the depth of penetration of the front relative to the diffusion/dispersion length parameter L_D, with $\alpha = \sqrt{x_p/L_D}$.

In the limiting case of $L_D = 0$, the situation reverts to the one discussed previously: for the chosen condition of a linear exchange isotherm and a step feed, the front remains a step front, located at $\xi = 0$, i.e. at x_p. If L_D is finite, the front present in the (hypothetical) infinite system follows the error function complement according to (9.53), see the curve marked O. It should be noted here that since ξ is expressed in units of $2\sqrt{D^*t} \equiv 2\sqrt{L_D x_p}$, the actual shape of the front in terms of $(x - v^*t)$ flattens with increasing time, or increasing x_p. Indicating the front "width" covering the range $0.16 < \bar{c} < 0.84$ as w (i.e. about 70% of the total variation of c), one finds $w = 2\sqrt{2L_D x_p}$. Thus for $L_D = 1$ cm, this width equals about 9 cm for $x_p = 10$ cm — the front stretching from $x = (10 - 4.5)$ to $x = (10 + 4.5)$ cm in the column — and about 30 cm for $x_p = 1$ m.

Using the boundary condition specified under (A), i.e. $\bar{c} = 1$ at $x = 0$ for all times, the "reduced" front shape changes from $\bar{c} = \text{erfc}\,\xi$ at $x_p = 0$ to $\bar{c}$ approaching $\frac{1}{2}$ erfc ξ for $x_p \to \infty$. The corresponding curves are indicated as the set A in Fig. 9.9 for different values of $\alpha = \sqrt{x_p/L_D}$ and may be back-translated in terms of $(x - v^*t)$ as indicated above for the curve marked O. It may be noted that the curves A differ considerably from the curve O for $\alpha < 2$. The area underneath these curves always exceeds the full area under curve O, as a result of the excess diffusion into the column when $\bar{c}$ is maintained at 1 for $x = 0$. As was pointed out before, the boundary condition (A) is actually not realistic as it requires the addition of solute into the feed solution in the early stages.

In contrast, the boundary condition (B), i.e. constant flux at $x = 0$, is about the best general guess for situations in the field. Even if in actuality this condition is violated slightly at the beginning, it ensures that the correct amount of solute has entered the column. The rewarding conclusion from Fig. 9.9 is then that the curves marked B soon become very close to O; that is, even at $\alpha = 1$, or $x_p = L_D$, the difference between the two is negligible compared to the accuracy of measurements. As, furthermore, any deviations from B would be in the direction of a slightly increased flux in the early stages, the use of the extremely simple equation (9.53) corresponding to curve O *but truncated* at the appropriate value of ξ seems absolutely warranted once α exceeds unity, or $x_p > L_D$. As was discussed in the previous section, this often implies that curve O and (9.53) are valid once x_p exceeds a few cm.

Another situation which lends itself to an analytical solution arises in the case where the diffusion term is counteracted by the steepening effect of non-linear *favorable* exchange. Then a steady state develops with respect to the moving coordinate $(x - v^*t)$, i.e. all points of the solute front travel with the same velocity which must then be equal to the mean rate of propagation v^*. Making use of (9.24) and (9.25), and replacing L_{dif} by L_D, this gives:

$$\left(\frac{\partial x}{\partial t}\right)_c = \frac{J^V}{q' + \theta}\left\{1 - L_D\,\frac{\partial(\partial c/\partial x)}{\partial c}\right\}$$

$$= \text{constant} = \frac{J^V}{\theta + \Delta q/\Delta c} \qquad (9.59)$$

with $\Delta q = (q_f - q_i)$ and $\Delta c = (c_f - c_i)$, as before. Replacing then $(\partial c/\partial x)_t$ by $dc/d(x - v^*t)$ one finds:

$$-L_D \frac{d\{dc/d(x - v^*t)\}}{dc} = \frac{(dq/dc) - (\Delta q/\Delta c)}{\theta + (\Delta q/\Delta c)}$$

or:

$$-L_D\, dc/d(x - v^*t) = B\Delta c \int (dq/\Delta q - dc/\Delta c) + \text{constant} \tag{9.60}$$

in which $B = \{(\Delta q/\Delta c)/(\theta + \Delta q/\Delta c)\}$. The system constant B equals $R_D/(R_D + 1)$ for the concentration range considered. In soil systems it is often close to unity, as R_D tends to be large (see chapter 7 of part A of this text).

Using as a boundary condition $\{dc/d(x - v^*t)\} = 0$ for $\bar{c} = 0$, (9.60) is conveniently expressed in terms of $\bar{c}$ and the corresponding fractional value $\bar{q}$ according to:

$$-L_D \frac{d\bar{c}}{d(x - v^*t)} = B\left[\int_0 d\bar{q} - \int_0 d\bar{c}\right] = B(\bar{q} - \bar{c}) \tag{9.61}$$

indicating that the slope of the front goes from zero at $c = c_i$, via finite negative values ($\bar{q} > \bar{c}$ for favorable exchange!) back to zero at $c = c_f$. The slope of the steady state front is then found from:

$$\frac{B(x - v^*t + s)}{L_D} = -\int_{\bar{c}_s} \frac{d\bar{c}}{\{\bar{q}(\bar{c}) - \bar{c}\}} = F(\bar{c}, \bar{c}_s) \tag{9.62}$$

where s is an (unknown) integration constant determined by the arbitrarily, but conveniently, chosen starting point for the integration of the RHS of (9.62), e.g. $\bar{c}_s = 0.5$. The limiting values $\bar{c} = 0$ and $\bar{c} = 1$ are not convenient for this purpose as the denominator approaches a zero value at either end. The shape of the solution front is, however, completely defined by (9.62) and may be constructed by graphical or numerical integration for any arbitrary adsorption isotherm. Indicating the above coordinate as $\tilde{x} \equiv (x - v^*t + s)$ one may thus construct $\bar{c}(\tilde{x})$ and also $\bar{q}(\tilde{x})$, i.e. the fronts in solution and adsorption phase relative to the position where $\bar{c} = \bar{c}_s$. In order to determine the location of these fronts with respect to x_p it suffices to locate graphically the mean position of the front in the solution phase, $\langle \tilde{x}_c \rangle$. Equation (9.61) then shows, on integration over $\tilde{x}$, while covering the full range of $\bar{c}$ from 0 to 1:

$$\langle \tilde{x}_q \rangle - \langle \tilde{x}_c \rangle \equiv \int_{-\infty}^{\infty} \bar{q}\, d\tilde{x} - \int_{-\infty}^{\infty} \bar{c}\, d\tilde{x} = L_D/B \tag{9.63}$$

i.e. $\langle \tilde{x}_q \rangle$ is located ahead from $\langle \tilde{x}_c \rangle$. Furthermore the weighted mean location of the combined fronts is found in the chosen coordinate system as:

$$\langle \tilde{x} \rangle \equiv \frac{\Delta q \langle \tilde{x}_q \rangle + \theta \Delta c \langle \tilde{x}_c \rangle}{\Delta q + \theta \Delta c} = B\langle \tilde{x}_q \rangle + (1-B)\langle \tilde{x}_c \rangle \tag{9.64}$$

which together with (9.63) gives:

$$\langle \tilde{x} \rangle = \langle \tilde{x}_c \rangle + L_D \tag{9.65}$$

This mean location, positioned at a distance L_D ahead of $\langle \tilde{x}_c \rangle$ as determined graphically, may then be identified as x_p (where $x = v^*t$). In Fig. 9.10 an example of the above procedure is shown, based on the convex lower part of the exchange isotherm given in Fig. 9.1, viz. for the range $f = 0$ to $f = 0.25$.

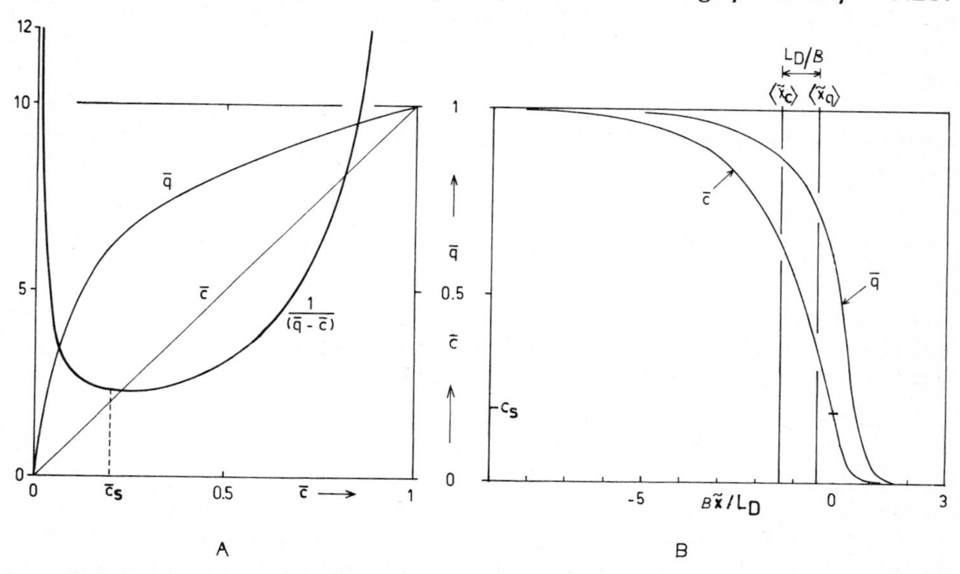

Fig. 9.10. Graphical construction of the steady state front arising for a favorable exchange isotherm with diffusion/dispersion.
A: $\bar{q}$ (right-hand scale) and $1/(\bar{q}-\bar{c})$ (left-hand scale) as a function of $\bar{c}$;
B: shape of the fronts in solution and adsorbed phase plotted against the coordinate $\tilde{x} = (x - v^*t + s)$ in units of L_D/B. At $\bar{c} = \bar{c}_s$, $\tilde{x} \equiv 0$.

If the exchange isotherm is given in terms of an exchange equation, the graphical or numerical procedure suggested above may be replaced by an analytical solution of (9.62). Reiniger and Bolt (1972) derived such an expression on the basis of the Gapon equation applied to complete exchange from 0% to 100% of the divalent cation. As the validity of the Gapon equation is highly questionable for a high percentage of monovalent cations (cf. chapter A5), while in practice one is usually concerned with the reclamation of sodic soils with gypsum starting from a sodium percentage of less than 40, the solution of (9.62) covering a range from q_i to Q will be given here. Restating the Gapon equation in terms of N_{2+}, f_{2+} and C according to equation (A7.5) in part A and using $\chi \equiv K_G\sqrt{2C}$ as a dimensionless exchange constant, one finds for the above range of q from q_i to Q, or from N_i to unity:

$$\bar{q} = (N - N_i)/(1 - N_i)$$

$$= \frac{\sqrt{f}}{\chi(1-f)+\sqrt{f}} \quad \frac{\chi(1-f_i)+\sqrt{f_i}}{\chi(1-f_i)} - \frac{\sqrt{f_i}}{\chi(1-f_i)} \tag{9.66a}$$

$$\bar{c} = \frac{f - f_i}{(1-f_i)} \tag{9.66b}$$

$$d\bar{c} = df/(1-f_i) \tag{9.66c}$$

Equation (9.62) then becomes:

$$B\tilde{x}/L_D = -\int \frac{\{\chi(1-f)+\sqrt{f}\}df}{(\sqrt{f}-\sqrt{f_i})(1-f)\{1-\chi(\sqrt{f}+\sqrt{f_i})\}} \tag{9.67}$$

which yields a solution of the type:

$$B\tilde{x}/L_D = -a\ln(\sqrt{f}-\sqrt{f_i}) - b\ln\{1-\chi(\sqrt{f}+\sqrt{f_i})\} + c\ln(1+\sqrt{f}) + d\ln(1-\sqrt{f})$$
$$\equiv \quad I \quad + \quad II \quad + \quad III \quad + \quad IV \tag{9.68}$$

with $$a = \frac{2\sqrt{f_i}(\chi+\sqrt{f_i})}{(1-f_i)(1-2\chi\sqrt{f_i})}$$

$$b = \frac{2\chi[\chi+\sqrt{f_i}\{1-\chi\sqrt{f_i})^2-\chi^2\}]}{(1-2\chi\sqrt{f_i})\{(1-\chi\sqrt{f_i})^2-\chi^2\}}$$

$$c = 1/(1+\chi-\chi\sqrt{f_i})$$

$$d = 1/(1-\chi-\chi\sqrt{f_i})$$

In order to determine $\langle\tilde{x}_c\rangle$, (9.68) is first integrated over f from f_i to 1 and divided by $(1-f_i)$ to give the mean value of the four terms over the chosen range of f. In order to sum up these four contributions, a common reference is introduced by calculating the value of each of the terms of (9.68) for a convenient value of $f = f_s$. The differences between the mean value of each term and its reference value may then be added giving $B\langle\tilde{x}_c\rangle/L_D$, and thus the location of $\langle\tilde{x}_c\rangle$ with respect to the chosen reference at $f = f_s$. After adding L_D one finds $\langle\tilde{x}\rangle$, which is identified with $x = v^*t$. A practical example may elucidate this procedure. A soil column containing about 17% exchangeable Na in addition to divalent cations is reclaimed with a solution of a calcium salt with a total concentration C = 0.02 N. Using K_G equal to 0.5 $(\text{mole/l})^{-1/2}$, this gives $\chi \equiv K_G\sqrt{2C} = 0.10$, and the Gapon equation reads $(1-N)/N = 0.1(1-f)/\sqrt{f}$. Taking $f_i = 0.16$, or $\sqrt{f_i} = 0.40$ one finds $N_i = 0.83$ (i.e. 83% divalent ions on the adsorption complex). This gives $a = 0.518$, $b = 0.111$, $c = 0.942$, and $d = 1.16$. Next the integrals of the four terms are determined over the range from $f_i = 0.16$ to $f = 1$ and divided by $(1-f_i)$ in order to find the mean value of the terms. Indicating these integrals, after division, by $\langle I\rangle$, $\langle II\rangle$, etc. one finds, respectively:

$$\langle I\rangle/a = -\ln(1-\sqrt{f_i}) + (1+3\sqrt{f_i})/2(1+\sqrt{f_i})$$

$$(1-f_i)\langle II\rangle/b = \{1/\chi-(1+\sqrt{f_i})\}\{1/\chi+(1-\sqrt{f_i})\}\ln\{1-\chi(1+\sqrt{f_i})\}$$
$$-\{1/\chi^2-(2\sqrt{f_i})/\chi\}\ln(1-2\chi\sqrt{f_i}) + \{1/\chi+(1-\sqrt{f_i})\}(1-\sqrt{f_i})$$

$$\langle \text{III} \rangle / c = \ln(1+\sqrt{f_i}) + (1-\sqrt{f_i})/2(1+\sqrt{f_i})$$

$$\langle \text{IV} \rangle / d = \ln(1-\sqrt{f_i}) - (3+\sqrt{f_i})/2(1+\sqrt{f_i}) \quad (9.69)$$

The mean location of the solution front relative to the location of the arbitrarily chosen reference value f_s is then found as:

$$B\langle \tilde{x}_c \rangle / L_D = \langle \text{I} \rangle - \text{I}(f_s) + \langle \text{II} \rangle - \text{II}(f_s) + \langle \text{III} \rangle - \text{III}(f_s) + \langle \text{IV} \rangle - \text{IV}(f_s) \quad (9.70)$$

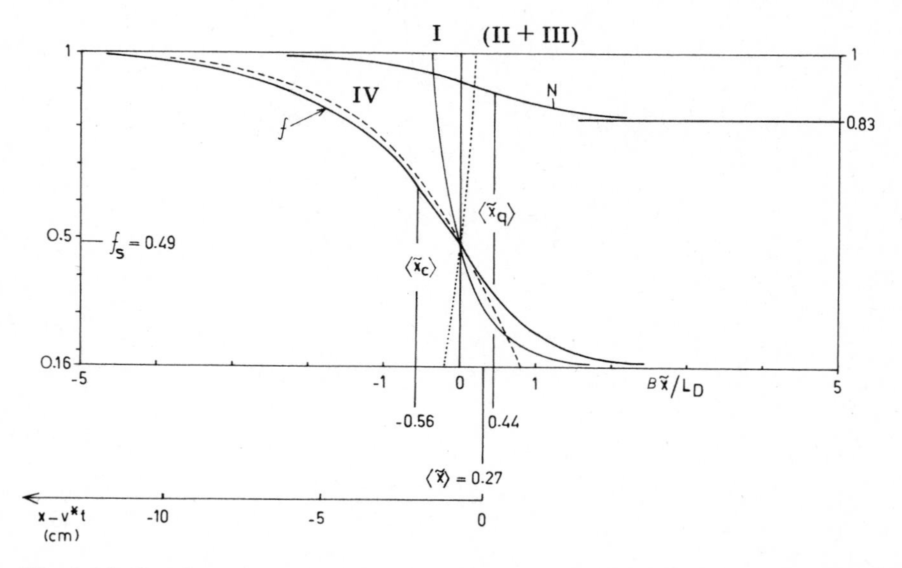

Fig. 9.11. Steady state front for Ca-ions fully replacing Na-ions in a soil column containing initially 17% of exchangeable Na-ions and 83% Ca-ions, as calculated by the Gapon equation with $K_G = \frac{1}{2}$ (kmol m^{-3})$^{-1/2}$ at 0.02 N electrolyte. Heavy lines show f_{Ca} and N_{Ca}, other lines show the contributions of the different terms specified in (9.68).

Using in the present case $f_s = 0.49$, or $\sqrt{f_s} = 0.70$, (9.70) gives: $B\langle \tilde{x}_c \rangle / L_D = 0.048 - 0.016 + 0.019 - 0.606 = -0.56$. In Fig. 9.11 the different terms constituting the complete front according to (9.68) are indicated, together with the above-calculated location of $\langle \tilde{x}_c \rangle$. The corresponding front in the exchanger ranging from $N_{2+} = 0.83$ to $N_{2+} = 1$, with its mean location at $\langle \tilde{x}_q \rangle = \langle \tilde{x}_c \rangle + L_D/B$, is also given. The final calibration in the column coordinate system obviously depends on B. Taking Q at 0.25 keq/m^3 and θ at 0.50, one finds $\Delta q = 0.17$ Q $= 0.042$ keq/m^3 as against $\theta \Delta c = 0.84$ C/2 $= 0.0084$ keq/m^3. The parameter B then equals about 5/6 and $L_D/B = 1.2 L_D$. With L_D about 2 cm, the front width as shown is thus about 10 cm; its actual location in the column is finally found by identifying the point located at $-0.56 + B = 0.27$ as $x = v^*t$.

While the above situation involved the counteraction of the favorable exchange isotherm on the spreading due to diffusion/dispersion, in the case of unfavorable exchange both the exchange isotherm and diffusion give rise to spreading of the front. Analytical solutions of (9.50) are not available and one must take recourse to numerical methods if the precise shapes of the fronts in solution and in the exchange phase are wanted (see the examples given in section 10.2.1). If only a superficial insight is needed into the expected shape of the concentration front, it should first be ascertained that the spreading action of the unfavorable exchange induces a front of which the (variable) slope at any value of c decreases proportionally to v^*t. In contrast the diffusion/dispersion effect, if acting on linear exchange, gives rise to $\bar{c} = (1/2)\,\mathrm{erfc}\,x/2\sqrt{D^*t}$ and thus the slope of the concentration front decreases proportionally to $\sqrt{L_D v^* t}$. Accordingly the diffusion/dispersion contribution in the case of unfavorable exchange becomes weakened with progressing time, the concentration gradients falling off more rapidly with time than would occur if the exchange isotherm were linear. Addition of the two effects as calculated separately, i.e. the spreading attributable to the non-linear isotherm in the absence of diffusion/dispersion and the diffusive spreading for a linear isotherm, respectively, would thus certainly constitute an overestimate of the combined effect. Such an estimate will often be satisfactory once considerable time has elapsed, as then the spreading due to unfavorable exchange becomes the dominant term.

Treating the spreading action of diffusion/dispersion on the solute front as a perturbation term to be added to the "base front" as found from the solution of the conservation equation in the absence of diffusion/dispersion, one may write formally:

$$x_{\bar{c}} = x_{\bar{c}}^0 + \delta x_{\bar{c}} \tag{9.71}$$

in which $x_{\bar{c}}^0$ is given by (9.11). Using a step feed at $t = 0$, this gives with the help of (9.12) (cf. also section A7.3.3 in part A):

$$x_{\bar{c}}^0 = \frac{J^V t}{\bar{q}'(\bar{c}) + \theta} = x_p \frac{R_D^\Delta + 1}{[\bar{q}'(\bar{c})R_D^\Delta + 1]} \tag{9.72}$$

in which R_D^Δ signifies the distribution ratio for the concentration increment covered by the step feed and $\bar{q}'(\bar{c})$ is the slope of the reduced exchange isotherm for the same range. The multiplier of x_p in (9.72) is thus a shape factor which may be plotted on the basis of a given exchange isotherm. It signifies the ratio of the integral storage capacity of the system (for the relevant concentration range) to the differential storage capacity and it will be indicated henceforth as $S_{\bar{c}}$.

A first estimate of the second term of (9.71) may be found by inverting (9.53) according to:

$$(x - v^*t) = 2\sqrt{D^*t}\,\mathrm{inverfc}\,(2\bar{c}) \tag{9.73}$$

where $(x - v^*t) \equiv (x - x_p)$ is the spreading caused by diffusion/dispersion when acting upon a step front, as arises in case of a linear isotherm. Replacing D^*t by $L_D v^* t = L_D x_p$, one may thus write:

$$\delta_0 x_{\bar{c}} = 2\sqrt{L_D x_p}\,\mathrm{inverfc}\,(2\bar{c}) \tag{9.74}$$

as valid for the step front. Using this as a first approximation to $\delta x_{\bar{c}}$, one finds:

$$x_{\bar{c}} \approx x_p S_{\bar{c}} + 2\sqrt{L_D x_p}\ \text{inverfc}\ (2\bar{c}) \tag{9.75}$$

which becomes exact for a linear isotherm as then $S_{\bar{c}} = 1$ over the entire concentration range.

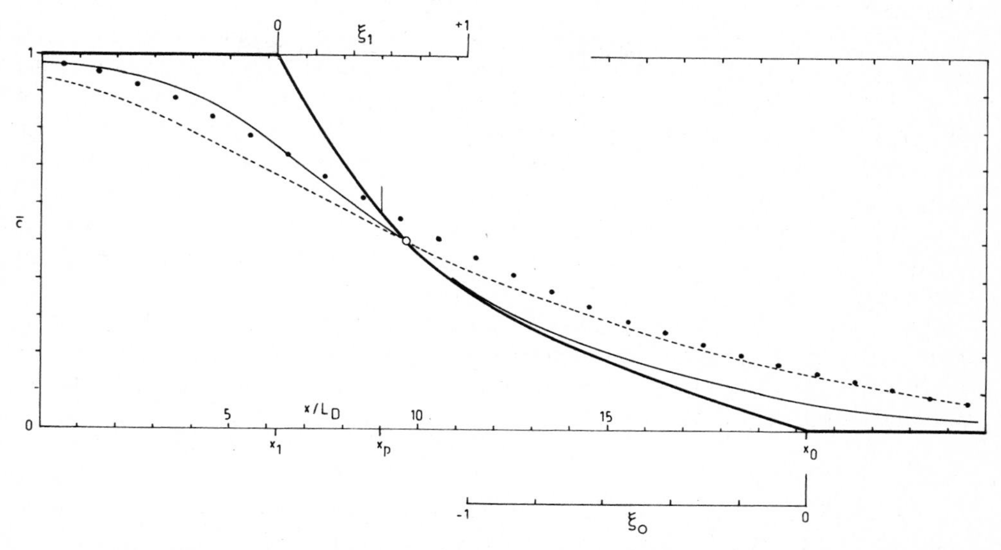

Fig. 9.12. Effect of diffusion/dispersion in the case of unfavorable exchange (arbitrary isotherm). Heavy line: "base front" formed in the absence of diffusion/dispersion. Heavy dots: exact location of the front as computed numerically. Dashed line: approximation following (9.75). Light-line: approximation following (9.77), using only the slope of the base front at its extremes $\bar{c} = 0$ and $\bar{c} = 1$ (cf. text). The corresponding calibrations for $\xi_0 = (x - x_p S_0)/2\sqrt{L_D x_p S_0}$ and $\xi_1 = (x - x_p S_1)/2\sqrt{L_D x_p S_1}$ are indicated.

As was pointed out above, (9.75) tends to overestimate the combined spreading attributable to a non-linear isotherm and diffusion/dispersion, because the concentration gradient in the front is always less negative than in the case of a front governed by linear exchange. Upon closer consideration, however, this statement must be modified somewhat. Thus the value of $D^* = L_D v^*$ appearing in (9.74) has been based on the integral capacity of the exchanger, while actually the effective value of D should refer to the differential capacity pertaining to the chosen value of $\bar{c}$. Translated into the parameters of (9.75) this means that the factor in front of the inverse error-function should be multiplied by $\sqrt{S_{\bar{c}}}$. This then leads to the interesting conclusion that, in the forward section of the concentration front, i.e. $x_{\bar{c}}$ for $S_{\bar{c}} > 1$, the second term contains compensating errors: the inverse error function is an overestimate (because of reduced slopes) while its multiplier is an underestimate. Accordingly (9.75) should be fairly dependable in this forward section. This is the most interesting part of the front as it determines the depth to which the solute has penetrated. Conversely both parts of the second term are overestimates of the spreading taking place at high values of $\bar{c}$.

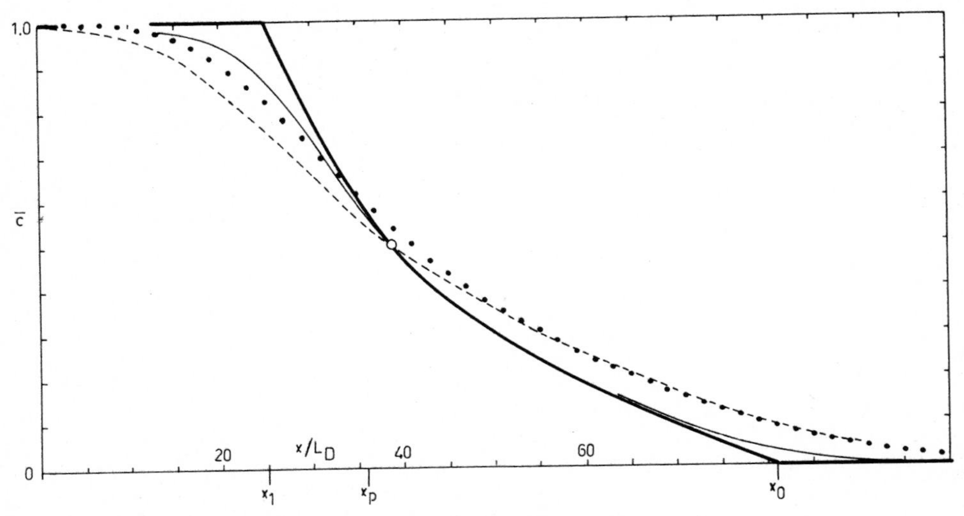

Fig. 9.13. Same as Fig. 9.12, after a four-fold increase of $x_p = v^*t$.

The above expectations are borne out when comparing the estimate according to (9.75) with the actual front computed numerically as is illustrated in Figs. 9.12, 9.13 and 9.14.

Figure 9.12 shows the "base front" (following an arbitrarily chosen concave isotherm) as a heavy line. Expressing the position coordinate for convenience in units of x/L_D, this front stretches at the chosen time from $x = 6.25L_D$ to $x = 20.25L_D$. The penetration depth of the combined front in the solid phase and solution is situated at $x_p = 9L_D$. The curve calculated according to (9.75) (dashed line) follows closely the computer calculated curve (separate dots) for $x > x_p$ and overestimates the spreading at low values of x. Figure 9.13 shows the same situation after a four-fold increase of the elapsed time, x_p being $36L_D$. In Fig. 9.14 the situation is pictured for unfavorable exchange following the Gapon equation, which leads to a convex exchange front. In this case the same numbers were used as those for the favorable exchange pictured in Fig. 9.11, i.e. $C = 0.02\,N$, $\theta = 0.5$, $Q = 0.25$ keq/m^3, $f_{+,f} = 0.84$ and $N_{+,f} = 0.17$, $f_{+,i} = 0.0$ and one finds that $x_p = J^V t/(\theta + \Delta q/\Delta c) = 0.324\,vt$. Selecting arbitrarily $x_p/L_D = 10$, the "base front" following $x^0_{f_+} = J^V t/\{\theta + q'(f_+)\}$ is plotted again as the heavy line, together with the front in the exchanger phase, N_+. Again the approximation following (9.75) (dashed line) is fairly close to the numerical solution (separate dots). The situation at $x_p = 40L_D$ is also shown (open circles)

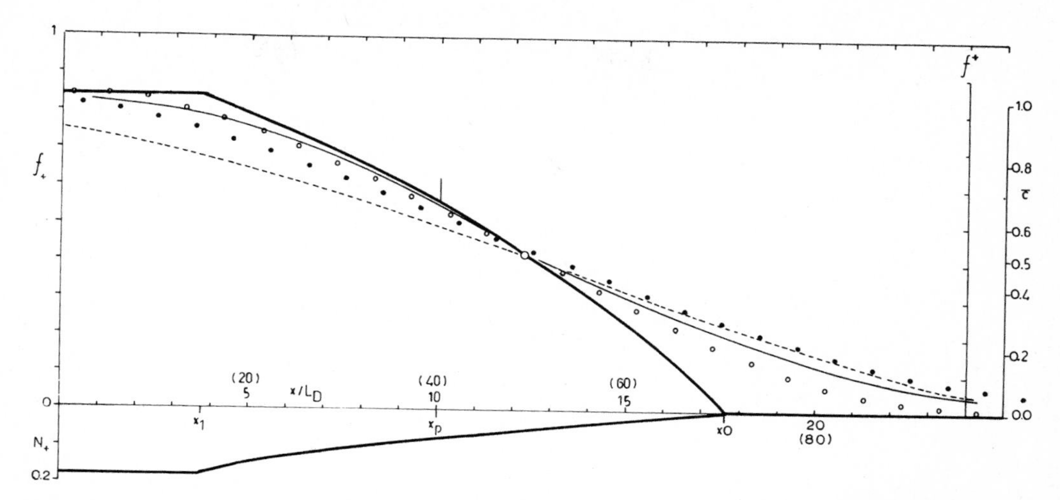

Fig. 9.14. Same as Fig. 9.12 but for an exchange isotherm following the Gapon equation in the range from 0 to 17% exchangeable Na-ions (indicated as N_+), i.e. the reverse of the situation shown in Fig. 9.11. Heavy dots and open circles indicate the exact location of the front for $x_p = 10L_D$ and $x_p = 40L_D$, respectively. Dashed line: approximation following (9.75) for $x_p = 10L_D$. Light line: approximation following (9.77) employing again the limiting slopes of the base curve, for $x_p = 10L_D$. At $x_p = 40L_D$ both approximations fall halfway between curves for $x_p = 10L_D$ and the base curve.

where the curve corresponding to the approximate solution should fall at one-half the distance from the base curve, i.e. half-way between the dashed line and the heavy line. Referring to section 9.3, the front pictured in Fig. 9.11 would follow the present one as a backward front if a block feed (of sufficient width) of a solution containing 0.0032 N NaCl and 0.0168 N $CaCl_2$ were brought into a column containing calcium exchanger and subsequently washed down with 0.02 N $CaCl_2$.

Further attempts to refine the estimation of $\delta x_{\bar{c}}$ are served by considering the spreading resulting from diffusion/dispersion of a sloping front. Solving, for this purpose, (9.51) for the conditions:

(C) infinite system stretching from $x \to -\infty$ to $x \to \infty$,

with $\bar{c} = 0$ for $x > w/2$ at $t = 0$, and for $t > 0$ at $x \to \infty$

$\bar{c} = 1$ for $x < -w/2$ at $t = 0$, and for $t > 0$ at $x \to -\infty$

$\bar{c} = \frac{1}{2} - x/w$ for $-w/2 < x < w/2$ at $t = 0$

one finds the well-known expression:

$$\bar{c} = \frac{\text{ierfc}\,(\xi_0 + \upsilon) - \text{ierfc}\,(\xi_0)}{2\upsilon} \tag{9.76}$$

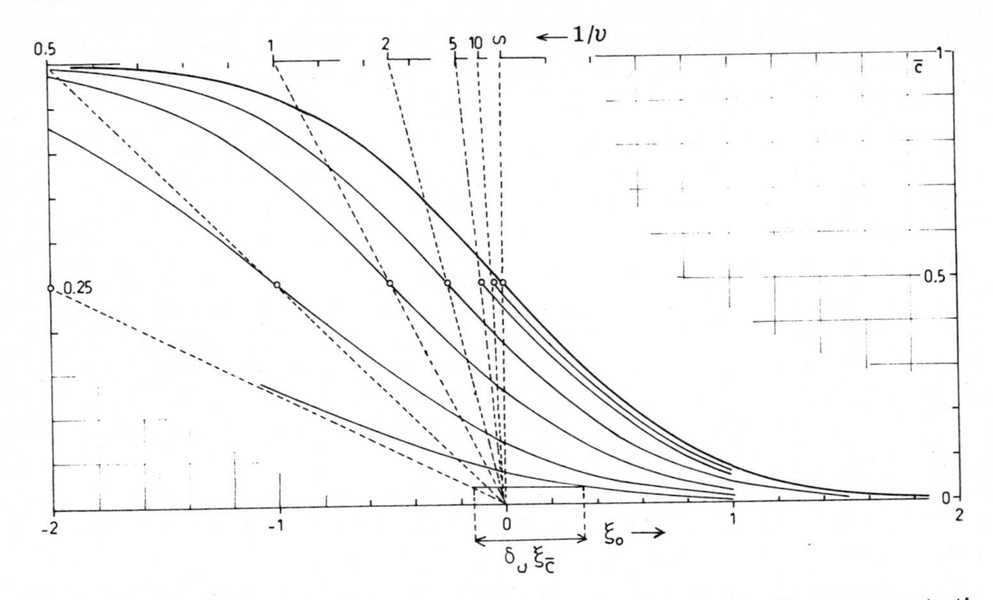

Fig. 9.15. Diffusion/dispersion front developing from an initially linear concentration distribution with finite slope (dashed lines), according to (9.76). The reduced slope of the initial distribution is indicated by the value of $1/\upsilon = -2\sqrt{D^*t}(d\bar{c}/dx)$.

in which $\upsilon = w/2\sqrt{D^*t}$ and $\xi_0 = (x - w/2 - v^*t)/2\sqrt{D^*t}$, i.e. the reduced moving coordinate with reference to the location of $\bar{c} = 0$ as found in the absence of diffusion/dispersion. In Fig. 9.15 this function is plotted for different values of the reduced initial slope of the front, $1/\upsilon$. It must be noted that these concentration fronts still refer to systems with a linear exchange isotherm, implying that the initial sloping front (cf. dotted lines in Fig. 9.15) would pass unchanged through the column in the absence of diffusion/dispersion. Such fronts could be formed by using an appropriate feed function $V_0(\bar{c})$.

Inverting the above relation one may write for the present system:

$$x_{\bar{c}} = x^0_{\bar{c}} + 2\sqrt{L_D x^0_{\bar{c}}}\,\delta_\upsilon \xi_{\bar{c}} \tag{9.77}$$

in which $x^0_{\bar{c}}$ is again the position coordinate in the absence of diffusion/dispersion, while $\delta_\upsilon \xi_{\bar{c}} \equiv (\xi_0 - \xi^0_0)$ may be read from Fig. 9.15 as indicated. To give an insight into the trajectory of $\delta_\upsilon \xi$ this parameter has been plotted against $\bar{c}$ for different values of $1/\upsilon$ in Fig. 9.16. Obviously the line indicated by $1/\upsilon = \infty$ simply equals $\delta_0 \xi = \text{inverfc}\,(2\bar{c})$ as used before, while $\delta_\upsilon \xi$ decreases with decreasing slope.

Whereas (9.77) is exact for linear exchange acting on a front with a finite initial slope, one might use the same equation as an approximation for non-linear exchange acting on a step feed. This could be accomplished graphically by first constructing the "base front" corresponding to $x^0_{\bar{c}} = x_p S_{\bar{c}}$. By drawing the tangents to this curve for different values of $\bar{c}$ one may read $w_{\bar{c}}$, which yields $\upsilon_{\bar{c}}$ as $w_{\bar{c}}/2\sqrt{L_D x_p S_{\bar{c}}}$. From Fig. 9.16 one may then determine $\delta_\upsilon \xi_{\bar{c}}$ which is entered into (9.77) to give the approximation for $x_{\bar{c}}$. In contrast to (9.75) this must lead to an underestimate of the spreading effect as the value of $1/\upsilon_{\bar{c}}$ used is a minimum value reached only at the chosen time. In practice it might be somewhat better to employ a mean value of the slope of the "base front", e.g. by using $w_{\bar{c}}/2$, i.e. twice the value of $1/\upsilon_{\bar{c}}$ indicated above. However, the fairly elaborate procedure described above will often hardly be warranted in view of the fair approximation obtained with (9.75). Using a mean slope (i.e. the chord) for the entire front might be worth considering. In Figs. 9.12—9.14 the outcome of a similar approach is plotted. As in this case it was attempted to investigate the behavior of convex and concave "base fronts", the construction was performed by using two extreme values of υ, the one corresponding to $(\partial\bar{c}/\partial x)$ at $\bar{c} = 0$ for the range $\bar{c} = 0$ to $\bar{c} = 0.5$ and the other corresponding to $(\partial\bar{c}/\partial x)$ at $\bar{c} = 1$ for the upper half of the curve. As the *minimum* values of $1/\upsilon$, were used, viz. those reached *at* the chosen time, the corresponding spread is indeed an underestimate.

Summarizing the discussion of the present section it appears that, in view of the many uncertainties as to the precise value of L_D in soil systems, the effect of diffusion/dispersion on cation exchange chromatography may be satisfactorily accounted for in most cases by applying graphically suitable corrections to the "base curve" given by (9.72). Using for this purpose the correction equation $\delta_0 x_{\bar{c}} = 2\sqrt{L_D\, x_p}$ inverfc $(2\bar{c})$, one finds the correct curve in the case of a linear isotherm, a slight overestimate in the case of unfavorable exchange, and an overestimate which increases rapidly with x_p for favorable exchange, as then the spreading effect is delimited by its steady state value as constructed in Figs. 9.10 and 9.11.

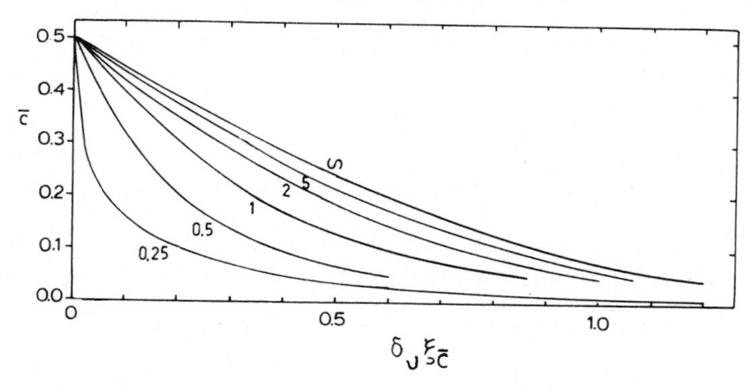

Fig. 9.16. Spreading effect of diffusion/dispersion upon an initially linear concentration distribution with finite slope $1/\upsilon$, plotted as $\delta_\upsilon\xi(\bar{c})$ for different values of $1/\upsilon$ (cf. Fig. 9.15).

9.6. NON-EQUILIBRIUM CHROMATOGRAPHY

In the previous sections the discussion was guided by the presumed existence of local equilibrium between exchanger and the (moving) liquid phase. Actually this standpoint was mitigated somewhat in the course of the attempt to obtain an estimate of the dispersion coefficient, when near-equilibrium with a stagnant phase, usually containing the exchange sites, was accepted as a satisfactory condition. Accordingly, in all cases treated till now the composition of moving liquid phase at any point supposedly implied the composition of the exchanger phase via the exchange isotherm.

Whether or not local equilibrium or near-equilibrium indeed exists will obviously depend on the relative magnitude of the rate of exchange between moving solution and exchanger, and the rate of change of the concentration in the solution passing by the exchanger. The latter is proportional to J^V, while the exchange reaction will, at least in the initial stages, follow a first-order rate law characterized by some rate coefficient, k, specified in sec^{-1}. Accordingly one may distinguish again a characteristic length parameter, equal to J^V/k, which determines to what degree local equilibrium is reached during the flow process.

Admitting beforehand that the addition of even a simple rate process to the other factors determining the shape of the exchange front leads to equations which must be solved numerically (see chapter 10), it will be attempted here to make some estimates on the basis of a simplified system. Following the procedure used in the previous section when discussing the effect of a diffusion/dispersion term, here again the step front arising in the case of a linear isotherm in the absence of diffusion/dispersion will be the point of departure. It may also be ascertained that in aggregated soils the accessibility of exchange or adsorption sites via a diffusion process is probably always the rate limiting step for the reaction. In such a case one may write:

$$\frac{\partial(q_a + \theta_a c_a)}{\partial t} = k_a \theta_m (c_m - c_a) \tag{9.78}$$

in which the subscript 'a' has been used to indicate a stagnant phase, presumably located inside an aggregate, while 'm' refers to the mobile phase. The rate constant k_a has been defined here somewhat arbitrarily by including θ_m in the RHS of (9.78). Assuming now that *inside* the aggregate equilibrium between adsorption sites and solution is reached instantaneously, dq_a is related to dc_a via the exchange isotherm by:

$$dq_a = q' dc_a \tag{9.79}$$

where q' is taken equal to the constant $\Delta q/\Delta c$ for the present case involving a linear isotherm. Introducing a distribution ratio between aggregate and mobile phase as:

$$R_D^{am} \equiv (\Delta q/\Delta c + \theta_a)/\theta_m \tag{9.80}$$

one may replace (9.78) by:

$$\frac{\partial c_a}{\partial t} = (k_a/R_D^{am})(c_m - c_a) \tag{9.81}$$

Next the conservation equation (9.4) may be rewritten for the present case, which gives, in the absence of a significant longitudinal diffusion/dispersion:

$$\theta_m R_D^{am} \frac{\partial c_a}{\partial t} + \theta_m \frac{\partial c_m}{\partial t} + J^V \frac{\partial c_m}{\partial x} = 0 \tag{9.82}$$

As appears from (9.82), all adsorption sites have been assigned to the stagnant phase, i.e. inside the aggregate. Although there is no necessity to do so, the present purpose of exposing the effect of a finite rate of equilibration is served by this simplification. In the following chapter (section 10.3.2) the effect of instantaneous equilibrium between the mobile phase and part of the adsorption sites is discussed. Recently, Cameron and Klute (1977) supplied evidence for the presence of such sites in soil.

The last two terms of (9.82) may now be combined to:

$$\theta_m \left(\frac{\partial c_m}{\partial t}\right)_x + J^V \left(\frac{\partial c_m}{\partial x}\right)_t \equiv J^V \left(\frac{\partial c_m}{\partial x}\right)_{t - x\theta_m/J^V}$$

whereas:

$$\left(\frac{\partial c_a}{\partial t}\right)_x \equiv \left[\frac{\partial c_a}{\partial (t - x\theta_m/J^V)}\right]_x$$

Introducing now the scaled variables:

$$\bar{c} = (c - c_i)/(c_f - c_i)$$

$$\hat{x} = k_a\ x\theta_m/J^V$$

$$\hat{t} = (k_a/R_D^{am})(t - x\theta_m/J^V)$$

equations (9.81) and (9.82) may be written as:

$$\left(\frac{\partial \bar{c}_a}{\partial \hat{t}}\right)_{\hat{x}} = (\bar{c}_m - \bar{c}_a) \tag{9.83}$$

$$\left(\frac{\partial \bar{c}_a}{\partial \hat{t}}\right)_{\hat{x}} = -\left(\frac{\partial \bar{c}_m}{\partial \hat{x}}\right)_{\hat{t}} \tag{9.84}$$

This set of partial differential equations is identical with the one describing cross-flow heat exchange. Using as boundary conditions $\bar{c}_a = 0$ at $\hat{t} = 0$ and $\bar{c}_m = 1$ at $\hat{x} = 0$, its solution has been given repeatedly by various workers in different fields. An early reference is Nusselt (1911) which provides tables for low values of $\hat{t}$ and $\hat{x}$ and a perspective drawing of $\bar{c}(\hat{t}, \hat{x})$. This solution is commonly expressed in terms of a function J(x, y) sometimes referred to as the Goldstein J-function because of the extensive discussion in Goldstein (1953). This function comprises an integral of a zero-order modified Bessel function, I_0, and has been tabulated by Brinkley (unpublished, as indicated in Brinkley et al., 1952)*.

The application of the above treatment to ion exchange in columns was introduced by Thomas (1944) and was followed up by Hiester and Vermeulen (1952). Actually, Thomas (1944) referred especially to the more involved case of second-order reversible kinetics. Granted that at high flow velocities around small exchanger particles the actual exchange reaction may be rate limiting, for the present case of relatively low flow velocities around particles of occasionally considerable size the second-order reaction seems of rather remote significance. Hiester and Vermeulen (1952) covered both the second-order and first-order reaction extensively, pointing out specifically the importance of diffusion control of the reaction. In contrast Klinkenberg (1948) covered solely the first-order case and produced a nomogram allowing an easy evaluation of J(x, y) once x and y exceed the value of 2, based on a rather convenient approximation equation. In Klinkenberg (1954) a comparison is made between the different approximations for J(x, y).

In short, the solution of (9.83) and (9.84) for the stated boundary conditions reads:

$$\overline{c}_m = J(\hat{x}, \hat{t})$$

$$= 1 - \int_0^{\hat{x}} \{\exp(-\hat{t} - \hat{x}')\} I_0(2\sqrt{\hat{t}\hat{x}'}) d\hat{x}' \qquad (9.85)$$

When $\hat{x}$ and $\hat{t}$ exceed the value of 2, a very good approximation is given by (cf. Klinkenberg, 1954):

$$\overline{c}_m \approx \tfrac{1}{2} + \tfrac{1}{2} \operatorname{erf}\{\sqrt{\hat{t}}(1 - \sqrt{\hat{x}/\hat{t}}) + (1 + \sqrt{\hat{t}/\hat{x}})/8\sqrt{\hat{t}}\} \qquad (9.85a)$$

The concentration inside the aggregate — and thus for the present linear isotherm also the value of the amount adsorbed, $\overline{q}_a$ — is given by:

$$\overline{q}_a \equiv \overline{c}_a = 1 - J(\hat{t}, \hat{x}) \qquad (9.86)$$

$$\approx \tfrac{1}{2} + \tfrac{1}{2} \operatorname{erf}\{\sqrt{\hat{t}}(1 - \sqrt{\hat{x}/\hat{t}}) - (1 + \sqrt{\hat{t}/\hat{x}})/8\sqrt{\hat{t}}\} \qquad (9.86a)$$

Recalling now the approximation used before in connection with the derivation of (9.46) and (9.48) — expressing the effect of a stagnant phase in terms of an equivalent length parameter L_r assuming near-equilibrium between the two phases — it may be investigated under what conditions such an approximation is warranted. For this purpose one must first express the rate coefficient k_r in the same parameters as used to calculate the flux into the stagnant phase leading to (9.42). For spherical aggregates this gives, in the absence of adsorption (cf. Crank, 1964):

* Computer program of the J-function available from the senior author of chapter 10.

$$\frac{\partial c_a}{\partial t} = \frac{15D}{R_a^2}(c_m - c_a) \tag{9.87}$$

which specifies the total flux into a unit volume of the stagnant phase. As the soil column contains only a fraction θ_a of the stagnant phase, the diffusion flux into the stagnant phase residing in a unit volume of column equals $\theta_a(\partial c_a/\partial t)$ in the absence of the adsorption sites. As this flux is solely determined by the concentration difference and the diffusion coefficient in the liquid phase (corrected for the tortuosity inside the aggregate, cf. (9.42)), it remains the same if adsorption sites are present inside the aggregate, so in general:

$$\frac{\partial(q_a + \theta_a c_a)}{\partial t} \equiv (\theta_a + q')\frac{\partial c_a}{\partial t}$$

$$= \frac{15D}{R_a^2}(c_m - c_a)\theta_a \tag{9.88}$$

It is pointed out in passing that (9.88) implies that $\partial c_a/\partial t$ still follows (9.87) provided D is replaced by $\theta_a D/(\theta_a + q')$, cf. the discussion in relation to (9.47).

Comparing now (9.88) and (9.78) one finds the link between the rate coefficient k_a and the rate of diffusion into a spherical stagnant phase as:

$$k_a = \left(\frac{\theta_a}{\theta_m}\right)\frac{15D}{R_a^2} \tag{9.89}$$

Returning to the situation in the absence of adsorption sites, one may use (9.48), with $A = 1$, to express L_r in terms of k_a, which gives:

$$L_r = \left(\frac{\theta_a}{\theta}\right)^2 \frac{J^V}{\theta_m k_a} \tag{9.90}$$

containing the ratio J^V/k_a indicated earlier. As in the absence of adsorption sites $J^V t/\theta = v^*t \equiv x_p$, the above relation may be used to express the solutions pertaining to near-equilibrium as given by (9.53), (9.54) and (9.56) in $k_a t$. Thus one finds in this case:

$$\xi \equiv (x - v^*t)/2\sqrt{D^*t}$$

$$= (x - v^*t)/2\sqrt{L_r v^*t}$$

$$= (x/v^*t - 1)\sqrt{\theta\theta_m k_a t}/2\theta_a \tag{9.91}$$

$$\alpha_r = \sqrt{v^*t/L_r} = \sqrt{\theta\theta_m k_a t}/\theta_a \tag{9.92}$$

Similarly the variables $\hat{x}$ and $\hat{t}$ are in the present case:

$$\hat{x} = (\theta_m/\theta)\, k_a t\, x/v^*t \tag{9.93}$$

$$\hat{t} = (\theta_m/\theta_a) k_a t(1 - x\theta_m/v^*t\theta) \tag{9.94}$$

With the help of the above one may follow the approach towards equilibrium between stagnant and mobile phases, as is demonstrated in Fig. 9.17. Selecting again $\theta_a = \theta_m = 0.25$, the distribution of $\bar{c}_a$ and $\bar{c}_m$ in a column has been plotted as a function of x/v*t according to (9.85) and (9.86) for different values of α_r, here equal to $\sqrt{2k_a t}$.

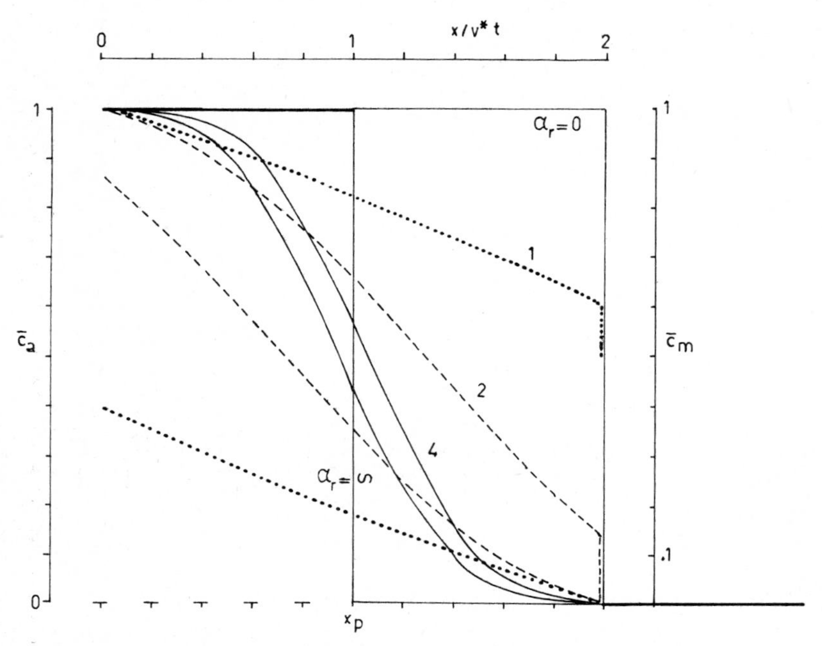

Fig. 9.17. Concentration fronts in a mobile phase ($\bar{c}_m$) and a stagnant phase ($\bar{c}_a$), as influenced by the rate of equilibrium relative to the rate of propagation of the front, α_r (cf. (9.85) and (9.86)0. For the present case, with $\theta_a = \theta_m = \frac{1}{2}\theta = 0.25$, the parameter $\alpha_r \equiv \sqrt{x_p/L_r}$ equals $\sqrt{x_p k_a/J^V}$ cf. (9.90). With $\alpha_r = 0$ the stagnant phase remains free from solute and $\bar{c}_m$ shows a step front at $x = J^V t/\theta_m = 2v^*t$. In contrast $\alpha_r \to \infty$ gives instantaneous equilibration leading to a step front for both $\bar{c}_a$ and $\bar{c}_m$ at $x = v^*t$.

Figure 9.17 shows the full transition of the concentration front $\bar{c}_m$ from $\alpha_r = 0$ — implying $D^* = 0$, i.e. the stagnant phase does not participate at all and a step front is formed at $x = J^V/\theta_m = 2v^*t$ — to $\alpha_r \to \infty$ (instantaneous equilibration giving a step front at $x = v^*t$). At intermediate values of α_r the fronts in the mobile and stagnant phases remain anti-symmetric, gradually approaching each other for increasing α_r. Now the near-equilibrium approach used in conjunction with the construction of L_r in Fig. 9.8 referred to the mean concentration, $\langle c \rangle$, which was presumably following a complementary error function. One may thus investigate the validity of this approach by comparing the trajectory of $\langle c \rangle$ according to (9.85) and (9.86) with the one predicted by (9.56). This has been done in Fig. 9.18 for $\alpha_r = 2$.

As the chosen boundary conditions for the solution in terms of $J(\hat{x}, \hat{t})$ imply a constant influx at the entrance boundary, the particular solution according to (9.56) was chosen. As was shown in Fig. 9.9, however, this solution differs only slightly from (9.53).

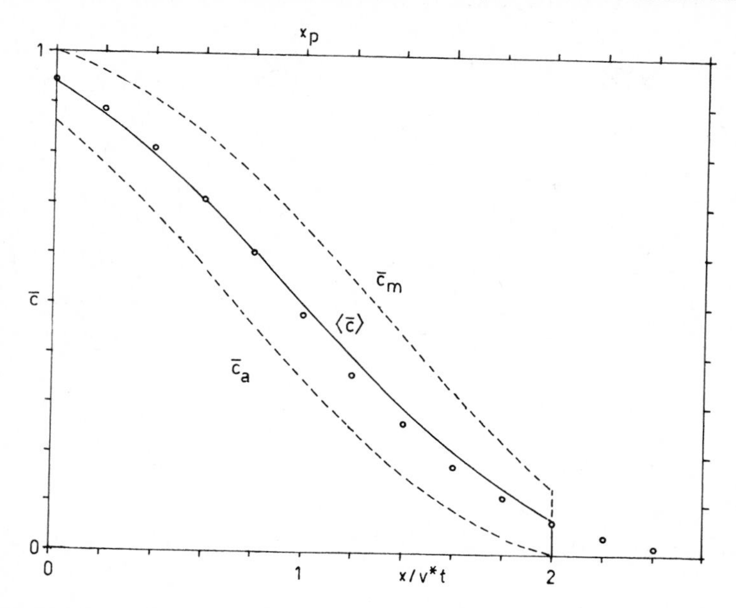

Fig. 9.18. Comparison between the mean value $\langle\bar{c}\rangle$ of the separate fronts of $\bar{c}_a$ and $\bar{c}_m$ as shown in Fig. 9.17 for $\alpha_r = 2$ (dashed lines) and the approximation obtained by assuming near-equilibrium between the phases (open circles, following the curve $\underline{B}$ with $\alpha = 2$ as given in Fig. 9.9).

The agreement between the two is remarkably good, taking into account that the cut-off at $x/v^*t = 2$ for the non-equilibrium case results from the boundary condition used in conjunction with (9.85) and (9.86), where $\bar{c}_a$ was fixed at zero at that point. In contrast (9.56) allowed for "leakage" across the boundary at $x/v^*t = 2$.

The above exposure warrants the conclusion that, in the absence of adsorption sites, a satisfactory condition for the validity of the near-equilibrium approach is given by $\alpha_r > 2$. This implies that the concentration front calculated with (9.53) on the basis of L_r as read from Fig. 9.8 should be a satisfactory approximation once the penetration depth, x_p, is at least four times the value of L_r (cf. footnote on Table 9.1, p. 335).

In order to express the above results in terms of permissible flow rates the following example may serve. Using aggregates of 2 cm diameter one finds, with D equal to 10^{-9} m^2/sec, the value of k_a as $15 \cdot 10^{-5}$ sec^{-1}. With $J^V = 1.5 \cdot 10^{-5}$ m/sec one finds a penetration depth, v^*t, at 0.4 m with $t = 1.33 \cdot 10^4$ sec ≈ 3.7 h. In that case $k_a t = 2$ and $\alpha = 2$, so $L_r = 0.1$ m, as may also be verified from Fig. 9.8. Thus the substantial flow rate of 1.5 m per day would be about the limiting value for 2-cm aggregates if one wants to apply the near-equilibrium solution for a penetration depth of about 0.5 m.

Coats and Smith (1964) presented a thorough analysis of dispersion resulting from a stagnant phase in the absence of adsorption, in which they arrived at the connection between k_a and L_r in a reverse manner. The pair of curves following (9.85) and (9.86) must coalesce for high values of $k_a t$ to a step front at the same position as follows from (9.53) with very low values of $D^* = L_r v^* t$. In that case both $\hat{t}$ and $\hat{x}$ reach very high values and the approximation (9.85a) may be further simplified by dropping the second term of the argument to give:

$$\bar{c}_m \approx \tfrac{1}{2} \operatorname{erfc} [\sqrt{\hat{x}} \{1 - \sqrt{1 + (\hat{t} - \hat{x})/\hat{x}}\}] \tag{9.95}$$

As in the neighborhood of $x = v^* t$, $\hat{t}$ approaches $\hat{x}$, expansion gives:

$$\bar{c}_m \approx \tfrac{1}{2} \operatorname{erfc} \{(\hat{x} - \hat{t})/2\sqrt{\hat{x}}\}$$

$$= \tfrac{1}{2} \operatorname{erfc} \{k_a(\theta_m/\theta_a)(\theta x/j^V - t)/2\sqrt{k_a \theta_m x/J^V}\} \tag{9.96}$$

Comparing this equation with (9.53), which in the absence of adsorption sites reads:

$$\bar{c} = \tfrac{1}{2} \operatorname{erfc} \{(x - J^V t/\theta)/2\sqrt{L_r J^V t/\theta}\} \tag{9.97}$$

one finds immediately:

$$L_r = \frac{\theta_a}{\theta} \frac{1}{k_a} \frac{\theta_a}{\theta_m} \frac{x}{t} \tag{9.98}$$

In the neighborhood of $x = J^V t/\theta = x_p$, equation (9.98) becomes identical with (9.90), indicating that the coalescence of $\bar{c}_a$ and $\bar{c}_m$ to a step front at $x = x_p$ via (9.53) requires indeed that L_r follows (9.90).

Generalizing this approach to systems involving adsorption sites inside the aggregate, (9.96) becomes:

$$\bar{c}_m \approx \tfrac{1}{2} \operatorname{erfc} \left[\frac{k_a\{\theta_m(R_D^{am} + 1)x/J^V - t\}}{2R_D^{am}\sqrt{k_a \theta_m x/J^V}} \right] \tag{9.99}$$

Comparing with (9.53), with $D^* t = L_r v^* t$ and $\theta_m (R_D^{am} + 1)/J^V = 1/v^*$, this gives:

$$L_r = \frac{(R_D^{am})^2}{(R_D^{am} + 1)} \frac{1}{k_a} \frac{x}{t} \tag{9.100}$$

Matching of the two curves at $x = x_p = J^V t/\theta(R_D + 1)$ then requires:

$$L_r = A^2 \left(\frac{\theta_a}{\theta}\right)^2 \frac{J^V}{\theta_m k_a} \tag{9.101}$$

in which:

$$A \equiv (1 + \Delta q/\theta_a \Delta c)/(1 + \Delta q/\theta \Delta c) \tag{9.102}$$

and where all exchange sites have again been assigned to the intra-aggregate space. In the denominator of (9.102) one recognizes $(R_D + 1) \equiv R_f$ (see chapter 10), the retardation factor of the formation for the passing front. The numerator is a similar retardation factor, but in this case it refers to the interior composition of the stagnant phase. As $\theta > \theta_a$ one finds $A > 1$; it may be visualized as a correction factor which allows for the fact that the exchange sites must be supplied with ions via the interior of the aggregates. As long as θ_a constitutes 50% or more of the total liquid content, which would seem probable in the case of soils with a high exchange capacity, the factor A^2 is rather insensitive towards the actual value of R_D, as it varies between unity for $\Delta q/\Delta c = 0$ to $(\theta/\theta_a)^2$ for $\Delta q/\Delta c \to \infty$. The fortunate conclusion is then that the estimate for L_r as obtained from Fig. 9.8 remains roughly valid, independent of the exchange capacity of the soil.

In this respect it may be pointed out that in practice the ratio θ_a/θ for a particular soil is often neither constant nor known as a function of the moisture status. As furthermore $\Delta q/\Delta c$ varies with electrolyte content of the system it might be best to take $A^2(\theta_a/\theta)$ in (9.48) equal to unity for a general guess, the uncertainty then amounting to the factor $\theta_a/\theta \to \theta/\theta_a$. If Fig. 9.8 is used as a guide, this implies that L_r, i.e. the ascending branch of L_D, should be taken at twice the value as plotted.

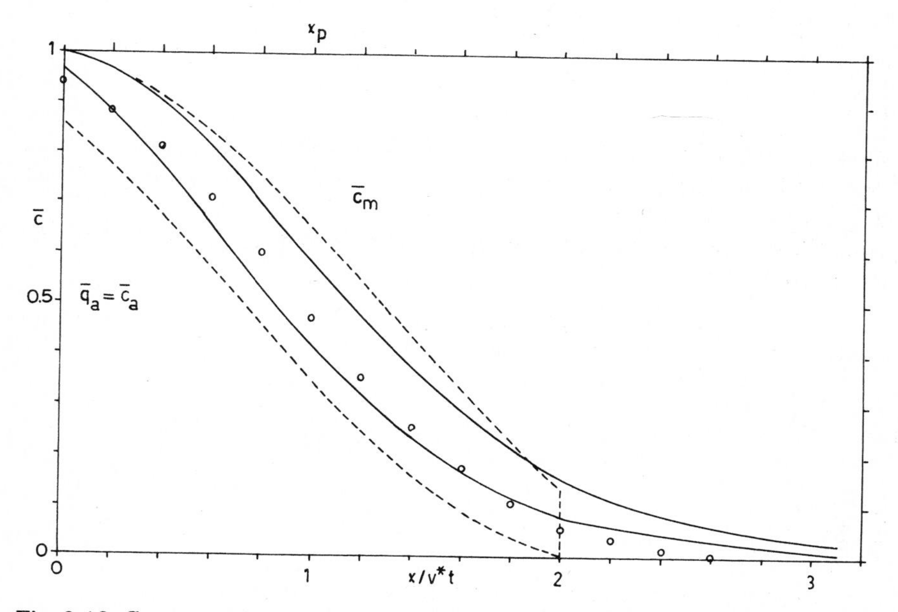

Fig. 9.19. Concentration fronts in stagnant and mobile phases in the case of a finite rate of equilibration. Dashed lines: $\alpha_r = 2$, $R_D^{am} = \theta_a/\theta_m$, i.e. no adsorption sites in the stagnant phase. Solid lines: $\alpha_r = 2$, $R_D^{am} = 9$, i.e. adsorption sites are present in the stagnant phase (obeying a linear isotherm). Open circles: average concentration in the column assuming near-equilibrium (same as in Fig. 9.18).

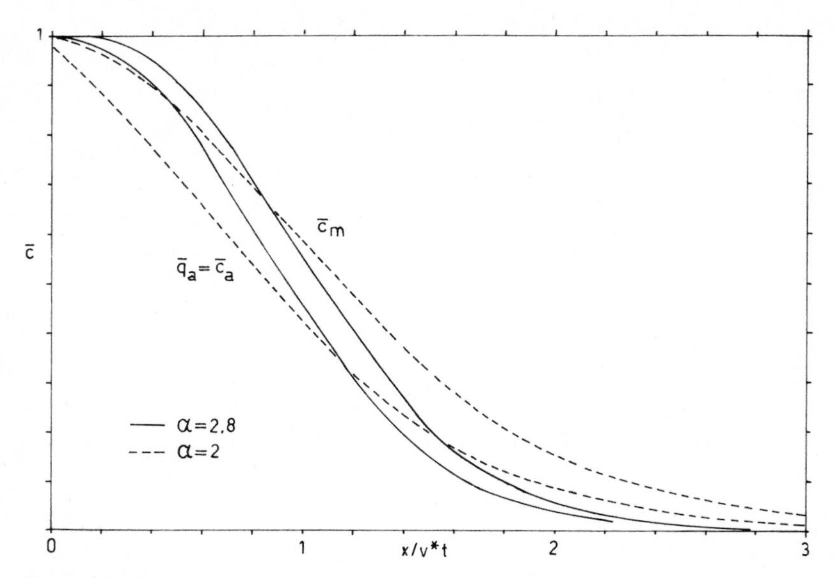

Fig. 9.20. Concentration fronts in stagnant and mobile phases in the case of a finite rate of equilibration for two values of α_r (cf. Fig. 9.17). The stagnant phase contains adsorption sites with a capacity $\Delta q/\Delta c = 4$, or $R_D = 8$ and $R_D^{am} = 19$.

Accepting the correction factor A^2 for the calculation of L_r, it may now be verified whether the approach from non-equilibrium towards near-equilibrium follows the same pattern as pictured in Fig. 9.17. For this purpose the system chosen before, i.e. $\theta_a = \theta_m$, was modified by introducing different values of $\Delta q/\Delta c$ equal to 2, 4.5 and 12, respectively. In Table 9.1 the corresponding values of the relevant parameters are given for $\alpha_r = 2$, i.e. $x_p = 4L_r$, which was found to be a satisfactory condition for near-equilibrium in the previous case. The results as given in Figs. 9.19 and 9.20 show that this condition remains satisfactory also in the presence of adsorption sites. It may be noticed that $\bar{q}_a$ and $\bar{c}_m$ actually approach each other upon increasing the amount of charged sites while keeping x, and thus L_r, constant. This is caused by the effect of a lower value of J^V, together with the increased total volume of solution, $J^V t$, necessary to reach the same penetration depth, resulting in an increase in contact time at, for example, $x = x_p$, which is proportional to R_D^{am} to a power larger than 1 (cf. $\hat{t}$ and $\hat{x}$ at $x = x_p$). At the same time the average concentration in the column is now found as a weighted mean, i.e. $\langle \bar{c} \rangle = (R_D^{am} \bar{c}_a + \bar{c}_m)/(R_D^{am} + 1)$. In Fig. 9.19, showing $\bar{c}_a$ and $\bar{c}_m$ for $\Delta q/\Delta c = 2$, or $R_D^{am} = 9$, this weighted mean value (not shown) should thus be compared with the near-equilibrium solution indicated by open circles. Especially because the chosen boundary condition, i.e. $\bar{q}_a = \bar{c}_a = 0$ at $x = (R_D^{am} + 1)v^*t$, implies that now the solute has moved beyond $x = 2v^*t$, the overall agreement between the exact calculation and the near-equilibrium approximation is already quite good at $\alpha_r = 2$.

TABLE 9.1

Distribution ratios R_D and R_D^{am} and the correction factor A as a function of the adsorber capacity, $\Delta q/\Delta c$, for $\theta_a = \theta_m = 0.25$

$\Delta q/\Delta c$	R_D	R_D^{am}	A	Corresponding values of $k_a t$ and ranges of $\hat{x}$ and $\hat{t}$ for $\alpha_r \equiv v^* t/\sqrt{D^* t} = 2$, or $v^* t/L_r = 4$*						Corresponding values of t and J^V for $k_a = 15 \cdot 10^{-5}$ sec^{-1} (2-cm aggregates) and $v^* t = 0.4$m		
					at $x = 0$		at $x = x_p$	at $x = J^V t/\theta_m$				
				$k_a t$	$\hat{x}$	$\hat{t}$	$\hat{x} = \hat{t}$	$\hat{x}$	$\hat{t}$	t (10^5 sec)	J^V ($m/10^5$ sec)	$J^V t$ (m)
0	0	1	1	2	0	2	1	2	0	0.13	1.5	0.2
2	4	9	9/5	32.4	0	3.6	3.24	32.4	0	2.16	0.46	1.0
4.5	9	19	19/10	72.2	0	3.8	3.61	72.2	0	4.81	0.42	2.0
12	24	49	49/25	192.1	0	3.92	3.84	192.1	0	12.81	0.39	5.0

* Using (9.100) and (9.102) one finds: $\hat{x} = A^2(\theta_a/\theta)^2 \alpha_r^2 (x/x_p)$ and $\hat{t} = A(\theta_a/\theta)\alpha_r^2 |1 - (x/x_{pc})|$, where $x_{pc} \equiv J^V t/\theta_m$ indicates the penetration depth of the carrier liquid. The bracketed expression reverts to $A(\theta_a/\theta)$ for $x = x_p$, then given $\hat{t} = \hat{x}$. For given values of θ_a/θ and $A > 1$, $\hat{x}$ and $\hat{t}$ thus become proportional to α_r^2; $\alpha_r > 2$ appears to be a satisfactory condition for the use of the near-equilibrium approach discussed before.

As is already obvious from Table 9.1, a further increase of $\Delta q/\Delta c$ has hardly any influence as $\hat{t}$ and $\hat{x}$ at $x = x_p$ rapidly approach their limiting value, equal to 4, for $\theta/\theta_a = 2$. Accordingly the curves given in Fig. 9.20 for $R_D^{am} = 19$, or $R_D = 9$, do not differ significantly from those given in Fig. 9.19 for $R_D^{am} = 9$, or $R_D = 4$. The curves for $\Delta q/\Delta c = 12$, or $R_D = 24$, are indistinguishable from the previous ones within the scale of plotting. Finally the effect of an increase of α_r to $2\sqrt{2}$, corresponding to $v^*t/L_r = 8$, is also shown in Fig. 9.20.

The above exposition is incomplete, as neither the simultaneous presence of longitudinal diffusion and convective dispersion nor non-linear adsorption isotherms were taken into account. Actually the well-known publication of Lapidus and Amundson (1952) presented a solution which also took longitudinal diffusion and/or dispersion into account. Their solution, however, is given in the form of double integrals of tabulated functions, which in practice require computer assistance. As, on the other hand, the exchange isotherm was still assumed to be linear, this analytical solution is, in practice, surpassed by direct application of a computer to the numerical solution of the differential equation itself, employing any suitable exchange isotherm.

It should also be pointed out that, in the above treatment, no specific attention was given to the shape of the break-through curve. This curve refers solely to $\bar{c}_m$ as a function of $\hat{t}$ for a given value of $\hat{x}$ and is more sensitive to lack of equilibrium in the sense that experimentally an early break-through is more easily detected. At the same time the corresponding "tailing" effect is more impressive in a break-through curve than if one plots an estimate of the front in the column. Actually most of the literature concerned with non-equilibrium behavior has been directed particularly at the shape of the break-through curve (cf. e.g. Thomas, 1944; Hiester and Vermeulen, 1952; Hashimoto et al., 1964; Passioura and Rose, 1971; Van Genuchten et al., 1974, 1977). In the following chapter particular attention is given to the relative importance of non-equilibrium effects in more involved situations, using computer-simulated solutions of the conservation equation.

9.7. EFFECT OF A (SIMPLE) PRODUCTION TERM

A second issue treated in the literature is non-equilibrium involving a production term. Nitrification of NH_4^+-ions and denitrification of NO_3^--ions are particular examples (Cho, 1971, Misra et al., 1974). Also, pesticides when present in soil will usually follow a conservation equation including a (negative) production term accounting for microbiological decay. Numerical solutions for such problems are becoming available, but analytical solutions may still be found if linear adsorption isotherms and first-order reaction kinetics are again assumed. Actually the situation is quite similar to the one treated previously and as such the solution of Lapidus and Amundson (1952) implies this case also.

Rewriting (9.23) with a linear production term one finds:

$$\frac{\partial(q+\theta c)}{\partial t} + J^V \frac{\partial c}{\partial x} - \theta D \frac{\partial^2 c}{\partial x^2} + k\theta c = 0 \qquad (9.103)$$

where a positive value of k indicates a negative production, as in case of breakdown or, for example, radioactive decay. If the adsorption isotherm is linear this gives:

$$\frac{\partial c}{\partial t} + v^* \frac{\partial c}{\partial x} - D^* \frac{\partial^2 c}{\partial x^2} + k^* c = 0 \qquad (9.104)$$

in which $k^* = k\theta/(\theta + q') = \text{constant}$.

In contrast to the previous case of first-order reversible kinetics this equation may be solved directly via Laplace transformation, at least for the boundary condition corresponding to a constant concentration at the column entrance (cf. Cho, 1971; Misra et al., 1974). In addition, the condition of a constant initial concentration is, because the decay usually acted over an unknown time period in the column, preferably put as $c_i = 0$. Introducing another length parameter, $L_d \equiv v^*/k^* = J^V/k\theta$, characterizing the decay effect, and defining a dimensionless "decay-diffusion/dispersion interaction parameter", ι, according to:

$$\iota \equiv \sqrt{1 + 4L_D/L_d} - 1 \qquad (9.105)$$

this solution may be written as:

$$\bar{c} = \exp\left\{-\frac{\iota x}{2L_D}\right\}$$

$$\cdot[\text{erfc}(\xi_\iota) + \exp(-\xi_\iota^2)\exp(\xi_\iota + \alpha_\iota)^2 \,\text{erfc}(\xi_\iota + \alpha_\iota)]/2 \qquad (9.106)$$

with $\bar{c} \equiv (c - c_i)/(c_f - c_i) = c/c_f$

$\xi_\iota = (x - v^*t - \iota v^*t)/2\sqrt{D^*t}$

$\alpha_\iota = (v^*t + \iota v^*t)/\sqrt{D^*t}$

The part of this equation between square brackets represents the solution as found in (9.54), apart from a shift in the position coordinate equal to ιx_p. This part is then multiplied with a decay factor as given by the first exponential term. It is important to note that this factor is solely dependent upon the position in the column. For a given value of ι (characteristic for the system) a plot of the above solution is thus obtained simply by first constructing the solute front according to Fig. 9.9 (or equation (9.54)), pertaining to the appropriate value of α_ι. Next, the exponential decay factor may be calculated as a function of the distance to the origin of the column and the previous curve is corrected accordingly.

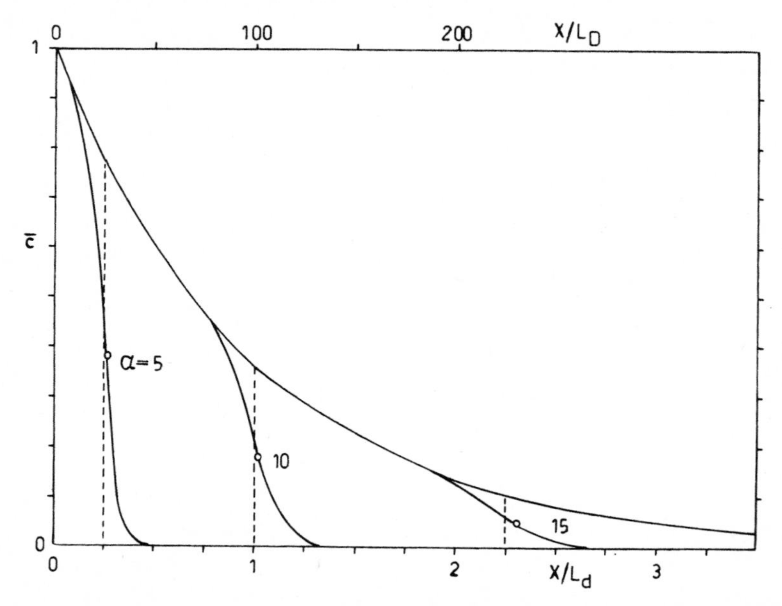

Fig. 9.21. Concentration front in a column of a solute following a linear adsorption isotherm and subject to a first-order decay process in combination with diffusion/dispersion. The distance axis is calibrated both in units of L_D and in units of $L_d \equiv J^V/\theta k$, corresponding to $L_d/L_D = 100$ or $\iota = 0.02$. (i.e. a very low value of the decay-diffusion interaction parameter). The parameter $\alpha = \sqrt{x_p/L_D}$ determines the location of x_p on the upper scale (dashed lines). The continuous decay curve constitutes the steady state concentration distribution reached at large values of t or α.

The development of the resulting solute front as a function of progressing time is demonstrated schematically in Fig. 9.21, using a rather small value of ι. Calibrating the distance axis in units of x/L_d, one may observe that the sigmoid diffusion front progresses proportionally with time, its maximum value being delimited by the time-independent decay curve. The latter curve then forms the steady state concentration-profile applying when v^*t has reached a sufficiently large value. It should be noted that the mid-point of the sigmoid curve lies slightly ahead of x_p (viz., at $(1 + \iota)x_p$), which is caused by the increased contribution of the diffusion flux owing to the concentration gradient resulting from the decay.

It should be pointed out here that the corresponding excess flux entering the column (i.e. in addition to $J^V c_f$) is not necessarily available in practice. In the above calculation it was implicitly present because of the chosen boundary condition, i.e. $\bar{c} = 1$ at $x = 0$. This argument is analogous to the one used in selecting the solution given in (9.56) in favor of (9.54). Although also here the solution of (9.104) with a constant flux at the entrance boundary would be preferable, the latter solution appears to be rather complicated. The difference between both solutions appears to be easily corrected for at sufficiently small values of ι (see below).

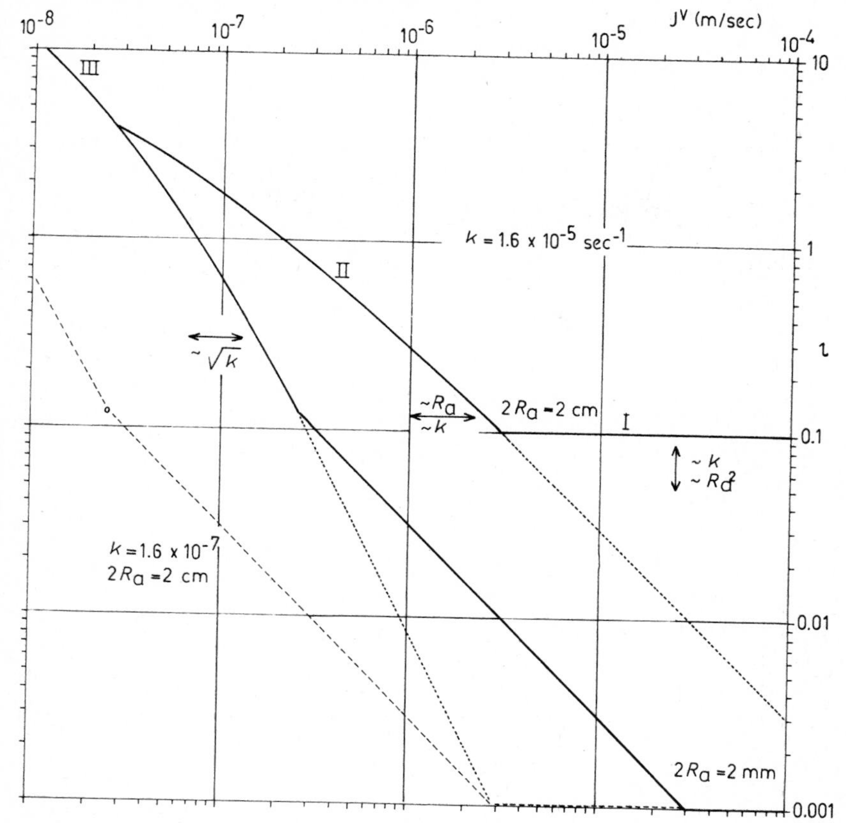

Fig. 9.22. Estimate of the decay-diffusion/dispersion interaction parameter ι as a function of the volume flux J^V. Upper solid line, consisting of three branches marked I, II, III, applies to aggregates of 2 cm diameter and is obtained by combining $L_d = J^V/k\theta$ with L_D as found from Fig. 9.8, for $k = 1.6 \cdot 10^{-5}$ sec^{-1}. For other values of k and R_a the three branches may be translated as indicated (see also text).

The magnitude of ι in soil columns depends particularly on the decay reaction involved. In an effort to delimit its value it is pointed out that one of the faster reactions known to occur in soils is the nitrification of NH_4-ions. In that case values of k have been suggested amounting to 0.2—0.8 $\cdot 10^{-5}$ sec^{-1} (cf. Cho, 1971; Misra et al., 1974). Presumably microbiological breakdown of pesticides is at least an order of magnitude smaller. Using as a maximum value $k = 1.6 \cdot 10^{-5}$ sec^{-1}, one finds, for J^V varying between 10^{-5} and 10^{-8} m/sec and at $\theta = \frac{1}{2}$, that L_d ranges for the above fast reaction from about 1 m down to 1 mm, under field conditions. In order to find the corresponding values of ι, the above range must be compared with L_D as given in Fig. 9.8. Using this information a composite graph may be constructed as shown in Fig. 9.22. Whereas two sets of lines are given corresponding to the above value of k in conjunction with aggregates of 2 cm and 2 mm diameter, respectively, the values applying for other values of k and R_a may be found by simple translations as shown. In this graph one recognizes the three ranges of L_D as distinguished in Fig. 9.8 and here identified by roman numerals. Obviously ι will be independent of J^V when the stagnant-phase effect dominates (range I), both L_D and L_d then being proportional to J^V. In that case ι equals 0.1 for 2-cm aggregates for the chosen high value of k, and becomes inversely proportional to both k and R_a^2. If convective dispersion is the controlling factor, L_D becomes constant (range II). If it amounts to a few cm, as is often found in practice, ι will be a function of kL_D/J^V, which varies strongly in the range of interest of J^V. The upper line shown corresponds to $\theta k L_D = 1.6 \cdot 10^{-7}$ m/sec, and the appropriate value of ι for other combinations of these variables may be read by horizontal translation, i.e. multiplying by a factor proportional to $\theta k L_D$.

In very homogeneous soils, where the convective dispersion parameter, L_{dis}, could be smaller than a few cm, longitudinal diffusion could become the controlling factor of L_D (range III). This situation corresponds with the curve on the left-hand side. This curve is independent of the aggregate size and ι is now a function of $\theta 2kD/(J^V)^2$. As plotted, $\theta k D$ was taken at $4 \cdot 10^{-15}$ m^2/sec^2, and it may be adjusted again for other values by horizontal translation, i.e. multiplying by a factor proportional to $\theta\sqrt{kD}$.

Returning now to the influence of the value of ι on the shape of the solute front, one may distinguish roughly three ranges. Focussing attention particularly on the stage where x_p/L_D exceeds the value 4 (i.e. $\alpha > 2$), one finds from (9.106), with $\iota > 1$, that $\xi_\iota < -1$ at $x = x_p$, such that $\frac{1}{2}$ erfc $\xi_\iota >$ 0.92. In other words, for $\iota > 1$ and $\alpha > 2$, a substantial descent of the sigmoid front starts at a point beyond x_p. Calculating the value of the exponential decay term gives, for the above conditions, $\exp(-\iota x/2L_D) < 0.14$, however, so it may be safely stated that the exponential decay term is acting independently of the bracketed second term of (9.106). Thus for $\iota > 1$ the front has practically reached its steady state (exponential) shape, once x_p exceeds $4L_D$. Accordingly it suffices that ι exceeds unity to warrant neglecting the bracketed term of (9.106) once x_p exceeds $4L_D$ (cf. Fig. 9.23).

In contrast, if $\iota < 0.1$, the shift of the half-way point of erfc ξ_ι soon becomes negligible, while the decay term may be expanded to $\exp(-x/L_D)$. In other words, $\bar{c}$ may be conveniently approximated by the product of the error-function solution given by (9.54) — which depends solely on x_p/L_D — and a decay term depending solely on L_d. The two effects, viz. diffusion/dispersion spreading and decay, then act independently, as is implied by the small value of the interaction parameter ι. In that case it would actually be preferable to use (9.56) or (9.53) to calculate the base curve.

Finally there remains the range $0.1 < \iota < 1$, where the spreading and decay truly interact. In this range, (9.106) must be applied. In Figs. 9.23 and

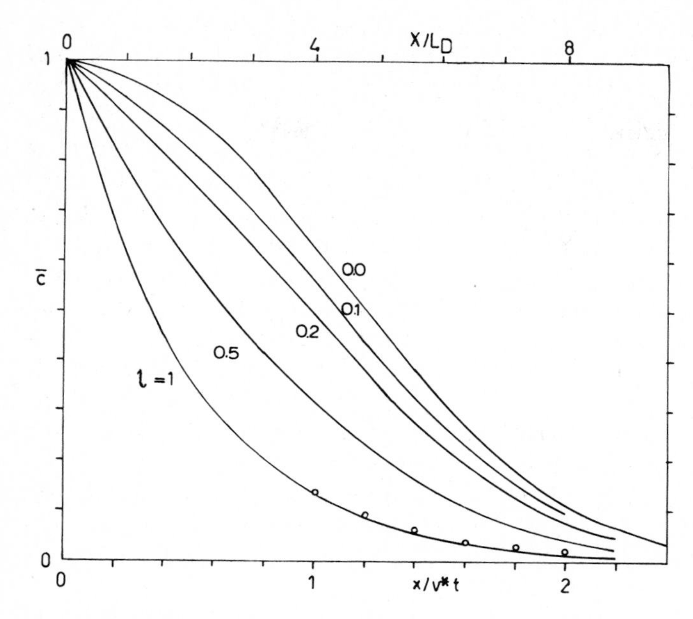

Fig. 9.23. Influence of a decay term on the concentration front in a column (subject to diffusion/dispersion and a linear adsorption isotherm) at shallow penetration of the front, i.e. $x_p/L_D = \alpha^2 = 4$, for different values of the interaction parameter ι. Open circles: approximation, neglecting diffusion/dispersion, for $\iota = 1$.

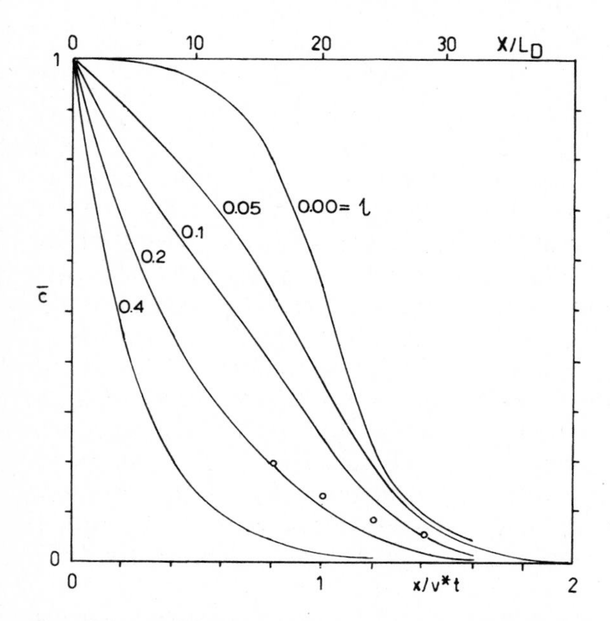

Fig. 9.24. As Fig. 9.23, at deep penetration of the front, i.e. $x_p/L_D = \alpha^2 = 20$. Open circles: approximation neglecting diffusion/dispersion, for $\iota = 0.2$.

9.24 some examples are plotted for different values of ι and x_p/L_D. It should be noted that for $x_p/L_D = 20$ the dominance of the decay term is actually reached at $\iota > 0.4$, while at $\iota = 0.2$ the trajectory of the decay term is an acceptable approximation up to $x = x_p$. It should also be noted that the curve for $\iota = 0.4$ with $x_p/L_D = 20$ coincides with the one for $\iota = 1$ with $x_p/L_D = 8$: once the decay term dominates, the product of ι and x_p/L_D determines the trajectory if $\bar{c}$ is plotted against x/x_p.

A warning note should be sounded against the boundary condition keeping $\bar{c}$ at unity at the entrance. Whereas in the absence of a decay term the total amount of solute needed to be brought in at the entrance — in addition to the amount carried by the convective flux, $J^V t c_f$ — remains finite, this is not so when decay is taking place. This is most easily demonstrated by calculating the influx needed to sustain the steady state condition as given by the first term of (9.106). This gives $J^V c_f(1 + \iota/2)$ which implies that, in order to maintain the steady state with $c = c_f$ at the entrance, one must continually add extra solute in an amount equal to $J^V c_f \iota/2$ keq/m^2 sec. In practice this will hardly ever be the case as the total influx is usually fixed at $J^V c_f$. This implies that the steady state solution with constant influx reads(cf. also (10.14) given in Ch. 10):

$$\bar{c} = \frac{1}{1 + \iota/2} \exp\{-\iota x/2L_D\} \tag{9.107}$$

It is easy to see that for $\iota < 0.1$ this effect may be incorporated into the exponential term by replacing x by $(x + L_D)$. In the above range of ι it will thus suffice to shift the curves as calculated in Figs. 9.21, 9.23 and 9.24 to the left over a distance equal to L_D. As for these small values of ι, the diffusion-spreading acts more or less independently of the decay, also the transient state preceeding the steady state could be corrected by applying the above shift.

Summarizing the situation it appears that the value of k used for the calculation of Fig. 9.22 presents a safe maximum for most decay processes to be expected in soil. Granted that, if a particular component is decaying as rapidly in the adsorbed state as in solution, the last term of (9.103) would have to be replaced by $k(\theta + \Delta q/\Delta c)c$ — such that L_d would decrease by a factor $(R_D + 1)$ — the numbers given above in relation to nitrification of NH_4-ions are presumably effective decay constants in soil containing an adsorption complex (cf. Misra et al., 1974). It thus seems reasonable to assume that in practice k will not exceed the maximum as used in Fig. 9.21. Instead, k will often be one or perhaps two orders smaller; decay constants mentioned in the literature for pesticides are often only about 10^{-6}—10^{-7} sec^{-1} (cf. Hamaker, 1972). Looking at Fig. 9.22 this implies that then ι will only rise significantly above 0.1 at flow rates of a few mm per day. Under such conditions one may safely construct the solute front in the absence of decay following the earlier part of this chapter including non-linear isotherms etc. and then apply to this front a decay correction equal to $\exp(-\iota x/2L_D)$.

Not nearly as common as decay processes are situations with a positive production of a solute, at least if simple decay products like CO_2 and H_2O are not included. A notable exception is the formation of NO_3^--ions from the nitrification reaction discussed before. Aside from denitrification of NO_3^--ions, which can occur simultaneously with the nitrification process, it can be established first that, in the absence of adsorption, the total amount of NO_3^--ions found in the column must equal the difference between the two curves obtained for the distribution of the NH_4^+-ions when using the appropriate value of ι and $\iota = 0$, respectively. If (linear) adsorption is present, the mentioned difference should be multiplied by $(R_D + 1)$ to match the total amount of NO_3^- in the column. The actual distribution of the nitrate has little relation to the "deficit" of the NH_4^+-curve. While maximum production takes place at high concentrations of NH_4^+, the NO_3^- produced is then transported through the column by convection and redistributed by diffusion/dispersion. An exception to this is the steady state distribution. In the absence of a significant contribution of a nitrite intermediate, and of denitrification, the two steady state curves of NH_4^+ and NO_3^- should be mirror images, as then the adsorbed NH_4^+, being constant, does not play any role. In the transient stage preceding, the fact that NH_4^+ is adsorbed — such that the advancing front moves with a velocity $J^V/\theta(R_D + 1)$ — while the NO_3^- produced moves with a velocity J^V/θ, causes the nitrate front to move far ahead of the ammonium front. In the literature cited above (Cho, 1971; Misra et al., 1974) some examples of curves are given covering the transient stage under varying conditions.

9.8. SUMMARY, WITH A GUIDE LINE FOR GUESSING

As was already inferred in chapter 7 of part A of this text, the solute front penetrating a soil column following the introduction of a step feed has in general a (distorted) sigmoid shape and it moves with a velocity proportional — but not equal — to the filter flux of the carrier liquid, J^V. Distinguishing between the (mean) location and the shape of the front, the first is usually of prime concern, while the second characteristic becomes important only if either the deviation of the location of a certain portion of the front from its mean location is large or the mean location is close to a critical point in the soil (e.g. the ground-water level). Indicating this mean location of the front as the penetration depth, x_p, its position should always be estimated first. For this purpose one needs only to know the value of the distribution ratio for the increment of concentration of the step feed, $R_D^\Delta \equiv \Delta q/\theta\Delta c$. If near-equilibrium prevails during the front passage, one finds x_p according to equation (A7.3) of part A, chapter 7 as:

$$x_p(\Delta c) = \frac{J^V t}{\theta} \frac{1}{R_D^\Delta + 1}$$

in which $J^V t$ actually stands for $\int_0^t J^V dt$, such that J^V signifies the mean filter flux during the relevant time period. Before going into further attempts to specify the shape of the front, it is advisable to guess the accuracy of the value of x_p thus obtained, taking into account the likely existing inhomogeneities of the soil profile.

Next one should verify whether the assumed existence of near-equilibrium is warranted. Referring to the relevant discussion in connection with Figs. 9.17—9.19, this depends on the value of $\alpha_r = \sqrt{x_p/L_r}$. Unless direct information is available about the value of the rate constant, k_r, of the adsorption reaction (in which case one uses (9.101) and (9.102)), one should take into account the delay caused by diffusion into a stagnant phase, presumably inside the "aggregates". Using an appropriate value of the (mean) diameter of the aggregates, $2R_a$, one may employ Fig. 9.8 for a first estimate of L_r. For this purpose one must use *only* the lines ascending proportionally to J^V, located on the LHS of Fig. 9.8, if necessary extrapolating the latter in the downward direction. As was pointed out in the small print paragraph on p. 333, the best general guess on L_r may be to take twice the value as read from Fig. 9.8, in order to account for the influence of R_D^Δ on L_r in a system for which the distribution of the liquid content θ over the stagnant phase inside the aggregate, θ_a, and the inter-aggregate mobile phase, θ_m, is not known. If the latter is known, L_r may be calculated from (9.48) together with (9.102).

Once L_r is found, the value of α_r can be calculated. As was shown, the assumption of near-equilibrium becomes acceptable if $\alpha_r > 2$. If this condition is not fulfilled one would expect a sizable deviation from equilibrium (cf. Fig. 9.17), such that $\bar{c}$ exceeds the value as found from equilibrium theory (while $\bar{q} < \bar{q}_{eq}$). Provided L_r is known with some accuracy a computer calculation could be used to find the actual shape of the front.

Assuming that near-equilibrium has been established, one may attempt to specify the shape of the front. First L_{dif} is read from Fig. 9.8 (line descending proportionally to J^V) and L_{dis}, covering convective dispersion, is introduced as, for example, the aggregate diameter. Summing up these parameters with L_r estimated above then gives L_D and $\alpha \equiv \sqrt{x_p/L_D}$ to be used later. The front shape is then determined by the adsorption isotherm, q(c), in the appropriate range and direction of the adsorption process. Distinguishing (a) linear, (b) un-favorable, and (c) favorable exchange (or adsorption), the following procedure may be applied:

(a) In the case of linear adsorption the front is situated around x_p following (9.53), which may be written as:

$$\frac{x(\bar{c}) - x_p}{x_p} = \frac{2}{\sqrt{\alpha}} \text{ inverfc } (2\bar{c})$$

provided $\alpha > 2$. In the case this condition is not met, the boundary conditions (e.g. constant values of the concentration, or of the flux, at $x = 0$)

will determine whether corrections according to Fig. 9.9 are worth considering.

(b) In the case of un-favorable adsorption (isotherm concave upwards) one should first construct the front as it would develop for $L_D = 0$. For this purpose one must construct the slope of the isotherm as $q'(c) = (dq/dc) = q'(\bar{c})\, R_D^{\Delta}$. The solution front then follows from (9.11) as:

$$x^0(\bar{c}) = J^V t/\{q'(c) + \theta\}$$

Using the adsorption isotherm one finds the corresponding front of the adsorbed phase. If now $\alpha > 3$, a reasonable correction for the diffusion/dispersion effect may be applied following:

$$\frac{\delta_0 x(\bar{c})}{x_p} = \frac{2}{\sqrt{\alpha}} \text{ inverfc } (2\bar{c})$$

such that $x(\bar{c}) \approx x^0(\bar{c}) + \delta_0 x(\bar{c})$. For $\alpha < 3$ computer calculations might be necessary.

(c) In the case of favorable adsorption one must first ascertain that the front is always situated between the one constructed in case (a) and a step front at $x = x_p$. Often this will be a satisfactory description of the front shape, particularly for small values of α. If more precision is wanted the graphical procedure discussed in connection with Fig. 9.10 could be attempted if computer calculation is not feasable, which then gives the steady state profile.

If, finally, a decay process is known to take place with a given value of the equivalent first-order rate constant, k, one can assess the decay length parameter as $L_d = J^V/k\theta$. From this the decay-diffusion/dispersion interaction parameter is found as $\iota \equiv (\sqrt{1 + 4L_D/L_d} - 1)$. If $\iota < 0.1$ and the adsorption isotherm is linear, the expected front may be constructed as outlined above, after which it is multiplied with a decay factor equal to $\exp\{-\iota(x + L_D)/2L_D\}$. The trajectory of this factor then also indicates the steady state profile. For $\iota > 1$ the steady state profile will be established within a shallow depth of the column and is usually all that matters.

Having outlined above a rapid guessing procedure based on the findings of the foregoing sections, some remarks should be made about its prospects with regard to applicability under field conditions. As was pointed out in the footnote on p. 287, the present treatise has been based on the prevalence of steady state leaching. Similarly, homogeneity of the soil column was used throughout. Granted that both conditions are hardly met in practice, known deviations from this "assumed homogeneity in place and time" can, however, be incorporated when solving the relevant conservation equation(s) numerically. In practice, prediction is limited by precise information about the parameters involved rather than by the available facilities for solving a conservation equation.

A cautioning remark is made here with respect to indiscriminate use of an assumed value of L_D in computor simulations, more particularly of its component covering convective dispersion, L_{dis}. In Fig. 9.8 it is suggested to take this component equal to the grain diameter, following the considerations starting with equation (9.26). It then follows that the validity of this approach is delimited by the condition: $c - L_{dis}(dc/dx) > 0$; the combined solute flux resulting from convection and convective dispersion, $J^V[c - L_{dis}(dc/dx)]$, must be in the same direction as the convective flux, $J^V c$.

Against this background attempts towards prediction of c(x, t) for a given solute in a given soil profile must necessarily be based on a critical evaluation of the significance of the different terms of the *complete* conservation equation in a particular situation, followed by a *suitable* simplification. The above treatment should be helpful in this respect, giving the relative importance of the terms in the case of a homogeneous, steady state system. Applying it for different (extreme) values of the parameters involved — as, for example, high and low values of J^V, θ, k_r, etc. — one should be able to gain an impression whether or not fluctuations of these in time and place are likely to play a dominant role. It should then be decided which terms are of secondary importance and which parameters should perhaps be determined experimentally, if possible. In certain cases this might lead to the conclusion that non-steady flow of the carrier is likely to exercise a far greater influence on the movement and spreading of a solute front than a diffusion/dispersion term as used in the above steady state solutions. Similarly, the prediction of the displacement of solutes in a cracked soil will probably require primarily a specification of the ensuing laterally inhomogeneous flow pattern of the carrier and the corresponding diffusion of solute in a direction perpendicular to the main direction of carrier flow.

A particular example of the first case mentioned is given in Addiscott (1977) and Addiscott et al. (1978). These papers discuss the winter leaching of anions from soil, following intermittent rainfall and evaporation. For a chosen thickness of consecutive soil compartments and assuming instantaneous equilibration between a stagnant and a mobile phase during periods of zero or upward flow of the carrier, while disconnecting the two phases during rapid downward flow, the observed displacement of a chloride tracer under field conditions could be simulated satisfactorily.

REFERENCES

Addiscott, T.M., 1977. A simple computer model for leaching in structured soils. J. Soil. Sci., 28:554—563.

Addiscott, T.M., Rose, D.A. and Bolton, J., 1978. Chloride leaching in the Rothamsted drain gauges. Influence of rainfall pattern and soil structure. J. Soil Sci., 29: 305—315.

Bear, J., 1972. Dynamics of Fluids in Porous Media. American Elsevier., New York, N.Y., 764 pp.

Bear, J., Zaslavsky, D. and Irmay, S., 1968. Physical Principles of Water Percolation and Seepage. Unesco, Paris, 465 pp.

Biggar, J.W. and Nielsen, D.R., 1962. Miscible displacement: II. Behavior of tracers. Soil Sci. Soc. Am. Proc., 26: 125—128.

Biggar, J.W. and Nielsen, D.R., 1963. Miscible displacement: V. Exchange processes. Soil Sci. Soc. Am. Proc., 27: 623—627.

Boucher, D.F. and Alves, G.E., 1959. Dimensionless numbers. Chem. Eng. Progr., 55: 55—64.

Boucher, D.F. and Alves, G.E., 1963. Dimensionless numbers — 2. Chem. Eng. Progr., 59: 75—83.

Bower, C.A., Gardner, W.R. and Goertzen, J.O., 1957. Dynamics of cation exchange in soil columns. Soil Sci. Soc. Am. Proc., 21: 20—24.

Brinkley, S.R., Edwards. H.E. and Smith, R.W., 1952. Table of the temperature distribution function for heat exchange between a fluid and a porous solid. Unpublished table announced in: Math. Tables and other Aids to Computation, 7: 40.

Cameron, D.R. and Klute, A., 1977. Convective—dispersive solute transport with a combined equilibrium and kinetic adsorption model. Water Resour. Res., 13: 183—188.

Carslaw, H.S. and Jaeger, J.C., 1959. Conduction of Heat in Solids. Clarendon Press, Oxford, 510 pp.

Cho, C.M., 1971. Convective transport of ammonium with nitrification in soil. Can. J. Soil Sci., 51: 339—350.

Coats, K.H. and Smith, B.D., 1964. Dead-end pore volume and dispersion in porous media. Soc. Pet. Eng. J., : 73—84.

Crank, J., 1964. The Mathematics of Diffusion. Clarendon Press, Oxford.

De Vault, D., 1943. The theory of chromatography. J. Am. Chem. Soc., 65: 534—540.

Feynman, R.P., Leighton, R.B. and Sands, M., 1963. The Feynman Lectures on Physics. Addison Wesley , New York, N.Y., pp. 43.1—43.10.

Frissel, M.J. and Poelstra, P., 1967. Chromatographic transport through soils: I. Theoretical evaluations. Plant and Soil, 26: 285—302.

Frissel, M.J. and Reiniger, P., 1974. Simulation of Accumulation and Leaching in Soils. Pudoc, Wageningen, 116 pp.

Frissel, M.J., Poelstra, P. and Reiniger, P., 1970. Chromatographic transport through soils: III. A simulation model for the evaluation of the apparent diffusion coefficient in undisturbed soils with tritrated water. Plant and Soil, 33: 161—176.

Gershon, N.D. and Nir, A., 1969. Effects of boundary conditions of models on tracer distribution in flow through porous media. Water Resour. Res., 5: 830—839.

Glueckauf, E., 1955a. Principles of operation of ion-exchange columns. In: Ion Exchange and Its Applications. Soc. Chem. Ind. London, pp. 34—46.

Glueckauf, E., 1955b. Theory of chromatography: 10. Formulae for diffusion into spheres and their application to chromatography. Trans. Faraday Soc., 51: 1540—1551.

Glueckauf, E. and Coates, J.J., 1947. Theory of chromatography: II, III, IV. J. Chem. Soc., 1302—1329.

Goldstein, S., 1953. On the mathematics of exchange processes in fixed columns: I. Mathematical solutions and asymptotic expansions. Proc. R. Soc. Lond., A 219: 151—185.

Hamaker, J.W., 1972. Quantitative aspects of decomposition. In: C.A.I. Goring and J.W. Hamaker (Editors), Organic Chemicals in the Soil Environment. Marcel Dekker, New York, N.Y., pp. 253—340.

Hashimoto, I., Deshpande, K.B. and Thomas, H.C., 1964. Péclet numbers and retardation factors for ion exchange columns. Ind. Eng. Chem. Fund., 3: 213—218.

Heller, J.P., 1960. An "unmixing" demonstration. Am. J. Phys., 28: 348—353.

Heller, J.P., 1972. Observations of mixing and diffusion in porous media. Proc. 2nd Symp. IAHR—ISSS on the fundamentals of Transport Phenomena in Porous Media. Univ. of Guelph, Canada: 1—26.

Hiester, N.K. and Vermeulen, T., 1952. Saturation performance of ion-exchange columns and adsorption columns. Chem. Eng. Prog., 48: 505—516.

Klinkenberg, A., 1948. Numerical evaluation of equations describing transient heat and mass transfer in packed solids. Ind. Eng. Chem., 40: 1992—1994.

Klinkenberg, A., 1954. Heat transfer in cross-flow heat exchangers and packed beds. Ind. Eng. Chem., 46: 2285—2289.

Lapidus, L. and Amundson, N.R., 1952. The effect of longitudinal diffusion in ion exchanges and chromatographic columns. J. Phys. Chem., 56: 984—988.

Misra, C., Nielsen, D.R. and Biggar, J.W., 1974. Nitrogen transformations in soil: I, II and III. Soil Sci. Soc. Am. Proc., 38: 289—304.

Nielsen, D.R. and Biggar, J.W., 1961. Miscible displacement in soils: I. Experimental information. Soil Sci. Soc. Am. Proc., 25: 1—5.

Nielsen, D.R. and Biggar, J.W., 1962. Miscible displacement in soils: II. Theoretical considerations. Soil Sci. Soc. Am. Proc., 26: 216—221.

Nusselt, W., 1911. Der Wärmeübergang im Kreuzstrom. Z. Ver. Dtsch. Ing., 55: 2021—2024.

Passioura, J.B., 1971. Hydrodynamic dispersion in aggregated media: 1. Theory. Soil Sci., 111:339—344.

Passioura, J.B. and Rose, D.A., 1971. Hydrodynamic dispersion in aggregated media: 2. Effects of velocity and aggregate size. Soil Sci., 111: 345—351.

Pfannkuch, H.O., 1963. Contribution à l'étude des déplacements de fluides miscibles dans un milieu poreux. Rev. Inst. Fr. Pét., 18: 215—270.

Reiniger, P. and Bolt, G.H., 1972. Theory of chromatography and its application to cation exchange in soils. Neth. J. Agric. Sci., 20: 301—313.

Rible, J.M. and Davis, L.E., 1955. Ion exchange in soil columns. Soil Sci., 79: 41—47.

Ritchie, A.S., 1966. Chromatography as a natural process in geology. Adv. Chromatogr., 3: 119—134.

Rose, D.A., 1977. Hydrodynamic dispersion in porous materials. Soil Sci., 123: 277—283.

Rose, D.A. and Passioura, J.B., 1971. The analysis of experiments on hydrodynamic dispersion. Soil Sci., 111: 252—257.

Scotter, D.R. and Raats, P.A.C., 1968. Dispersion in porous mediums due to oscillating flow. Water Resour. Res., 4: 1201—1206.

Taylor, G.I., 1953. Dispersion of soluble matter in solvent flowing slowly through a tube. Proc. R. Soc. Lond., A, 219: 186—203.

Thomas, H.C., 1944. Heterogeneous ion exchange in a flowing system. J. Am. Chem. Soc., 66: 1664—1666.

Thornthwaite, C.W., Masher, J.R. and Nakamura, J.K., 1960. Movement of radiostrontium in soils. Science, 131: 1015—1019.

Van der Molen, W.H., 1957. The exchangeable cations in soils flooded with sea water. Agric. Res. Rep., 63.17, p. 167.

Van Genuchten, M.Th., Davidson, J.M. and Wierenga, P.J. 1974. An evaluation of kinetic and equilibrium equations for the prediction of pesticide movement through porous media. Soil Sci. Soc. Am. Proc., 38: 29—35.

Van Genuchten, M.Th. and Wierenga, P.J., 1976. Mass transfer studies in sorbing porous media: I. Analytical solutions. Soil Sci. Soc. Am. J., 40: 473—480.

Van Genuchten, M.Th. and Wierenga, P.J., 1977. Mass transfer studies in sorbing porous media: II. Experimental evaluation with tritium (^3H_2O). Soil Sci. Soc. Am. J., 41: 272—278.

Van Genuchten, M.Th., Wierenga, P.J. and O'Connor, G.A., 1977. Mass transfer studies in sorbing porous media: III. Experimental evaluation with 2,4,5-T. Soil Sci. Soc. Am. J., 4: 278—285.

CHAPTER 10

MOVEMENT OF SOLUTES IN SOIL: COMPUTER-SIMULATED AND LABORATORY RESULTS

M. Th. van Genuchten and R.W. Cleary*

Many of the mechanisms which determine the rate at which chemicals move through soil have been discussed in the previous chapter. These mechanisms include such processes as diffusion/dispersion, adsorption, decay, and intra-aggregate diffusion. Exact analytical solutions can be derived for some of these processes, such as for linear adsorption and linear decay (Chapter 9). However, for non-linear cases analytical methods cannot be used to obtain exact solutions of the transport equations and approximate methods must be employed. The use of numerical techniques can provide a useful and often the only alternative in modeling the transport of solutes in such cases. Obviously, numerical and analytical approaches can and should complement and augment each other. For example, an analytical solution may be used to check the accuracy of a numerical program. On the other hand, a numerical solution may be used to demonstrate the appropriateness (or shortcoming) of a particular assumption necessary in the development of an analytical solution.

In this chapter some of the transport equations used at present to describe the movement of chemicals in soils are restated and solutions based on both numerical and analytical techniques are presented. The influence of several mechanisms affecting solute transport phenomena are studied, such as non-linear adsorption, hysteresis in the equilibrium isotherms, and decay. Special consideration is given to the occurrence of non-equilibrium conditions between a given chemical and its adsorbent, the soil.

10.1. MATHEMATICAL DESCRIPTION

The unsteady-state, one-dimensional convective-dispersive mass transport equation which describes the concentration distribution of a chemical undergoing adsorption and linear decay in an unsaturated or saturated soil is given by:

---

* Present address: U.S. Salinity Laboratory, 4500 Glenwood Dr., Riverside, Calif. 92501, U.S.A.

$$\frac{\partial(\theta c)}{\partial t} + \frac{\partial(J^V c)}{\partial x} = \frac{\partial}{\partial x}\left(\theta D \frac{\partial c}{\partial x}\right) - \rho_b \frac{\partial\, {}^w q}{\partial t} - k\theta c \qquad (10.1)$$

where c is the solute concentration (ML^{-3}), ${}^w q$ is the sorbed concentration ($= q/\rho_b$), ρ_b is the (dry) bulk density of the soil (ML^{-3}), J^V is the Darcy volumetric flux of the feed solution (LT^{-1}), θ is the volumetric moisture content, D is the dispersion coefficient (L^2T^{-1}), k is a first-order rate coefficient (T^{-1}), t is the time (T) and x is the distance (L). The boundary and initial conditions for a semi-infinite system undergoing infiltration with a pulse of water containing a solute of concentration c_f, followed by solute-free water may be stated as:

$$-\theta D \frac{\partial c}{\partial x} + J^V c = \begin{cases} J^V c_f & x = 0 \qquad 0 < t \leqslant t_o \\ 0 & x = 0 \qquad t > t_o \end{cases}$$

$$\frac{\partial c}{\partial x} = 0 \qquad x \to \infty \qquad t \geqslant 0$$

$$c = 0 \text{ in} \qquad 0 \leqslant x < \infty \qquad t = 0 \qquad (10.1a)$$

where t_o is the duration of the solute pulse applied to the soil surface.

If D is assumed constant, (10.1) for steady, saturated conditions (J^V and θ become constants), or for unsaturated, constant moisture content conditions (Warrick et al., 1971), reduces to:

$$\frac{\partial c}{\partial t} + v \frac{\partial c}{\partial x} = D \frac{\partial^2 c}{\partial x^2} - \frac{\rho_b}{\theta} \frac{\partial\, {}^w q}{\partial t} - kc \qquad (10.2)$$

and the boundary and initial conditions become:

$$-D \frac{\partial c}{\partial x} + vc = \begin{cases} vc_f & x = 0 \qquad 0 < t \leqslant t_o \\ 0 & x = 0 \qquad t > t_o \end{cases}$$

$$\frac{\partial c}{\partial x} = 0 \qquad x \to \infty \qquad t \geqslant 0$$

$$c = 0 \text{ in} \qquad 0 \leqslant x < \infty \qquad t = 0 \qquad (10.2a)$$

where v is the average interstitial or pore-water velocity ($v = J^V/\theta$).

Solutions of (10.2) can be obtained after specifying the adsorption rate $\partial\, {}^w q/\partial t$ in the transport equation. Several models for adsorption or ion exchange may be used for this purpose. These may be divided into two broad categories: equilibrium models which assume instantaneous adsorption of the chemical, and kinetic models which consider the rate of approach towards equilibrium. Table 10.1 presents examples of some of the adsorption models at present available. Not included in table 10.1 are those models

TABLE 10.1.

Some equations available to describe the adsorption of chemical on soil

Model	Equation	Reference
1. Equilibrium		
1.1 (linear)	${}^{w}q = k_1 c + k_2$	Lapidus and Amundson (1952) Lindstrom et al. (1967)
1.2 (Langmuir)	${}^{w}q = \dfrac{k_1 c}{1 + k_2 c}$	Tanji (1970) Ballaux and Peaslee (1975)
1.3 (Freundlich)	${}^{w}q = k_1 c^{k_2}$	Lindstrom and Boersma (1970) Swanson and Dutt (1973)
1.4	${}^{w}q = k_1 c \exp(-2k_2 {}^{w}q)$	Lindstrom et al. (1971) Van Genuchten et al. (1974)
1.5 (modified Kielland)	${}^{w}q/{}^{w}q_f = c[c + k_1(c_f - c) \exp\{k_2(c_f - 2c)\}]^{-1}$	Lai and Jurinak (1971)
2. Non-equilibrium		
2.1 (linear)	$\dfrac{\partial {}^{w}q}{\partial t} = k_r(k_1 c + k_2 - {}^{w}q)$	Lapidus and Amundson (1952) Oddson et al. (1970)
2.2 (Langmuir)	$\dfrac{\partial {}^{w}q}{\partial t} = k_r\left(\dfrac{k_1 c}{1 + k_2 c} - {}^{w}q\right)$	Hendricks (1972)
2.3 (Freundlich)	$\dfrac{\partial {}^{w}q}{\partial t} = k_r(k_1 c^{k_2} - {}^{w}q)$	Hornsby and Davidson (1973) Van Genuchten et al. (1974)
2.4	$\dfrac{\partial {}^{w}q}{\partial t} = k_r \exp(k_2 {}^{w}q)\{k_1 c \exp(-2k_2 {}^{w}q) - {}^{w}q\}$	Lindstrom et al. (1971)
2.5	$\dfrac{\partial {}^{w}q}{\partial t} = k_r({}^{w}q_f - {}^{w}q) \sinh\left(k_1 \dfrac{{}^{w}q_f - {}^{w}q}{{}^{w}q_f - {}^{w}q_i}\right)$	Fava and Eyring (1956) Leenheer and Ahlrichs (1971)
2.6	$\dfrac{\partial {}^{w}q}{\partial t} = k_r(c)^{k_1}({}^{w}q)^{k_2}$	Enfield et al. (1976)

which describe the competition between two ionic species, such as the commonly used cation exchange equations. Except for a few cases (cf. Lai and Jurinak, 1971), generally two transport equations (i.e. of the type given by (10.1)) have to be solved for such multi-ion problems. Only single-ion transport models will be considered here.

Most of the equilibrium models listed in Table 10.1 are special cases of the non-equilibrium models. Model 1.4 in particular follows directly from the rate equation used by Lindstrom et al. (1971) (model 2.4) by setting the time derivative, $\partial^{w}q/\partial t$, to zero. All adsorption models further represent reversible adsorption reactions, except for model. 2.6. The rate equation proposed by Enfield et al. (1976) is an irreversible adsorption equation because it does not allow for any chemical desorption (the adsorption rate is always positive). Model 2.6 was used by Enfield et al. to describe the adsorption of phosphorus on several soils.

For illustrative purposes this chapter will consider only models 1.3 and 2.3, i.e. the Freundlich-type equilibrium and non-equilibrium adsorption equations. Both models will be discussed separately.

10.2. EQUILIBRIUM ADSORPTION

It has been shown by several authors (cf. Kay and Elrick, 1967; Bailey et al., 1968; Davidson and Chang, 1972) that for many organic chemicals the relationship, at equilibrium, between adsorbed (${}^{w}q$) and solution concentration (c) can be described by a general Freundlich isotherm (model 1.3):

$$^{w}q = kc^{n} \tag{10.3}$$

where k and n are temperature-dependent constants. Although (10.3) could be derived on physicochemical grounds (cf. Rideal, 1930), it suffices for the present objective to treat k and n as just empirical constants which may be determined experimentally, for example using batch equilibration methods.

Equation (10.3) assumes that once the chemical and the porous matrix are brought sufficiently close together, adsorption will be an instantaneous process. However, the equation does not specify the exact mode in which the chemical and the soil adsorption sites may be brought together (e.g. convective transport, diffusion, a shaking process, etc.).

Equation (10.3) may be differentiated with respect to time and substituted into (10.2). This allows the transport equation to be written in terms of one dependent variable, c:

$$\frac{\partial c}{\partial t} = \frac{1}{R_f}\left(D\frac{\partial^2 c}{\partial x^2} - v\frac{\partial c}{\partial x} - kc\right) \tag{10.4}$$

where the retardation factor R_f (introduced by Hashimoto et al., 1964) is defined as:

$$R_f \equiv 1 + q'/\theta \quad \text{(cf. (9.7) and (9.9))}$$
$$= 1 + \rho_b n k c^{n-1}/\theta \tag{10.5}$$

The retardation factor defines the mean velocity of the moving liquid relative to the mean velocity at which the chemical itself moves through the soil.

Exact analytical solutions of (10.4) are available only when a linear adsorption isotherm is present ($n = 1$). The retardation factor then becomes independent of the solute concentration, i.e.:

$$R_f = 1 + \rho_b k/\theta \tag{10.5a}$$

The solution of (10.4) subject to a first-type boundary condition at the soil surface and for a semi-infinite profile has been given in Gershon and Nir (1969). Cleary and Adrian (1973) derived the analytical solution for the case of a finite profile with linear adsorption and a first-type boundary condition. Selim and Mansell (1976) extended Cleary and Adrian's work by including the effects of linear decay. The direct solution of (10.4) subject to the initial and third-type boundary conditions given by (10.2a) is apparently not in the literature. Without giving the lengthy integral transform derivation, we present this solution for the first time here:

$$\frac{c}{c_f} = \begin{cases} c_1(x,t) & 0 < t \leqslant t_0 \\ c_1(x,t) - c_1(x, t - t_0) & t > t_0 \end{cases} \tag{10.6}$$

where:

$$\begin{aligned} c_1(x,t) = {} & \frac{v}{v_\iota} \exp(vx/2D) \left[\exp(-v_\iota x/2D) \operatorname{erfc}\left\{\frac{R_f x - v_\iota t}{(4DR_f t)^{1/2}}\right\} \right. \\ & \left. - \exp(v_\iota x/2D) \operatorname{erfc}\left\{\frac{R_f x + v_\iota t}{(4DR_f t)^{1/2}}\right\} \right] \\ & + \frac{v^2}{4kD} \exp(vx/D) \left[2 \exp(-kt/R_f) \operatorname{erfc}\left\{\frac{R_f x + vt}{(4DR_f t)^{1/2}}\right\} \right. \\ & + \left(\frac{v}{v_\iota} - 1\right) \exp\left\{-\frac{x}{2D}(v + v_\iota)\right\} \operatorname{erfc}\left\{\frac{R_f x - v_\iota t}{(4DR_f t)^{1/2}}\right\} \\ & \left. - \left(\frac{v}{v_\iota} + 1\right) \exp\left\{-\frac{x}{2D}(v - v_\iota)\right\} \operatorname{erfc}\left\{\frac{R_f x + v_\iota t}{(4DR_f t)^{1/2}}\right\} \right] \end{aligned} \tag{10.7}$$

and:

$$v_\iota = v\sqrt{1 + 4kD/v^2}$$

Equation (10.7) may be simplified for the particular case when no decay processes are present ($k = 0$). By applying L'Hospital's rule to the second term of (10.7) it may be shown that the equation reduces to the solution derived by Lindstrom et al. (1967) for no decay:

$$c_1(x, t) = \frac{1}{2}\,\mathrm{erfc}\left\{\frac{R_f x - vt}{(4DR_f t)^{1/2}}\right\} + \left(\frac{v^2 t}{\pi D R_f}\right)^{1/2} \exp\left\{-\frac{(R_f x - vt)^2}{4DR_f t}\right\}$$

$$-\frac{1}{2}\exp(vx/D)\left(1 + \frac{vx}{D} + \frac{v^2 t}{DR_f}\right)\mathrm{erfc}\left\{\frac{R_f x + vt}{(4DR_f t)^{1/2}}\right\} \qquad (10.8)$$

Referring to section 9.7 of the previous chapter it is pointed out that in the above equation (10.7) the decay-diffusion/dispersion interaction is expressed in terms of the variable $v_\iota = v(1 + \iota)$, whereas previously the parameter ι was used for this purpose (cf. (9.105)). Since the present case is based on a linear adsorption isotherm, the retardation factor R_f is a constant, so that $v/R_f = v^*$ and $D/R_f = D^*$ (9.104). Formulating (10.7) in terms of the variables used previously:

$$L_D = D/v = D^*/v^*$$

$$\xi = \tfrac{1}{2}(x - v^*t)/\sqrt{D^*t}; \qquad \xi_\iota = \tfrac{1}{2}\{x - (\iota + 1)v^*t\}/\sqrt{D^*t}$$

$$\alpha = v^*t/\sqrt{D^*t}; \qquad \alpha_\iota = (\iota + 1)v^*t/\sqrt{D^*t}$$

one obtains:

$$\frac{c}{c_f} = \frac{\exp(-\tfrac{1}{2}\iota x/L_D)}{(\iota + 2)}\left[\mathrm{erfc}(\xi_\iota) - \exp\{(\iota + 1)x/L_D\}\,\mathrm{erfc}(\xi_\iota + \alpha)\right]$$

$$+ \frac{2\exp(x/L_D)}{\iota(\iota + 2)}\left[\exp\{-\tfrac{1}{4}\iota(\iota + 2)\alpha^2\}\,\mathrm{erfc}(\xi + \alpha)\right.$$

$$\left. - \exp(\tfrac{1}{2}\iota x/L_D)\,\mathrm{erfc}(\xi_\iota + \alpha_\iota)\right] \qquad (10.7a)$$

where $x/L_D \equiv (\xi + \alpha)^2 - \xi^2$ has been left intact. For $\iota \to 0$, $\xi_\iota \to \xi$, and $\alpha_\iota \to \alpha$; expanding the last term to cancel ι appearing in the denominator then leads immediately to the previous equation (9.56), which is thus identical with (10.8) above.

Unfortunately, no exact analytical solutions are available when the adsorption isotherm is non-linear. Approximate solutions may still be derived for some special cases as was shown in chapter 9 for favorable exchange ($n < 1$). Generally, however, one must resort to numerical techniques in order to obtain results (Van Genuchten and Wierenga, 1974; Gupta and Greenkorn, 1976). The numerical solutions presented in this chapter were obtained using a finite difference approach, and were programmed in the

IBM S/360 CSMP-language as discussed extensively elsewhere (Van Genuchten and Wierenga, 1974, 1976a). In the CSMP approach, the spatial derivatives are approximated with suitable finite differences, while the integration in time is performed using a predictor—corrector Runge-Kutta type algorithm (IBM, 1967). This approach has proven to be a relatively simple way of obtaining solutions for rather complex distributive systems, without sacrificing accuracy in the results.

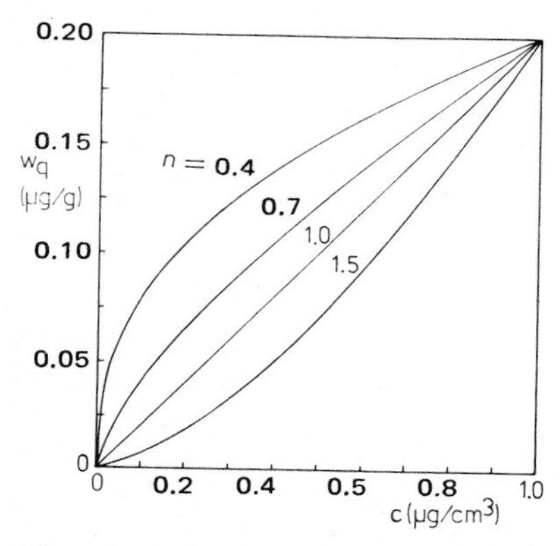

Fig. 10.1. Freundlich equilibrium plots for k = 0.2 and four different values for the exponent n.

10.2.1. Non-linear adsorption effects

First the influence of a non-linear adsorption isotherm on the shape and position of a solute pulse travelling through a soil column will be studied. For illustrative purposes the constant k in (10.3) was chosen to be 0.2 and the value of n was allowed to vary between 0.4 and 1.5, as shown in Fig. 10.1. The isotherms are clearly much more non-linear near the origin, but become approximately linear as the maximum value is approached. The maximum concentration, c_f, of the feed solution was given an arbitrary value of 1.0 $\mu g/cm^3$. The curves plotted in Fig. 10.1 were subsequently used in conjunction with (10.4) ($k = 0$) to calculate the distribution of the chemical versus depth, both in the solution and adsorbed phases. Results are shown in Fig. 10.2. A total of 20 cm^3 of a chemical solution was applied to the soil surface and leached out with a solute-free solution at a rate (J^V) of 16 cm/day. The chemical hence is in the feed solution for only 1.25 days (t_0). The calculated profiles in Fig. 10.2 are plotted after three days. Other data used to calculate the curves are given with Fig. 10.2.

Recalling the extensive discussion on the shape of the "base front", i.e. the front in the absence of diffusion/dispersion, as influenced by concave, linear, and convex adsorption isotherms (see section 9.3), one finds that the curves based on the linear isotherm ($n = 1$) exhibit a nearly symmetrical

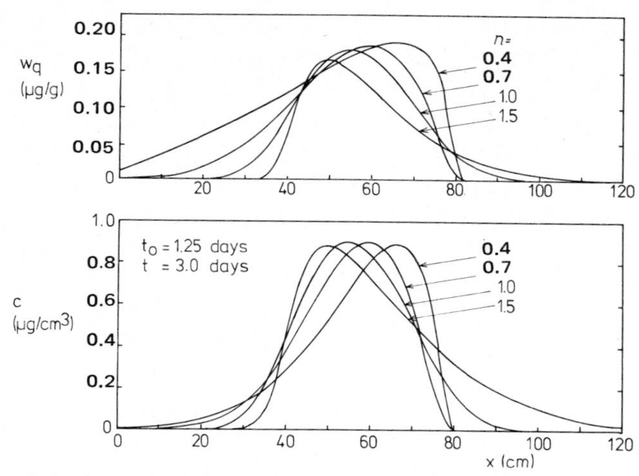

Fig. 10.2. Solution (c) and adsorbed ($^w q$) concentration distributions versus distance as obtained with the four equilibrium isotherms of Fig. 10.1, and the following soil-physical data: $\rho_b = 1.40\ g/cm^3$, $J^V = 16\ cm/day$, $\theta = 0.40\ cm^3/cm^3$ and $D = 30\ cm^2/day$.

shape for the present block feed, i.e. a block front as modified by diffusion/dispersion. This is not true, however, when n deviates from one: the more non-linear the adsorption isotherm, the more asymmetrical the resulting solute distribution becomes. When $n = 0.4$, for example, a very steep solute front develops while the distribution near the soil surface becomes much more dispersed. The appearance of a sharp front may be explained by considering the ratio v/R_f in (10.4). Since R_f increases rapidly with decreasing concentration ($n = 0.4$), the ratio v/R_f will decrease when the solute concentration becomes smaller, i.e. ahead of the solute front. This in turn slows down the movement of the chemical before the front. The ratio v/R_f on the other hand is much higher at the higher concentrations. The higher the concentration, the faster the apparent velocity of the chemical will be; the higher concentrations now tend to override the lower concentrations and a step front will develop. Owing to the developing high concentration gradients, however, the front will never attain a block form. The diffusion forces become very strong here and will always maintain a smooth curve. It should be noted that the retardation factor also influences the steepness of the front through the ratio D/R_f in (10.4). Since R_f is very large near the toe of the concentration front, the apparent dispersion of the solute distribution will be much less here than at the higher concentration (cf. the relevant discussion on p.321 in the previous chapter).

The sharpening of the solute front when $n < 1$ may also be explained by considering the differential capacity for adsorption, $d^w q/dc$, i.e. the slope of the equilibrium isotherms. When $n < 1$, $d^w q/dc$ is relatively large at the lower concentrations. This means that the rate of adsorption at lower concentrations is much higher than at higher concentrations. Because more material will be drawn from the soil solution towards the adsorption sites located at the downstream side of the solute front, the solution concentration initially will increase only very slowly. With each increase in concentration, however, d^wq/dc decreases, resulting in the adsorption rate becoming smaller. This in turn forces the solution concentration to increase faster, leading to the observed steep front. When the bulk of the chemical has passed, i.e. near the soil surface, a small decrease in concentration initially will result in the release of only a small amount of material adsorbed (d^wq/dc is small at the higher concentrations). With each decrease in concentration, the desorption rate of the chemical increases, bringing more and more chemical into solution and hence resulting in a slower decrease in the solution concentration. This then explains the highly dispersed solute distributions near the entrance of the soil column.

When $n > 1$, the reverse occurs. Adsorption at the lower concentrations is now relatively minor and the toe of the front propagates with a velocity nearly equal to that of the liquid. At the higher concentration, however, adsorption is now much more extensive, resulting in a slower rate of movement. Hence the concentration front becomes increasingly more dispersed in appearance as time progresses. After most of the chemical has passed and the concentration starts decreasing, initially much of the chemical which was adsorbed will go into solution (d^wq/dc is large) and hence the concentration will remain relatively high. Again, with each decrease in concentration, less material will go into solution (d^wq/dc decreases rapidly) and the solution concentration will drop quicker, resulting in the steep front upstream of the solute pulse. Some of these observations have also been discussed in the previous chapter. The curves presented in Fig. 10.2 are representative of the processes of desodication ($n < 1$) and sodication ($n > 1$), although the adsorption isotherms may be somewhat different than the Freundlich ones used here (cf. sections A7.3.3 and A7.4 in volume A of this text, and the discussion of Figs. 9.11 and 9.14).

10.2.2. Chemical hysteresis effects

Several studies recently have revealed the possibility of a hysteresis phenomenon when studying the adsorption and desorption of different pesticides and soils. Hornsby and Davidson (1973), Swanson and Dutt (1973), and Van Genuchten et al. (1974) all encountered hysteresis in the equilibrium isotherms. In these studies the coefficients k and n in the Freundlich isotherm (10.3) were found to depend on the sorption direction, i.e. whether *ad*sorption ($\partial^wq/\partial t > 0$) or *de*sorption ($\partial^wq/\partial t < 0$) occurred. Figure 10.3 shows an example where adsorption and desorption follow different curves. These isotherms were used by Van Genuchten et al. (1974) to describe the movement of the herbicide picloram (4-amino-3,4,6-trichloropicolinic acid) through a Norge loam soil. Other examples of

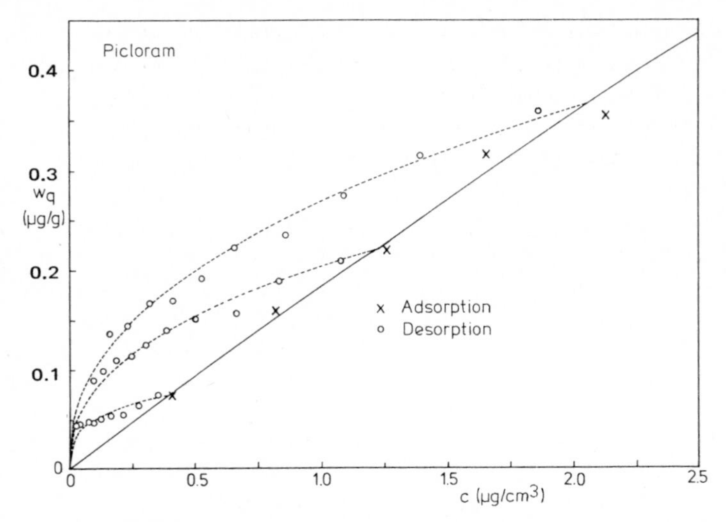

Fig. 10.3. Equilibrium adsorption (${}^{w}q_a$) and desorption (${}^{w}q_d$) isotherms for Picloram sorption on Norge loam. The adsorption isotherm is given by ${}^{w}q_a = 0.180\ c^{0.94}$. The desorption isotherms may be calculated from equations 10.10 and 10.11.

hysteresis can be found in the studies quoted above. The desorption parameters of the curves plotted in Fig. 10.3 may be determined by equating the Freundlich equations for adsorption and desorption, as follows:

$$ {}^{w}q_r = k_a(c_r)^{n_a} = k_d(c_r)^{n_d} \tag{10.9} $$

where the subscripts a and d refer to adsorption and desorption, respectively, and where c_r and ${}^{w}q_r$ represent the solution and sorbed concentrations when the sorption direction is reversed from adsorption to desorption, i.e. where $\partial {}^{w}q/\partial t = 0$. Solving (10.9) for k_d yields:

$$ k_d = {}^{w}q_r \left\{ \frac{k_a}{{}^{w}q_r} \right\}^{n_d/n_a} \tag{10.10} $$

The ratio n_a/n_d was found to be a function of ${}^{w}q_r$:

$$ n_a/n_d = 2.105 + 0.062\,{}^{w}q_r^{-1.076} \tag{10.11} $$

For the three curves shown in Fig. 10.3 the values of n_a/n_d are 2.22 (${}^{w}q_r = 0.365$), 2.34 (${}^{w}q_r = 0.223$) and 2.98 (${}^{w}q_r = 0.0767$). These values are close to those reported by Swanson and Dutt (1973), who observed an average value of 2.3 for the adsorption and desorption of atrazine using different soils. Similar values were obtained also by Hornsby and Davidson (1973), and Van Genuchten et al. (1977). Smaller values, however, were obtained by Wood and Davidson (1975), and Farmer and Aochi (1974) for the desorption of different organic chemicals from different soils.

The isotherms shown in Fig. 10.3 were used by Van Genuchten et al. (1974) to calculate effluent concentration distributions of picloram from 30-cm long columns. Results for one experiment (fig. 3 in the original study) are shown in Fig. 10.4 and are compared with the experimental effluent data. The concentrations in the Figure are plotted versus the number of pore volumes leached through the soil column.

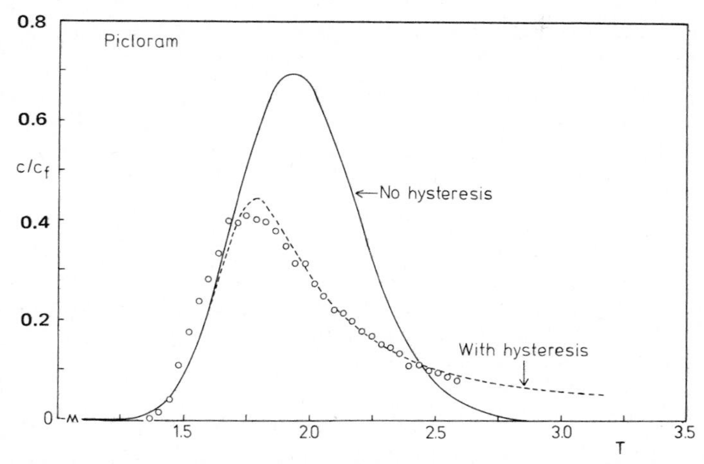

Fig. 10.4. Calculated and observed effluent curves for Picloram movement through Norge loam. The calculated curves were obtained with the following soil-physical data: $\rho_b = 1.53$ g/cm^3, $J^V = 14.2$ cm/day, $\theta = 0.363$ cm^3/cm^3, $D = 2.8$ cm^2/day, $t_0 = 0.896$ days ($T_0 = 1.17$) and $L = 30$ cm.

The number of pore volumes may be defined as the amount of liquid leached through the column, divided by the liquid capacity of the column, i.e.:

$$T \equiv \frac{J^V t}{\theta L} = \frac{vt}{L} \tag{10.12}$$

where L is the length of the column. When no adsorption occurs and diffusion/dispersion processes are neglected, the solute in the column would travel with a velocity equal to that of the liquid in which it is dissolved. The retardation factor R_f in that case would be 1.0 and the chemical will appear in the effluent after exactly one pore volume. When adsorption occurs, however, the chemical will travel slower than the liquid and hence will appear in the effluent after more than one pore volume of solution is leached through the soil column. Assuming a linear approximation of the *ad*sorption isotherm ($k_a = 0.18$, $n_a = 1.0$), and using the data of Fig. 10.4, one may calculate a retardation factor R_f of 1.76 ((10.5a)). The chemical will now travel with a mean velocity equal to $(1.76)^{-1}$ or 0.57-times the velocity of the liquid, i.e. the chemical will appear in the effluent after 1.76 pore volumes. Owing to the presence of diffusion and dispersion processes and the slight non-linear character of the adsorption isotherm, the exact break-through of the chemical may be somewhat different, but should still be centered around 1.76 pore volumes (Fig. 10.4). By plotting the relative concentration versus pore volumes, the adsorption characteristics of the chemical and the soil are more clearly shown.

Figure 10.4 illustrates the importance of using both adsorption and desorption characteristics. The solid line is calculated when the hysteresis phenomena are ignored and adsorption and desorption are assumed to follow the same (adsorption) isotherm. The effluent curve in this case acquires a fairly symmetrical shape. When hysteresis is included in the calculations a much better description of the data results. The front side of the calculated curves is not affected by hysteresis, since the chemical processes are determined here by the *ad*sorption isotherm only. During desorption, however, the isotherm is convex upwards ($n < 1$) and a long "tail" develops during elution of the chemical. The development of the tailing in Fig. 10.4 may be explained in a similar way as was done in the previous section for a general non-linear isotherm, which resulted in the dispersed appearance of the chemical near the entrance of the soil column (Fig. 10.2, $n = 0.4$).

Figure 10.4 not only shows that the occurrence of hysteresis in the equilibrium isotherms induces heavy tailing, but also that the peak concentration decreases significantly. From the Figure it is clear that excluding hysteresis effects when calculating effluent concentration distributions may result in serious disagreements between observed and calculated distributions.

10.2.3. Decay effects

The two previous examples were obtained with the assumption that the chemical undergoes no decay reactions (biological or chemical transformation). In many systems which contain such compounds as pesticides, ammoniacal fertilizers, and radioactive materials, decay processes are usually present. A thorough analytical analysis of decay was given in section 9.7, and only one example will be briefly discussed here.

To illustrate the influence of a decay sink on an effluent curve, the experimental parameters of the last example were used (assuming no hysteresis), together with a first-order decay coefficient k (cf. (10.2)) which was allowed to vary between 0.0 and 0.5 day^{-1}. Results were obtained for both a short pulse ($t_0 = 0.896$ day) and a continuous feed solution, the latter case resulting in a complete break-through of the chemical (Fig. 10.5; solid and dashed lines, respectively). Owing to the increase in degradation with increasing values of k, the maximum concentration in the effluent clearly decreases. The complete break-through curves (dashed lines) approach an equilibrium value which depends on the value of k used in the calculations. It should be noted that a slightly non-linear isotherm was used to obtain the curves plotted in Fig. 10.5 ($n = 0.94$). When the isotherm is approximated by a linear relation ($\text{k} = 0.18$, $n = 1.0$), the concentration distributions are given exactly by the analytical solution of (10.7). The steady-state value of the concentration during break-through (Fig. 10.5) for this case is given by (see for example Gershon and Nir, 1969):

$$\frac{c}{c_f} = \frac{2\exp\left\{\dfrac{vx}{2D}\left(1-\sqrt{1+4Dk/v^2}\right)\right\}}{\left(1+\sqrt{1+4Dk/v^2}\right)} \tag{10.13}$$

$$\left(= \frac{2\exp\left(-\frac{1}{2}\,v\iota x/D\right)}{(\iota+2)}\ ;\ \text{cf. } (9.107)\right) \tag{10.14}$$

It is interesting to note that this steady-state solution ($\partial c/\partial t = \partial\,{}^{w}q/\partial t = 0$) holds regardless of whether or not the chemical is adsorbed by the medium, and hence is independent of the particular form of the adsorption isotherm. The final concentrations in Fig. 10.5, reached after approximately 2.3 pore volumes, can, therefore, also be calculated with (10.13).

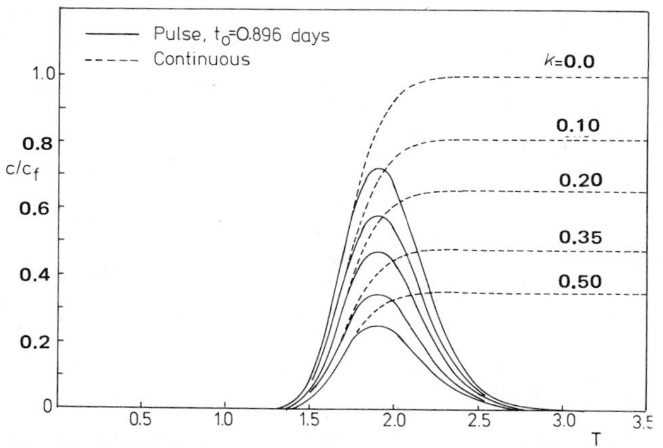

Fig. 10.5. Effect of first-order decay on Picloram movement through Norge loam. The soil-physical data are the same as used in Fig. 10.4.

10.3. NON-EQUILIBRIUM ADSORPTION

In Fig. 10.4 experimental effluent concentrations were compared with calculated curves which were obtained with the assumption that at any time equilibrium exists between sorbed (${}^{w}q$) and solution (c) concentrations. Although an excellent description of the experimental data was obtained, provided the observed hysteresis phenomenon was included in the numerical calculations, some small deviations are apparent at the break-through side of the effluent curves. The chemical seems to travel somewhat faster than predicted with the equilibrium model. Similar and more severe deviations between experimental and equilibrium curves have been observed in a number of studies (Kay and Elrick, 1967; Davidson and Chang, 1972; Green et al., 1972; Van Genuchten et al., 1974, 1977; Wood and Davidson, 1975).

The deviations mentioned here are presumably caused by the inability of the chemical to be adsorbed instantaneously by the medium during its transport through the soil.

Many explanations have been given to account for this inability to reach equilibrium. First there is the possibility that a true kinetic adsorption mechanism is present, i.e. the chemical will be adsorbed slowly, even when it is assured that the chemical molecule and soil adsorption site are physically in close contact. Hence a finite rate of adsorption is present. When this is the case and it is the only reason for non-equilibrium, the movement of the chemical in a soil should be described completely by coupling the transport equation (10.2) with one of the kinetic rate equations listed in Table 10.1 (or with any other appropriate rate equation).

A second possible reason for non-equilibrium may be the significant physical resistances encountered by a chemical in trying to reach the sorption sites of the porous matrix during its movement through the soil. This may be especially true for aggregated (Green et al., 1972) and/or unsaturated soils (Nielsen and Biggar, 1961). In these cases the chemical has to diffuse first out of the larger, liquid-filled pores (mobile liquid) into an immobile liquid region before it can be adsorbed by those sorption sites located inside the aggregates or along the walls of blind pores. This situation is now merely a physical problem in that the physical make-up of the medium is responsible for non-equilibrium (*physical* vs. *kinetic* non-equilibrium). Adsorption in the immobile region of the soil becomes now a diffusion-controlled mechanism. It should be noted that this physical non-equilibrium may be demonstrated also with chemicals which are not adsorbed by the medium. Material still has to diffuse into the immobile zone, especially when the soil is highly aggregated (see for example the studies by Nielsen and Biggar, 1961; McMahon and Thomas, 1974; Van de Pol, 1974). Both mechanisms for non-equilibrium will now be discussed separately.

10.3.1. Kinetic non-equilibrium

As mentioned previously, serious deviations have been observed between experimental data and predictions based on equilibrium models. Figure 10.6 shows an example of how serious these deviations can be. The example is taken from Van Genuchten et al. (1977), who studied the adsorption and movement of the herbicide 2,4,5-T (2,4,5-Trichlorophenoxyacetic acid) through 30-cm long columns containing Glendale clay loam soil. The equilibrium adsorption and desorption isotherms for this herbicide and soil are given in Fig. 10.7. Note again the hysteresis between adsorption and desorption. The isotherms, both for adsorption and desorption, are described with Freundlich-type equations. The ratio n_a/n_d was found to be 2.3 ((10.10)), resulting in desorption curves of the form:

$$^{w}q = k_d c^{0.344} \tag{10.15}$$

The curves plotted in Fig. 10.6 were obtained with and without the inclusion of hysteresis effects in the calculations. Once again one can see that the elution side of the experimental effluent curve is reasonably well described when hysteresis is taken into account. The break-through part of the effluent curve, however, is poorly predicted.

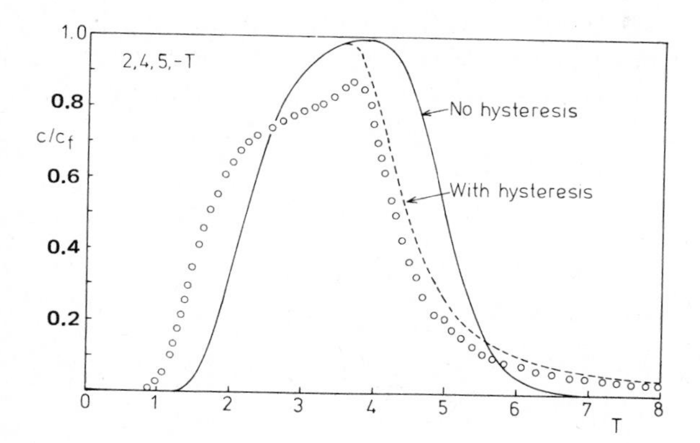

Fig. 10.6. Calculated and observed effluent curves for 2,4,5-T movement through Glendale clay loam. The following soil-physical data were used in the calculations: $\rho_b = 1.36$ g/cm^3, $J^V = 5.11$ cm/day, $\theta = 0.473$ cm^3/cm^3, $T_0 = 2.761$, $D = 8.4$ cm^2/day and $L = 30$ cm.

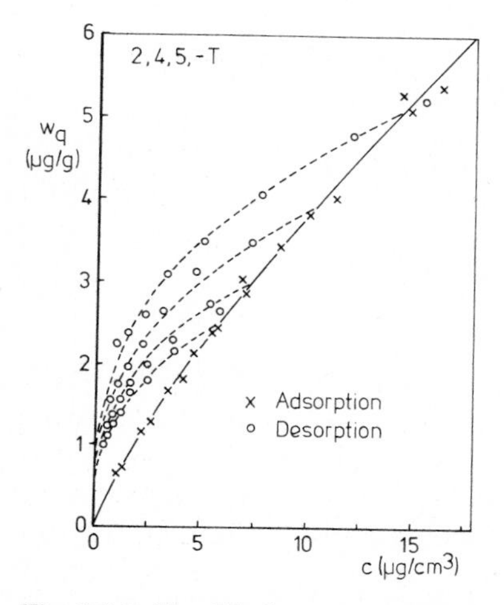

Fig. 10.7. Equilibrium adsorption ($^{w}q_a$) and desorption ($^{w}q_d$) isotherms for 2,4,5-T sorption on Glendale clay loam. The adsorption isotherm is given by $^{w}q_a = 0.616\ c^{0.792}$, the desorption curves by $^{w}q_d = k_d c^{0.344}$.

Several authors have suggested the use of a kinetic model, rather than an equilibrium one, to improve the predictions. A rate equation frequently used is based on model 2.3 (Table 10.1). In the present notation, this equation may be written as:

$$\frac{\partial\, {}^{w}q}{\partial t} = k_r(kc^n - {}^{w}q) \tag{10.16}$$

where k_r is a kinetic rate coefficient with units of day^{-1}. Several analytical solutions of (10.2) and (10.16) may be derived, provided the adsorption isotherm is linear ($n = 1$) and hysteresis is neglected (Van Genuchten, 1974, appendix B). Subject to the particular initial and boundary conditions of (10.2a), the solution may be expressed in the following dimensionless form:

$$\frac{c}{c_f} = \begin{cases} c_1(s, t) & 0 < t \leqslant t_0 \\ c_1(s, t) - c_1(s, t - t_0) & t > t_0 \end{cases} \tag{10.17}$$

with:

$$c_1(s, t) = G(s, t) \exp(-rR_D t) + r \int_0^t G(s, \tau)\, [R_D I_0(\zeta) + \sqrt{R_D \tau/(t-\tau)}\, I_1(\zeta)] \exp[-r(R_D\tau + t - \tau)]\, d\tau \tag{10.18}$$

$$G(s, t) = \frac{1}{2} \operatorname{erfc} \sqrt{\frac{s(1-t)^2}{4t}} + \sqrt{\left(\frac{st}{\pi}\right)} \exp\left\{\frac{-s(1-t)^2}{4t}\right\} - \frac{1}{2}(1 + s + st)e^s \operatorname{erfc} \sqrt{\frac{s(1+t)^2}{4t}} \tag{10.19}$$

$$\zeta = 2r\sqrt{R_D \tau (t - \tau)}; \qquad R_D = \frac{\rho_b k}{\theta}; \qquad r = \frac{k_r x}{v} \tag{10.20}$$

$$s = \frac{vx}{D}; \qquad t = \frac{vt}{x}; \qquad t_0 = \frac{vt_0}{x} \tag{10.21}$$

where I_0 and I_1 are modified Bessel functions of the first kind and of order 0 and 1, respectively. To maximize generality, the solution is expressed in four dimensionless variables. The variable R_D is a system constant since it is completely defined by properties of the chemical and medium. The variable r defines to what degree local equilibrium between solution concentration and adsorbed concentration is reached during the transfer process, and is essentially the same as the scaled variable $\hat{x}$, introduced in section 9.6 of the previous chapter. The *scaled* variables s and t reduce to the often used Péclet number, Pé ($= vL/D$), and the number of pore volumes, T ((10.12)), when effluent concentrations from a finite soil column are considered.

It should be mentioned here that all solutions given in this chapter are derived for a semi-infinite medium. When these solutions are used to calculate effluent concentrations by setting $x = L$, some small errors may be introduced. However, owing to the uncertainty of the exact physical processes at the exit of the column as well as the relatively small influence of the mathematical boundary conditions generally imposed on the system, i.e. the condition $\partial c/\partial x = 0$ at $x = L$ (see for example the discussions by Wehner and Wilhelm, 1956; Pearson, 1959; Van Genuchten and Wierenga, 1974), the solutions obtained for a semi-infinite medium should provide close approximations for those of a finite medium.

The analytical solution given above will now be used to investigate whether or not the kinetic model represents an improvement over the equilibrium model in describing the experimental data of Fig. 10.6. To be able to use this solution, the observed hysteresis effects have to be ignored. This presents no objection, however, since interest for the moment is only in the description of the break-through side of the effluent curve (hysteresis has no effect on this part of the curve as was shown in Figs. 10.4 and 10.6). It is further necessary to linearize the slightly non-linear adsorption isotherm of Fig. 10.7. The following procedure was used by Van Genuchten et al. (1977) to obtain this linearization. Denoting the linearized isotherm by ${}^{w}q = k_a^l c$, and requiring that the areas under the isotherm over the range 0—10 ppm (the column was leached with a 10-ppm 2,4,5-T solution) for both the linearized and the non-linear Freundlich equations be the same, one has from Fig. 10.7:

$$\int_0^{10} k_a^l c\,dc = \int_0^{10} 0.616\, c^{0.792}\,dc \qquad (10.22)$$

from which a value of 0.426 for the linearized adsorption constant, k_a^l, may be calculated.

Figure 10.8 compares the experimental data of Fig. 10.6 with curves based on (10.17)—(10.21), using values of the dimensionless rate coefficient, r, ranging from 0 to ∞. The curve labeled $r = 0$ represents the solution when no adsorption takes place (no exchange of material), while the curve for which $r \to \infty$ represents the limiting case of equilibrium adsorption. It is evident from Fig. 10.8 that no one value of r results in an acceptable description of the experimental data. It was found that the predictions could not be improved by either varying the Peclet number (Pé), or by including the non-linearity of the adsorption isotherm into a numerical solution of the same kinetic model. Similar plots were prepared, based on either model 2.4 or 2.5 (Table 10.1) and these showed only minor differences from those given in Fig. 10.8. Hence the kinetic model seems unable to fit the experimental data of Fig. 10.6.

Several other authors have compared predictions based on kinetic models with experimental data. Hornsby and Davidson (1973) compared the same kinetic model as discussed above with experimental data on the movement of Fluometuron through Norge loam. Concentration distributions versus depth were presented for two pore-water velocities (v): 0.59 and 5.50 cm/h. The values of the ratio k_r/v for these two experiments were 5.93 and 0.64, respectively. By comparison, the ratio k_r/v for the curve labeled 7.0 in Fig. 10.8 equals 7.0/30, or 0.23 cm^{-1}. The rather high values of k_r/v used by Hornsby and Davidson (1973) suggest that their calculated curves would be the same or very close to those obtained with equilibrium solutions. The kinetic model used by these authors was not compared with the equilibrium model; hence no conclusions could be drawn on whether the kinetic model actually improved the description of the data or not.

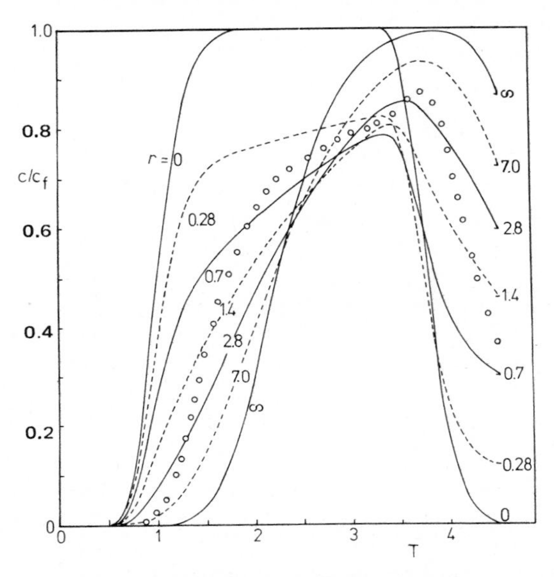

Fig. 10.8. Effect of the dimensionless rate coefficient, r, on calculated effluent curves for 2,4,5-T movement through Glendale clay loam. The open circles represent observed data points, and are the same as those shown in Fig. 10.6.

Van Genuchten et al. (1974) compared observed and calculated effluent concentrations for the movement of picloram through Norge loam. The calculated curves were based on both the equilibrium model and kinetic models 2.3 and 2.4 (Table 10.1). No satisfactory description of the experimental data could be obtained with any of the three models, except for a few experiments where the experimental data approached the equilibrium model. In that case the predictions of the kinetic and equilibrium models were about the same, since again rather high rate coefficients had to be used. The two kinetic models hence did not substantially improve the predictions of the data. The authors suggested that physical rather than kinetic phenomena produced the observed deviations between experiment and calculations.

Further experimental and calculated concentration profiles in soil columns were presented by Huggenberger et al. (1972) for the movement of lindane through different soils. These authors used the model derived by Oddson et al. (1970), i.e. similar to (10.17) but with negligible dispersion. The solution of Oddson et al. (1970) is rather conveniently expressed with Goldstein's (1953) J-function, as follows:

$$\frac{c}{c_f} = \begin{cases} c_1(t) & 0 < t \leqslant t_0 \\ c_1(t) - c_1(t - t_0) & t > t_0 \end{cases} \tag{10.23}$$

where

$$c_1(t) = J(u, w) = \begin{cases} 0 & t \leqslant 1 \\ 1 - e^{-w} \int_0^u e^{-\tau} I_0(2\sqrt{w\tau}) d\tau & t > 1 \end{cases} \tag{10.24}$$

and

$$u = rR_D; \quad w = r(t-1) \tag{10.25}$$

The use of this model by Huggenberger et al. (1972) resulted in a rather poor prediction of the experimental data. The chemical not only traveled much faster into the column than predicted by their model (the ratios k_r/v used by these authors ranged from 0.1 to 1.0 cm^{-1}), but the concentration distribution also appeared much more dispersed in the column than could be described by the no-dispersion model.

From the discussions above it appears that as yet no conclusive evidence exists that kinetic rate models are able to describe the often observed faster movement of the chemical than that predicted by equilibrium models. Obviously, more experimental evidence should be produced before definitive conclusions can be reached. However, the limited amount of experimental and theoretical investigations to date suggest that other, maybe more important, physical soil phenomena may have to be theoretically included if one is to describe satisfactorily the observed experimental data.

10.3.2. Physical non-equilibrium

Nearly all attempts to describe solute movement in soils have been based thus far on the convective—dispersive equation, appropriately modified to include such processes as adsorption and decay ((10.2)). Without adsorption and decay, this equation becomes:

$$\frac{\partial c}{\partial t} = D\frac{\partial^2 c}{\partial x^2} - v\frac{\partial c}{\partial x} \tag{10.26}$$

The general features of (10.26) have been discussed at length in section 9.5 ((9.51)). Except for small values of the dimensionless group vx/D, solute distributions based on (10.26), whether plotted versus depth or as effluent curves, appear either sigmoidal in shape or more or less symmetrical (depending on the boundary conditions imposed at the soil surface).

As early as in 1961, Nielsen and Biggar (1961) observed serious deviations between the sigmoid-type break-through curves predicted with (10.26) and experimental curves for the movement of chloride through Oakley sand and glass beads. The lower the water content during their experiments, the more serious these deviations became. The solute appeared earlier in the effluent than could be predicted with the equation, while considerably more water was needed to reach a relative concentration of one. The early break-through of the chemical and the tailing towards complete displacement was attributed to the unsaturated conditions during their experiments. The authors argued that, when the water content decreases, more and more of the larger pores are eliminated for transport, resulting in a larger proportion of the water being located in relatively slow moving or immobile (stagnant or dead) zones. They suggested further that these almost stagnant or immobile water zones would act as sinks to ionic diffusion during the displacement process, leading to a nearly incomplete displacement. The occurrence of this stagnant water has been the subject of many other studies (Turner, 1958; Aris, 1959; Coats and Smith, 1964; Villermaux and Van Swaay, 1969; Gupta et al., 1973).

Several other conditions seem to influence the occurrence of immobile water, such as the presence of aggregates (Biggar and Nielsen, 1962; Green et al., 1972), or the use of undisturbed soil cores (McMahon and Thomas, 1974). The magnitude of the flow rate has also been linked with the presence of immobile water (Biggar and Nielsen, 1962; Gaudet et al., 1977).

Several theoretical studies have dealt with the presence of immobile liquid zones (see also chapter 9). Deans (1963) modified (10.26) to include the transfer by diffusion from mobile to immobile zones. His model was later used by Coats and Smith (1964), Villermaux and Van Swaay (1969), and Bennet and Goodridge (1970), and may be written in the following form:

$$\theta_m \frac{\partial c_m}{\partial t} + \theta_a \frac{\partial c_a}{\partial t} = \theta_m D \frac{\partial^2 c_m}{\partial x^2} - v_m \theta_m \frac{\partial c_m}{\partial x} \tag{10.27}$$

$$\theta_a \frac{\partial c_a}{\partial t} = k_a (c_m - c_a) \tag{10.28}$$

where the subscripts m and a refer to mobile and immobile liquid regions, respectively, and v_m is the average interstitial velocity of the liquid in the mobile region of the soil ($v_m = J^V/\theta_m$). The mass transfer coefficient k_a (day^{-1}) determines the rate of exchange between the two liquid regions. The model assumes that this rate of exchange is proportional to the concentration difference between the two liquid regions.

Equations (10.27) and (10.28) assume that no adsorption is present.

Van Genuchten and Wierenga (1976b) modified the model to include adsorption. The following set of differential equations resulted:

$$\theta_m \frac{\partial c_m}{\partial t} + \rho_b p \frac{\partial\, {}^w q_m}{\partial t} + \theta_a \frac{\partial c_a}{\partial t} + \rho_b(1-p)\frac{\partial\, {}^w q_a}{\partial t}$$

$$= \theta_m D \frac{\partial^2 c_m}{\partial x^2} - v_m \theta_m \frac{\partial c_m}{\partial x} \qquad (10.29)$$

$$\rho_b(1-p)\frac{\partial\, {}^w q_a}{\partial t} + \theta_a \frac{\partial c_a}{\partial t} = k_a(c_m - c_a) \qquad (10.30)$$

where ${}^w q_m$ and ${}^w q_a$ represent the amounts adsorbed in the "mobile" and stagnant regions, respectively, both specified per unit mass of soil assigned to these two regions. The mass fraction of solid phase assigned to the mobile region on the basis of the corresponding adsorption capacity, Q, is indicated with the parameter p, such that ${}^w Q_m = p\,{}^w Q$, and ${}^w Q_a = (1-p)\,{}^w Q$. Equations (10.29) and (10.30) were derived by assuming that adsorption around the larger liquid-filled pores would not necessarily be the same as the adsorption around the micropores in the immobile zone. When a chemical moves through an aggregated and/or unsaturated soil, only part of the adsorption sites may be readily accessible for the invading solution. These sites are located around the larger pores and in direct contact with the mobile liquid. However, when an immobile liquid zone is present, some of the material can only be adsorbed by the stagnant part of the medium after it has diffused into the immobile zone, made up of dead (blind) pores and micropores inside aggregates. The division of the adsorption sites into two fractions, one fraction located in the 'mobile' region and one fraction in the stagnant region of the soil is characterized by the parameter p. When $p = 0$, all adsorption occurs away from the mobile liquid inside the stagnant region of the medium. In that case the model reduces to the one elaborated upon in section 9.6.

When it is assumed that equilibrium adsorption occurs, and the Freundlich isotherm is again used to describe the relation between sorbed and solution concentrations ((10.3)), the number of dependent variables may be decreased from four to two. Substitution of (10.3) into (10.29) and (10.30) leads then to:

$$[\theta_m + \rho_b p k n c_m^{n-1}] \frac{\partial c_m}{\partial t} + [\theta_a + \rho_b(1-p) k n c_a^{n-1}] \frac{\partial c_a}{\partial t}$$

$$= \theta_m D \frac{\partial^2 c_m}{\partial x^2} - v_m \theta_m \frac{\partial c_m}{\partial x} \qquad (10.31)$$

$$[\theta_a + \rho_b(1-p) k n c_a^{n-1}] \frac{\partial c_a}{\partial t} = k_a(c_m - c_a) \qquad (10.32)$$

Several analytical solutions of (10.31) and (10.32) are available, provided, again, that the adsorption isotherm is linear ($n = 1$) and that no hysteresis between adsorption and desorption is present (Appendix B of Van Genuchten, 1974). For initial and boundary conditions equivalent to (10.2a) (c has to be replaced by c_m), the dimensionless solution is:

$$\frac{c_m}{c_f} = \begin{cases} c_1(s, t) & 0 < t \leqslant t_0 \\ c_1(s, t) - c_1(s, t - t_0) & t > t_0 \end{cases} \qquad (10.33)$$

$$\frac{c_a}{c_f} = \begin{cases} c_2(s, t) & 0 < t \leqslant t_0 \\ c_2(s, t) - c_2(s, t - t_0) & t > t_0 \end{cases} \qquad (10.34)$$

with:

$$c_1(s, t) = G(s, t) \exp(-rt/zR_f) + \frac{r}{R_f} \int_0^t G(s, \tau)\, H_1(t, \tau) d\tau \qquad (10.35)$$

$$c_2(s, t) = r \int_0^t G(s, \tau)\, H_2(t, \tau) d\tau \qquad (10.36)$$

$$G(s, t) = \frac{1}{2} \operatorname{erfc}\left\{\frac{s^{1/2}(zR_f - t)}{(4zR_f t)^{1/2}}\right\} + \left(\frac{st}{\pi z R_f}\right)^{1/2} \exp\left\{\frac{-s(zR_f - t)^2}{4zR_f t}\right\}$$

$$- \frac{1}{2}\left(1 + s + \frac{st}{zR_f}\right) e^s \operatorname{erfc}\left\{\frac{s^{1/2}(zR_f + t)}{(4zR_f t)^{1/2}}\right\} \qquad (10.37)$$

$$H_1(t, \tau) = e^{-u-w}\left\{I_0(y)/z + I_1(y)/(1-z)\left(\frac{u}{w}\right)^{1/2}\right\} \qquad (10.38)$$

$$H_2(t, \tau) = e^{-u-w}\left\{I_0(y)/(1-z) + I_1(y)/z\left(\frac{w}{u}\right)^{1/2}\right\} \qquad (10.39)$$

$$y = 2\sqrt{uw}; \qquad u = \frac{r\tau}{zR_f}; \qquad w = \frac{r(t-\tau)}{(1-z)R_f} \qquad (10.40)$$

The (italic) dimensionless variables are defined as:

$$z = \frac{\theta_m R_{f,m}}{\theta R_f}; \qquad s = \frac{v_m}{D}; \qquad t = \frac{vt}{x}; \qquad r = \frac{k_a x}{\theta v} \qquad (10.41)$$

where:

$$\theta_m R_{f,m} = \theta_m + \rho_b pk; \quad \theta R_f = \theta + \rho_b k; \quad v = \frac{v_m \theta_m}{\theta} = \frac{J^V}{\theta} \qquad (10.42)$$

Note that when effluent concentration distributions are considered ($x = L$), s and t, respectively reduce to the column Peclet number Pé ($= v_m L/D$) and the number of pore volumes leached through the column, $T(= vt/L)$. In that case the solutions are expressed in four independent parameters, R_f, Pé, z, and r. Of these four parameters, the retardation factor is the most easily determined independently, for example, from batch equilibration studies. The remaining three parameters however are less easy to quantify. The parameter z, for example, depends on the fraction of mobile liquid in the medium $\phi_m (= \theta_m/\theta)$ and the fraction of the sorption sites located in the dynamic region of the soil (p). These quantities are difficult to quantify a priori. Generally elaborate curve-fitting techniques have to be used to determine their values (Van Genuchten and Wierenga, 1977).

Referring to (9.48) and (9.102) of the previous chapter it is pointed out that for $p = 0$ (all adsorption sites assumed to be located within the stagnant region) one obtains $z = \theta_m/(\theta + \rho_b k)$, such that $(1 - z) = A\phi_a$.

Before continuing the discussion of the 2,4,5-T effluent data given in the previous two sections, a sensitivity analysis will be carried out for each of the four dimensionless parameters z, R_f, Pé, and r. Each variable was allowed to vary while keeping the remaining three constant. Results are shown in Figs. 10.9—10.12. The concentration curves were plotted versus pore volumes, assuming that the chemical was in the feed solution during the first three pore volumes.

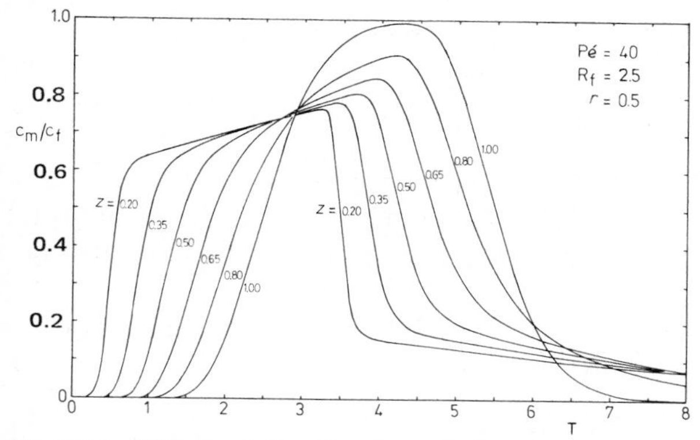

Fig. 10.9. Effect of the dimensionless variable z on calculated effluent curves from an aggregated sorbing medium.

The influence of the parameter z is shown in Fig. 10.9. This parameter defines the relative contribution of the mobile phase to the overall retardation of a solute front in the system. In the limiting case when $z = 1$, no immobile water is present ($\phi_m = 1$) and all sorption sites are immediately accessible for the displacing chemical solution ($p = 1$). The model then reduces to the convective—dispersive equation, modified for linear adsorption ((9.45)), and the effluent curve becomes nearly symmetrical in shape

(Fig. 10.9). When a large immobile liquid zone is present ($\theta_m \ll \theta$), the velocity of the liquid in the mobile region will increase because the convective transfer processes are confined to a smaller cross-sectional area ($v_m = J^V/\theta_m$). Particularly the combination of $\phi_m \ll 1$ and $p \ll 1$ (stagnant regions containing a substantial fraction of the sorption sites) leads to an early break-through of the chemical (a "faster" moving liquid with relatively less retardation in the mobile region). At the same time the chemical may diffuse slowly from the mobile into the stagnant region, where part of it will be adsorbed. This diffusion process will continuously remove material from the mobile region, resulting in a "tailing" phenomenon as manifested by the slow increase in concentration after the first appearance of the chemical in the effluent (Fig. 10.9). The extreme case when $z = 0.4$ may occur when the medium is highly aggregated or exhibits severe channeling (also for certain fractured media). In that case the effluent curves become very distorted (asymmetrical). It should be noted that the analytical solution ((10.33)–10.42)) breaks down when $\theta_m = \theta$ and $p < 1$ (some sorption sites are located in "dry" stagnant regions, possibly in densely compacted aggregates or between closely packed soil particles). Some of the sites are then assumed to be completely inaccessible for the displacing fluid. Total adsorption as seen by the "undisturbed" soil will now become less than that inferred from static (batch) equilibrium measurements on 'disturbed' soil ($R_{f,m} < R_f$). The solute distribution in this case will retain its nearly symmetrical shape as predicted by the linear equilibrium model (section 10.2.1), but the resulting effluent curve will be displaced to the left of the curve labeled 1.0 in Fig. 10.9. This particular approach was used by Van Genuchten et al. (1974) and Wood and Davidson (1975) to model the relatively faster movement of the chemical through the soil.

The influence of the retardation factor is shown in Fig. 10.10. As expected, the chemical appears later in the effluent when R_f increases (i.e. when sorption increases). Also the peak concentration seems to decrease

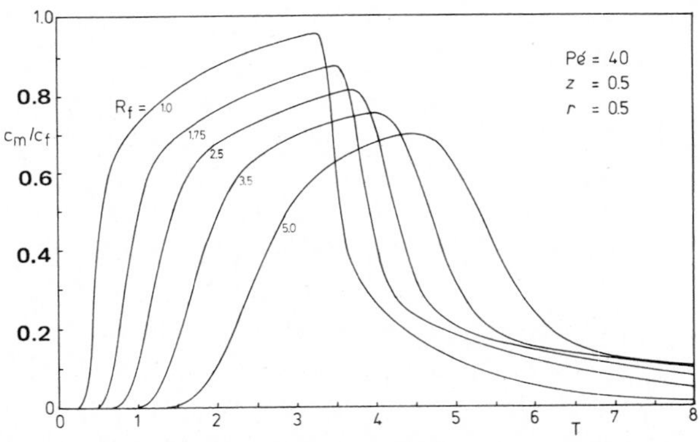

Fig. 10.10. Effect of the retardation factor, R_f, on calculated effluent curves from an aggregated sorbing medium.

somewhat with increasing adsorption. When $R_f = 1$, no adsorption occurs and the model simulates the movement of an inert chemical through an aggregated medium. Tailing in this case becomes less pronounced (Fig. 10.10), because no adsorption can take place in the stagnant region and the chemical can be stored only in the immobile liquid of the stagnant region. Hence relatively less chemical has to diffuse into the immobile zone.

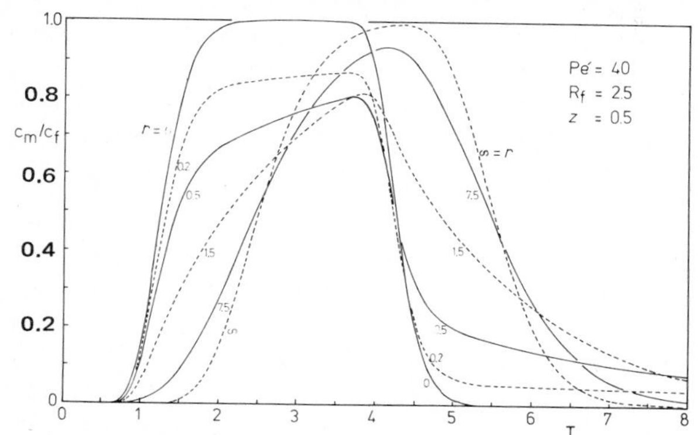

Fig. 10.11. Effect of the dimensionless mass transfer coefficient, r, on calculated effluent curves from an aggregated sorbing medium.

Fig. 10.11 shows the effect of the mass transfer coefficient r on the shape and position of the effluent concentration distributions. When $r = 0$ no material can diffuse into the stagnant region of the soil, and adsorption is confined to the dynamic region of the soil. In this case, the effluent curve again acquires a nearly symmetrical shape. When r is small, the exchange by diffusion between dynamic and stagnant regions is still very slow, resulting in the bulk of the chemical traveling through the medium nearly unaffected by the exchange. However, the stagnant zone keeps exchanging material with the mobile liquid, resulting in extensive tailing, both during break-through and elution. With increasing values of the mass transfer coefficient, the rate of exchange increases, eventually leading to a new equilibrium where the concentrations in mobile and immobile liquid are identical ($c_m = c_a$). This limiting case is represented by the curve $r \to \infty$. The curves given in Fig. 10.11 are nearly identical to those shown in Fig. 10.8. This is not surprising because the intra-aggregate diffusion model ((10.31) and (10.32)) is mathematically similar to the kinetic adsorption model ((10.2) and (10.16)).

Finally, the influence of the column Peclet number, Pé, is shown in Fig. 10.12. The influence of Pé on the shape of the calculated curves is relatively small. Note the discontinuities of the curve for which Pé $= \infty$, at both 1.3, and 4.3 pore volumes. From Fig. 10.12 one may conclude that an approximate value of the dispersion coefficient generally is sufficient when describing the leaching of chemical through aggregated, unsaturated soils. An interesting observation may be made when comparing Figs. 10.11 and 10.12. When r is large, a small change in the value of this coefficient seems to have

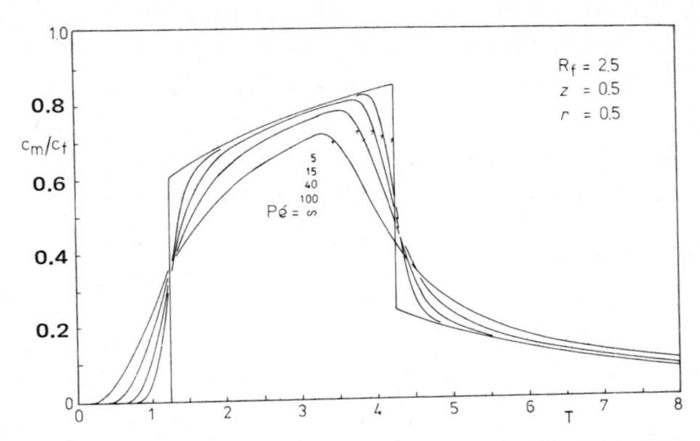

Fig. 10.12. Effect of the Peclet number, Pé, on calculated effluent curves from an aggregated sorbing medium.

the same effect as a change in the dispersion coefficient. Also, from Fig. 10.8, when the kinetic rate coefficient is large, i.e. near equilibrium adsorption, a small change in this coefficient results in effects which could also be obtained in an approximate way by a change in the dispersion coefficient (see the relevant discussion of Fig. 9.17). These are rather important observations. It seems highly probable that when the dispersion coefficient is fitted to experimentally obtained concentration profiles, it will also include any small kinetic or intra-aggregate diffusion effects which may be present.

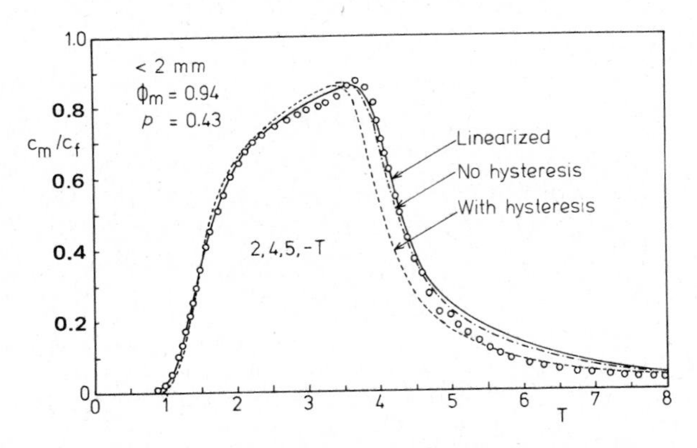

Fig. 10.13. Calculated and observed effluent curves for 2,4,5-T movement through Glendale clay loam. The open circles represent observed data points, and are the same as those shown in Fig. 10.6.

The question remains whether the intra-aggregate diffusion model is able to better describe the previously shown experimental 2,4,5-T effluent data than was possible with either the equilibrium model (Fig. 10.6) or the kinetic model (Fig. 10.8). Figure 10.13 shows a best-fit comparison of the experimental data with the diffusion model, using a value of 0.426 for the linearized adsorption constant k_a^l. Values of the three curve-fitted parameters, z, Pé, and r, are shown in the Figure. An important improvement in the description of the data is obtained; especially the front side of the effluent curve is well described by the model. The same Figure also shows the effects of hysteresis and non-linear adsorption on the calculated results. It appears that these influences are rather small. The slope of the break-through side of the curve increases slightly when non-linear adsorption is considered (section 10.2.1). The inclusion of hysteresis in the model again does not affect the front part of the curve, but it results in an earlier decrease in concentration during elution of the chemical, as well as more prolonged tailing at the higher pore volumes. However, from Fig. 10.13 it is clear that hysteresis does not affect the results to any great extent. Such a conclusion is not possible when intra-aggregate diffusion is neglected, as was shown in Figs. 10.4 and 10.6. Thus for the conditions of this study, intra-aggregate exchange of 2,4,5-T between dynamic and stagnant regions seems to be far more important than the observed hysteresis in the adsorption—desorption process.

The curves representing non-linear adsorption, with and without hysteresis (Fig. 10.13), were obtained with a numerical solution of (10.31) and (10.32) (Van Genuchten and Wierenga, 1976a). To be able to use this solution, one needs separate estimates of the parameters p and θ_m. These may be obtained by leaching the same column with an inert material such as tritium. Figure 10.14a shows effluent distributions of tritiated water which were obtained from the same column for which the 2,4,5-T data were obtained. The effluent data of tritium were also fitted to the linearized intra-aggregate diffusion model, resulting in the parameter values shown in the Figure. A small retardation of the tritium was observed during its displacement through the column. The linearized adsorption constant k_a^l in the Freundlich isotherm was found to be 0.009, equivalent to a retardation factor R_f of 1.027. Estimates of the parameters p and $\phi_m (= \theta_m/\theta)$ may then be obtained by assuming that they are the same for both tritium and 2,4,5-T movement. From the definitions of z ((10.41)), $R_{f,m}$, and R_f, one obtains:

$$zR_f = \phi_m + p(R_f - 1) \tag{10.43}$$

Using the values for 2,4,5-T (Fig. 10.13; $R_f = 2.225$; $z = 0.661$) and tritiated water (Fig. 10.14a; $R_f = 1.027$; $z = 0.926$), (10.43) leads then to:

$$\begin{aligned} (0.661)(2.225) &= \phi_m + p(1.225) \qquad (2,4,5\text{-T}) \\ (0.926)(1.027) &= \phi_m + p(0.027) \qquad (^3H_2O) \end{aligned} \tag{10.44}$$

or:

$$\begin{aligned} \phi_m &= \frac{\theta_m}{\theta} = 0.94 \\ p &= 0.43 \end{aligned} \tag{10.45}$$

The above derivation assumes that the ratio of mobile to total water (θ_m/θ) for the two experiments is the same. This assumption seems reasonable since the two displacements were carried out at approximately the same flow velocities (J^V equals 5.11 and 5.09 cm/day, respectively for 2,4,5-T and tritium) and at the same water contents (θ equaled 0.473 and 0.460, respectively). The assumption that p is also the same for the two chemicals, however, may not be entirely true owing to the different adsorption mechanisms for 2,4,5-T and tritium. The herbicide 2,4,5-T is adsorbed by the organic matter fraction of the soil (O'Connor and Anderson, 1974), while the observed tritium retardation may be caused by some isotopic exchange of tritium with crystal-lattice hydroxyls of the clay fraction of the soil, or by replacement of some exchangeable cations by $^3H^+$. However, by changing the value of p for tritium slightly and considering the small value of the retardation factor for tritium, it is easily demonstrated that the values of ϕ_m and p (for 2,4,5-T) will remain essentially the same.

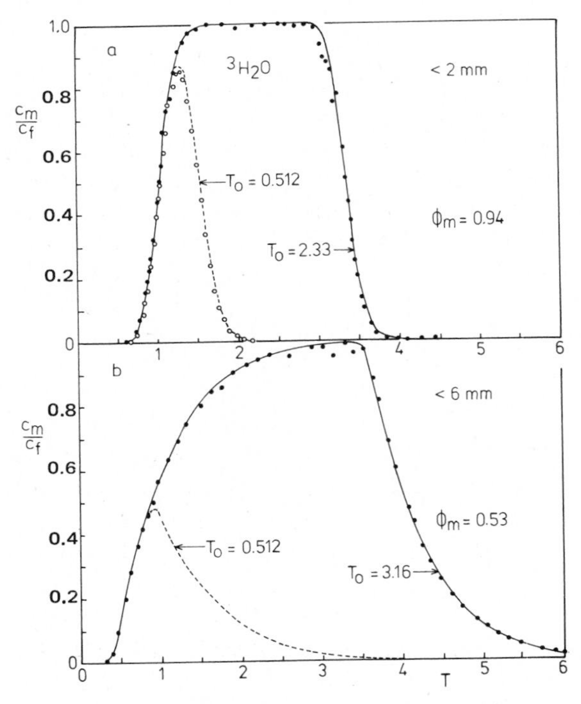

Fig. 10.14. Calculated and observed effluent curves for tritium movement through Glendale clay loam. Results were obtained with two different aggregate sizes and bulk densities, but with approximately the same pore-water velocities. a: $<$ 2 mm aggregates, $\rho_b = 1.36\ g/cm^3$, $J^V = 5.09$ cm/day, $\theta = 0.460\ cm^3/cm^3$, $z = 0.926$, Pé = 95, $r = 1.47$ and $R_f = 1.027$. b: $<$ 6 mm aggregates, $\rho_b = 1.13$, $J^V = 5.54$ cm/day, $\theta = 0.399\ cm^3/cm^3$, $z = 0.531$, Pé = 35, $r = 1.54$, and $R_f = 1.025$.

The value of the parameter ϕ_m as derived above indicates that about 94% of the water in the column may be considered mobile water, and that only 6% is immobile. This small amount of immobile water has little effect on the calculated results for tritium (Fig. 10.14a). In fact, the effluent distribution of this example could be described equally well with appropriate solutions of the convective—dispersive equation (10.26). When the amount of immobile water increases, however, such a description becomes exceedingly difficult. Figure 10.14b represents a more extreme case where nearly 50% of the water in the column is immobile. Note that the effluent curves, for both a short and long solute pulse, become much more asymmetrical than the approximately symmetrical ones shown in Fig. 10.14a. The curves in Fig. 10.14b were obtained using larger aggregates in the soil column, as well as a lower bulk density. Both conditions seem to favor the presence of stagnant liquid regions in the soil. When larger aggregates are present, the pore-size distribution is likely to become much broader owing to the creation of larger interaggregate pores and the existence of a significant amount of small intra-aggregate (micro) pores. An increase in aggregate size also increases the diffusion length between mobile and immobile regions, leading to a slower exchange of material between the two liquid regions. The bulk density may also affect the amount of immobile water present. When the bulk density decreases, the pore-size distribution is also likely to become broader, owing to the creation of larger interaggregate pores (given the same aggregate size). The broader the pore-size distribution, the broader also the pore-water velocity distribution will become: some water will move faster in the soil, and some water will not move at all or only very slowly compared to the average pore-water velocity. Hence, the diffusion processes between fast and slowly moving liquid zones become more important. The curves shown in Fig. 10.14b may be typical for the leaching processes in undisturbed aggregated field soils.

The influence of aggregate size on the effluent distributions for 2,4,5-T seems to be less than for tritium. Figure 10.15 shows the effluent concentration profiles obtained from aggregates of less than 6 mm (experiment 4-1 of Van Genuchten et al., 1977). The chemical appears slightly earlier in the effluent compared to the experimental data shown in Fig. 10.13, but this is primarily a result of the presence of more immobile liquid. The value of p found for the larger aggregates is only slightly smaller than for the smaller (< 2 mm) aggregates: 0.40 vs. 0.43. This value was not significantly affected by either aggregate size or bulk density (Van Genuchten et al., 1977). The average value determined was 0.40, indicating that 40% of the sorption sites are located around the larger, liquid-filled pores of the dynamic region, while 60% of the sites are located either inside aggregates or along blind (dead-end) pores.

Figure 10.16 further shows the effluent concentration distributions of boron (H_3BO_4) obtained from 30-cm long columns packed with the same

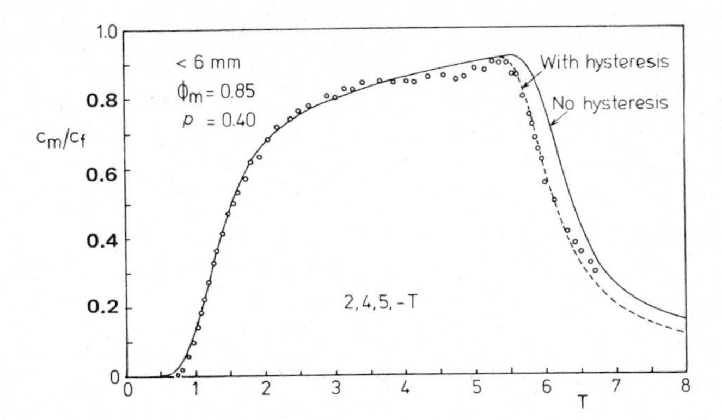

Fig. 10.15. Calculated and observed effluent curves for 2,4,5-T movement through Glendale clay loam. The following soil-physical data were used in the calculations: $\rho_b = 1.31$ g/cm^3, $J^V = 16.81$ cm/day, $\theta = 0.456$ cm^3/cm^3, $t_0 = 4.028$ days and $L = 30$ cm.

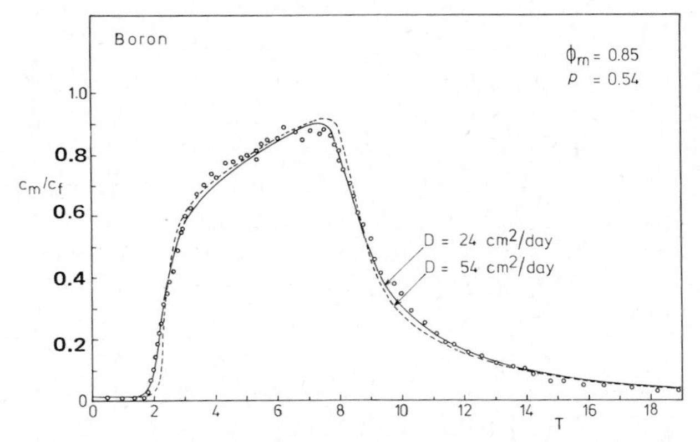

Fig. 10.16. Calculated and observed effluent curves for boron movement through Glendale clay loam. The following soil-physical data were used in the calculations: $\rho_b = 1.22$ g/cm^3, $J^V = 17.12$ cm/day, $\theta = 0.445$ cm^3/cm^3, $t_0 = 5.06$ days and $L = 30$ cm.

(6-mm aggregate-sized) Glendale clay loam soil material. The equilibrium adsorption isotherm for this chemical and soil was adequately described with the non-linear isotherm ${}^wq = 2.77\ C^{0.672}$ (Van Genuchten, 1974). No hysteresis was observed for this chemical. In a separate experiment with tritium, the fraction of mobile liquid, ϕ_m, was found to be 0.85, and the Peclet number, Pé, was found to be 56. The dashed line in Fig. 10.16 is based on these parameter values. Although the description of the experimental data is reasonable, it may be improved by increasing the value of the dispersion coefficient (solid line). Boron adsorption during batch equilibration was found to be complete only after shaking the chemical and soil for about 1—5 h. Hence, some kinetic effects are likely to be present during boron movement at a volumetric flux, J^V, of 17.12 cm/day. It was shown earlier (Fig. 10.8) that the kinetic adsorption mechanism will smear the concentration front in a way which is very much similar to an increase in dispersive transport (Fig. 10.12). Because the present study assumes equilibrium adsorption, the observed small kinetic effects may be included in the model by increasing the dispersion coefficient.

The fraction of the sorption sites located in the dynamic region of the soil (p) for the boron experiment is higher than the value found for 2,4,5-T (0.54 vs. 0.40). This indicates that relatively more boron is adsorbed in the vicinity of the main flowing paths of the moving liquid. Apparently more boron is adsorbed at the exterior of the aggregates, while relatively more 2,4,5-T is adsorbed inside the aggregates. The differences in p for boron and 2,4,5-T are probably indicative of the different adsorption mechanisms for the two chemicals. Boron adsorption presumably takes place on the mineral fraction of the soil (Rhoades et al., 1970), while 2,4,5-T adsorption occurs on the organic fraction.

10.3.3. Mobile and immobile concentration distributions in soils

The success of the intra-aggregate diffusion model in describing experimental data for the movement of tritium, 2,4,5-T, and boron suggests the importance of physical processes associated with the leaching of chemicals in the field. Many field soils exhibit some kind of aggregation or channeling; such conditions seem to favor the existence of large zones of immobile water. In addition, leaching processes in the field frequently occur under unsaturated conditions and often at relatively low flow rates, both of which also favor more immobile liquid. Wild (1972) studied the movement of water and mineralized nitrate (formed from soil organic matter) under a bare field site, and found that the leaching of the nitrate was much more gradual than could be predicted by miscible displacement theories. He suggested that water was moving mainly through the larger cracks and channels, and hence was unable to leach the nitrate that was contained largely in the relatively immobile regions of the aggregates (peds). Gumbs and Warkentin (1975) also concluded that water was moving through the largest interaggregate pores during infiltration into an aggregated clay soil under zero pressure. The wetting front apparently bypassed some of the aggregates, which then remained relatively dry for some period of time. These two examples suggest that the infiltrating water may not necessarily displace the moisture inside the aggregates, but may leave this immobile water relatively undisturbed during its movement through the soil. When the invading liquid contains a solute, displacement in the aggregates and away from the larger cracks and channels will become strongly dependent upon diffusion processes, and should be incomplete for a considerable amount of time after the water/solute front has passed.

Considerable evidence for the existence of immobile liquid zones in the field may be found in a recent study by Quisenberry and Phillips (1976). Small pulses of water and chloride were applied to two field soils and the downward propagation of these pulses was followed in time. Both the water and solute moved rapidly, even though the moisture content during the experiments was generally less than the field capacity of the soil. The rapid migration of both water and solute beyond the lowest (90 cm) sampling depth suggests that it was the applied water which moved through the soil and not the initial soil water which could have been displaced by the feed solution. If the surface-applied water had displaced the initial soil water during its movement downwards, the chloride should have been measured near the soil surface (as was the case in studies reported by Warrick et al. (1971), and Kirda et al. (1973)). Apparently channeling, especially beneath the Ap horizon, was responsible for the rapid movement of both water and chloride, resulting in little or no convective displacement away from the channels. It seems evident that in this case a considerable amount of dead water is present in the soil, probably much higher than the 50% observed for the data in Fig. 10.14 (see also the recent study by Wild and Babiker, 1976).

When such large percentages of stagnant water are present, it is evident that the conventional convective—dispersive equation may be inadequate in describing the leaching process. The intra-aggregate diffusion model may prove to be a better model for such conditions. An important feature of this model is its ability to differentiate between such mobile and immobile liquid zones. Figure 10.17 gives an example of the type of distributions one may obtain with this model. The solute profiles inside a 30-cm long soil column were obtained with (10.33) and (10.34), using the same parameter values as were used for the calculated curve shown in Fig. 10.14b. It was assumed, however, that the solute was in the feed solution for only 1 day ($t_0 = 1$). Figure 10.17 shows that the concentration in the immobile liquid (c_a) is considerably different from that in the mobile liquid (c_m). Owing to the diffusion of the solute into and out of the immobile zone, the immobile concentration lags behind the mobile concentration. The weighted average concentration, $\langle c \rangle$, at any point in the column is given by:

$$\langle c \rangle = \phi_m c_m + (1 - \phi_m) c_a \qquad (10.46)$$

Since the fraction of mobile water was 0.53 for this experiment it follows that the average concentration in the profile is approximately the average of c_m and c_a.

When a sorbing medium is considered, the differences between mobile and immobile concentrations become even more pronounced. Figure 10.18 shows the distributions of 2,4,5-T, inside a 30-cm long soil column, obtained with the same data as given in Fig. 10.15. Also in this case 60% of the adsorption occurs in stagnant regions, but because of the increased sorptive capacity of the soil much more material can diffuse into the stagnant region before adsorption in this part of the soil will be complete. As a result, the

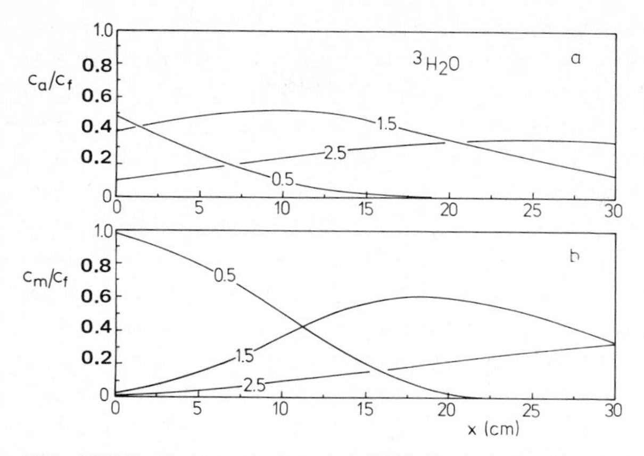

Fig. 10.17. Concentration profiles inside a 30 cm long soil column at different times during leaching with a short pulse ($t_0 = 1$ day) of tritiated water. Numbers on the curves indicate times (days) after leaching is initiated. Values of the parameters used to construct the curves are given in Fig. 10.14b.

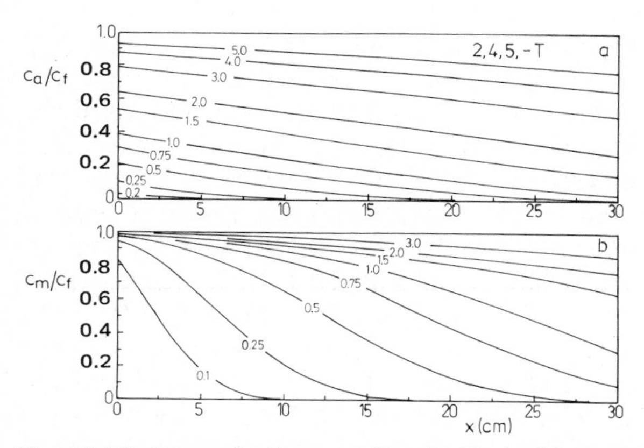

Fig. 10.18. Concentration profiles inside a 30 cm long soil column at different times during leaching with a continuous feed solution of 2,4,5-T. Numbers on the curves indicate times (days) after leaching is initiated. Values of the parameters used to construct the curves are given in Figure 10.15.

concentration in the immobile region will increase much more slowly than when tritium movement is considered. For example, at 20 cm from the soil surface, the relative concentration in the mobile phase reaches 0.5 after 0.8 day, while at this depth it takes about 2.4 days or three-times as long for the immobile liquid to reach a relative concentration of 0.5. That fraction of the chemical which diffuses into the stagnant region of the soil and which will remain in solution may be calculated by considering transport equation (10.32), which governs the exchange between mobile and stagnant regions of the soil. The first term on the left of (10.32) represents the material which will be stored in the immobile liquid, while the second term gives the material which is actually adsorbed by the stagnant region. Assuming linear adsorption ($n = 1$), the fraction remaining in the solution is found to be equal to $\theta_a / \{\theta_a + (1 - p)\rho_b k\}$.

For the data given in Fig. 10.15, one may calculate that only 17% of the chemical which diffuses into the stagnant region of the soil will remain in solution, the remaining 83% being adsorbed by this part of the soil. For tritium, however, nearly 100% will remain in solution. Thus, when sorption increases, relatively more material can diffuse into the stagnant region. Since this diffusion process is very slow compared to the rate at which the bulk of the chemical moves through the soil, the differences between mobile and immobile concentrations are likely to increase when sorption further increases.

Solute distributions like those shown in Figs. 10.17 and 10.18 may also be helpful in understanding at least some of the factors controlling the variability of sampling results obtained when suction cups are used for solute measurements in the field (England, 1974; Hansen and Harris, 1975). A suction cup essentially represents a point source of suction in the soil matrix, and the suction applied through the cups to the soil will determine, in part, the minimum pore size that will be drained in the immediate vicinity of the cup. It is obvious that the physical make-up of the cup as well as the way they are installed, play important roles in the performance of the cups (Parizek and Lane, 1970). The microscopic location of the porous cup in the medium, however, seems to be of prime importance. When the cup is installed inside a large soil aggregate (or ped), or away from a channel or large interaggregate pore, it is likely that mainly immobile liquid will be withdrawn through the sampler device. This will also be the case when the cup, during its installment, is pressed into the soil too strongly, so that the soil becomes locally compacted. The immobile liquid which will be withdrawn from the soil in such cases, may have a concentration quite different from that of the mobile liquid, as evidenced by the distributions shown in Figs. 10.17 and 10.18.

One may also expect a concentration—time effect when soil water is withdrawn through suction cups. This effect will depend upon the history of the solute distribution. For example, when a suction cup is installed such that it is in contact with both macropores and micropores, the sampler is likely to first withdraw mainly mobile liquid. With increasing time, and depending upon the conductivity of the individual pores, the mobile liquid around the porous cup will drain slowly away, leading to a lower rate of water uptake by the sampler. When a sufficiently large pressure is applied, more and more liquid will subsequently be withdrawn from the immobile regions through the smaller micropores. When this is the case, the solute concentration sampled is likely to change with time. Referring to the distributions of Fig. 10.18, for example, one may always expect a decrease in concentration with time, since the immobile concentration lags behind the mobile concentration. For the distributions given in Fig. 10.17, however, the concentration will decrease with increasing sampling length only when the mobile concentration is higher than the immobile concentration (e.g. in the upper 15^{cm} after 1 day). After 1.5 days this situation is reversed and the concentration of the liquid withdrawn through the cups is likely to increase during sampling in the upper part of the column.

Obviously any of several factors may account for the serious concentration variability sometimes observed in the field during suction cup cosampling. The intra-aggregate diffusion model results of this chapter, however, appear to provide a quantitative description and explanation for the important soil-physical phenomena which no doubt are a major contributor to this observed variability for many soils.

ACKNOWLEDGEMENTS

This work was supported, in part, by funds obtained from EPA Grant No. R803827-01, the Northeastern Forest Experiment Station of the USDA, Grant No. 23-586, and NSF Grant No. ENG-75-19184.

REFERENCES

Aris, R., 1959. Diffusion and reaction in flow systems of Turner's structures. Chem. Eng. Sci., 10: 80—87.

Bailey, G.W., White, J.L. and Rothberg, T., 1968. Adsorption of organic herbicides by montmorillonite. Role of pH and chemical character of adsorbate. Soil Sci. Soc. Am. Proc., 32: 222—234.

Ballaux, J.C. and Peaslee, D.E., 1975. Relationships between sorption and desorption of phosphorous by soils. Soil Sci. Soc. Am. Proc., 39: 275—278.

Bennet, A. and Goodridge, F., 1970. Hydrodynamic and mass transfer studies in packed adsorption columns: I. Axial liquid dispersion. Trans. Inst. Chem. Eng., 48: 232—240.

Biggar, J.W. and Nielsen, D.R., 1962. Miscible displacement: II. Behavior of tracers. Soil Sci. Soc. Am. Proc., 26: 125—128.

Cleary, R.W. and Adrian, D.D., 1973. Analytical solution of the convective—dispersive equation for cation adsorption in soils. Soil Sci. Soc. Am. Proc., 37: 197—199.

Coats, K.H. and Smith, B.D., 1964. Dead-end pore volume and dispersion in porous media. Soc. Pet. Eng. J., 4: 73—84.

Davidson, J.M. and Chang, R.K., 1972. Transport of picloram in relation to soil-physical conditions and pore-water velocity. Soil Sci. Soc. Am. Proc., 36: 257—261.

Deans, H.A., 1963. A mathematical model for dispersion in the direction of flow in porous media. Soc. Pet. Eng. J., 3: 49—52.

Enfield, C.G., Harlin, C.C. Jr., and Bledsoe, B.E., 1976. Comparison of five kinetic models for orthophosphate reactions in mineral soils. Soil Sci. Soc. Am. J., 40: 243—249.

England, C.B., 1974. Comments on "A technique using porous cups for water sampling at any depth in the unsaturated zone", by Warren W. Wood. Water Resour. Res., 10: 1049.

Farmer, W.J. and Aochi, Y., 1974. Picloram sorption by soils. Soil Sci. Soc. Am. Proc., 38: 418—423.

Fava, A. and Eyring, H., 1956. Equilibrium and kinetics of detergent adsorption — A generalized equilibration theory. J. Phys. Chem., 65: 890—898.

Gaudet, J.P., Jégat, H., Vachaud, G. and Wierenga, P., 1977. Solute transfer, with exchange between mobile and stagnant water, through unsaturated sand. Soil Sci. Soc. Am. J., 41: 665—671.

Gershon, N.D. and Nir, A., 1969. Effects of boundary conditions of models on tracer distribution in flow through porous mediums. Water Resour. Res., 5: 830—839.

Goldstein, F.R.S., 1953. On the mathematics of exchange processes in fixed columns: I. Mathematical solutions and asymptotic expansions. Proc. R. Soc., Lond., 219: 151—185.

Green, R.E., Rao, P.S.C. and Corey, J.C., 1972. Solute transport in aggregated soils: tracer zone shape in relation to pore-velocity distribution and adsorption. Proc. 2nd Symp. Fundamentals of Transport Phenomena in Porous Media. Guelph, August 7—11, IAHR—ISSS, 2: 732—752.

Gumbs, F.A. and Warkentin, B.P., 1975. Prediction of infiltration of water into aggregated clay soil samples. Soil Sci. Soc. Am. Proc., 39: 255—263.

Gupta, R.K., Millington, R.J. and Klute, A., 1973. Hydrodynamic dispersion in unsaturated porous media. I. Concentration distribution during dispersion. J. Indian Soc. Soil Sci., 21: 1—7.

Gupta, S.P. and Greenkorn, R.A., 1976. Solution for radial-flow with non-linear adsorption. J. Environ. Eng. Div., Am. Soc. Cir. Eng., 102: 87—94.

Hansen, E.A. and Harris, A.R., 1975. Validity of soil-water samples collected with porous ceramic cups. Soil Sci. Soc. Am., Proc., 39: 528—536.

Hashimoto, I., Deshpande, K.B. and Thomas, H.C., 1964. Péclet numbers and retardation factors for ion exchange columns. Ind. Eng. Chem. Fund., 3: 213—218.

Hendricks, D.W., 1972. Sorption in flow through porous media. In: Fundamentals of Transport Phenomena in Porous Media. Developments in Soil Science, 2. Elsevier, Amsterdam, pp. 384—392.

Hornsby, A.G. and Davidson, J.M., 1973. Solution and adsorbed fluometuron concentration distribution in a water-saturated soil: experimental and predicted evaluation. Soil. Sci. Soc. Am. Proc., 37: 823—828.

Huggenberger, F., Letey, J. and Farmer, W.J., 1972. Observed and calculated distribution of lindane in soil columns as influenced by water movement. Soil Sci. Soc. Am. Proc., 36: 544—548.

IBM, 1967. System/360 Continuous System Modeling Program (360A-CX-16X), User's Manual. Data Processing Division, 122 East Post Road, White Plains, New York, N.Y.

Kay, B.D. and Elrick, D.E., 1967. Adsorption and movement of lindane in soils. Soil Sci., 104: 314—322.

Kirda, C., Nielsen, D.R. and Biggar, J.W., 1973. Simultaneous transport of chloride and water during infiltration. Soil Sci. Soc. Am. Proc., 37: 339—345.

Lai, Sung-Ho and Jurinak, J.J., 1971. Numerical approximation of cation exchange in miscible displacement through soil columns. Soil Sci. Soc. Am. Proc., 35: 894—899.

Lapidus, L. and Amundson, N.R., 1952. Mathematics of adsorption in beds: VI. The effect of longitudinal diffusion in ion exchange and chromatographic columns. J. Phys. Chem., 56: 984—988.

Leenheer, J.A. and Ahlrichs, J.L., 1971. A kinetic and equilibrium study of the adsorption of carbaryl and parathion upon soil organic matter surfaces. Soil Sci. Soc. Am., Proc., 35: 700—705.

Lindstrom, F.T., Haque, R., Freed, V.H. and Boersma, L., 1967. Theory on the movement of some herbicides in soils. Environ. Sci. Technol., 1: 561—565.

Lindstrom, F.T. and Boersma, L., 1970. Theory of chemical transport with simultaneous sorption in a water saturated porous medium. Soil Sci., 110: 1—9.

Lindstrom, F.T., Boersma, L. and Stockard, D., 1971. A theory on the mass transport of previously distributed chemicals in a water saturated sorbing porous medium. Isothermal cases. Soil Sci., 112: 291—300.

McMahon, M.A. and Thomas, G.W., 1974. Chloride and tritiated water flow in disturbed and undisturbed soil cores. Soil Sci. Soc. Am. Proc., 38: 727—732.

Nielsen, D.R. and Biggar, J.W., 1961. Miscible displacement in soils: I. Experimental information. Soil. Sci. Soc. Am. Proc., 25: 1—5.

O'Connor, G.A. and Anderson, J.U., 1974. Soil factors affecting the adsorption of 2,4,5-T. Soil Sci. Soc. Am. Proc., 38: 433—436.

Oddson, J.K., Letey, J. and Weeks, L.V., 1970. Predicted distribution of organic chemicals in solution and adsorbed as a function of position and time for various chemical and soil properties. Soil Sci. Soc. Am. Proc., 34: 412—417.

Parizek, R.R. and Lane, B.E., 1970. Soil-water sampling using pan and deep pressure—vacuum Iysimeters. J. Hydrol., 11: 1—21.

Pearson, J.R.A., 1959. A note on the 'Danckwerts' boundary conditions for continuous flow reactors. Chem. Eng. Sci., 10: 281—284.

Quisenberry, V.L. and Phillips, R.E., 1976. Percolation of surface-applied water in the field. Soil Sci. Soc. Am., J., 40: 484—489.

Rhoades, J.D., Ingvalson, R.D. and Hatcher, J.T., 1970. Laboratory determination of leachable soil boron. Soil Sci. Soc. Am., Proc., 34: 871—875.

Rideal, E.K., 1930. Surface Chemistry. Cambridge University Press.

Selim, H.M. and Mansell, R.S., 1976. Analytical solution of the equation for transport of reactive solutes through soils. Water Resour. Res., 12: 528—532.

Swanson, R.A. and Dutt, G.R., 1973. Chemical and physical processes that affect atrazine movement and distribution in soil systems. Soil Sci. Soc. Am. Proc., 37: 872—876.

Tanji, K.K., 1970. A computer analysis on the leaching of boron from stratified soil columns. Soil Sci., 110: 44—51.

Turner, G.A., 1958. The flow-structure in packed beds. Chem. Eng. Sci., 7: 156—165.

Van de Pol, R.M., 1974. Solute Movement in a Layered Field Soil. MS thesis, New Mexico State University, Las Cruces, New Mexico.

Van Genuchten, M.Th., 1974. Mass Transfer Studies in Sorbing Porous Media, Ph.D. thesis, New Mexico State Univ., Las Cruces, New Mexico.

Van Genuchten, M.Th. and Wierenga, P.J., 1974. Simulation of one-dimensional solute transfer in porous media. New Mexico Agric. Exp. Sta. Bull., 628.

Van Genuchten, M.Th. and Wierenga, P.J., 1976a. Numerical solution for convective—dispersion with intra-aggregate diffusion and non-linear adsorption. In: G.C. van Steenkiste (Editor), System Simulation in Water Resources. North-Holland, Amsterdam.

Van Genuchten, M.Th. and Wierenga, P.J., 1976b. Mass transfer studies in sorbing porous media: I. Analytical solutions. Soil Sci. Soc. Am. J., 40: 473—480.

Van Genuchten, M.Th. and Wierenga, P.J., 1977. Mass transfer studies in sorbing porous media: II. Experimental evaluation with tritium (^3H_2O). Soil Sci. Soc. Am. J., 41: 272—278.

Van Genuchten, M.Th., Davidson, J.M. and Wierenga, P.J., 1974. An evaluation of kinetic and equilibrium equations for the prediction of pesticide movement through porous media. Soil Sci. Soc. Am. Proc., 38: 29—35.

Van Genuchten, M.Th., Wierenga, P.J. and O'Connor, G.A., 1977. Mass transfer studies in sorbing porous media: III. Experimental evaluation with 2,4,5-T. Soil Sci. Soc. Am. J., 41: 278—285.

Villermaux, J. and Van Swaay, W.P.M., 1969. Modèle représentatif de la distribution des temps de séjour dans un réacteur semi-infini à dispersion axiale avec zones stagnantes. Chem. Eng. Sci., 24: 1097—1111.

Warrick, A.W., Biggar, J.W. and Nielsen, D.R., 1971. Simultaneous solute and water transfer for an unsaturated soil. Water Resour. Res., 7: 1216—1225.

Wehner, J.F. and Wilhelm, R.H., 1956. Boundary conditions of flow reactor. Chem. Eng. Sci., 6: 89—93.

Wild, A., 1972. Nitrate leaching under bare fallow at a site in Northern Nigeria. J. Soil Sci., 23: 315—324.

Wild, A. and Babiker, I.A., 1976. The asymmetric leaching pattern of nitrate and chloride in a loamy sand under field conditions. J. Soil Sci., 27: 460—466.

Wood, A.L. and Davidson, J.M., 1975. Fluometuron and water content distribution during infiltration: measured and calculated. Soil Sci. Soc. Am. Proc., 39: 820—825.

CHAPTER 11

ELECTROCHEMICAL PHENOMENA IN SOIL AND CLAY SYSTEMS

G.H. Bolt

In chapters 1, 2, 3 and 7 considerable attention was given to the existence of electrostatic fields near charged surfaces as abundantly present in soil systems. In these chapters the static aspects of such systems were the prime concern and the discussion was directed towards obtaining an insight into the equilibrium distribution of ions in soils. In chapters 9 and 10 transport processes were introduced, but here the discussion was limited to the net outcome in terms of displacement and transformations of electrically neutral compounds, such as the transformation of A-exchanger into B-exchanger upon the passage of B-salt. As in this case the final outcome depends primarily on the equilibrium distribution ratio of the adsorbed species, R_D, the actual extent of the electrostatic fields mentioned did not enter the discussion.

A notable exception to this could be the occurrence of an intra-aggregate stagnant phase connected via narrow voids with the mobile phase (cf. Blackmore, 1976). The rate constant k_r governing anion equilibration would then depend on the extent of the DDL in such voids.

In the present chapter attention will be given to electro-chemical phenomena accompanying irreversible (i.c. transport) processes in soils and clays. As will be shown, such phenomena are caused by the motion of counter-ions and co-ions relative to the solid matrix. In turn this ion movement interacts with the motion of the liquid phase, which exhibits a velocity distribution as a function of the distance from the solid surface. Accordingly the actual distribution of the ions in the liquid layer adjoining the solid surface will be of decisive importance.

11.1. SPECIFICATION OF FLUXES AND FORCES IN ISOTHERMAL POROUS MEDIA

Before a discussion of particular phenomena, as for example electro-osmosis, suspension effect, etc., the *group* of phenomena will be described in a comprehensive, though severely simplified, manner, using the framework as supplied by "Thermodynamics of irreversible processes". Reference is made to existing literature (De Groot and Mazur, 1962; Katchalsky and Curran, 1965; Groenevelt and Bolt, 1969) for the underlying derivations and it suffices here to reiterate a general expression for the energy dissipation during irreversible processes comprising only the isothermal, steady motion of (ionic) species under the influence of existing gradients of their thermodynamic potential $\tilde{\mu}_k$, viz.:

$$T \frac{d_i {}^v S}{dt} = \Sigma_{k=1}^{n} {}^m j_k \nabla(-\tilde{\mu}_k)_T \tag{11.1}$$

Applied here to a macroscopically one-dimensional flow system, ${}^m j_k$ indicates the macroscopic flux (density) in the direction of flow, specified in kmol/m^2 sec. In that case $\nabla(-\tilde{\mu}_k)_T$ signifies the rate of decrease of $\tilde{\mu}_k$ in the direction of flow, $-\Delta(\tilde{\mu}_k)/\Delta x$. In the LHS of (11.1) $d_i({}^v S)/dt$ indicates the entropy "produced" per unit volume of the system, per unit time, i.e. the energy dissipated by the moving species k upon traversing this unit volume, subject to the "driving force" $\nabla(-\tilde{\mu}_k)$. Viewing the motion of the species k as a simple frictional process, the LHS of (11.1) indicates the heat of friction generated by the moving species. At constant temperature this then corresponds to the loss in free enthalpy of the species.

The requirement that the entropy production be positive for all spontaneous processes then leads in the present case — i.e. one-dimensional steady flow of a number of species k — to the conclusion that there must exist multiple functional relationships between all j_k and all $\nabla(-\tilde{\mu}_k)_T$. The *simplest* acceptable relationship to the above purpose is of the type:

$${}^m j_k = \Sigma_{l=1}^{n} L_{k,l} \nabla(-\tilde{\mu}_l)_T \tag{11.2}$$

i.e. each flux is a linear homogeneous function of all "driving forces" acting on the n different components. Together with (11.1) this gives for the present isothermal transport systems the basic condition:

$$\Sigma_{k=1}^{n} \nabla(-\tilde{\mu}_k)_T \, \Sigma_{l=1}^{n} L_{k,l} \nabla(-\tilde{\mu}_l)_T > 0 \tag{11.3}$$

in which the coefficients $L_{k,l}$ constitute a n × n matrix transforming the n forces $\nabla(-\tilde{\mu}_k)_T$ into the corresponding set of n fluxes.

Lest the significance of the above reasoning be under-estimated it is pointed out that it provides a necessary mitigation of the common understanding that "in nature spontaneous processes involving certain constituents imply the running downhill of these" (energetically speaking). The latter statement, which could be formulated in the present context as $j_k \nabla(-\tilde{\mu}_k) > 0$, is not necessarily correct in multicomponent systems: certain components may spontaneously "run uphill". A trivial example of such a situation is the "leaky" osmometer: the direction of flow of water into an osmometer compartment containing a solution with an osmotic pressure Π may be reversed upon piercing the membrane, the water then flowing in the direction of increasing $\tilde{\mu}_w$ (cf. the relevant discussions by Corey, Kemper and Low (1961)). Generally speaking such behavior may be described as the dragging along of one component (against its own potential gradient) with another one running downhill. Equation (11.1) thus specifies the bounds of spontaneity as these apply to an entire system: running uphill is possible for specific components as long as the overall production of entropy remains positive.

Translated into practical consequences, (11.3) implies that the matrix L must fulfil the conditions $L_{kk} > 0$ and $L_{kk}L_{ll} > L_{kl}^2$, i.e. the diagonal coefficients and the determinant of the matrix are non-negative. Furthermore, for the present simplified system, the so-called "Onsager law" requires the matrix to be symmetric, which implies that only $n(n+1)/2$ coefficients L_{kl} are independent parameters.

Equation (11.1), which comprises per definition a complete set of n independent fluxes and their conjugated forces, may now be manipulated to deliver any other set of n independent fluxes and forces (cf. Katchalsky and Curran, 1965; Groenevelt and Bolt, 1969), provided the new fluxes and forces are obtained by suitable combinations of the previous ones, leaving (11.1) algebraically intact. Such a reapportioning of fluxes is of importance if it is attempted to relate the coefficients L to known physical processes. The appropriate choice of independent fluxes in a particular system is then dictated, at least in part, by a priori notions about the driving forces and the controlling reactive forces for the motion of the different components. Following Katchalsky and Curran (1965), it will be shown below how a set of fluxes and forces particularly suitable for the present system may be derived from (11.1).

For a system comprising fairly dilute aqueous solutions moving through a charged porous medium it appears logical to regard the volume flux of the solution as largely controlled by frictional forces between the solid matrix and the solution. Furthermore the motion of solute molecules with respect to the solvent should be controlled by a molecular diffusion resistance. As the solutes involved are ionic it is also convenient to distinguish between fluxes of electrically neutral combinations (i.e. salts) and excess fluxes of either cations or anions, which constitute an electric current.

The above may be illustrated for a simple three-component system containing one type of cation with valence z_+, one type of anion with valence z_- and the solvent water, w. The corresponding three independent molar fluxes of these components, ${}^m j_k$, may now be combined to give a volume flux for the entire solution, J^V, a diffusion flux for the neutral salt relative to the solvent, J_s^D, and finally a flux of charge, corresponding to an electric current (density), I, according to:

$$J^V \equiv {}^m j_w \bar{V}_w + {}^m j_+ \bar{V}_+ + {}^m j_- \bar{V}_- \qquad (11.4a)$$

$$J_s^D \equiv {}^m j_- / {}^m c_- - {}^m j_w / {}^m c_w \qquad (11.4b)$$

$$I \equiv (z_+ {}^m j_+ + z_- {}^m j_-)F = ({}^m j_+ / {}^m c_+ - {}^m j_- / {}^m c_-) z_+ {}^m c_+ F \qquad (11.4c)$$

In these equations $\bar{V}_k$ indicates the molar volume of the species concerned, while the salt flux has been equated (arbitrarily) to the flux of the anionic species.

It must be stressed that the concentration ${}^m c_-$ appearing here in the definition of J_s^D refers to the *local equilibrium* concentration of the system (cf. the remark on p. 393). Accordingly $J_s^D + {}^m j_w / {}^m c_w$ is an "equivalent linear velocity" of the anions in the system which would deliver the actual value of the anion flux, ${}^m j_-$, on multiplication by *this* equilibrium concentration. The actual mean linear velocity of the anions would exceed this value if the actual concentration were below its local equilibrium value for example, because of anion exclusion within the porous medium. Also the concentration c_+ to be used later refers to this local equilibrium solution.

Next one may write out $\nabla(\tilde{\mu}_k)_T$ as:

$$\nabla(\tilde{\mu}_k)_T = \bar{V}_k \nabla P + \nabla \mu_k^c + z_k F \nabla \psi \tag{11.5}$$

in which μ_k^c indicates the concentration-dependent part of the thermodynamic potential $\tilde{\mu}_k$. Introduction of (11.4) and (11.5) into (11.1) then gives:

$$T \frac{d_i{}^v S}{dt} = J^V \nabla(-P) + J_s^D \phi_w \nabla(-\Pi) + I\nabla(-E_+) \tag{11.6}$$

in which $\phi_w \equiv {}^m c_w \bar{V}_w$, $E_+ = (\psi + \mu_+^c / z_+ F)$, Π = osmotic pressure. The multiplier of I appearing in the last term of (11.6) may now be identified with the gradient of the potential registered by an electrode *reversible* to the cation species involved. This equation then identifies the forces conjugate to the chosen fluxes as, respectively, $\nabla(-P)$ for the volume flux J^V, $\phi_w \nabla(-\Pi)$ for the diffusion flux relative to the solvent and $\nabla(-E_+)$ for the current density.

The above transformation of (11.1) is effected as follows. For a neutral salt solution at constant T, like the *local equilibrium* solution of the system, one may write:

$$^m c_w \nabla(-\tilde{\mu}_w) = \phi_w \nabla(-P) + {}^m c_w \nabla(-\mu_w^c) \tag{11.7a}$$

$$^m c_- \nabla(-\tilde{\mu}_-) = \phi_- \nabla(-P) + {}^m c_- \nabla(-\mu_-^c) + z_- {}^m c_- F \nabla(-\psi) \tag{11.7b}$$

$$^m c_+ \nabla(-\tilde{\mu}_+) = \phi_+ \nabla(-P) + {}^m c_+ \nabla(-\mu_+^c) + z_+ {}^m c_+ F \nabla(-\psi) \tag{11.7c}$$

in which $\phi_k \equiv {}^m c_k \bar{V}_k$, i.e. the volume fraction occupied by component k in the equilibrium solution. The Gibbs-Duhem relation for the system reads:

$$\Sigma_k {}^m c_k \nabla(-\mu_k^c) = 0 \tag{11.8}$$

which allows one to define the osmotic pressure, Π, of the equilibrium solution by:

$$^m c_w \nabla(-\mu_w^c) \equiv -\phi_w \nabla(-\Pi) = -\{{}^m c_- \nabla(-\mu_-^c) + {}^m c_+ \nabla(-\mu_+^c)\}$$

$$= -{}^m c_- / \nu_- \{\nabla(-\mu_s^c)\} \tag{11.9}$$

where ${}^m c_- / \nu_- = {}^m c_+ / \nu_+ = {}^m c_s$, the subscript s referring to the salt present. Combining now the last two terms of (11.7c) into one corresponding to the gradient of the electrochemical potential of the cation as read with an electrode reversible to this species one finds:

$$^m c_+ \nabla(-\mu_+^c) + z_+ {}^m c_+ F \nabla(-\psi) \equiv z_+ {}^m c_+ F \nabla(-E_+) = -z_- {}^m c_- F \nabla(-E_+) \tag{11.10}$$

Entering (11.9) and (11.10) into (11.7b) then gives:

$$^m c_- \nabla - \tilde{\mu}_- = \phi_- \nabla(-P) + {}^m c_- \nabla(-\mu_-^c) + {}^m c_+ \nabla(-\mu_+^c) - {}^m c_+ \nabla(-\mu_+^c) - z_+ {}^m c_+ F \nabla(-\psi)$$

$$= \phi_- \nabla(-P) + \phi_w \nabla(-\Pi) - z_+ {}^m c_+ F \nabla(-E_+) \tag{11.11}$$

Introducing now the equivalent "filter" velocity of the components as:

$$v_k \equiv {}^m j_k / {}^m c_k \; (= {}^m j_k \bar{V}_k / \phi_k) \tag{11.12}$$

i.e. the velocity which upon multiplication by the *local equilibrium* concentration delivers the actual flux, one may write:

$$v_w {}^m c_w \nabla(-\tilde{\mu}_w) = v_w \phi_w \nabla(-P) - v_w \phi_w \nabla(-\Pi) \tag{11.13a}$$

$$v_- {}^m c_- \nabla(-\tilde{\mu}_-) = v_- \phi_- \nabla(-P) + v_- \phi_w \nabla(-\Pi) - v_- z_+ {}^m c_+ F \nabla(-E_+) \tag{11.13b}$$

$$v_+ {}^m c_+ \nabla(-\tilde{\mu}_+) = v_+ \phi_+ \nabla(-P) \qquad + v_+ z_+ {}^m c_+ F \nabla(-E_+) \tag{11.13c}$$

Upon summation one thus finds:

$$\Sigma j_k \nabla(-\tilde{\mu}_k)_T = (v_w \phi_w + v_- \phi_- + v_+ \phi_+) \nabla(-P) + \phi_w (v_- - v_w) \nabla(-\Pi) + (v_+ - v_-) z_+ {}^m c_+ F \nabla(-E_+) \tag{11.14}$$

which identifies the fluxes J^V, J_s^D and I — defined in (11.4) — as the fluxes conjugate to the forces $\nabla(-P)$, $\phi_w \nabla(-\Pi)$ and $\nabla(-E_+)$

The convenience of the present choice of fluxes and forces, as identified by (11.6), becomes apparent upon consideration of a limiting case. If the porous medium consists of a chemically inert and uncharged solid phase with a pore size far in excess of molecular dimensions, then the presence of a pressure gradient causes only a flow of the solution as a whole, which is then controlled by the flow resistance of the medium for the solution. Furthermore, a gradient of the osmotic pressure (or solute concentration) causes only a diffusion flux, J_s^D, at zero flux of the solution, J^V. This diffusion flux is then controlled by a resistance against molecular diffusion within the liquid phase. Finally, a gradient of the electric potential causes only motion of cations and anions in opposite direction, again at zero flux of the solution.

Accordingly the "coupling" coefficients connecting J^V to $\phi_w \nabla(-\Pi)$ and $\nabla(-E_+)$, J_s^D to $\nabla(-P)$ and I to $\nabla(-P)$ all vanish in the limiting case of a porous medium not exhibiting permiselectivity for the components of the liquid phase. In short, the matrix $L_{k,1}$ attains maximum simplicity in the limiting case, for the chosen set of fluxes and forces. Conversely, the coefficients corresponding to this chosen set may be related directly to the permiselectivity of the porous medium.

Admittedly the present choice of $\nabla(-E_+)$ as the force conjugate to I is arbitrary, and could be replaced by $\nabla(-E_-)$, the potential gradient as measured with electrodes reversible to the anion. In that case, however, the salt flux must be equated to the cation flux (cf. Groenevelt and Elrick, 1976; Groenevelt et al., 1978). In principle one might also employ the mean salt flux defined as $(j_-/c_- + j_+/c_+)/2$, together with $\nabla(-E_0)$, the potential gradient as measured with irreversible electrodes. In that case, however, measurements are subject to the uncertainty of junction potentials of often unknown magnitude (at least within charged porous media, c.f. section 11.8).

In the section following, the relation of the coefficients $L_{k,1}$ with the charge density of the solid phase of the porous medium will be demonstrated, assuming a system of extreme geometric simplicity. For this purpose it appears worthwhile to redefine the current density in terms of the equivalent linear velocity of current-carrying ions, as then all coefficients attain the same dimension. For this purpose one may introduce $J_{\pm}$ as:

$$J_{\pm} \equiv I/z_{+}{}^{m}c_{+}F = {}^{m}j_{+}/{}^{m}c_{+} - {}^{m}j_{-}/{}^{m}c_{-} \tag{11.15}$$

i.e. the excess velocity (relative to the anions) with which cations present at the local equilibrium concentration, $z_{+}{}^{m}c_{+}$, would have to move in order to give the required value of I. Obviously the chosen equivalent of the electric current then implies an adjustment of the conjugated force as:

$$\nabla(-p_{e_+}) \equiv z_{+}{}^{m}c_{+}F\nabla(-E_{+}) \tag{11.16}$$

which has the dimension of a pressure gradient, i.e. the force acting on the cationic charge present in a unit volume of the local equilibrium solution.

In analogy with (11.2) the chosen fluxes and forces may be connected via a matrix of transport coefficients as shown in Table 11.1.

TABLE 11.1

Transport coefficients in a single salt system

Fluxes	Forces		
	$\nabla(-P)$	$\phi_w\nabla(-\Pi)$	$\nabla(-p_{e_+})$
J^V	L_V	L_{VD}	L_{VE}
J_s^D	L_{DV}	L_D	L_{DE}
$J_{\pm}$	L_{EV}	L_{ED}	L_E

11.2. THE TRANSPORT COEFFICIENTS EXPRESSED IN COEFFICIENTS OF FRICTION

In the above matrix all coefficients are specified in terms of a velocity per unit force (acting on the unit volume) and may thus be seen as the reciprocal of a coefficient of friction per unit volume of transported material. The singly subscripted coefficients appearing in the main diagonal of the matrix connect the fluxes to their conjugated force and will be termed straight coefficients. The remainder are then coupling coefficients. Referring to Kedem and Katchalsky (1963), Katschalsky and Curran (1965), Groenevelt and Bolt (1969), Bolt and Groenevelt (1972) and Groenevelt and Elrick (1976) for details of more general derivations, it will be tempted here to discuss schematically the *interpretation* of the coefficients using the following simplifying conditions.

(a1) The steady motion of cations and anions relative to the water and the matrix is subject to "frictional forces" between ions and water and between ions and the matrix. Accepting Stokes law as a guide line, the corresponding coefficients of friction are constants and will be specified as:

b_+ = coefficient of friction between cation and water, in N per m^3 of ionic volume present in excess water, at unit velocity difference;

b_- = same for anion;

b = same for cation and anion, between ion and solid matrix.

The first two coefficients are obviously determined largely by the (hydrated) ionic size and are directly related to ionic mobilities in dilute aqueous solutions. The third coefficient, in contrast, is determined by the average contact area between ion and matrix, which depends primarily on the pore size of the porous matrix. All three coefficients will finally contain the "system peculiar" ratio of pore tortuosity and liquid content, λ/θ, necessary because velocities are to be expressed as filter velocities, ${}^{m}j_k/{}^{m}c_k$.

(a2) The motion of the water in the system is subject to frictional forces between water and matrix and between water and ions. The corresponding coefficients are, respectively:

b = coefficient of friction between water and matrix, again per unit volume of water as present in the porous matrix;

$b_+\phi_+/\phi_w$ = coefficient of friction between water and cation, per unit volume of water;

$b_-\phi_-/\phi_w$ = same between water and anion.

The first coefficient is again largely determined by the pore size of the matrix and may, to a first approximation, be taken equal to the coefficient for ion against matrix, implying for the time being that the volume fraction of the ions in contact with the matrix is of the same order of magnitude as that of the water. The coefficients of friction of water against the ions follow directly from the "action equals reaction" principle, with $\phi_k \equiv c_k \overline{V}_k$.

(b) The ionic distribution in the, supposedly charged, porous medium is characterized by accumulation factors according to:

$$c'_w = c_w k \tag{11.17a}$$

$$c'_- = c_- \bar{u}_- \tag{11.17b}$$

$$c'_+ = c_+ \bar{u}_+ \tag{11.17c}$$

in which the primes denote the actual (average) concentration within the porous medium. The concentrations c_+ and c_- are then the "local equilibrium concentration", i.e. the concentrations that would be found in neutral salt solutions that were locally brought to equilibrium with the liquid phase inside the porous matrix.

It is of importance to note once more that the introduction of the (hypothetical) local equilibrium solution is necessary because the local values of $\tilde{\mu}_k$ are related to the composition of this equilibrium solution. In fact the entire operation of splitting the force into its components in (11.5) refers to this local equilibrium solution. Accordingly P, Π and $\underline{p}_+$ are defined similarly as pertaining to this equilibrium solution, which checks with their experimental evaluation via a locally inserted measuring compartment.

Making use of the parameters of friction and accumulation mentioned above one may write out the force balance for steady state motion of the components through the porous matrix. Specifying per unit volume of the three components moving with *local* velocities relative to the matrix equal to v'_w, v'_+ and v'_-, respectively, this gives for the frictional forces:

$$-F_w = v'_w b + (v'_w - v'_-)\phi'_- b_- / \phi'_w + (v'_w - v'_+)\phi'_+ b_+ / \phi'_w \tag{11.18a}$$

$$-F_- = v'_- b + (v'_- - v'_w) b_- \tag{11.18b}$$

$$-F_+ = v'_+ b + (v'_+ - v'_w) b_+ \tag{11.18c}$$

The above actual local velocities may be translated into the chosen fluxes J^V, J^D_- and $J_\pm$ by inverting (11.4) and (11.15) and using the accumulation factors defined by (11.17). This gives:

$$k v'_w = J^V - \phi_s J^D_s - \phi_+ J_\pm \tag{11.19a}$$

$$v'_- / \bar{u}_- = J^V + \phi_w J^D_s - \phi_+ J_\pm \tag{11.19b}$$

$$\bar{u}_+ v'_+ = J^V + \phi_w J^D_s + (1 - \phi_+) J_\pm \tag{11.19c}$$

with $\phi_s \equiv (\phi_+ + \phi_-)$, i.e. the volume fraction of salt in the local equilibrium solution. Substitution of (11.19) into (11.18) then gives the frictional forces on the different components in terms of the chosen fluxes. For each component the force balance under steady state conditions then requires that the frictional force (per unit volume of the component) cancels the driving force (per unit volume), which gives:

for $\nabla(-\Pi) = \nabla(-\underline{p}_+) = 0$	for $\nabla(-P) = \nabla(-\underline{p}_+) = 0$	for $\nabla(-P) = \nabla(-\Pi) = 0$	
$-F_w = \nabla(-P)$	$-F_w = -\nabla(-\Pi)$	$-F_w = 0$	(11.20a)
$-F_- = \nabla(-P)$	$-F_- = (\phi_w/\phi_-)\nabla(-\Pi)$	$-F_- = -(1/\phi_-)\nabla(-\underline{p}_+)$	(11.20b)
$-F_+ = \nabla(-P)$	$-F_+ = 0$	$-F_+ = (1/\phi_+)\nabla(-\underline{p}_+)$	(11.20c)

Combination of the sets (11.20) with (11.19) and (11.18) then gives three sets of three equations relating the fluxes J^V, J^D_s and $J_\pm$ to $\nabla(-P)$ — in the absence of other driving forces —, to $\phi_w \nabla(-\Pi)$ and to $\nabla(-\underline{p}_+)$, respectively. These sets thus allow one to express the coefficients L of Table 11.1 in terms of the frictional and accumulation parameters introduced previously.

Although straightforward, the above procedure if fully executed is rather cumbersome, even for the present, simplified system containing only three coefficients of friction and three accumulation factors. In contrast the first row of coefficients of Table 11.1 may be found straight away by summing up the equations (11.18) to give the frictional force acting on a unit volume of the entire liquid phase, according to:

$$-[\phi'_w F_w + \phi'_- F_- + \phi'_+ F_+] = [\phi'_w v'_w + \phi'_- v'_- + \phi'_+ v'_+]b \qquad (11.21)$$

With $\phi'_k v'_k = \phi_k v_k$ and $\Sigma_k \phi'_k = 1$ this gives:

$$-F_t = J^V b \qquad (11.21a)$$

so,

$$\text{for } \nabla(-\Pi) = \nabla(-p_{e+}) = 0 : J^V = \frac{1}{b}\nabla(-P) \qquad (11.22)$$

$$\text{for } \nabla(-P) = \nabla(-p_{e+}) = 0 : J^V = \frac{(\phi'_- \phi_w/\phi_-) - \phi'_w}{b}\nabla(-\Pi)$$

$$= -\frac{(k - 1/\bar{u}_-)}{b}\phi_w \nabla(-\Pi) \qquad (11.23)$$

$$\text{for } \nabla(-P) = \nabla(-\Pi) = 0 : J^V = \frac{(\phi'_+/\phi_+) - (\phi'_-/\phi_-)}{b}\nabla(-p_{e+})$$

$$= \frac{\bar{u}_+ - 1/\bar{u}_-}{b}\nabla(-p_{e+}) \qquad (11.24)$$

The coefficients L_V, L_{VD} and L_{VE} thus having been calculated, one may express the members (b) and (c) of the sets (11.18) and (11.20) in terms of J_s^D and $J_\pm$ by substituting J^V via (11.22)—(11.24). This leads to the expected result that $L_{DV} = L_{VD}$ and $L_{EV} = L_{VE}$.

For this chosen simple friction model the Onsager relations thus appear as the result of the assumption used previously: viz. the frictional force exerted by the solute on the water is equal and opposite to the frictional force exerted by the water on the solute.

The four coefficients left become rather complicated expressions which tend to obscure their main features. Introducing some minor approximations, however, leads to a somewhat clearer picture. Thus if $\phi_w \approx 1$ and $\phi_+ \ll 1$, all coefficients are readily found, as given in Table 11.2 (cf. Bolt and Groenevelt, 1972).

11.3. SIGNIFICANCE OF THE TRANSPORT COEFFICIENTS

The chosen (over)simplified model allows easy identification of the different coefficients. Thus one recognizes L_V as the hydraulic conductivity of the porous medium when a dilute electrolyte solution is pressed through the medium while maintaining $\nabla(-\Pi)$ and $\nabla(-p_{e+})$ at zero value. While $\nabla(-\Pi) = 0$ implies the passage of a solution with a constant equilibrium concentration — which is not too difficult to realize for a limited time period by flushing the end-compartments — the condition $\nabla(-p_{e+}) = 0$ needs some

TABLE 11.2

Transport coefficients in a three-component system with $\phi_w \approx 1$ and $\phi_+ \ll 1$

	$\nabla(-P)$	$\phi_w \nabla(-\Pi)$	$\nabla(-p_{e+})$
J^V	$\frac{1}{b}$	$-\frac{(1-1/\bar{u}_-)}{b}$	$\frac{(\bar{u}_+ - 1/\bar{u}_-)}{b}$
J_s^D	$-\frac{(1-1/\bar{u}_-)}{b}$	$\frac{(1-1/\bar{u}_-)^2}{b} + \frac{1}{\phi_-\bar{u}_-(b+b_-)}$	$-\frac{(\bar{u}_+ - 1/\bar{u}_-)(1-1/\bar{u}_-)}{b} - \frac{1}{\phi_-\bar{u}_-(b+b_-)}$
$J_\pm$	$\frac{(\bar{u}_+ - 1/\bar{u}_-)}{b}$	$-\frac{(\bar{u}_+ - 1/\bar{u}_-)(1-1/\bar{u}_-)}{b} - \frac{1}{\phi_-\bar{u}_-(b+b_-)}$	$\frac{(\bar{u}_+ - 1/\bar{u}_-)^2}{b} + \frac{1}{\phi_-\bar{u}_-(b+b_-)} + \frac{\bar{u}_+}{\phi_+(b+b_+)}$

special provisions. Noting that for $\nabla(-\Pi) = 0$, $\nabla(-p_{e+}) = -\nabla(-p_{e-}) = z_+{}^m c_+ F\nabla(-\psi)$, this second condition implies shorting the medium with suitable electrodes. Thus $L_V = 1/b$ is the hydraulic conductivity for the short-circuited medium. In essence this implies the transport of the solution with unhindered motion of *all* ions — including the countercharge — by suppressing the streaming potential (gradient) and the osmotic pressure gradient.

In practice this is not as simple as it may seem at first sight. Thinking in terms of a column into which a neutral electrolyte solution is fed (with a composition equal to the position-invariant local equilibrium solution) this implies that the excess counterions needed to sustain unhindered motion must be fed into the system via an electrode reversible to this counterion. Simple short-circuiting with metal electrodes would imply the occurrence of electrode reactions presumably giving off a mixture of H-ions and metal ions and leading subsequently to (proper) diffusion fluxes (cf. Groenevelt et al., 1978, for the effect of short-circuiting electrodes reversible to the co-ion).

The coefficients L_{DV} and L_{EV} in the same column are typical examples of simple drag terms. Thus in the absence of $\nabla(-\Pi)$ and $\nabla(-p_{e+})$ the ions do not experience a direct driving force and autonomous motion is precluded. The volume flux simply implies the motion of the salt deficit, $c_s(1 - 1/\bar{u}_-)$, and the excess volume charge, $z_+{}^m c_+(\bar{u}_+ - 1/\bar{u}_-)$, which constitute a backward "diffusion flux" and a forward charge-flux according to the chosen definitions of J_s^D and $J_\pm$. The occurrence of this backward diffusion flux is often indicated as a salt-sieving or salt-reflection effect of a charged matrix.

The ratio L_{DV}/L_V is then termed the reflection coefficient, σ (Staverman, 1952), such that:

$$J_s^D = -\sigma L_V \nabla(-P) \tag{11.25}$$

Similarly L_{EV} expresses the streaming current in terms of $\nabla(-P)$, with $L_{EV}/L_V = (\bar{u}_+ - 1/\bar{u}_-)$, the distribution ratio of the mobile countercharge to the salt present at equilibrium concentration.

The coefficients L_{VD} and L_{VE} of the first row indicate the volume flux induced by $\nabla(-\Pi)$ and $\nabla(-\underset{e+}{p})$, respectively, in the absence of the other two driving forces. Referring to (11.20), second column, one finds in this case the driving force per unit volume of water as $-\nabla(-\Pi)$ and per unit volume of anion as $(\phi_w/\phi_-)\nabla(-\Pi)$, while the driving force on the cations vanishes in the absence of $\nabla(-\underset{e+}{p})$. Specifying for the systems in situ, containing $\phi_w' \approx 1$ and $\phi_-' = \phi_-/\bar{u}_-$, this gives for the total driving force acting on a unit liquid phase:

$$F_t = -(1 - 1/\bar{u}_-)\nabla(-\Pi) \tag{11.26}$$

giving L_{VD}/L_V again as $-(1 - 1/\bar{u}_-) = -\sigma$. The coefficient L_{VD} may be termed the coefficient of capillary osmosis, which is the counterpart to salt-sieving. Similar reasoning explains $L_{VE} = (\bar{u}_+ - 1/\bar{u}_-)L_V$, the coefficient of electro-osmosis and the counterpart to L_{EV}, the coefficient governing the streaming current.

The interpretation of L_D hinges on the recognition of the simultaneous occurrence of a drag flux and an autonomous diffusion flux, composing together the diffusion flux $J_s^D = ({}^m j_-/{}^m c_- - {}^m j_w/{}^m c_w)$. The first term then gives the relative salt deficit, $(1 - 1/\bar{u}_-)$, dragged along with the volume flux induced by $\nabla(-\Pi)$, giving $(1 - 1/\bar{u}_-)^2 L_V$. The second term specifies the autonomous salt flux (i.e. the salt flux in excess of the water flux, cf. (11.18b) and (11.20b), second column, giving:

$$v_- = v_-'/\bar{u}_- \approx \nabla(-\Pi)/\phi_-\bar{u}_-(b + b_-) \tag{11.27}$$

where second-order terms have been neglected. Similarly L_{ED} contains the corresponding drag term for the mobile charge, $-(\bar{u}_+ - 1/\bar{u}_-)(1 - 1/\bar{u}_-)L_V$, plus the autonomous charge flux. The latter equals minus the autonomous diffusion flux, as only the anions are subject to a driving force if $\nabla(-\underset{e+}{p}) = 0$. The coefficient L_{DE} follows directly as containing a drag component equal to $-(1 - 1/\bar{u}_-)J^V$, the volume flux being determined by $L_{EV}\nabla(-\underset{e+}{p})$. The autonomous diffusion flux (of the anions!) is found from (11.27), using $-\nabla(-\underset{e+}{p})/\phi_-$ as the driving force per unit volume of anions (cf. (11.20b), third column). The coefficients L_{DE} and L_{ED} govern the salt transport induced by the application of a gradient of the electric potential in systems with unequal transport numbers, and the diffusion current accompanying salt diffusion under the same condition.

Finally L_E contains two terms governing the autonomous anion and cation fluxes, in addition to the drag flux of the mobile charge. The first one is found as equal and opposite to the autonomous (salt) diffusion flux in the coefficient L_{DE}: a negative anion flux corresponds to a positive charge flux. The second one is the autonomous cation flux, with $\bar{u}_+$, b_+ and ϕ_+ replacing $1/\bar{u}_-$, b_- and ϕ_- in the first one.

Apart from the present set of nine coefficients governing the case where three independent forces are operative, one may use certain combinations to describe the situation arising when one of the fluxes is maintained at zero value. In this respect it is pointed out that ordinarily the flow of salt solutions through porous media takes place in the absence of electrodes imposing a chosen gradient of $\underset{\sim}{p}_{e+}$. In that case the flow of charge is prohibited (no return flow of electric current via the electrodes). As a result a gradient of $\underset{\sim}{p}_{e+}$ arises ("streaming and/or diffusion potential"), which may be expressed in terms of $\nabla(-P)$ and $\nabla(-\Pi)$ by solving the last row in Table 11.1 for $J_\pm = 0$. This gives:

$$-\nabla(-\underset{\sim}{p}_{e+}) = (L_{EV}/L_E)\nabla(-P) + (L_{ED}/L_E)\nabla(-\Pi) \qquad (11.28)$$

which could be evaluated experimentally by measuring the difference in potential between reversible electrodes inserted at certain positions in the column. Aside from whether or not such electrodes are present (without delivering an electric current), the coefficients commanding the system with two independent forces at zero electric current are found as given in Table 11.3.

TABLE 11.3

Transport coefficients at zero current

	$\nabla(-P)$	$\phi_w \nabla(-\Pi)$
J^V	$(L_V - L_{VE}L_{EV}/L_E)$	$(L_{VD} - L_{VE}L_{ED}/L_E)$
J_s^D	$(L_{DV} - L_{DE}L_{EV}/L_E)$	$(L_D - L_{DE}L_{ED}/L_E)$

11.4. SOME NUMERICAL ESTIMATES

The model described in the previous section served primarily to indicate the *origin* of the coupling coefficients, viz. the presence of a salt deficit and of a countercharge in the liquid phase. The pressure-gradient-induced liquid flux J^V will drag these along, giving rise to coupling between the fluxes J_s^D and $J_\pm$ and the force $\nabla(-P)$. Conversely the forces $\nabla(-\Pi)$ or $\nabla(-\underset{\sim}{p}_{e+})$, if acting on the salt deficit and countercharge, respectively, will be transmitted to the liquid phase via drag forces, thus coupling J^V to the above forces, etc.

The difficulty one faces when applying these principles in order to obtain numerical estimates of the coefficients involved is the non-homogeneous distribution of both the ions and the liquid velocity inside a pore, in addition to the existence of a pore-size distribution within the porous medium. While now the *average* values of the salt deficit and countercharge are usually known with sufficient accuracy, as well as for example the *average* liquid velocity for a given value of $\nabla(-P)$, the simple multiplication of these averages as suggested in Table 11.2 is a severe and often untenable simplification.

For lack of precise information on the necessary distribution functions in actual porous media it is of interest to look at the outcome if the homogeneous distribution of ions (cf. Table 11.2) is replaced by the "ideal" diffuse distribution in a planar double layer, according to the Gouy theory (cf. chapter 1). For this purpose the liquid velocity in such a planar layer is assumed to obey the one-dimensional, steady state Navier-Stokes equation with constant viscosity, η, according to:

$$\eta\, d^2v/(dx)^2 = -F \qquad (11.29)$$

with $v(x)$ indicating the local velocity at a distance x from the solid surface and $F(x)$ specifying the local driving force in N/m^3. Following Groenevelt and Bolt (1969), one may then construct the velocity pattern arising in a liquid layer typified by a concentration distribution according to:

$$\phi'_+(x) = \phi_+ u_+(x) \qquad (11.30a)$$

$$\phi'_-(x) = \phi_- / u_-(x) \qquad (11.30b)$$

where $u_+(x)$ and $1/u_-(x)$ are the local values of the accumulation factors pertaining to the ions concerned as for example found with the help of the relevant equations in chapter 1. This gives:

(a) for the condition $\nabla(-\Pi) = \nabla(-p_{e+}) = 0$

$$v(x) = \frac{\nabla(-P)}{\eta} \int_0^x dx \int_x^{d_1} dx \qquad (11.31a)$$

(b) for $\nabla(-P) = \nabla(-p_{e+}) = 0$

$$v(x) = -\frac{\phi_w \nabla(-\Pi)}{\eta} \int_0^x dx \int_x^{d_1} (1 - 1/u_-) dx \qquad (11.31b)$$

(c) for $\nabla(-P) = \nabla(-\Pi) = 0$

$$v(x) = \frac{\nabla(-p_{e+})}{\eta} \int_0^x dx \int_x^{d_1} (u_+ - 1/u_-) dx \qquad (11.31c)$$

The liquid velocity pattern $v(x)$ given in (11.31) may now be combined with the distribution functions 1, $(1 - 1/u_-)$ and $(u_+ - 1/u_-)$, and integrated over the liquid layer stretching from $x = 0$ to $x = d_1$, respectively. This gives:

$$J^V = \frac{\theta/\lambda}{d_1} \int_0^{d_1} v(x)dx \tag{11.32}$$

and the drag terms:

$$J_s^D(\text{drag}) = \frac{\theta/\lambda}{d_1} \int_0^{d_1} \{1 - 1/u_-(x)\} v(x) dx \tag{11.33}$$

$$J_\pm(\text{drag}) = \frac{\theta/\lambda}{d_1} \int_0^{d_1} \{u_+(x) - 1/u_-(x)\} v(x) dx \tag{11.34}$$

In cases where the distribution functions of u appearing in (11.31b) and (11.31c) vary only slightly across the liquid layer, they may be replaced by the average values $(1 - 1/\bar{u}_-)$ and $(\bar{u}_+ - 1/\bar{u}_-)$. Then the velocity pattern becomes parabolic in all cases and (11.33) and (11.34) deliver the drag components of the coefficients as shown in Table 11.2. The model based on this condition will be indicated henceforth as model I. If the distribution functions are taken as those derived from the Gouy theory for the freely extended double layer, the integrals appearing in the above equations may still be solved analytically (see the sections following) and the ensuing values of the different coefficients are readily obtained. This model (II) will now be used to assess the effect of a diffuse ion distribution on the coupling coefficients, using a system with given values for the total (mobile) countercharge and salt deficit. Thinking in terms of a solid phase with a surface charge density, Γ, of 10^{-9} keq/m^2 (e.g. montmorillonite), liquid layers with thickness d_1 of 400 Å units with a (local) equilibrium concentration of 10^{-3} NaCl, and ionic mobilities corresponding to an equivalent conductance Λ of $5.5 \cdot 10^{-3}\ \Omega^{-1}\text{m}^2$ per equivalent of single ion, one finds the coefficients as given in Table 11.4 (still to be multiplied by an appropriate value for θ/λ).

TABLE 11.4

Influence of the allowance for a distributed countercharge (model II) on the calculated magnitude of the coefficients L (in units of 10^{-13} m/sec per N/m^3); $d_1 = 400$ Å, $c_0 = 10^{-3}$ keq/m^3, $\Gamma = 10^{-9}$ keq/m^2, $\theta/\lambda = 1$

	Model I (homogeneous charge)	Model II (Gouy distribution)
L_V	5.3	5.3
$L_{VD} = L_{DV}$	−2.7	−1.8
L_D	(1.3 + 2.9) = 4.3	(0.8 + 2.9) = 3.7
$L_{VE} = L_{EV}$	133	13.0
$L_{DE} = L_{ED}$	−(67 + 3) = −70	−(5.9 + 2.9) = −8.8
L_E	(3333 + 147 + 3) = 3483	(96 + 147 + 3) = 246

For this table the value of $(\bar{u}_+ - 1/\bar{u}_-)$ was taken equal to $\Gamma/d_l c_0 = 25$. Using $d_{ex,Cl}$ at 200 Å (cf. chapter 7), one finds $(1 - 1/\bar{u}_-)$ as $d_{ex}/d_l = 0.5$. Translation of the frictional coefficient of the ions, $b_+ = b_-$, into Λ gives $\phi_- b_- = \phi_+ b_+ = z_+{}^m c_{0,+} F^2/\Lambda_+ = 1.7 \cdot 10^{12}$ N/m^3 per m/sec at the chosen value of $c_0 = 1$ keq/m^3. The coefficients pertaining to model II were calculated according to Groenevelt and Bolt (1969).

As expected, the effect is most striking for the transport of charge: $(u_+ - 1/u_-)$ steeply declines away from the surface, while in this region the liquid velocity is still far below its average value as given by J^V. In contrast L_{DV} is influenced much less as $(1 - 1/u_-)$ declines slowly away from the surface and a major part of the anion exclusion reaches into the region of fairly high liquid velocities.

The overall conclusion is that model I has only practical significance if the thickness of the liquid layer does not exceed a few tens of Å units, i.e. where severe double layer truncation may lead to a fairly homogeneous distribution of the ions. As will be commented on in the following sections, model II offers somewhat better perspectives. In fact this model tends to present an upper limit for calculations of the type suggested above: most, if not all, deviations from the assumptions underlying Gouy distributions in conjunction with laminar flow will cause contraction of the diffuse layer and reduction of liquid velocities close to the solid surface, thus diminishing the coupling coefficients. It should also be pointed out that if coupling coefficients tend to decrease with increasing value of the liquid layer thickness, the presence of a pore-size distribution will further decrease the "macroscopic" coupling coefficients. In that case a relatively large part of the liquid flow will be through the pores with the smallest coupling coefficients.

11.5. SALT-SIEVING AND OSMOTIC EFFICIENCY COEFFICIENTS IN SOIL

Salt-sieving or reverse osmosis is governed primarily by the reflection coefficient $\sigma = -L_{DV}/L_V$, as is its counterpart, the osmotic solvent flux induced by concentration gradients. Both phenomena have been studied extensively for clay pastes by Kemper et al. in the period from 1960 to 1970. An excellent review of this work is given by Kemper in Nielsen et al. (1972), covering the original papers: Kemper (1960, 1961a, b), Kemper and Evans (1963), Kemper and Maasland (1964), Kemper and Rollins (1966), Kemper and Letey (1968), Kemper and Quirk (1972), Letey and Kemper (1969) and Letey et al. (1969).

As was explained in the previous section, the coefficient $L_{VD} = L_{DV}$ is determined by the co-ion distribution in the liquid films multiplied by the local liquid velocity. Now the co-ion deficit itself is presumably already rather insensitive towards specific interactions between matrix surface and co-ions (cf. chapter 7). At the same time the liquid velocity is low close to the solid surface, which further minimizes the effect of such interactions.

Aside from a counter electro-osmotic disturbance it appears that salt-sieving lends itself well for experimental verification of a co-ion exclusion theory in systems with a narrow pore-size distribution.

In practice the situation is not quite that simple because the conditions $\nabla(-\Pi) = \nabla(-\underset{\sim}{p}_+) = 0$ are difficult to attain experimentally. Thus $\nabla(-\Pi) = 0$ implies that the end compartments bordering a clay plug through which the equilibrium solution is forced by means of a pressure gradient must be flushed with excess solution. In turn this makes an accurate determination of the salt "reflected" by the column entrance difficult. The condition $\nabla(-\underset{\sim}{p}_+) = 0$ implies the insertion of shorted electrodes which should in principle be able to deliver at the column entrance the excess counterions necessary to replace the freely moving counterions inside the liquid films, while removing these at the exit. In practice this condition is nearly impossible to meet. It may be noted in this respect that the insertion of shorted electrodes reversible to the co-ion (i.c. AgCl electrodes) would not only disturb the observation of the reduced passage of salt by liberating co-ions at the exit of the column but could also create concentration gradients by introducing some Ag-ions at the entrance side (cf. Elrick et al., 1976; Groenevelt et al., 1978).

The standard experiment consists then of forcing an equilibrium solution through a column of the porous material under the condition of zero electric current while flushing the entrance and exit compartments. Alternatively one may determine the volume flux arising if a concentration difference is maintained between entrance and exit compartments, again for zero electric current. Unfortunately the condition $J_\pm = 0$ yields information only about the reflection coefficient σ', defined as (cf. Table 11.3):

$$\begin{aligned}\sigma'(J_\pm = 0) &= -J_s^D/J^V \text{ for } \nabla(-\Pi) = 0, J_\pm = 0 \\ &= -(L_{DV} - L_{DE}L_{EV}/L_E)/(L_V - L_{VE}L_{EV}/L_E) \\ &\equiv -\frac{J^V/\nabla(-\Pi) \text{ for } \nabla(-P) = 0, J_\pm = 0}{J^V/\nabla(-P) \text{ for } \nabla(-\Pi) = 0, J_\pm = 0}\end{aligned} \qquad (11.35)$$

where the last line defines the osmotic efficiency at zero current as identical with the reflection coefficient at zero current. This then brings the uncertainty about the counterionic behavior in a charged matrix into the picture.

Keeping the above in mind it appears nevertheless of interest to compare observed reflection coefficients with those predicted on the basis of the expected distribution pattern of co-ions (cf. chapter 7). Particularly because the electro-osmotic effects tend to remain below the magnitude predicted from an assumed diffuse counterion distribution according to Gouy with ionic mobilities equal to those in solution, it seems of interest to plot experimental data against extreme assumptions. For this purpose the calculation of σ' may be used as typifying maximal counter electro-osmotic disturbance while the simple calculation of $\sigma = -L_{VD}/L_V$ could serve as the system with minimal electro-osmotic disturbance.

In the theoretical analysis accompanying some of his early measurements Kemper (1960) set up a layer-wise computation of σ' for interacting Gouy layers. The gradient of the streaming potential was obtained by iteration to zero current transport. The observed salt reflection ranged from 0.32 (at 0.01 N NaCl in a montmorillonite paste with liquid-filled porosity $\theta = 0.7$) to 0.10 (at 0.1 N NaCl in a paste with $\theta = 0.6$). Using the Darcy coefficient as measured and allowing for pore tortuosity, the estimated value of the liquid layer thickness in the above systems was taken as 95 and 40 Å, respectively, which appeared to compare well with the thickness needed to account for the observed salt reflection, viz. 100 and 43 Å, respectively. In a later paper Kemper and Rollins (1966) reported the observed osmotic efficiency in similar clay samples for varying conditions. Layer-wise calculations following the pattern described above again gave satisfactory matching for a number of cases (cf. Fig. 11.1). Encouraged by the above results Römkens (pers. comm., 1966) executed a computer calculation of σ' covering a fairly wide range of conditions of surface density of charge, electrolyte level and thickness of the liquid film. A summarizing reference to these calculations is contained in Groenevelt and Bolt (1972); in Fig. 11.1 the results covering three relevant cases are plotted for comparison with experimental data given in Kemper and Rollins (1966). In view of the uncertainties still involved in translating observed Darcy coefficients in terms of the "mean effective liquid layer thickness", the agreement between the experimental reflection coefficients as found by Kemper and Rollins (1966) for the NaCl systems and the calculated curves for σ' is rather good.

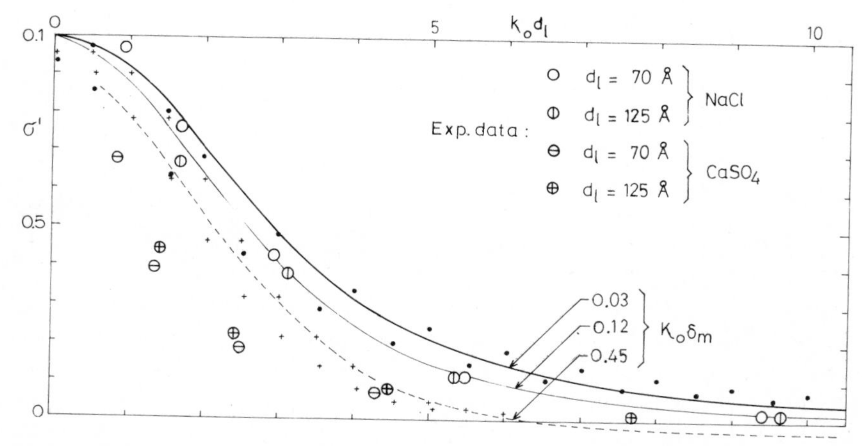

Fig. 11.1. The reflection coefficient at zero current, σ', plotted as a function of the reduced thickness of the mobile liquid layer, $\kappa_0 d_1$, for different values of $\kappa_0 \delta_m$ characterizing the magnitude of the mobile countercharge. The curves shown were calculated for a Gouy distribution taking into account counter electro-osmosis and double layer interaction (from Römkens, pers. comm., 1968). Large circles indicate experimental data given by Kemper and Rollins (1966). Small dots represent the reflection coefficient at zero streaming potential, σ, according to (11.39). Small crosses refer to the approximation equation (11.40).

Some additional remarks should be made with regard to the curves shown in Fig. 11.1. For any given value of the mobile liquid layer, d_l, the reflection coefficient will depend primarily on $d_{ex} = Q/\sqrt{\beta c_0} - \delta_m$ (cf. (7.8)) where δ_m is here related to the mobile countercharge, Γ_m, according to $\delta_m = 4/z_+\beta\Gamma_m$ (cf. (1.37)). A generalized plot of σ' against $\kappa_0 d_l$ for different values of $\kappa_0\delta_m$ as shown in Fig. 11.1 then represents the course of σ' as a function of d_l (at constant electrolyte level) for a constant value of Γ_m — which follows from the particular combination of $\kappa_0\delta_m$ and the electrolyte level chosen. If used as a plot of σ' as a function of the electrolyte level (at constant d_l) the corresponding value of Γ_m varies with the electrolyte level for the curves shown. Fortunately the three curves shown may be interpolated linearly for other values of $\kappa_0\delta_m$ in the range $0.03 < \kappa_0\delta_m < 0.2$. The experimental data from Kemper and Rollins (1966) as shown in Fig. 11.1 thus refer to varying values of $\kappa_0\delta_m$ even if Γ_m were a constant in their series of measurements at varying electrolyte levels.

The theoretical curves shown include fully the secondary effects caused by the electrokinetic phenomena. Here uncertainty prevails on three counts, however; viz.
(a) the drag flux of counterions varies strongly with the thickness of an immobile layer;
(b) the mobility of counterions present locally at high concentrations is probably different from the mobility in the equilibrium solution;
(c) the present model based on (11.29) and (11.31) does not allow for excess friction between the ions and the matrix as compared to solution and matrix.

While referring to the following section for further elaboration on this, it may be stated here that the contribution of the electrokinetic phenomena on σ' is fortunately often rather small. Moreover, it is seen from (11.35) that the electrokinetic effects affect J_s^D and J^V in a similar manner: both fluxes are reduced with comparable factors and thus the ratio J_s^D/J^V for $J_\pm = 0$, $\nabla(-\Pi) = 0$ is usually not very different from J_s^D/J^V for $\nabla(-\underline{p}_{e+}) = 0$, $\nabla(-\Pi) = 0$.

The above is illustrated by using the numbers given in Table 11.4. For model II this gives:

$$\sigma = -L_{DV}/L_V = 0.51$$

$$\sigma' = -(L_{DV} - L_{DE}L_{EV}/L_E)/(L_V - L_{VE}L_{EV}/L_E) = 0.48$$

Having thus characterized the electro-kinetic terms as usually having fairly small and compensating effects on the ratio of J_s^D and J^V, it seems worthwhile to devise an approximation equation for σ' based on the calculation of σ. Recalling the co-ionic distribution in truncated diffuse layers as calculated in chapter 7, it then appears warranted for the present purpose to use the equations for the freely extended double layer from $x = \delta$ to $x = x_t$ in order to circumvent the use of elliptic integrals. Although this will lead to a slight underestimate of the co-ion exclusion this may serve to compensate for the fact that $\sigma > \sigma'$.

Making use of the relevant equations in Table 1.1 of chapter 1 one finds for symmetric homoionic systems:

$$\kappa_0 x = \ln \coth(zy/4) - \ln \coth(zy_s/4)$$

$$= -\ln\left[\frac{\sqrt{u^z}-1}{\sqrt{u^z}+1}\right] + \ln\left[\frac{\sqrt{u_s^z}-1}{\sqrt{u_s^z}+1}\right] \tag{11.36}$$

where the present x-coordinate replaces the x-coordinate used in chapter 1, its zero point being taken at the beginning of the mobile layer where $u = u_s$. Replacing the argument of the ln terms by t and t_s, respectively, this gives:

$$\kappa_0 x = -\ln(t/t_s) \tag{11.37a}$$

$$\kappa_0 \delta = -\ln t_s \tag{11.37b}$$

$$\kappa_0 d_l = -\ln(t_d/t_s) \tag{11.37c}$$

$$(1 - 1/u) = 4t/(1+t)^2 \tag{11.37d}$$

Introducing the above into (11.31) and (11.32) one finds:

$$\sigma = -J_s^D/J^V \text{ for } \nabla(-\Pi) = \nabla(-p_+) = 0$$

$$= -\frac{\int (d_l x - x^2/2)(1-1/u)dx}{\int (d_l x - x^2/2)dx}$$

$$= \frac{3}{(\kappa_0 d_l)^3}\left[\kappa_0 d_l \int_{t_s}^{t_d} \ln(t/t_s)\frac{4t}{(1+t)^2}\, d(\ln t) + \tfrac{1}{2}\int_{t_s}^{t_d} \ln^2(t/t_s)\frac{4t}{(1+t)^2}\, d(\ln t)\right] \tag{11.38}$$

The solution of this equation is found as:

$$\sigma = \frac{12}{(\kappa_0 d_l)^3}\left[\kappa_0 d_l\left\{\ln(1+t_s) - \frac{\kappa_0 d_l}{2}\,\frac{t_d}{1+t_d}\right\} - \left| t - \frac{t^2}{4} + \frac{t^3}{9} - \ldots \right|_{t_s}^{t_d}\right] \tag{11.39}$$

where the appropriate values of t_s and t_d are found from $t_s = \exp(-\kappa_0\delta)$, $t_d = \exp\{-\kappa_0(d_l + \delta)\}$. The limiting curve of $\sigma(\kappa_0 d_l)$, valid for $\kappa_0 d_l > 5$ and $\kappa_0\delta < 0.01$, is then given by:

$$\sigma = \frac{12 \ln 2}{(\kappa_0 d_l)^2} - \frac{\pi^2}{(\kappa_0 d_l)^3} \tag{11.39a}$$

which serves to obtain a first estimate.

In Fig. 11.1 the values of σ from (11.39) derived above are shown as separate dots for $\kappa_0 \delta_m = 0$ (limiting case corresponding to a highly charged surface without an immobile layer) and for $\kappa_0 \delta_m = 0.45$. Comparison with the curves representing σ' — including the maximal effect of counter electro-osmosis — shows a satisfactory check for $\kappa_0 d_l < 4$. Beyond this value the relative difference between σ and σ' becomes considerable. As in practice electro-osmosis will hardly reach the maximum values as assumed for the calculation of the curves (cf. following section), (11.39) may still serve as an estimate for σ' if an absolute error of 0.05 is acceptable.

An even simpler approximation equation was proposed by Kemper (cf. Nielsen et al., 1972). This equation, based on the contention that the major part of the liquid transport takes place near to $x = d_l$, consists of applying model I (cf. Table 11.2) while using the value of u pertaining to $x = d_l$ for the constant $\bar{u}_-$.

To evaluate this estimate one may read U_t from graph 1.1 in chapter 1 at the appropriate value of $\kappa_0(\delta + d_l) \equiv \kappa_0 D$. As for symmetric systems $U_t = \cosh(zy_t)$, one then finds:

$$\sigma' \approx 1 - \exp(-\operatorname{argcosh} U_t) \tag{11.40}$$

In Fig. 11.1 the above approximation has also been indicated for $\kappa_0\delta = 0$ and $\kappa_0\delta = 0.45$, respectively. As would be expected, this equation gives an underestimate, particularly for large values of $\kappa_0 d_l$.

To summarize, it may be concluded that salt-sieving and osmotic efficiency factors in soils in situ may be estimated rather easily, provided a reasonable guess is available on the "effective" thickness of the liquid layers. In practice, knowledge about the liquid layer geometry is likely to be the limiting factor for such estimates. While experimental verification of the phenomenon in clay plugs has proved to be relatively simple, its quantitative significance with respect to soil behavior may be rather small. In this respect it should be recalled that the effect becomes large only in thin liquid layers, when flow rates of the soil solution tend to be low. It is pointed out that then the (relatively large) diffusion leakage (L_D) will soon lead to a steady state concentration gradient.

11.6. ELECTROKINETIC EFFECTS IN SOILS

Much in contrast to salt-sieving, electrokinetic effects in soils are difficult to predict quantitatively from model theories. At the same time they are well known and of definite practical significance. The streaming potential (gradient) accompanying liquid flow through a charged porous medium forms part of the interpretation of the so-called "Spontaneous Potential" observed in well logging as employed in oil explorations (SP logs). The information available on this subject should thus primarily be sought in petroleum engineering literature (cf. Smits, 1968; Lynch, 1962). The counterpart of the streaming potential is the electro-osmotic liquid flow arising when a gradient of the electric potential is applied over a soil column. This aspect has found an application in the electro-osmotic stabilization of weak clays, used to advantage in soil engineering. Although part of this "stabilization" may be traced to electrode reactions leading to local aluminization of soil, also short-term effects arise following the removal of liquid from the anode. A literature review covering early attempts towards practical utilization in soil engineering was given by Casagrande (1953). A short communication by Miller (1955) should be mentioned here as indicative of the numerous interfering effects which may arise if a soil column of some length is subjected to an applied electric field of several V/cm. Electrolytic conductance then induces gradients of electrolyte concentration which in turn cause a strong variation of the potential gradient along the column, the pore water pressure varying accordingly.

The difficulty in predicting the magnitude of the above electro-kinetic effects in soils and clay pastes arises partly because of the uncertainty about the mobilities of the counterions in the system. Even if the mobilities governing electrolytic conductance in these systems are known roughly (cf. following section), such figures are at best average values, while the contribution of the frictional forces between ions and matrix, although not known, is probably substantial. Briefly, a known electric force acting on *all* counterions is transmitted in part directly to the solid matrix, which part does not induce electro-osmotic flow. The remaining part is transmitted to the liquid phase, but this force is then largely concentrated near the solid matrix (where the majority of the counterions are situated), which is again the region where uncertainties exist about the precise distribution of the counterions, and presumably also about the local viscosity of the solvent. Finally the mentioned electric force acting on the ions may differ from the one calculated on the basis of the externally applied electric field, because of the variation in conductance near the charged surface.

In view of the uncertainties mentioned it is important to establish first that, for a given amount of mobile countercharge subject to an applied electric field of known magnitude, the electro-osmotic liquid flux tends to increase the further this countercharge extends into the liquid layer. This follows immediately from (11.29) if the driving force acting on a unit volume of liquid phase is taken to be proportional to the local value of the volume density of charge, $\rho_{\pm}(x)$. One then finds:

$$\eta(\mathrm{d}v/\mathrm{d}x) \sim \int_{x}^{\mathrm{d_l}} \rho_{\pm}(x)\mathrm{d}x \tag{11.41}$$

where the RHS represents the part of the mobile countercharge located beyond a chosen distance x from the solid surface, to be indicated as $\Gamma(x)$. Obviously $\Gamma(o)$ then represents the total countercharge (here considered as a given quantity). For any chosen distribution of the countercharge, $\Gamma(x)$ must be a curve situated between the extremes representing a homogeneous charge distribution (h) and a fully condensed distribution (c), as is shown in Fig. 11.2A. The corresponding velocity pattern is found by integrating over such a curve as shown in Fig. 11.2B. Clearly, both the maximum and the average value of the liquid velocity must increase with an increased spreading of the countercharge. This effect is reinforced if the viscosity increases close to the solid surface.

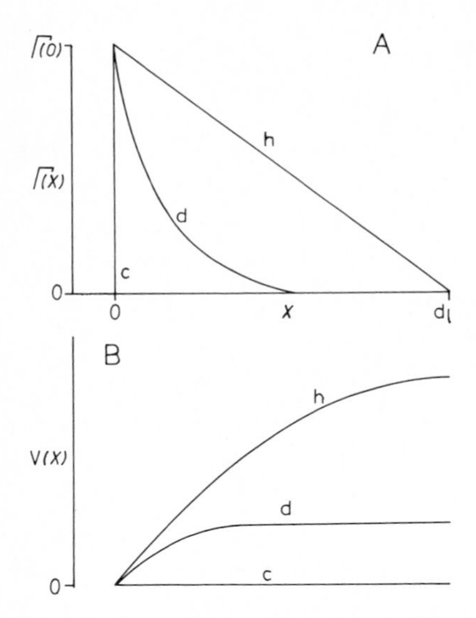

Fig. 11.2. Schematic representation of the electro-osmotic liquid velocity, v(x), in a liquid layer with thickness d_l and containing a total mobile countercharge $\Gamma(o)$, for different types of charge distribution. This distribution has been characterized by the amount of countercharge present beyond a distance x from the solid surface, $\Gamma(x)$, as shown; viz. h: homogeneous; d: diffuse; c: fully condensed.

As now an ionic distribution following the Gouy theory may be considered as the most "diffuse" type of distribution that could exist in a liquid layer, it follows that calculations based on the Gouy theory lead to the maximal possible values for the coefficients covering electro-kinetic phenomena. If experiments indicate values below these maximal ones calculated on the basis of a fully mobile countercharge, this may be formally accounted for by introducing a reduced value for the mobile countercharge, Γ_m, according to:

$$\Gamma_m \equiv \bar{m}\Gamma \qquad (11.42)$$

Values of $\bar{m}$ less than unity then imply that any or all of the effects reducing the electro-osmotic liquid flux may be operative, like double layer contraction for example owing to specific adsorption of certain cations, increased viscosity close to the surface, etc. Alternatively one may express the observed reduction of the mobile charge in terms of the equivalent thickness of a liquid layer adjacent to the solid surface which would accomodate the immobile part of the countercharge. Using again the Gouy theory for this purpose the latter can be found according to (1.37), relating a surface charge density to the cut-off distance δ, as:

$$\delta_{im} \equiv \delta_m - \delta = \frac{4}{z_+\beta}\left(\frac{1}{\Gamma_m} - \frac{1}{\Gamma}\right)$$

$$= \frac{1-\bar{m}}{\bar{m}} \frac{4}{z_+\beta\Gamma} \tag{11.43}$$

With the help of the equations given in chapter 1 one may now calculate the liquid velocity pattern arising if a planar diffuse double layer is exposed to a potential gradient $\nabla(-E)$ parallel to the solid surface. Introducing the Poisson equation (1.1) into (11.29) this gives the constituting equation as:

$$\eta \frac{d^2 v}{dx^2} = \frac{\epsilon'}{4\pi} \frac{d^2 \psi}{dx^2} \nabla(-E) \tag{11.44}$$

Integrating twice, taking $v = 0$ and $\psi = \psi_m$ at $x = 0$ this gives:

$$v(x) = -\frac{\epsilon'}{4\pi\eta}(\psi_m - \psi) \quad \nabla(-E) \tag{11.45}$$

It should be noted that the velocity pattern thus follows the course of the electric potential in the double layer, the maximum velocity being reached once ψ approaches zero and then being equal to the velocity given by the well-known Helmholtz-Smoluchowski equation. The potential ψ_m is then referred to as the ζ-potential, i.e. the potential "at the plane of shear" in the liquid layer. Equation (11.45) may now be worked out with the help of Table 1.1 in chapter 1 provided the present position coordinate x taken relative to the plane of shear is translated into the coordinate x, according to:

$$x = \mathrm{x} - \delta_m \tag{11.46}$$

where $\delta_m \approx 4/z_+\beta\Gamma_m$ indicates the cut-off value used before (cf. (1.37)), now referring to the *mobile* countercharge, situated beyond the plane of shear. For a system containing one symmetric electrolyte, Table 1.1 gives for a freely extended double layer

$$v(x) = \frac{\epsilon' RT}{2\pi z_+ F\eta}[\ln \coth(\kappa_0 \delta_m/2) - \ln \coth(\kappa_0 \mathrm{x}/2)]\nabla(-E) \tag{11.47}$$

With $\Gamma_m = \bar{m}\Gamma$, the maximum value of v(x) reached at large values of x is then found as:

$$v(x \to \infty) = \frac{F\nabla(-E)}{\eta} \frac{4}{z_+\beta} \ln \coth(2\kappa_0/\bar{m}z_+\beta\Gamma) \tag{11.48}$$

For the present model (which was used before in calculating Table 11.4) the course of $v(x)$ is strongly dependent on the electrolyte level (via κ_0), which also determines the ζ-potential for a given value of $\bar{m}\Gamma$. At a given value of $\kappa_0\delta_m$, however, $v(x)$ approaches this maximum value asymptotically, showing little increase once $\kappa_0 x > 2$. This indicates that the velocity pattern is insensitive with respect to the thickness of the liquid layer once the latter exceeds the value of $2/\kappa_0$.

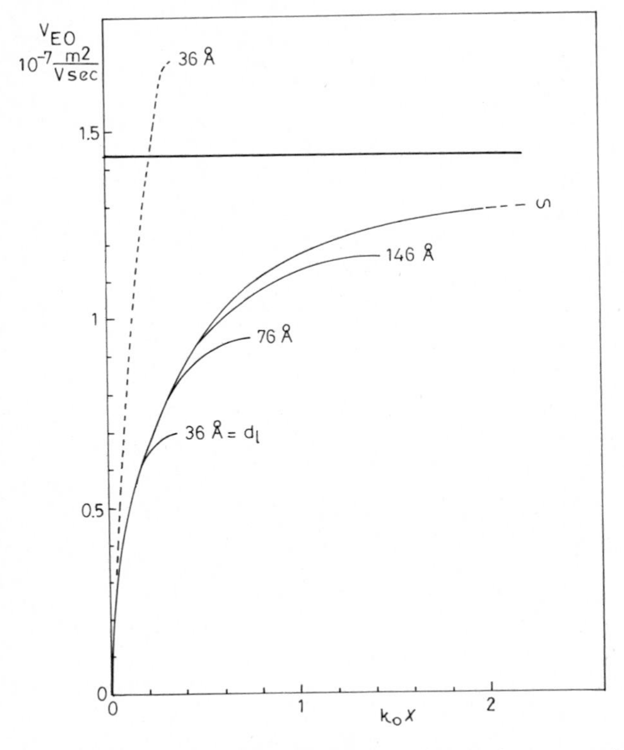

Fig. 11.3. The electro-osmotic liquid velocity, v_{EO}, in liquid films of thickness d_l as indicated. The curves shown correspond to a charged surface in equilibrium with mono—monovalent salt at 10^{-3} *N* concentration, with a total mobile charge of 10^{-9} keq/m^2, following a Gouy distribution. Heavy line represents the Helmholtz-Smoluchowski maximum velocity obtained for $d_l \rightarrow \infty$. Dashed line: parabolic velocity pattern arising for a homogeneously distributed charge of the same magnitude, shown only for $d_l = 36$ Å. In this case the maximum velocity is proportional to d_l^2.

If the thickness of the liquid layer characterized by $\kappa_0 d_l$ becomes smaller than about 2, double layer truncation will reduce the liquid velocities. This follows immediately from (11.45), indicating that for $\psi_t > 0$ the maximum value corresponding to the Helmholtz-Smoluchowski equation is never reached. Calculation of the velocity pattern in this case requires the use of elliptic integrals as given in (1.13). In Fig. 11.3 a few examples of the ensuing velocity pattern are shown. In this particular case $\kappa_0 \delta_m$ was taken at 0.04, corresponding to a mobile charge of 10^{-9} keq/m^2 and an equilbirium electrolyte level of 10^{-3} *N* mono—monovalent salt. As follows from the coinciding curves close to $\kappa_0 x = 0$, other cases may be found by simply recalibrating the vertical axis in terms of the appropriate value of $v(x \to \infty)$ as found from (11.48). For comparison also the parabolic velocity pattern arising in the case of a homogeneous charge distribution is shown for $d_1 = 36$ Å. In this case the maximum velocity is proportional to d_1^2. This velocity pattern is the basic ingredient of the theory of electro-osmotic flow proposed by Schmid (1951). It clearly offers no grounds for realistic estimates once d_1 exceeds a few tens of Å units (see also model I in Table 11.4).

The calculation of L_{VE} from the above velocity pattern implies another integration over the entire liquid layer. For truncated double layers this requires computer assistance, though a reasonable approximation equation may be derived (see below). If $\kappa_0 d_1 > 2$, equation (11.47) is a satisfactory approximation. Upon integration over the liquid layer this yields the equation given before by Groenevelt and Bolt (1972) valid for symmetric electrolytes:

$$L'_{VE} = H \ln u_m - \frac{2H/z}{\kappa_0 d_1}\left[\frac{\pi^2}{4} - \ln\sqrt{u_m^z}\ \ln\frac{\sqrt{u_m^z}+1}{\sqrt{u_m^z}-1} - 2\left\{\frac{1}{\sqrt{u_m^z}} + \frac{1}{3^2(\sqrt{u_m^z})^3} + \ldots\right\}\right] \tag{11.49}$$

where L'_{VE} is given in $m^2/Vsec$ with $H = 2F/\beta\eta$, $u_m^z = \coth^2(\kappa_0\delta_m/2)$, the Boltzmann factor applying to the plane of shear. The first term of this equation corresponds with the Helmholtz-Smoluchowski equation, the following ones signify the correction necessary for the reduced velocity close to the surface. It should be noted that $\ln u_m = F\psi_m/RT$ and is thus proportional to the ζ-potential.

In the case of severe truncation, i.e. $\kappa_0 d_1 < 0.5$, the approximation equations (1.27a and b) are satisfactory for describing $\psi(x)$, leading to:

$$L'_{VE} = -\frac{2H}{z}\ln\sin(\pi\delta_m/2x_t) - \frac{2H/z}{\kappa_0 d_1}\left|-\chi\ln\chi + \chi + \frac{\chi^3}{18} + \frac{\chi^5}{900} + \frac{\chi^7}{19845} + \ldots\right|_{\pi\delta_m/2x_t}^{\pi/2} \tag{11.50}$$

In this case the first term corresponds to the maximum velocity obtained in a truncated double layer, which relates to the Helmholtz-Smoluchowski equation provided the ζ-potential is replaced by $(\zeta - \psi_t)$. The second term again gives the reduction needed for the gradual increase of v.

In Fig. 11.4 L'_{VE} as calculated from (11.49) and (11.50) is plotted as a function of $\kappa_0\delta_m$ for different values of $\kappa_0 d_1$. As stated above, the first parameter is directly related to u_m (or the ζ-potential) and may also be regarded as a measure of the reduced value of the mobile charge, according to $\kappa_0\delta_m = 4/\Gamma_m\sqrt{z_+} = 4\sqrt{c_0}/\Gamma_m\sqrt{\beta z_+}$ (cf. equation (3.35)). The second parameter then specifies the thickness of the liquid layer in units of $1/\kappa_0$. Figure 11.4 allows a quick estimate of L'_{VE} for any chosen combination of Γ_m, κ_0 and d_l.

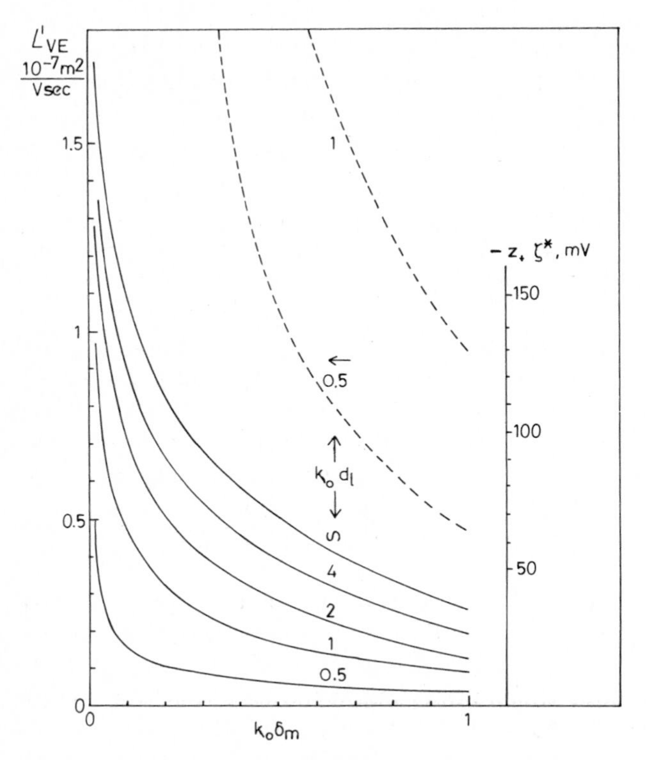

Fig. 11.4. Electro-osmotic coefficient L'_{VE} as a function of the reduced value of the mobile charge, characterized by $\kappa_0\delta_m = 4/\Gamma_m\sqrt{z_+}$, for different values of $\kappa_0 d_l$. Upper solid line corresponds to the Helmholtz-Smoluchowski value which may also be expressed in a ζ-potential. Dashed lines represent L'_{VE} as calculated according to model I (homogeneous charge distribution), full lines are calculated from Gouy theory.

As an example one may consider a solid surface with a mobile countercharge equal to $2 \cdot 10^{-10}$ keq/m^2 (20% of the total charge of montmorillonite), d_l equal to 30 Å and in equilibrium with 0.01 M $CaSO_4$. This gives with $1/\kappa_0 = 15$ Å, $\kappa_0\delta_m = 0.67$, $\kappa_0 d_l = 2$ and thus $z_+L'_{EV}$ at about $2 \cdot 10^{-8}$ m^2/Vsec. Taking into account a moisture content of 50% by volume and a tortuosity factor equal to 2, one finds the electro-osmotic conductivity of the porous medium as about $0.25 \cdot 10^{-8}$ m^2/Vsec, which is of the same order as the values observed in practice (see below).

For further orientation it is pointed out that the limiting line for $\kappa_0 d_l \to \infty$ corresponds to the Helmholtz-Smoluchowski equation, expressing L'_{VE} in terms of the ζ-potential, written in the present context as:

$$z_+L'_{VE} = -z_+\epsilon'\zeta/4\pi\eta \text{ m}^2/\text{Vsec} \tag{11.51}$$

Accordingly a second vertical axis has been added in Fig. 11.4, calibrated in mV. This axis then specifies the apparent ζ-potential, ζ^*, as back-calculated from observed electro-osmotic conductivities via the Helmholtz-Smoluchowski equation. Because of the reduction of the electro-osmotic flux owing to the gradual decrease of the potential in the double layer and double layer truncation, the actual value of the potential at the plane of shear may be much higher. As $\kappa_0 \delta_m$ is hardly affected by double layer truncation, one may estimate the actual value of ζ by reading the value on the $z_+\zeta^*$-axis via the limiting curve for $\kappa_0 d_l \rightarrow \infty$ (at $\kappa_0 \delta_m$ as found from $z_+ L'_{VE}$ at the *given* value of $\kappa_0 d_1$).

Thus in the example used previously the estimated value of $z_+ L'_{EV}$ at $2 \cdot 10^{-8}$ m^2/V sec would translate into an apparent potential $z_+\zeta^* = -30$ mV, i.e. $\zeta^* = -15$ mV. For the chosen value $\kappa_0 \delta_m = 0.67$, back-translation via the limiting curve gives the potential at the plane of shear as $\zeta = -55/z_+ = -28$ mV.

Turning to some experimental information, Ravina and Zaslavsky (1968), in an extensive review of the subject, inferred that the electro-osmotic conductivity of soils appears to center around $0.5 \cdot 10^{-8}$ m^2/V sec for a wide variety of soils. This number checks very well with data collected in the course of a study on electro-osmotic stabilization of soils conducted at Harvard University around 1955*, where the observed values for the electro-osmotic conductivity ranged from about $0.2 \cdot 10^{-8}$ for field clays to 10^{-8} m^2/Vsec in sodium kaolin.

Using again a factor of 4 to account for tortuosity and about 50% pore space, the above value checks also with apparent ζ-potentials around 25 mV which have often been reported. In view of the many uncertainties about the geometry of liquid films in soils this number might indicate that the present model gives a satisfactory description of the situation. Referring to the example discussed above in the small-print section, straightforward interpretation would then indicate that probably less than one-quarter of the total countercharge contributes to electro-osmotic transport, so $\bar{m} < 0.25$.

While likely to be correct, the above statement should be handled with care. Again the inhomogeneity of the applied field, arising in a heterocapillary system when most of the counterions are located in narrow pores, is mentioned. The surface conductance in these small pores would tend to lead to local low values of the applied electric field, interspersed with regions of higher potential gradients in wide pores with a low electric conductivity.

Regardless of the above limitations to predictions, it will be clear that the electro-osmotic conductivity (which is determined primarily by mobile charges situated close to the solid phase, presumably often in the smallest pores), has no relation to the hydraulic conductivity, which is largely determined by the widest pores. The fact that the ratio of hydraulic and electro-osmotic conductivity for different soils varies over many orders of magnitude is thus no surprise. In this context it is pointed out that, even for homogeneous

---

* Report to the Bureau of Yards and Docks, Soil Mechanics Laboratory, Harvard University, 1955.

capillary systems, this ratio is only indicative of the magnitude of the mobile charge if this charge were distributed homogeneously as was assumed in the above model I (and also in the theory put forward by Schmid (1951)). At best this may apply to charged membranes with pores of the order of 10 Å units.

Even more complicated is the prediction of streaming potentials in soils. This potential arises if the liquid phase is forced through a charged porous medium in the absence of electrodes, allowing for a return flow of charge external to the liquid phase. The net charge transport within the liquid phase then remains zero and a potential gradient must arise to balance the drag flux of the mobile countercharge with the liquid flux by conductance. Referring to Table 11.1 one thus finds:

$$J_{\pm} = L_{EV}\nabla(-P) + L_{E}\nabla(-p_{e+}) = 0$$

or

$$\nabla(-p_{e+})/\nabla(-P) = -L_{EV}/L_{E} \qquad (11.52)$$

assuming that $\nabla(-\Pi)$ is maintained at zero. As in that case $\nabla(-E_{+}) = \nabla(-E)$, the above ratio defines the streaming potential (gradient) in terms of the applied gradient of the hydraulic pressure. Where previously L'_{VE} was expressed in m/sec per V/m, L'_{EV} will now be defined in Coul/m^2 sec per N/m^3 and may then be derived in direct analogy with L'_{VE} for the present model as:

$$L'_{EV} = \frac{1}{d_1}\int_0^{d_1} \rho_{\pm} v dx$$

$$= -\frac{\epsilon'}{4\pi\eta}\int_0^{d_1}\left(\frac{d^2\psi}{dx^2}\right)\left(d_1 x - \frac{x^2}{2}\right) d(x/d_1)$$

$$= -\frac{\epsilon'}{4\pi\eta}\int_0^{d_1} (\psi_m - \psi)\, d(x/d_1) \qquad (11.53)$$

In addition to the uncertainties incurred in the calculation of L'_{EV} as regards the mobility of the countercharges, another set of assumptions is necessary to calculate the coefficient L'_{E}. These concern the contribution of all ions to the electric conductance of the liquid layer where now both the countercharge and the "neutral" salt contribute. Deferring a discussion on the electric conductance in charged porous systems to the section following, it will be accepted here that the specific conductance may be determined in situ. The coefficient L'_{E} is then defined as:

$$L'_{E} = I/\nabla(-E) = K'_{s} + K'_{a} \qquad (11.54)$$

the last two terms specifying the contributions to the conductivity of the salt and the adsorbed ions, respectively, in mho/m (see the section following). With $L'_{EV} = L'_{VE}$ the streaming potential is then found as:

$$\nabla(-E)/\nabla(-P) = -\frac{L'_{VE}}{K'_s + K'_a} \tag{11.55}$$

where the previous models for L'_{VE} could be used. Taking the Helmholtz-Smoluchowski model as the simplest, and not unreasonable, one this gives:

$$\nabla(-E)/\nabla(-P) = \frac{\epsilon'\zeta}{4\pi\eta(K'_s + K'_a)} \tag{11.56}$$

Whereas L'_{VE} was shown above to become smaller with increasing electrolyte level ($\kappa_0 \delta_m$ increasing for a fixed value of Γ_m or δ_m), this effect is obviously reinforced for the streaming potential because of the simultaneous increase of K'_s. Using the previous number of $L'_{VE} = 2 \cdot 10^{-8}$ m^2/Vsec as a guide number for L'_{EV} specified in Coul/sec per m^2 of *liquid phase* per V/m and assuming that tortuosity and pore volume influence the electric conductance in the same manner as L'_{EV}, one may deduce a maximum estimate for the streaming potential. Thus in the presence of 10^{-2} N electrolyte ($K_0 \approx 10^{-1}$ mho/m) one finds in the absence of a significant contribution of the adsorbed ions to the conductance:

$|\nabla(-E)/\nabla(-P)| < 2 \cdot 10^{-7}$ m^2 V/N (= 20 mV/bar).

In practice one finds rarely more than about 5 mV/bar at this concentration level, which is then rather easily accounted for by the conductance contribution of adsorbed ions.

An interesting set of data on streaming potentials was produced by Smits (1969), on cores of shaly sands and shales. At the lowest electrolyte level used (0.02 N NaCl) the observed values ranged from about 7 mV/bar in a shaly sand with a hydraulic conductivity of about 0.5 Darcy to around 1 mV/bar in shales with a hydraulic conductivity of about 1 milli Darcy. Back-calculation to the corresponding values of $L'_{VE} = L'_{EV}$, with L'_E as determined experimentally, gave in situ numbers around $0.1 \cdot 10^{-8}$ m^2/Vsec for the coarser samples. In view of the low porosities this seems still within the range discussed earlier.

Finally mention is made of the counter electro-osmosis induced by the streaming potential. Whereas in the present context the primary velocity distribution induced by the gradient of the hydraulic pressure is parabolic in shape, the electro-osmotic flow pattern follows the pattern of the electric potential across the liquid layer. For relatively wide films this somewhat resembles a plug flow as was pictured in Fig. 11.2. If combined in the opposite direction an involved flow pattern arises with a potential possibility for a net negative velocity close to the solid phase. In Fig. 11.5 an example of such a case has been pictured, based on the calculations by Römkens (pers. comm., 1966) used in section 11.5 for the calculation of the salt-sieving effect. As has become clear in the present section this effect, though undoubtedly present, is unlikely to develop to the full magnitude as assumed in these calculations (taking a very high mobile charge).

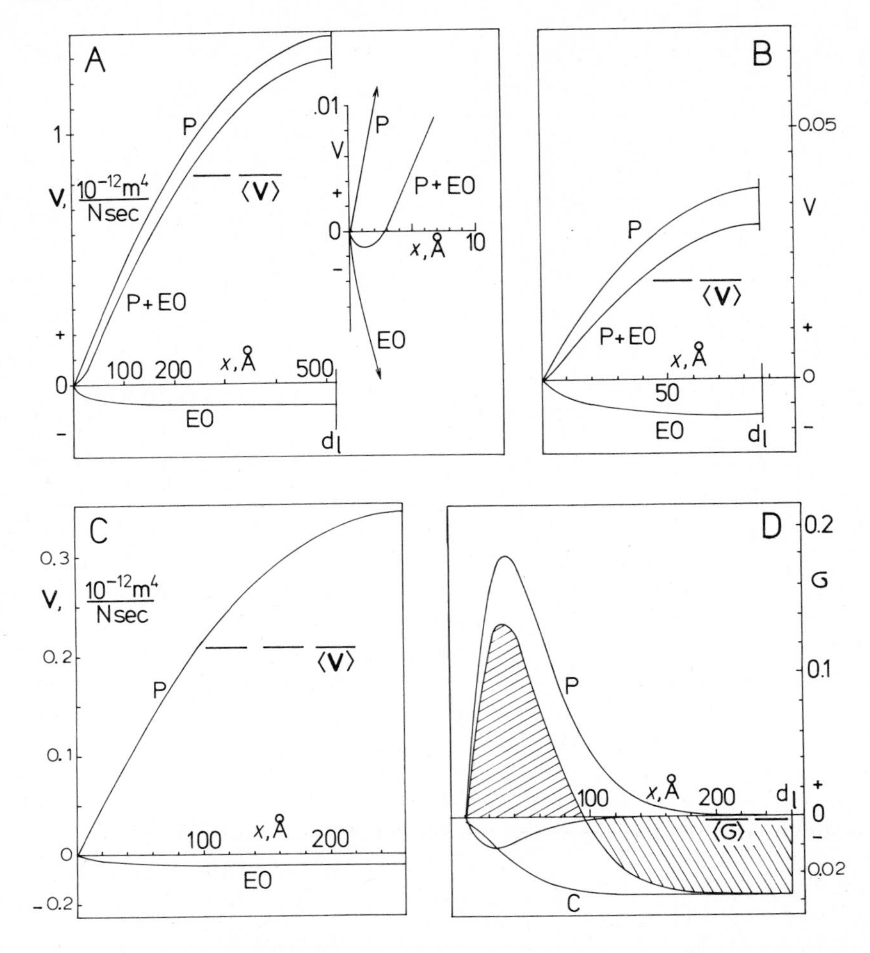

Fig. 11.5. Flow pattern as calculated for a liquid film of thickness d_l and carrying a mobile charge Γ_m, distributed according to the Gouy theory. Shown are the Poiseuille component (P), and the counter electro-osmotic component (EO) of the local valocity v given in 10^{-12} m/sec per N/m^3 (after Römkens, pers. comm., 1966).
A: $d_l = 520$ Å, $\Gamma_m = 1.4 \cdot 10^{-9}$ keq/m^2, $C_0 = 10^{-3}$ N NaCl; giving $\nabla(-E)/\nabla(-P) \approx$ 60 mV/bar. Insert: situation close to the solid surface, showing a net negative velocity.
B: $d_l = 87$ Å, $\Gamma_m = 1.4 \cdot 10^{-9}$ keq/m^2, $C_0 = 10^{-3}$ N NaCl; giving $\nabla(-E)/\nabla(-P) \approx$ 6 mV/bar.
C: $d_l = 260$ Å, $\Gamma_m = 0.3 \cdot 10^{-9}$ keq/m^2, $C_0 = 1.5 \cdot 10^{-2}$ N NaCl; giving $\nabla(-E)/\nabla(-P) \approx$ 20 mV/bar.
D: As C but the components of the liquid velocity have been multiplied by $(1 - 1/u_-)/\langle v \rangle$. Addition of the local conductance transport of anions (C), as induced by the streaming potential, gives the local contribution to the reflection coefficient, σ. In this case the net reflection for the entire liquid layer, $\langle \sigma \rangle$, is slightly negative, because the conductance transport just exceeds the salt deficit dragged along with the liquid flux (hatched area; cf. also Fig. 11.1 for $\kappa_0 d_l = 10$).

In summary, electrokinetic effects in soils have been verified to exist within the range to be expected, albeit definitely smaller than would be predicted on the basis of a high fraction mobile charge. The use of a corrected equation for the ion distribution in a diffuse layer would be of help, although probably the uncertainty about the precise geometry of the electric field as applied or arising would be the major obstacle to further improvement.

11.7. ELECTRIC CONDUCTANCE IN SOILS AND CLAY SUSPENSIONS

The transport coefficient, L_E, used in Table 11.1, specified the apparent velocity of motion of positive charge induced by a unit driving force (of electrical origin) corresponding to one Newton per m^3 of local equilibrium solution. As in the previous section the corresponding parameter L'_E will be used here, giving the charge transport in Coul/sec per m^2 liquid area under the influence of a unit gradient of the electric potential (in the absence of other driving forces). Allowing for geometry corrections comprising at least the liquid content θ and a tortuosity factor, L'_E may be regarded as a measure of the DC conductivity, $\bar{\bar{K}}$, of the porous medium as a whole. This parameter then comprises an electro-osmotic drag term, $\mathring{K}$, plus an AC conductivity, $\tilde{K}$, corresponding to the autonomous motion of the ions with respect to the liquid phase. These two contributions to $\bar{\bar{K}}$ may be identified in Table 11.2, lower RH quadrangle. The relative magnitude of the drag term obviously depends strongly on the particular distribution of the counterions in the porous medium. Referring to Table 11.4 one finds for the over-simplified model I (homogeneous distribution) that the drag term would soon become very dominant if the thickness of the liquid layer exceeds some tens of Å. The model II, which proved to be more dependable in the previous section, indicates a much smaller contribution from $\mathring{K}$, although it still constitutes 40% of the total conductivity for the example based on a fully mobile countercharge of a magnitude found in montmorillonite clay. An even smaller contribution from the drag term may be deduced from the value of L'_{EV} suggested in the previous section as characteristic for soils, i.e. around $2 \cdot 10^{-8}$ m^2/Vsec. While this number is again about 40% of the mobility of the regular ions in solution (e.g. $\Lambda' = 5 \cdot 10^{-8}$ m^2/Vsec), the velocity pattern resulting from electro-osmosis implies that only a small fraction of the countercharge will be dragged along with the *mean* liquid velocity as given by L'_{EV} (cf. (11.45)). It thus seems safe to conclude that $\mathring{K}$ will, at most, contribute 10% to $\bar{\bar{K}}$, while it should often be negligible.

It is of interest to note that in dilute suspensions the electro-osmotic contribution to $\bar{\bar{K}}$ is replaced by the electrophoretic motion of the colloid charge (cf. Van Olphen and Waxman, 1958).

The fortunate conclusion of the above is that in practice one may obtain a fair estimate of L'_E from the AC conductivity of the system. Considerable information on this subject is available in the literature, particularly on suspensions pastes of montmorillonite, e.g. Van Olphen (1957), Van Olphen and Waxman (1958), Low (1958), Letey and Klute (1960), Gast (1966), Dakshinamurti (1960), Cremers and Laudelout (1965, 1966), Shainberg and Kemper (1966), Shainberg and Levy (1975). Extensive data on cores of shaly sands are contained in Waxman and Smits (1968).

The specific conductance of a clay suspension or paste in an AC field, $\tilde{K}$, is determined by number and mobility of the ions present and by the geometrical conditions prevailing in the system. The total amount of ions present per unit volume of the system may be determined experimentally and comprises neutral salts plus an excess of counterions (c.q. a deficit of co-ions). Indicating these amounts by θc_0 and Q, both in keq per m^3 of system, one may state that, for a homoionic system containing only one type of salt at an equilibrium concentration c_0:

$$\tilde{K} < \theta c_0(\Lambda_+ + \Lambda_-) + Q\Lambda_+ \tag{11.57}$$

where Λ_+ and Λ_- are the ionic mobilities in bulk solution, expressed as an equivalent conductance in $\Omega^{-1}\,m^2$/keq. The above inequality follows from the consideration that both the tortuosity of the pores and the impaired mobility of at least part of the ions present lead to a decrease in the specific conductance. From this equation it follows that, for any given adsorbent/salt system, $\tilde{K}$ may be studied as a function of c_0, θ and Q.

Efforts to replace (11.57) by an equality with a predictive value have been largely concentrated on liquid-saturated systems and the following discussion will be primarily directed to these. In that particular case Q may be replaced by $(1-\theta)\rho_s\Gamma S$, where $\rho_s\Gamma S$ specifies the countercharge in keq per unit volume of solid phase, and the number of variables influencing $\tilde{K}$ is limited to θ and c_0. The variation of $\tilde{K}$ with c_0 for a given value of θ follows in general a curve of the type as pictured in Fig. 11.6, where $\tilde{K}$ has been plotted against $\tilde{K}_0 = c_0(\Lambda_+ + \Lambda_-)$, the specific conductance of the equilibrium solution. Expressing formally the specific conductance of the porous medium as the sum of the contributing conductances of salt-ions and of the adsorbed countercharge, K'_s and K'_a, one may write:

$$K \equiv K'_s + K'_a \tag{11.58}$$

where the tilde has been omitted as only AC conductance will be discussed henceforth. From Fig. 11.6 it may be observed that, upon increasing values of the equilibrium concentration, eventually K becomes linear with K_0, which checks with the expectation that at high concentrations of the permeating solution K'_a gives a small, constant contribution to K.

The reciprocal of the limiting slope of the $K(K_0)$-curve is usually termed the formation factor, Φ, which covers the effect of pore tortuosity and θ. Extrapolation of this limiting slope to the K-axis then gives $K'_{a,\infty}$, the

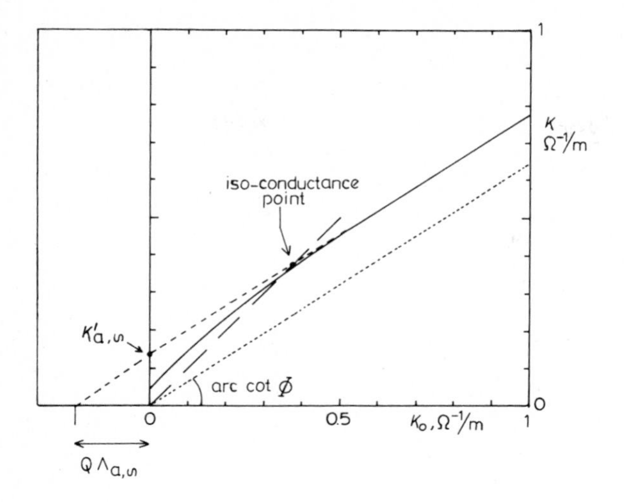

Fig. 11.6. Specific conductance of a colloid suspension, K, as a function of the specific conductance of the equilibrium solution, K_0 (3.75% Na-montmorillonite; data from Gast, 1966). The tangent of the limiting slope represents the reciprocal of the formation factor, $1/\Phi$. The construction of the iso-conductance point is shown.

(constant) contribution of the adsorbed countercharge at high electrolyte level. As indeed the amount of countercharge is constant, while presumably the geometry of the pore space and electric field become invariant with K_0 at high electrolyte level, the subdivision of K into its components seems beyond doubt at this stage.

Extension of this procedure to low electrolyte level, viz. defining:

$$K'_a \equiv K - K_0/\Phi \tag{11.59}$$

obviously contains an arbitrary element, as K'_s is related to K_0 via a constant value of Φ. Now there are several reasons for a change of Φ — as presumably acting on the conductance of the salt ions. Thus the growing extent of a diffuse double layer would cause increased co-ion exclusion, so the actual salt content of the system would decrease to $[(d_1 - d_{ex,-})/d_1]$-times the assumed amount θc_0, while geometry may change owing to swelling phenomena. Also the equilibrium salt concentration will often exceed that of the solution used, because of dissolution of soil constituents (cf. Shainberg and Levy, 1975). If such effects are taken into account there still remains an observable decrease of K'_a towards low values of K_0. The simplest explanation for this decrease is the one forwarded by Shainberg and Levy (1973), viz. the "cloud" of adsorbed ions, located in the vicinity of charged solid particles, becomes an isolated patch interspaced with solution of low conductance. In effect this means that the applied field becomes inhomogeneous, the adsorbed ions being subjected to a potential gradient less than the applied (mean) value. In short, the observed decrease of K'_a does not necessarily reflect a decreased mobility.

Further efforts to interpret K_a' in terms of the amount and mobility of the countercharge requires further assumptions. These have been guided primarily by the observed existence of an iso-conductivity *value* characteristic for a given solid phase. As follows from Fig. 11.6, all curves pertaining to a particular value of the liquid and solid content must exhibit an iso-conductance *point*, i.e. the intersection of the diagonal with the observed K-curve, or with its extrapolated slope. It has been established (Dakshinamurti, 1960; Cremers and Laudelout, 1966b; Shainberg and Levy, 1975) that the iso-conductivity points of a series of suspensions with different suspension concentrations tend to coincide rather closely over a range of concentrations between 0 and 25% by volume, thus giving a characteristic iso-conductivity value (e.g. $0.4\,\Omega^{-1}$/m for the Na-montmorillonite shown in Fig. 11.6). If now K_a' is expressed in terms of the mean excess concentration of countercharge in the liquid phase subject to the *same* formation factor Φ as derived for the salt contribution from the limiting slope, one finds at an iso-conductivity point:

$$K \equiv K_{iso} = \{K_{iso} + (1-\theta)\rho_s \Gamma S \Lambda_a/\theta\}/\Phi \tag{11.60}$$

where Λ_a is the apparent solution mobility of the adsorbed charge. If now K_{iso} is independent of θ, one finds the formation factor as:

$$\Phi = 1 + C(1-\theta)/\theta \tag{11.61}$$

with the relative volume conductance C, characteristic for the solid phase:

$$C = \rho_s \Gamma S \Lambda_a / K_{iso} \tag{11.62}$$

As was shown by Cremers and Laudelout (1966), different clays and clays saturated with different cations gave reasonably constant values of C for a wide range of clay concentrations. Presumably the particular value of C characteristic of a certain solid reflects the system geometry as influenced by particle anisometry. Values reported by Cremers and Laudelout (1966a, b) amounted to about 2 for an illitic soil in the range $0.38 < \theta < 0.46$, 4.3 for illite suspensions ranging from 10 to 60% by weight, 5 for zettlitz kaolinite, 9 for Camp Berteau montmorillonite, and about 14 for Na-Wyoming bentonite. For Camp Berteau saturated with Ca, C increased considerably above 9 for $\theta < 0.7$, indicating structural effects (e.g. plate condensation). In Fig. 11.7 the reciprocal of $\Phi(\theta)$ has been plotted for different values of C.

Although not giving solid proof of the necessity of using the same formation factor Φ for both K_s' and K_a', the existence of an iso-conductivity value lends considerable support to this practice. Thus if the assumption is made that K_0 and $\Lambda_a c_a$ should possibly be divided by two different formation factors, Φ_0 and Φ_a, respectively, both being functions of θ, the condition to be satisfied becomes:

$$(\Phi_0 - 1)(\Phi_a/\Phi_0) = C(1-\theta)/\theta$$

which hardly leaves room for a conclusion different from $\Phi_a/\Phi_0 = \text{constant}$ if also $K = K_0$ at $\theta = 1$ while K remains finite for $\theta \to 0$.

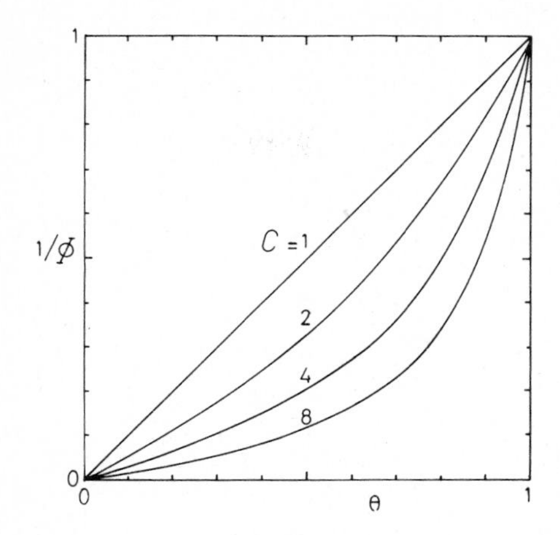

Fig. 11.7. The reciprocal of the formation factor, $1/\Phi$, as a function of the liquid content, θ, in liquid-saturated systems, for different values of the parameter C (cf. text).

Having accepted a common formation factor for both K_s' and K_a' (the latter term being expressed in terms of its contribution to the mean concentration in the liquid phase), one may calculate the equivalent conductance of the adsorbed ions from:

$$K_a' \equiv K - K_0/\Phi \;=\; Q\Lambda_a/\Phi \tag{11.63}$$

where $Q = (1-\theta)\rho_s\Gamma S$, the amount of countercharge per unit volume of system. Referring to Fig. 11.6 one thus finds $Q\Lambda_a$ as minus the intercept on the K_0-axis by running a line parallel to K_0/Φ from the point under consideration. Dividing this value by $Q\Lambda_+$, i.e. full countercharge multiplied by bulk equivalent conductance of the ion, gives the fractional mobility of the counterions, for AC conduction, $\tilde{m}$. This number may be interpreted as the fraction of the countercharge conducting at bulk mobility or as the mean relative mobility of the full charge. Using the limiting value pertaining to high electrolyte levels in Fig. 11.6 one finds $Q\Lambda_a = 0.2\,\Omega^{-1}$/m. At 3.75% Na-montmorillonite (as given) with a cation exchange capacity of around 10^{-3} keq/kg this gives $Q = 0.04$ keq/m^3 and thus $\tilde{m}$ would be around 0.9 at high salt levels. As may be seen in Fig. 11.6, $\tilde{m}$ decreases fairly gradually to about 0.3 at very low salt levels.

Making use of the iso-conductance point at about $0.35\,\Omega^{-1}$/m one may estimate C for the present system from (11.62). Using $\rho_s = 2600$ kg/m^3, $\Gamma S = 10^{-3}$ keq/kg this gives, with $\Lambda_a = 5\,\Omega^{-1}$ m^2/keq, a value of C at about 30. This is indeed consistent with the value of Φ as calculated for $\theta = 0.985$ (3.75% clay by weight) at $1 + (30 \times 0.015) = 1.45$ and found experimentally at 1.47. The above value of C back-calculated from Gast (1966) as shown in Fig. 11.6 is about twice the value given by Cremers and Laudelout (1966a) for a Na-Wyoming bentonite.

Gast (1966) also reported data on a 4.5% Sr-bentonite suspension, which lead to a value of $Q\Lambda_a$ at about $0.1\,\Omega^{-1}/m$, giving $\Lambda_a \approx 2\,\Omega^{-1}\,m^2/keq$, i.e. about two-thirds of the free ion mobility at high electrolyte level. Upon decreasing the salt level, Λ_a decreased considerably, i.e. to about $0.2\,\Omega^{-1}\,m^2/keq$ at very low salt levels. Shainberg and Levy (1975) analyzed the conductivity data for two dilute suspensions of Na-Wyoming bentonite over a wide range of electrolyte levels. The relative mobility of the adsorbed ions back-calculated from the actual iso-conductivity point, using a value of C at 25, came to about 0.6, and values from the limiting slope would possibly be slightly higher. Cremers and Laudelout (1966b) produced numbers for the limiting mobilities of the counterion (applicable for $\theta < 0.9$) in a Na-bentonite (giving $\tilde{m} = 0.8$) and a series of samples of Camp Berteau montmorillonite, with Na ($\tilde{m} = 0.6$), K (0.3), Rb (0.15), and Cs (0.08). This series indicates a sharp decrease with decreasing hydrated radius, inferring that an increasing fraction of the countercharge is effectively immobilized (cf. the negative adsorption data from Edwards et al. (1965) discussed in chapter 7).

As for the conductance at low porosity in soil systems, a large collection of data was presented by Waxman and Smits (1968). A series of sandstones and shaly sands with porosities ranging from 10 to 30% and a volume exchange capacity Q ranging from 0.05 to 0.5 keq/m^3 were studied with permeating solutions of NaCl from 0.02 to 0.5 N. The values of $\tilde{m}$ ranged from 0.9 at high concentrations down to about 0.3 for the extrapolated value for "zero" salt. At 0.02 N, $\tilde{m}$ was about 0.5.

Within the context of this chapter it appears sufficient to conclude that the specific conductance K and also L'_E may be estimated by taking into account the counterions with $\tilde{m}$ ranging from 0.3 to 0.9 for sodium ions, and perhaps about 0.1—0.5 for divalent cations, the maximum values arising only at high electrolyte levels ($>0.2\,N$).

In connection with the previous section this implies that the relative importance of the drag term in L'_E (which amounted to only, for example, 20% of the countercharge moving with a velocity smaller than the mean electro-osmotic liquid velocity — which in turn is only 40% of the free ion mobility) is probably truly negligible in systems akin to soil systems.

Rough estimates of K are probably most easily obtained from the equation:

$$K \approx c_0\{(\Lambda_- + \Lambda_+) + R_D\tilde{m}\Lambda_+\}/\Phi \tag{11.64}$$

where Φ is taken from (11.61) with $C = 2$ for soils. For very low porosities Waxman and Smits (1968) proposed to take $\Phi = \theta^{-n}$, with n ranging between 1.7 and 2.4 for the sandstones studied. Using $n = 1.7$ gives a reasonable transition between both expressions with Φ at about 3 for $\theta = 0.5$, which is a typical value for water-saturated field soils.

It is finally pointed out that the value of $\Lambda_a \equiv \tilde{m}\Lambda_+$, as used above to characterize the effective equivalent conductance of the counterions, may be combined with the surface density of charge to give the so-called specific surface conductance, according to:

$$K_S \equiv \Gamma\Lambda_a \tag{11.65}$$

This quantity is specified in units of Ω^{-1} and gives the contribution of the adsorbed ions present per unit surface area. For Na-montmorillonite with Λ_a at 3 Ω^{-1} m^2/keq and Γ at 10^{-9} keq/m^2 one thus finds K_S at about $3 \cdot 10^{-9}$ Ω^{-1}, which is a well-known number (cf. Shainberg and Levy, 1975). One may also use the weight conductance:

$$K_{wt} \equiv \Gamma S\Lambda_a \tag{11.66}$$

which for the above example gives about $2.5 \cdot 10^{-3}$ Ω^{-1} m^2/kg, as montmorillonite contains about 0.8 eq of countercharge per kg of clay.

Finally a few remarks will be made about the specific conductance of the non-saturated soil. This is of special interest if in situ observations are made in saline soils (cf. Rhoades and van Schilfgaarde, 1976; Rhoades et al., 1976). If the solid phase geometry is considered to be invariant with the liquid content it might be argued that the contribution of K_a' to K is fairly constant, particularly at salt levels where K_a' is only a fairly small fraction of K. In practice this amounts to the assumption that the formation factor, as dependent upon geometry, is satisfactorily constant with moisture content at least for the adsorbed ions, which could be defended by stating that the counterions are close to the surface where the liquid-phase geometry remains roughly unchanged upon drying out.

Starting from the saturated paste with θ equal to the porosity, p, one could thus write K according to (11.59) as:

$$K = K_0/\Phi(p) + K_a'(p) \tag{11.67}$$

If now the moisture content is decreased from p to $\theta = p\theta/p$ one would find:

$$K = \frac{\theta K_0}{p\Phi(p)} + K_a' \tag{11.68}$$

because the amount of salt present has decreased by a factor θ/p. Rhoades et al. (1976) investigated the ratio K/K_0 as a function of θ for a field soil at four values of K_0 and found a slightly concave relation, indicating that the formation factor as used in (11.68), $p\Phi(p)$, increases with decreasing θ. The relative increase of the co-ion, or salt exclusion upon decreasing θ might account for this observation. For the particular soil studied the above authors found that the reciprocal of the formation factor (indicated as transmission factor) was satisfactorily linear with θ over the full range of θ occurring in the field.

Referring to Tables 11.1 and 11.2 one finds that imposing a concentration gradient, $\nabla(-\Pi)$, leads to a current transport, $J_{\pm}$. Governed by the coefficient L_{ED}, this comprises again a drag term — proportional to $(\bar{u}_+ - 1/\bar{u}_-)(1 - 1/\bar{u}_-)$ — and a term caused by the autonomous motion of the salt owing to diffusion. Preventing a net transport of charge (as is the case in the absence of electrodes feeding ions into the system), a gradient of the electric potential arises, according to (cf. (11.28)):

$$\nabla(-\underset{e+}{p}) = -(L_{ED}/L_E)\nabla(-\Pi) \tag{11.69}$$

where $\nabla(-P)$ has been put at zero. Translating into the EMF gradient as measured with electrodes reversible to the cation this gives:

$$\nabla(-E_+) = -(L'_{ED}/L'_E)\nabla(-\Pi) \tag{11.69a}$$

Using Table 11.2 for a first orientation one finds:

$$\frac{z_+{}^m c_+ F\nabla(-E_+)}{\nabla(-\Pi)} = \frac{(\bar{u}_+ - 1/\bar{u}_-)(1 - 1/\bar{u}_-)/b + 1/\phi_-\bar{u}_-(b + b_-)}{(\bar{u}_+ - 1/\bar{u}_-)^2/b + \bar{u}_+/\phi_+(b + b_+) + 1/\phi_- u_-(b + b_-)} \tag{11.70}$$

which clearly shows the limits to be expected. Thus for $\bar{u}_+ \gg 1/\bar{u}_-$ the RHS of (11.70) approaches zero: if the porous medium does not pass the anions, equilibrium prevails and $\nabla(-E_+) = 0$. As $z_+{}^m c_+ F\nabla(-E_+) \equiv {}^m c_+\nabla(-\mu_+^c) + z_+{}^m c_+ F\nabla(-\psi)$ (cf. (11.10)), one finds $z_+\nabla(-\psi)$ as the well-known 60 mV per ten-fold increase in concentration ${}^m c_+$ across the medium. In this case the "diffusion" potential reverts to a "membrane" potential attributable to a perfect cation-selective membrane.

The other limit is found for $\bar{u}_+ = 1/\bar{u}_- \approx 1$ (no adsorbed countercharge). Then the drag terms disappear and one finds:

$$z_+{}^m c_+ F\nabla(-E_+) = T_-\nabla(-\Pi) \tag{11.71}$$

where, with $c_- = c_+$:

$$T_- = \frac{1/\phi_-(b + b_-)}{1/\phi_-(b + b_-) + 1/\phi_+(b + b_+)} = \frac{c_-\Lambda_-}{c_-\Lambda_- + c_+\Lambda_+} \tag{11.72}$$

indicates the Hittorf transport number of the anion in the system. Expressing (11.71) in terms of $\nabla(-\psi)$ and $\nabla(-\mu^c)$ with the help of (11.9) and (11.10) leads to the well-known expression for the liquid-junction potential in cells with transference according to:

$$F\nabla(-\psi) = -\frac{T_+}{z_+}\nabla(-\mu_+^c) - \frac{T_-}{z_-}\nabla(-\mu_-^c) \tag{11.73}$$

As is discussed in Katchalsky and Curran (1965), (11.73) may be retained for charged media provided J^V is held at zero (e.g. in an enclosed system). Because $J^V = 0$ usually implies that the solvent flux also becomes vanishingly small, the ions then move in accordance with their mobilities, Λ, through a liquid phase at rest.

As was shown in the classical paper by Staverman (1952), equation (11.70) may be written in a more general form by introducing "reduced" electrical transport numbers, T^*, according to:

$$T_k^* \equiv {}^m j_k/(I/F), \text{ at } \nabla(-P) = \nabla(-\Pi) = 0 \tag{11.74}$$

i.e. the number of moles of species k transported per Faraday passing through the system under the influence of an applied gradient of the electric potential in the absence of other driving forces. Applied to simple salt solutions one finds $T_k^* = T_k/z_k$, but in the case of transport through charged porous media T_k^* includes the contribution of drag phenomena, which in turn can be related to the reduced transport number of the solvent. This gives:

$$T_k^* = \frac{T_k}{z_k} + \frac{{}^m c_k}{c_w} T_w^* \tag{11.75}$$

In order to apply the above to (11.70) it must first be ascertained that (with $\phi_w \approx 1$):

$$z_+{}^m c_+ F(\bar{u}_+ - 1/\bar{u}_-)/b \equiv L'_{VE} (= L'_E T_w^*/Fc_w)$$

$$-z_-{}^m c_- F/\phi_-(b + b_-) = \Lambda'_-$$

both quantities being specified in $m^2/Vsec$. Introducing these into (11.70) gives after multiplying numerator and denominator by $(z_+{}^m c_+)^2 F$:

$$\frac{z_+{}^m c_+ F \nabla(-E_+)}{\nabla(-\Pi)} = \frac{(-z_-{}^m c_-/\bar{u}_-)\{\Lambda'_- - L'_{VE}\} - z_-{}^m c_- L'_{VE}}{(-z_-{}^m c_-/\bar{u}_-)\{\Lambda'_- - L'_{VE}\} + (z_+{}^m c_+\bar{u}_+)\{\Lambda'_+ + L'_{VE}\}} \tag{11.76}$$

The denominator then specifies I/F at unit gradient of the electric potential, while the numerator gives the equivalent flux of the anion caused by conduction and drag, plus the solvent flux multiplied by the equilibrium concentration of the solute. Using (11.74) then gives:

$$z_+{}^m c_+ F \nabla(-E_+) = z_- T_-^* \nabla(-\Pi) - z_-{}^m c_- T_w^* \nabla(-\Pi)/c_w \tag{11.77}$$

Using (11.9) this may be written, with $z_+{}^m c_+ = -z_-{}^m c_-$, as:

$$F\nabla(-E_+) = -T_-^* \nabla(-\mu_s^c)/\nu_- - T_w^* \nabla(-\mu_w^c) \tag{11.78}$$

The corresponding expression for the diffusion potential is easily memorized as:

$$F\nabla(-\psi) = -\Sigma T_k^* \nabla(-\mu_k^c) \tag{11.79}$$

where, in contrast to (11.73), the summation must include the solvent term $-T^*_w \nabla(-\mu^c_w)$. It is also pointed out that, if the system is explored by inserting electrodes reversible to the anionic species (e.g. Ag—AgCl electrodes), the expression (11.78) becomes:

$$F\nabla(-E_-) = -T^*_+\nabla(-\mu^c_s)/\nu_+ - T^*_w\nabla(-\mu^c_w) \tag{11.80}$$

Using (11.76) as starting point, one may attempt to estimate the value of T^*_- in soil systems. The (DC) conductance contribution of the anions specified in the numerator may be expressed as:

$$\bar{\bar{K}}'_- = F(\Lambda'_- - L'_{VE})(1-\sigma)c_0 \tag{11.81}$$

where Λ'_- is the mobility in bulk and where $1/\bar{u}_-$ has been replaced by $(1-\sigma)$ as a "passage" factor for salts subject to co-ion exclusion. Using Λ'_- at $5 \cdot 10^{-8}$ m^2/Vsec and the "maximum" value of L'_{VE} for systems containing NaCl at low concentration given in connection with (11.51) as $2 \cdot 10^{-8}$ m^2/Vsec, this gives $\bar{\bar{K}}'_-$ as $(0.3$ to $0.6) \times \Lambda'_-c_0$ for $(1-\sigma)$ ranging from 0.5 to 1.0. Similarly one finds for the counterions:

$$\bar{\bar{K}}'_+ \approx F(\Lambda'_+ + L'_{VE})(\tilde{m}R_D + 1)c_0 \tag{11.82}$$

where $\tilde{m}R_D$ is used to indicate the distribution ratio of mobile counterions. Using $\Lambda'_+ = \Lambda'_-$ this gives for $\tilde{m}R_D \gg 1$ the value of $\bar{\bar{K}}'_+$ as $1.4 \times \tilde{m}R_D\Lambda_-c_0$. Accordingly one finds:

$$z_-T^*_- \approx \frac{0.6\,(1-\sigma)}{1.4\,\tilde{m}R_D + 0.6\,(1-\sigma)} \approx \frac{0.4\,(1-\sigma)}{\tilde{m}R_D} \tag{11.83}$$

which will range from, say 0.02 at very low salt levels to 0.2 at moderate levels up to 0.1 N. For single salt systems the reduced transport number for the cations is then found as $1 - z_-T^*_-$, thus ranging from 0.8 at moderate salt level to 0.98 or so at low levels.

The reduced transport number for water is found from:

$$\frac{T^*_w}{Fc_w} \approx \frac{L'_{VE}}{L'_E} \tag{11.84}$$

where ϕ_w has been taken at unity. Using (11.54) and (11.55) this gives:

$$T^*_w \approx -Fc_w\nabla(-E)/\nabla(-P) \tag{11.85}$$

Having predicted previously that $\nabla(-E)/\nabla(-P)$ will rarely reach values of more than 5 mV/bar one may expect T^*_w to have values of 250 mol/96500 Coul or less.

From measured diffusion potentials and streaming potentials across cores of shaly sands (covering a considerable range of hydraulic conductivities and exchange capacities) Smits (1969) calculated the corresponding values of T^*_{Na} and T^*_w from equations of the type given above. In Fig. 11.8 and 11.9 these results are reproduced to provide an insight into the order of magni-

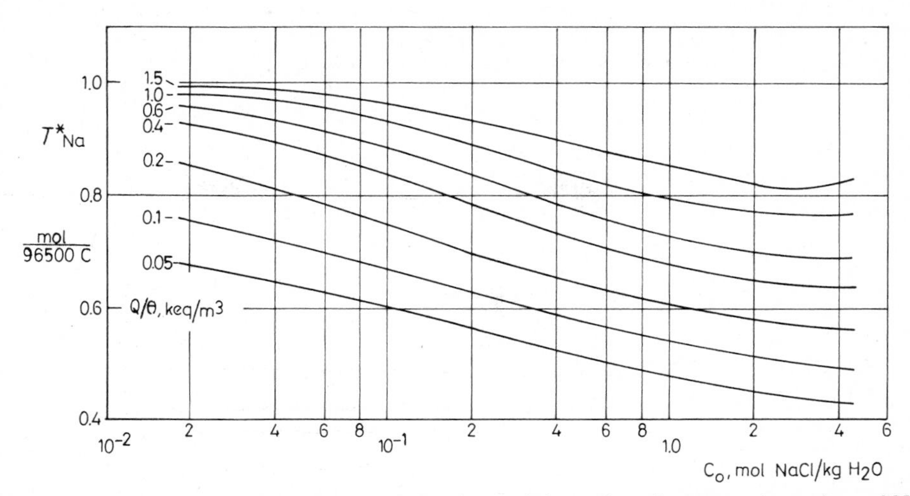

Fig. 11.8. Experimentally determined values of the reduced transport number of Na-ions, T^*_{Na}, as a function of the concentration of the permeating solution in shaly sands with various values of Q/θ, the concentration of counterions in the liquid phase. Data taken from Publ. 318, Kon. Shell EPL, Rijswijk, The Netherlands, by Smits (1968, in English), which served as a draft for Smits (1969).

tude to be expected. As was pointed out before, the interest in the electrochemical behavior of soil-like materials has been stimulated greatly by the need for interpretation of SP logs as used in oil exploration.

Quite a different aspect of diffusion potentials in charged porous media is their significance with regard to the interpretation of potentiometric determinations in such systems. Electrometric determination of pH in soil suspensions has been a routine from the early days of soil science. Around 1950 particular interest in such determinations was aroused by the becoming available of ion-specific electrodes which were expected to give a direct insight into the activities of cations in soil (cf. Marshall, 1964). Thus it was observed that a pair of electrodes — one reversible for a particular cation or the hydrogen ion, the other a reference calomel electrode — do not register the same potential difference in a (soil) suspension as that found in the equilibrium dialyzate of such a suspension. In the case of pH measurement the value calculated from the observed potential difference is usually consistently and significantly lower in the suspension than in its equilibrium dialyzate. The difference in pH between a suspension and its dialyzate has been termed the "Suspension Effect". In the soils literature around that period a discussion arose about the significance of these "increased activities" of cations in suspensions (e.g. Jenny et al., 1950; Coleman et al., 1951; Marshall, 1951, 1952; Peech and Scott, 1951; Peech et al., 1953). It could then be established by inserting pairs of electrodes as

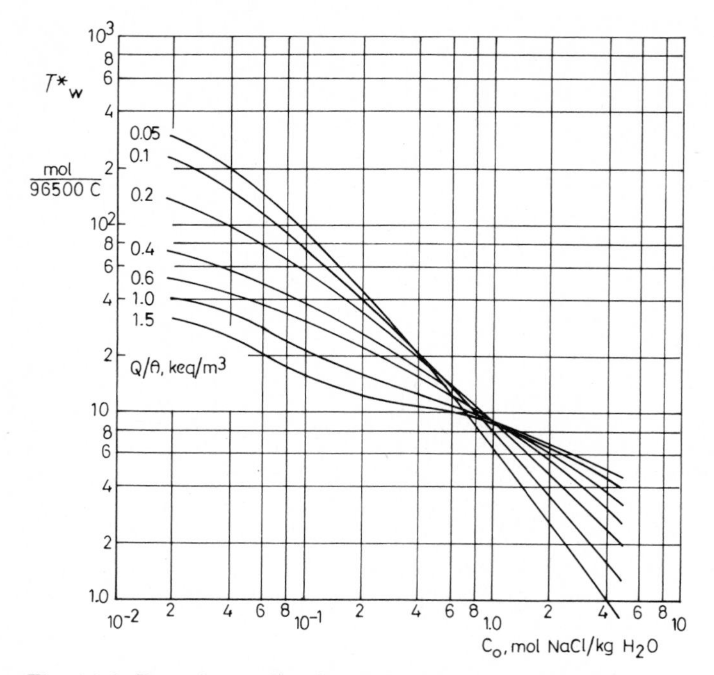

Fig. 11.9. Experimentally determined values of the reduced transport number of water, T^*_w, for the same system as in Fig. 11.8. Data taken from Publ. 318, Kon. Shell EPL, Rijswijk, The Netherlands, by Smits (1968, in English), which served as a draft for Smits (1969).

described above in a double system consisting of a suspension separated by a membrane from its equilibrium dialyzate, that the EMF between the two reversible electrodes situated in suspension and dialyzate would remain zero (as it should) if proper precautions for maintaining equilibrium were taken. Accordingly the suspension effect corresponds to the potential difference between the reference electrodes immersed in suspension and dialyzate, respectively. Overbeek (1953) presented a clear analysis of this system based on the application of the equation for the liquid-junction potential at zero volume flow as derived by Staverman (1952) and given above in (11.73). Using a suspension of K-clay in equilibrium with KCl at concentration c_0 as an example, the composite cell involved may be presented as:

$$\mathrm{Cal} \mid \mathrm{KCl_{sat}} \vdots\, \mathrm{KCl_{0}} \underset{\mathrm{eq}}{\mid} \mathrm{K\text{-}clay} \,\vdots\, \mathrm{KCl_{sat}} \mid \mathrm{Cal} \tag{11.86}$$

where the dotted lines indicate diffusion boundaries. Equation (11.73) may then be written as:

$$-F\nabla(-\psi) = T_{\mathrm{K}}\nabla(-\mu^{\mathrm{c}}_{\mathrm{K}}) - T_{\mathrm{Cl}}\nabla(-\mu^{\mathrm{c}}_{\mathrm{Cl}}) \tag{11.87}$$

With $T_K' = (1 - T_{Cl})$ this gives:

$$F\nabla(-\psi) = T_{Cl}\nabla(-\mu^c_{KCl}) - \nabla(-\mu^c_K) \qquad (11.88)$$

The EMF of the cell given in (11.86) is then found by integrating (11.88) over the three boundaries involved, viz. from KCl_{sat} via KCl_0 across the equilibrium boundary between dialyzate and suspension and back to KCl_{sat}. For this integration path the second term disappears regardless of the spatial variation of the single ionic potential, μ^c_K. At the same time the first term vanishes at the equilibrium boundary (presumably a membrane impermeable to the clay) as there $d\mu^c_{KCl} \equiv 0$. Accordingly one finds:

$$F\Delta(-\psi) = \int_{KCl_{sat}}^{KCl_0} [T_{Cl,sat} - T_{Cl,sus}]\, d(-\mu^c_{KCl}) \qquad (11.89)$$

While T_{Cl} is very close to $\frac{1}{2}$ across the solution boundary, quite different values would be expected across the boundary between suspension and saturated KCl. Thus at the "dilute" front of the diffusion boundary T_{Cl} will approach very low values as then the large excess of K^+ counterions will dominate the charge transport. In the direction of the saturated KCl bordering the suspension T_{Cl} will gradually approach again the value of $\frac{1}{2}$, the relative deficit of Cl ions decreasing with mounting electrolyte concentration. As a result one finds $F\Delta(-\psi) > 0$: the suspension registers a negative potential relative to its dialyzate if both sides are in contact with a saturated solution of KCl. If also cation reversible electrodes (e.g. glass electrodes) were inserted into suspension and its equilibrium dialyzate, the EMF registered by the electrode pair in the suspension must be "more positive" than the EMF indicated by the electrode pair in the equilibrium solution. This then implies pH(suspension) $<$ pH(equilibrium solution). As this conclusion remains valid also if suspension and equilibrium dialyzate are separated, it is obviously not warranted to interpret the observed suspension effect as associated in a quantitative manner with the existence of a phase potential difference between suspension and dialyzate, albeit that such a potential difference is likely to be present.

The situation is perhaps most easily summarized as follows. The use of the standard reference calomel electrode in solution chemistry is based on the necessity to use an electrode which attains the same potential with respect to a wide variety of solutions. Because of nearly identical mobilities of K- and Cl-ions, the saturated KCl solution junction appears suitable for this purpose. Thus equal diffusion fluxes of K and Cl — necessary under the condition of zero current flow — arise in the presence of a zero potential gradient. When immersed in a colloid suspension containing an excess of positively charged counterions, the combined flux of the positive ions will only become equal to that of the negative ions after a suitably retarding gradient of the electric potential has arisen (i.e. positive in the forward direction as compared to the diffusion). Such a gradient arises instantaneously following a minute forward shift of the positive ions: the mobile charge has become "polarized" to a slight extent.

If now a pair of KCl junctions are installed on both sides of a suspension-dialyzate system the condition of zero current flow is again satisfied following the initial "polarization" of the mobile charge, where it is immaterial whether or not part of this polarization takes place across the membrane separating suspension and dialyzate and part in the immediate environment of the KCl junction with the suspension. Accordingly any subdivision of the observed potential difference between the two calomel electrodes inserted in the KCl solutions forming the junctions contains elements which depend on the choice of the model used for the description of the boundaries involved. As was pointed out by Overbeek (1953), such models cannot be verified from the simple observation of the suspension effect. This follows also from the integrated equation (11.81), where the subdivision of $\Delta\psi$ over different junctions would imply the independent assessment of $\Delta(\mu_K^c)$ for these junctions, which is not possible experimentally.

It is finally pointed out that, for colloid suspensions containing counterions other than K^+, additional diffusion potentials will arise, depending on the relative mobilities of all cations involved. For the above simple KCl system, Overbeek (1953) constructed some limiting models. Of special interest in this respect is the approximation based on colloid particles with high charge density leading to:

$$F\Delta\psi \approx RT \ln \tilde{K}/\tilde{K}_0 \qquad (11.90)$$

where $\tilde{K}$ and $\tilde{K}_0$ indicate the specific conductance of suspension and equilibrium solution, respectively.

Recently Groenevelt et al. (1978) presented a detailed analysis of the development with time of the diffusion potential arising across a clay plug connecting two compartments at different initial salt concentrations. Using the condition $J^V = I = 0$, the coefficients observed experimentally were used to calculate the thickness of the liquid films with the help of a Gouy-type model, yielding a quite reasonable result. Of particular interest in this paper is the observed change in sign of the difference in hydrostatic pressure between the end-compartments upon shorting the system with electrodes reversible to the co-ion, corresponding to negative anomalous osmosis (cf. Katchalsky and Curran, 1965). The deficient magnitude of this pressure difference as compared to predicted values clearly illustrates the difficulties encountered when using shorted electrodes, as mentioned already on pp. 396, 402.

REFERENCES

Blackmore, A.V., 1976. Salt sieving within clay soil aggregates. Aust. J. Soil Res., 14: 149—158.

Bolt, G.H. and Groenevelt, P.H., 1972. Coupling between transport processes in porous media. Proc. 2nd IAHR—ISSS Symp. on Fundamentals of Transport Phenomena in Porous Media, Guelph, Canada, pp. 630—652.

Casagrande, L., 1953. Review of past and current work on electro-osmotic stabilization of soils. Harvard Soil Mech. Ser., 45: 83 pp.

Coleman, N.T., Williams, D.E., Nielsen, D.T. and Jenny, H., 1951. On the validity of interpretations of potentiometrically measured soil pH. Soil Sci. Soc. Am. Proc., 15: 106—114.

Corey, A.T. and Kemper, W.D., 1961. Concept of total potential in water and its limitations. Soil Sci., 91: 299—302.

Corey, A.T., Kemper, W.D. and Low, P.F., 1961. Concept of total potential in water and its limitations. Soil Sci., 92: 281—283.

Cremers, A. and Laudelout, H., 1965. On the "isoconductivity value" of clay gels. Soil Sci., 100: 298—299.

Cremers, A. and Laudelout, H., 1966a. Conductivité électrique des gels argileuse et anisométrie de leurs éléments. J. Chim. Phys., 10: 1155—1162.

Cremers, A. and Laudelout, H., 1966b. Surface mobilities of cations in clays. Soil Sci. Soc. Am. Proc., 30: 570—576.

Dakshinamurti, C., 1960. Study on the conductivity of clay systems. Soil Sci., 90: 302—306.

De Groot, S.R. and Mazur, P., 1962. Non-Equilibrium Thermodynamics. North-Holland, Amsterdam, 510 pp.

Edwards, D.G., Posner, A.M. and Quirk, J., 1965. Repulsion of chloride ions by negatively charged clay surfaces. Trans. Faraday Soc., 61: 2808—2823.

Elrick, D.E., Smiles, D.E., Baumgartner, N. and Groenevelt, P.H., 1976. Coupling phenomena in saturated homoionic montmorillonite. I. Experimental. Soil Sci. Soc. Am. J., 40: 490—491.

Gast, R.G., 1966. Applicability of models to predict rates of cation movement in clays. Soil Sci. Soc. Am. Proc., 30: 48—52.

Groenevelt, P.H. and Bolt, G.H., 1969. Non-equilibrium thermodynamics of the soil—water system. J. Hydrol., 7: 358—388.

Groenevelt, P.H. and Bolt, G.H., 1972. Permiselective properties of porous materials as calculated from diffuse double layer theory. Proc. 1st Symp. Fundamentals of Transport Phenomena in Porous Media, Haifa, Israel. Elsevier, Amsterdam, pp. 241—255.

Groenevelt, P.H. and Elrick, D.E., 1976. Coupling phenomena in saturated homo-ionic montmorillonite: II. Theoretical. Soil Sci. Soc. Am. J., 40: 820—823.

Groenevelt, P.H., Elrick, D.E. and Blom, T.J.M., 1978. Coupling phenomena in saturated homo-ionic montmorillonite: III. Analysis. Soil Sci. Soc. Am. J., 42: 671—674.

Jenny, H., Nielsen, D.R., Coleman, N.T. and Williams, D.E., 1950. Concerning the measurement of pH, ion activities, and membrane potentials in colloidal systems. Science, 112: 164—167.

Katchalsky, A. and Curran, P.F., 1965. Non-Equilibrium Thermodynamics in Biophysics. Harvard University Press, Cambridge, U.S.A.

Kedem, O. and Katchalsky, A., 1963. Permeability of composite membranes, parts 1, 2, 3. Trans. Faraday Soc., 59: 1918—1953.

Kemper, W.D., 1960. Water and ion movements in thin films as influenced by the electrostatic charge and diffuse layer of cations associated with clay mineral surfaces. Soil Sci. Soc. Am. Proc., 24: 10—16.

Kemper, W.D., 1961a. Movement of water as effected by free energy and pressure gradients: I. Application of classic equations for viscous and diffusive movements to the liquid phase in finely porous media. Soil Sci. Soc. Am. Proc., 25: 255—260.

Kemper, W.D., 1961b. Movement of water as effected by free energy and pressure gradients: II. Experimental analysis of porous systems in which free energy and pressure gradients act in opposite direction. Soil Sci. Soc. Am. Proc., 25: 260—265.

Kemper, W.D. and Evans, N.A., 1963. Movement of water as effected by free energy and pressure gradients: III. Restriction of solutes by membranes. Soil Sci. Soc. Am. Proc., 27: 485—490.

Kemper, W.D. and Letey, J., 1968. Solute and solvent flow as influenced and coupled by surface reactions. Trans. 9th Int. Congr. Soil Sci., 1: 233—261.

Kemper, W.D. and Maasland, D.E.L., 1964. Reduction in salt content of solution on passing through thin films adjacent to charged surfaces. Soil Sci. Soc. Am. Proc., 28: 318—323.

Kemper, W.D. and Quirk, J.P., 1972. Ion mobilities and electric charge of external clay surfaces inferred from potential differences and osmotic flow. Soil Sci. Soc. Am. Proc., 36: 426—433.

Kemper, W.D. and Rollins, J.B., 1966. Osmotic efficiency coefficients across compacted clays. Soil Sci. Soc. Am. Proc., 30: 529—534.

Letey, J. and Kemper, W.D., 1969. Movement of water and salt through a clay—water system. Experimental verification of the Onsager reciprocal relation. Soil Sci. Soc. Am. Proc., 33: 25—29.

Letey, J. and Klute, A., 1960. Apparent mobility of potassium and chloride ions in soil and clay pastes. Soil Sci., 90: 259—265.

Letey, J., Kemper, W.D. and Noonan, L., 1969. The effect of osmotic pressure gradients on water movement in unsaturated soil. Soil Sci. Soc. Am. Proc., 33: 15—18.

Low, P.F., 1958. The apparent mobilities of exchangeable alkali metal cations in bentonite water systems. Soil Sci. Soc. Am. Proc., 22: 395—398.

Low, P.F., 1961. Concept of total potential in water and its limitations: a critique. Soil Sci., 91: 303—305.

Lynch, E.J., 1962. Formation Evaluation. Harper and Row, New York, N.Y.

Marshall, C.E., 1951. Measurements in colloidal systems. Science, 113: 43—44.

Marshall, C.E., 1952. Potentiometric measurements in colloidal systems. Science, 115: 361—362.

Marshall, C.E., 1964. The Physical Chemistry and Mineralogy of Soils. Wiley, New York, N.Y., 388 pp.

Miller, R.D., 1955. Neglected aspects of electro-osmosis in porous bodies. Science, 122: 373—374.

Nielsen, D.R., Jackson, R.D., Cory, J.W. and Evans, D.D. (Editors), 1972. Soil Water. A.S.A.—SSSA, Madison, U.S.A., 175 pp.

Overbeek, J.Th.G., 1953. Donnan-EMF and suspension effect. J. Colloid Sci., 8: 593—605.

Peech, M. and Scott, A.D., 1951. Determinations of ionic activities in soil—water systems by means of the Donnan-membrane equilibrium. Soil Sci. Soc. Am. Proc., 15: 115—119.

Peech, M., Olsen, R.A. and Bolt, G.H., 1953. The significance of potentiometric measurements involving liquid junction in clay and soil suspensions. Soil Sci. Soc. Am. Proc., 17: 214—218.

Ravina, I. and Zaslavsky, D., 1968. Non-linear electrokinetic phenomena: I. Review of literature. Soil Sci., 106: 60—66.

Rhoades, J.D. and Van Schilfgaarde, J., 1976. An electrical conductivity probe for determining soil salinity. Soil Sci. Soc. Am. J., 40: 647—651.

Rhoades, J.D., Raats, P.A.C. and Prather, R.J., 1976. Effects of liquid-phase electrical conductivity, water content, and surface conductivity on bulk soil electrical conductivity. Soil Sci. Soc. Am. J., 40: 651—655.

Schmid, G., 1951. Zur Elektrochemie feinporiger Kapillar Systemen: II. Elektro-osmose. Z. Elektrochem., 55: 229—237.

Shainberg, I. and Kemper, W.D., 1966. Conductance of adsorbed alkali cations in aqueous and alcoholic bentonite pastes. Soil Sci. Soc. Am. Proc., 30: 700—706.

Shainberg, I. and Levy, R., 1975. Electrical conductivity of Na-montmorillonite suspensions. Clays and Clay Minerals, 23: 205—210.

Smits, L.J.M., 1968. SP log interpretation in shaly sands. Soc. Pet. Eng. Trans. AIME, 243: 123—136.

Smits, L.J.M., 1969. Propriétés électrochimiques de sables argileux et de chistes argileux. Rev. Inst. Fr. Pet.: 91—120.

Staverman, A.J., 1952. Non-equilibrium thermodynamics of membrane processes. Trans. Faraday Soc., 48: 176—185.

Van Olphen, H., 1957. Surface conductance of various ion forms of bentonite in water and the electrical double layer. J. Phys. Chem., 61: 1276—1280.

Van Olphen, H. and Waxman, M.H., 1958. Surface conductance of sodium bentonite in water. Proc. 5th Natl. Conf. Clays and Clay Minerals, pp. 61—80.

Waxman, M.H. and Smits, L.J.M., 1968. Electrical conductivities in oil-bearing shaly sands. Soc. Pet. Eng. Trans. AIME, 243: 107—122.

CHAPTER 12

CLAY TRANSFORMATIONS: ASPECTS OF EQUILIBRIUM AND KINETICS

Robert Brinkman

Much of the present volume deals with processes in soils in terms of local, partial equilibrium between a mobile and a stationary phase, the latter being considered as an exchanger with variable proportions of adsorbed ions, but otherwise invariant with time. This approach is valid where the solid phase is thermodynamically stable under the conditions considered, and often is a good approximation even where the solid phases are not thermodynamically stable. Exchange reactions are generally rapid, approaching equilibrium in seconds or minutes, whereas partial equilibrium between clay minerals and the surrounding solution may be approached over periods of weeks or months. The approach of the solution to equilibrium conditions normally requires very small amounts of solid phases to be dissolved.

Over these longer periods, the composition of the soil solution in an open system is the resultant of reactions toward equilibrium with the more reactive minerals (mainly the clay minerals) and of other factors. These include the solutes from rain and any ground water or surface water entering the soil, solutes formed by biotic activity, as well as the degree of concentration by evapotranspiration. This concentration factor over a given period may be approximated by the ratio (precipitation minus net run-off)/(precipitation minus net run-off minus actual evapotranspiration).

A steady state near to equilibrium between all clay phases present and the mobile liquid phase is not reached in most soils. Some old soils on ancient stable surfaces in the humid tropics have approached this condition. The clay minerals in most soils are at least in part inherited from the sediment, or brought in as a sedimentary admixture, or formed by weathering of sand- or silt-sized minerals from the parent rock under conditions different from those prevailing in the soil at present. Soil clays may have been formed near the weathering front of a rock, for example, or in sea water, at higher pH values and cation activities than in the present soil, or under a different past climate.

Clay minerals in soils may become unstable and may be decomposed after a change in environmental conditions that lowers or raises activities of solutes, for example by increased rainfall or evapotranspiration. Many clay minerals in soils are presumed to have formed at earth-surface temperatures and pressure, and their transformations take place under similar conditions.

In equilibrium calculations, normally constant temperature and pressure (25°C and 1 bar) are assumed. On the one hand, this simplifies studies in soil science compared with those undertaken in geology, where a range of temperatures and pressures may need to be considered. On the other hand, equilibrium compositions in geological studies may often be approximated by a closed system (a constant amount of material with a given chemical composition), in which minerals and an aqueous solution reach equilibrium at a certain temperature and pressure, whereas soils are not closed systems and the solid phases are normally not in overall equilibrium.

In spite of these difficulties, much information can be obtained about the direction and sequence of clay mineral transformations under different soil-forming processes by simple partial equilibrium considerations.

Very slow rates of dissolution (and formation) inhibit equilibrium between sand- and silt-sized minerals and the soil solution in many soils. This is often due to their low surface area per unit soil mass, not because of a slow rate per unit surface. Clay minerals on the other hand, having a much higher surface area, are often near equilibrium with the soil solution. Hence, the rate-determining step in clay transformation is generally the rate of replacement of the soil solution. This makes it possible in some cases to estimate rates of soil formation (clay mineral transformation) on the basis of external data (e.g., effective rainfall and solute concentrations) by calculations based on an assumption of partial equilibrium.

The different processes of clay transformation in soils are summarized in section 12.1, below, and discussed individually in sections 12.4—12.8 (in a slightly different sequence for simplicity of presentation). Sections 12.2 and 12.3 deal with some general aspects needed for that discussion. Activity diagrams used are described in section 12.2. Activity coefficients at high concentrations, ion association and the existence of polynuclear Al species, which complicate equilibrium calculations, are discussed in section 12.3. Some of the different weathering sequences occurring under different external conditions are mentioned in section 12.9, and some rates of clay transformation are indicated in the last section.

12.1. NATURE OF THE DIFFERENT PROCESSES AFFECTING CLAYS IN SOILS

External conditions determine the nature and, normally, the rates of the clay transformation processes in soils, whereas the actual clay mineral suite largely determines the composition of the soil solution by local, partial equilibria. The wide range of external conditions may be summarized by the description of a small number of clay transformation processes, and most of the important processes may be discussed with reference to simple two-dimensional activity diagrams.

The main processes of clay transformation are presumably the following five, in approximate order of decreasing areal importance.

(a) Hydrolysis by water containing carbon dioxide is a dominant process of clay formation and transformation particularly under humid tropical conditions, above as well as below a ground-water table, but also in part of the temperate zone. Even locally in a glacial environment this process may dominate soil formation, where water is abundant and plant activity limited (Reynolds, 1971). The process involves desilication as well as loss of basic cations.

(b) Cheluviation, the dissolution and removal of especially aluminum and iron by chelating organic acids (Swindale and Jackson, 1956), overrules hydrolysis mainly in humid temperate climates but in parts of the tropics as well. This occurs particularly where the activities of divalent ions and the availability of nitrogen are very low (in some cases enhanced by low temperatures or periodic reduction). These conditions slow down the complete decomposition of organic matter to CO_2 and water as well as the polymerization to virtually insoluble humic acids with high molecular weights, and favor the formation of soluble, chelating fulvic acids, with many carboxylic and phenolic groups.

(c) Ferrolysis, a cyclic process of clay transformation and dissolution based on the alternate reduction and oxidation of ferric oxides and ferrous ions (Brinkman, 1970), may be the dominant process of clay transformation wherever soils are alternately wet and dry and where there is some net leaching, in a wide range of climates.

(d) Dissolution by strong acids occurs in the acid sulfate soils in which pyrites are oxidized after drainage, where rain or snow is polluted by sulfuric (and other) acids from fossil fuels or industrial activity (as over much of Europe and the eastern United States), and locally in volcanic areas.

(e) Clay transformation under strongly alkaline conditions takes place, for example, in sodic (formerly "alkali") soils in arid climates, and in sediments in and near natron lakes. Under these conditions, the main aluminum species in solution is tetrahedral $Al(OH)_4^-$, in contrast to the octahedral (6-coordinated) aluminum in most clay minerals and in soil solutions at low pH. For this reason, the second activity diagram discussed below has a double vertical scale, involving both Al^{3+} and $Al(OH)_4^-$ activities.

12.2. ACTIVITY DIAGRAMS AND STABILITY RELATIONS

These clay transformation processes under earth-surface conditions may be summarized in a simple model with the aid of activity diagrams (Figs. 12.1 and 12.2). These Figures show $\log(H_4SiO_4)$ (parentheses represent activities) on the abscissa, and $\log(Al^{3+}) + 3\,pH$ on the ordinate. The

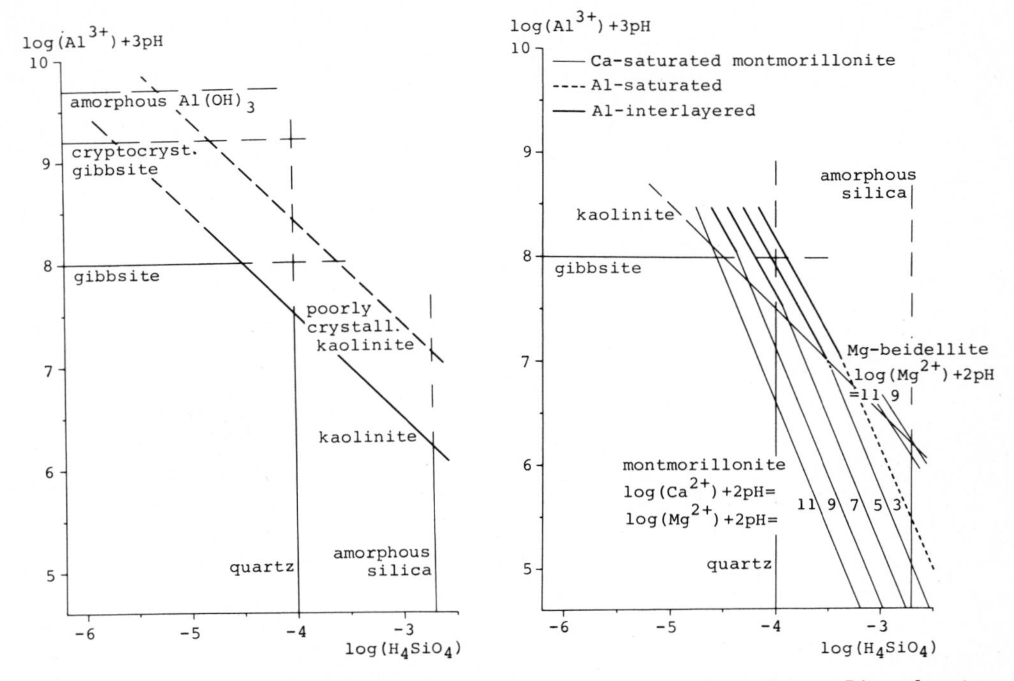

Fig. 12.1. Activity diagrams of some minerals in neutral to acid conditions. Lines denote equilibrium; regions above and to the right of lines, supersaturation; regions below and to the left of lines, undersaturation.

variable $\log(M^{z+}) + z\,pH$, in which M represents a metal such as Al, Mg or K, may be used instead of the separate variables $\log(M^{z+})$ and pH because metal and hydrogen ions occur in this ratio in dissolution equilibria. The diagrams might be considered as cross-sections*, because the activity products of hydroxides of cations not shown on abscissa or ordinate are held constant along each line of equilibrium.

Although the same equilibrium equations remain valid at low and at high pH, it is desirable to consider mineral equilibria in acidic conditions in terms of $\log(Al^{3+}) + 3\,pH$, and in alkaline conditions in terms of the dominant Al species above pH 7, $Al(OH)_4^-$. In dissolution equilibria this appears in the combination $\log(Al(OH)_4^-) - pH$. As a consequence of the linear relationship

$$\log(Al(OH)_4^-) - pH = \log(Al^{3+}) + 3\,pH - 23.6$$

(Sillén and Martell, 1964), both variables can be shown along the same ordinate (Fig. 12.2).

---

* The diagrams shown in Volume A of this series (Figs. A 8.3 and 4, pp. 150 and 151) and by some other authors might be considered as projections, because in such diagrams, $\log(Al^{3+}) + 3\,pH$ (not shown on an axis) is assumed to vary on the basis of continual equilibrium with the different minerals indicated.

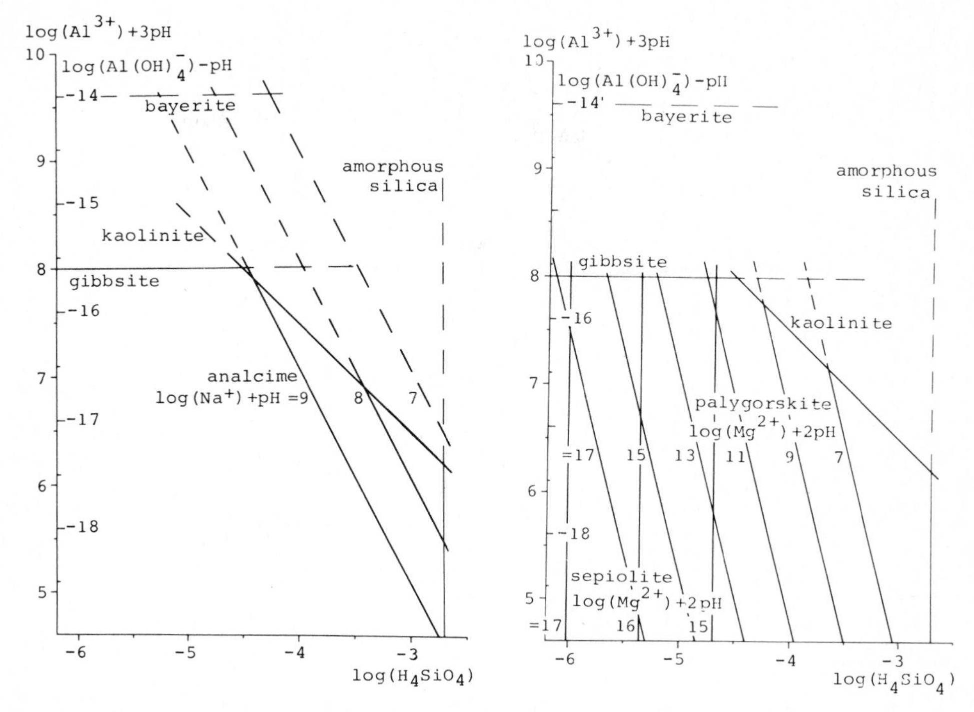

Fig. 12.2. Activity diagrams of some minerals in alkaline conditions.

Experimental data do not yet cover adequately all aspects of the equilibrium model. For example, the nature and the formation constants of aluminum chelates or of partially hydrolyzed simple or polynuclear aluminum ions in soils are not known in detail, and there is still considerable uncertainty about the values of the formation constants of different minerals, particularly those with a range in chemical composition.

The lower left-hand part in Figs. 12.1 and 12.2 represents soil solutions undersaturated with respect to the different minerals indicated. The lines represent solutions in equilibrium with a mineral; their intersections represent solutions in equilibrium with the two minerals. Regions above and to the right of lines shown denote supersaturation with respect to the corresponding minerals. Lines continuing in such regions represent metastable states. These may be realized in cases where the rate of formation of the stable phase is (much) slower than the rates of formation, transformation or dissolution of the metastable phases.

The lines of equilibrium for (well crystallized) gibbsite, kaolinite and quartz, for example, are invariant. In contrast, the lines for most clay minerals are displaced with changes in the activities in solution of their constituent ions other than Si or Al. This is illustrated by the several parallel

TABLE 12.1

Equations and constants*1 used in construction of Figs. 12.1 and 12.2

Mineral		Source
Quartz	$\log(H_4SiO_4) = -4.0$	Helgeson (1969)
Amorphous silica	$\log(H_4SiO_4) = -2.7$	Stumm and Morgan (1970)
Gibbsite	$\log(Al^{3+}) + 3\,pH = 8.0$	Helgeson (1969)
Cryptocryst. gibbsite	$\log(Al^{3+}) + 3\,pH = 9.2$	Helgeson (1969)
Amorphous $Al(OH)_3$	$\log(Al^{3+}) + 3\,pH = 9.7$	Sillén and Martell (1964)
Bayerite	$\log(Al(OH)_4^-) - pH = (-13.5\ to) -14$	Hayden and Rubin (1974)
	$\log(Al^{3+}) + 3\,pH = \log(Al(OH_4^-) - pH + 23.6$	Sillén and Martell (1964)
Kaolinite	$\log(Al^{3+}) + 3\,pH + \log(H_4SiO_4) = 3.5$	Kittrick (1966)
Poorly cryst. kaolinite	$\log(Al^{3+}) + 3\,pH + \log(H_4SiO_4) = 4.4$	Kittrick (1966)
Montmorillonite, $(Al_{3.03}Mg_{0.58}FeIII_{0.45})(Si_{7.87}Al_{0.13})O_{20}(OH)_4^{0.56-}$		Kittrick (1971)
Ca-montmorillonite	$3.16[\log(Al^{3+}) + 3\,pH] + 0.28[\log(Ca^{2+}) + 2\,pH] + 0.58[\log(Mg^{2+}) + 2\,pH] + 7.87\log(H_4SiO_4) = -1.18$	
Al-montmorillonite	$3.35\,[\log(Al^{3+}) + 3\,pH] + 0.58\,[\log(Mg^{2+}) + 2\,pH] + 7.87\log(H_4SiO_4) = -1.25$	
Al-interlayered montmorillonite	$4.28\,[\log(Al^{3+}) + 3\,pH] + 0.58\,[\log(Mg^{2+}) + 2\,pH] + 7.87\log(H_4SiO_4) = 5.26$	Van Breemen*2
Mg-beidellite, $Al_4(Si_{7.33}Al_{0.67})O_{20}(OH)_4Mg_{0.33}$	$\log(Al^{3+}) + 3\,pH + 0.07[\log(Mg^{2+}) + 2\,pH] + 1.58\log(H_4SiO_4) = 2.61$	Helgeson (1969)

Analcime, $NaAlSi_2O_6.H_2O$	$\log(Al^{3+}) + 3\,pH + \log(Na^+) + pH + 2\log(H_4SiO_4) = 8.1$	Thompson (1973)
Palygorskite*3, $Al_{0.96}Mg_{0.91}Fe_{0.10}Si_{4.00}O_{10.5}$	$0.96[\log(Al^{3+}) + 3\,pH] + 0.91[\log(Mg^{2+}) + 2\,pH] + 4.00\log(H_4SiO_4) = -1.4$	Singer (1977)
Sepiolite, $Mg_2Si_3O_4(OH)_8$	$2[\log(Mg^{2+}) + 2\,pH] + 3\log(H_4SiO_4) = 15.9$	Christ et al. (1973)

*1 Values of constants are approximate, and vary with the crystallinity of the mineral as indicated, for example, by the data for gibbsite and kaolinite. Constants for minerals with other compositions and modified or more accurate values of constants continue to appear in the literature.

*2 Recalculated by Van Breemen (pers. comm.) from data by Kittrick (1971) assuming equilibrium with hematite and Al-saturation under Kittrick's experimental conditions, and assuming the following cation exchange relationships:

$$\text{ex-}Ca_3 + 2\,Al^{3+} \leftrightarrows \text{ex-}Al_2 + 3\,Ca^{2+} \qquad \log K = 0.78$$

$$\text{ex-}Al + 5\,Al^{3+} + 15H_2O \leftrightarrows \text{ex-}[Al_2(OH)_5]_3 + 15\,H^+ \qquad \log K = 35.$$

The latter constant implies stability of interlayers with respect to gibbsite owing to a degree of solid solution (not shown).

*3 Recalculated assuming equilibrium with an iron oxide.

lines in Fig. 12.1 for montmorillonite at different Mg activities. The slope of each line, and its displacement with change in activities of other ions, depend on the chemical composition of the mineral, which in many cases is known with reasonable accuracy. However, the position of a line (at given activities of other constituent ions) depends on the formation constant, which for many minerals is not known with an accuracy better than one order of magnitude.

The position of a line is strongly dependent on the degree of order (crystallinity) of a phase, as illustrated by the lines for gibbsite and (metastable) cryptocrystalline gibbsite and amorphous Al $(OH)_3$ in Fig. 12.1.

The montmorillonite shown in this diagram is only an example. Lines for other smectites or for illites could have been shown, each with its own slope and with a position depending on its composition and structure, as well as on the activities in solution of different ions such as Mg, Fe^{2+} or K. The beidellite in Helgeson (1969), for example, only has a small range of stability, close to the intersection of the lines for kaolinite and amorphous silica (Fig. 12.1). Equations and constants used in the construction of Figs. 12.1 and 12.2 are listed in Table 12.1.

The stability relations for many different clay minerals containing several cations in different proportions may be compared more conveniently by computer than by hand calculation or stability diagram. Computer programs for this purpose have been developed, for example, by Brown and Skinner (1974), and Helgeson et al. (1969, 1970), the latter partly modified by Droubi et al. (1976). The former is designed to calculate equilibrium mineral assemblages of a given chemical composition (at a given pressure and temperature) in a closed system. Soils, however, are open systems in which local, partial equilibrium is common, but overall equilibrium very rare. The latter program was designed to calculate the sequence and amounts of minerals dissolving or precipitating in contact with a solution initially not in equilibrium with the solid phases, or during dilution or concentration of a solution. It also calculates the changes in solution composition.

Both programs are relatively simple as long as minerals with a constant composition (e.g. kaolinite) are considered. Several clay minerals, however, have a variable composition, depending on the conditions during their formation. In the program by Brown and Skinner, these are dealt with as ideal solid solutions of end-members with fixed compositions, which may be mixed in all proportions. The stability of solid solutions is higher than of the coexisting end-members, owing to entropy of mixing. This stability difference is calculated as a function of the composition.

In the program by Droubi et al., the stability constant of a phase with variable composition is approximated by including in the data a large number of similar minerals (24 montmorillonites), each with its own formation constant.

Examples of the practical importance of solid solution may be found in the Al chlorites. Pure Al interlayers from a chlorite synthesized under laboratory conditions tend to recrystallize in the form of gibbsite, with liberation of the original smectite, over periods of less than a year, and even incomplete interlayers, that persist for more than 8 months, are metastable with respect to gibbsite (Turner and Brydon, 1967). On the other hand, Al chlorites in many soils (presumably with less pure interlayers than their man-made equivalents) appear to persist for long periods (summary by Blum, 1976).

There are indications that in an open system (with leaching) in the tropics, Al-interlayered clay minerals are slowly transformed into kaolinite, whereas similar clay minerals containing some ferrous iron in the Al interlayers appear to persist without further transformation (Brinkman, 1977a).

12.3. ACTIVITIES AND CONCENTRATIONS, ION ASSOCIATION, Al SPECIES

Before going into the more detailed discussion of clay transformation processes in the following sections, some attention should be paid to relations between activities and concentrations, ion association and the existence of polynuclear Al species.

At very low ionic strengths, e.g. prevailing during hydrolysis by water containing carbon dioxide, activity corrections are small in comparison with the uncertainties in formation constants of the minerals, and activities and concentrations in solution may be considered equal. At intermediate ionic strengths, up to about 0.1, the extended Debye-Hückel formula may be used to relate activities and concentrations. Data for this are listed in part A of this volume, tables A 2.1 and A 2.2, pp. 18 and 19. At higher ionic strengths, which may occur, for example, during decomposition by strong acid in acid sulfate soils and during clay transformation in strongly alkaline conditions, the Debye-Hückel approximation breaks down. Single-ion activity coefficients may then be calculated from experimentally determined mean activity coefficients ($f_{\pm}$) for salts (the mean salt method). This calculation is based on two postulates:

$$f_{\pm,KCl} = f_{+,K} = f_{-,Cl}$$

$$(f_{\pm,M_mA_a})^{m+a} = (f_M)^m \times (f_A)^a$$

The latter presupposes that ion association is negligible during the determination of $f_{\pm}$. Some examples of single-ion activity coefficients at high ionic strengths are shown in Fig. 12.3.

Ion association (ion pair formation) also needs to be taken into account at moderate or high ionic strengths, as suggested, for example, by the unusually low values for $f_{SO_4^{2-}}$ in Fig. 12.3. Ion association may be important between sulface and most cations, between carbonate and bicarbonate and divalent cations, between hydroxyl and trivalent iron and aluminum at low pH, and between hydroxyl and some divalent cations at high pH. This, in combination with the activity/concentration relationships that depend on

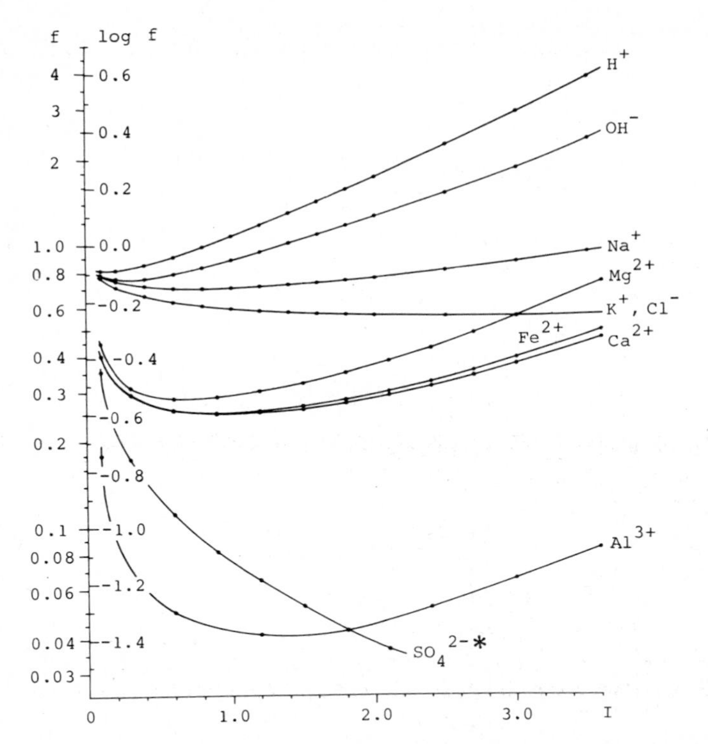

Fig. 12.3. Single-ion activity coefficients for some common ions. Lines for the cations were calculated from activity coefficients of the chlorides, and for the anions from activity coefficients of the potassium salts, listed in Robinson and Stokes (1968).
*The unusually low line for SO_4^{2-} includes the effect of ion-pair formation.

the ionic strength, makes hand calculation of ionic equilibria and activities at high ionic strength impracticable. Computer programs have been developed to perform the necessary iterative calculations in a reasonable time.

Van Breemen (1973a), for example, extended the Garrels and Thompson (1962) model for sea water to include 33 dissolved species and programmed the calculations for computer treatment. His paper lists the dissociation constants of ion pairs used, quotes sources for the calculation of individual activity coefficients of several ions by the mean salt method, and makes simple assumptions for a few ions. Satisfactory results were obtained for ionic strengths up to 1 and pH values between 1.8 and 8.2. The species $Al(OH)_4^-$, as well as dissociation constants for ion pairs between Ca or Mg and OH, would need to be included for use of the program in strongly alkaline conditions.

One of the (more extensive) computer programs by Helgeson et al. (1969, 1970), adapted by Droubi et al. (1976) is applicable particularly to alkaline conditions, but uses an extended Debye-Hückel approximation for ionic activity coefficients even at relatively high ionic strengths.

It is difficult to assign confidence limits to these activity calculations, but probably the relative errors in activities are within about 10% at ionic strengths up to about 0.1, and might be up to about a factor 2 at high ionic strengths (1 or more) with use of the D-H approximation or, even with activity coefficients by the mean salt method, in solutions with "unusual" compositions (relatively far from that of sea water).

The former approaches are based on the assumption that the activity coefficients depend only on the ionic strength of the solution, but not on its composition.

Quite different is the method given by Pitzer (1973, 1975), Pitzer and Mayorga (1973, 1974), and Pitzer and Kim (1974) for the calculation of mean activity coefficients of salts in mixed solutions. This is developed on the basis of theoretical insights from improved analysis of the Debye-Hückel model. The specific close-distance interactions of each ion with all other ions in the solution are taken into account. The formation of ion pairs is incorporated in the parameters for close-distance interaction. The interaction coefficients have been tabulated for 227 electrolytes. The authors report the method to be of high accuracy in pure electrolytes and mixtures up to ionic strengths of several molal.

Some special attention needs to be paid to the different Al species in solution, because of the central importance of Al in weathering and the existence, in certain conditions, of soluble polynuclear Al species. Total aluminum concentrations in solution may be considerably higher than Al^{3+} concentrations, particularly at high values of $\log(Al^{3+}) + 3\,pH$, near the equilibrium value for gibbsite. This may be due to the presence of soluble polymeric species as suggested, for example, by Richburg and Adams (1970). These authors calculated the solubility product $\log(Al^{3+}) + 3\,pH$ for a set of acid soils, and found a reasonably constant value if they assumed the presence of polynuclear aluminum species according to the reaction:

$$6\,Al^{3+} + 15H_2O \rightleftharpoons Al_6(OH)_{15}^{3+} + 15H^+$$

in which log K could be chosen between -30 and -42.

Hayden and Rubin (1974) tested several hydrolysis schemes in aqueous solution, involving monomeric and larger Al species, by comparing the fit of calculated pH-neutralization curves to slow titration data. The best fit (least squares of differences) was achieved by postulating the species $AlOH^{2+}$ and $Al_8(OH)_{20}^{4+}$. Equilibrium constants were calculated by successive approximation. They were expressed as mixed (practical) constants, incorporating concentrations instead of activities except for hydrogen activity, at a constant ionic strength of 0.15 M (in $NaNO_3$ solution). On the basis of estimated activity coefficients these practical constants may be converted to the "true" log K values (applicable to activities) as listed below:

$$\log(AlOH^{2+}) - pH - \log(Al^{3+}) = -5.05$$

$$\log(Al_8(OH)_{20}^{4+}) - 20\,pH - 8\log(Al^{3+}) = -63.1$$

These should only be used in calculations if Hayden and Rubin's value for the solubility product of gibbsite (10.05) is used as well.

It is probably correct to assume that such polymeric aluminum species are metastable, with "half-lives" of several months, rather than stable species.

Metastable aluminum polymers were found in aqueous solutions by Turner (1968) and Smith (1971), and in soil extracts by Bache and Sharp (1976). The soluble polynuclear Al species may comprise one (or possibly several) rings, with OH/Al ratios ranging from about 2.3 to 2.5. Structures with higher OH/Al ratios appear to be colloidal solids (Smith, 1971; Hayden and Rubin, 1974). Because the nature of the aluminum species in solution is time-dependent, the assumption of equilibrium is not valid, and calculations involving a single polynuclear ion can only be approximate at best. Nevertheless, the equations above could be used as an approximation to the relationships prevailing in a soil system in which the soil solution is weakly to moderately acid and contact times between solid phases and solution are of the order of a few weeks to a few months.

12.4. HYDROLYSIS BY WATER AND CARBON DIOXIDE

Water in equilibrium with a CO_2 pressure of $10^{-3.5}$ bar, as normally in the atmosphere, has a pH of about 5.6. Initial hydrolysis of a clay mineral by such water proceeds congruently (without a solid residue) until the solubility product of gibbsite is reached at an extremely low Al^{3+} activity, of the order of 10^{-9} mol/l. Dissolution then continues incongruently with the formation of gibbsite, while activities in solution of silica and other ions from the clay mineral continue to rise. Under these conditions, virtually all iron from the dissolving clay mineral is similarly retained in solid form, as goethite or hematite or a more hydrated phase.

If only kaolinite is present in the system, or if leaching is sufficiently rapid to renew the solution at this point in the process, i.e. before the silica activity reaches 10^{-4} mol/l (about 6 ppm SiO_2), the clay minerals decompose with production of secondary gibbsite and iron oxides, their proportions depending on the composition of the original clays. Even quartz is unstable at these low silica activities, and may slowly dissolve.

This process of clay decomposition with formation of gibbsite was demonstrated by Pedro (1970) and Pedro et al. (1970) in their accelerated leaching experiments. It occurs naturally in well drained, perhumid or humid tropical environments where much of the organic matter from vegetation decays on the soil surface, releasing carbon dioxide to the atmosphere rather than within the soil.

Rain water is not distilled water, however. Estimated concentrations in "average" rain, mainly from Garrels and Mackenzie (1971), are given in Table 12.2. Soluble aluminum in rain water is of the order of 0.1 mg/l for rain at pH 3.93 (Ruppert, 1975) to 0.01 mg/l for rain presumably at about pH 5 (Cawse, 1974), i.e. 3—0.3 μmol/l. This is 10—20% of the total (soluble plus particulate) Al concentration (Cawse, 1974). The silica concentration in "average" rain, about 5 μmol/l, is close to the equilibrium activity between cryptocrystalline gibbsite and kaolinite. In cases where soil-derived aerosol is present in rain, a considerable part of the initial, congruent hydrolysis (and, conceivably, some incongruent

TABLE 12.2

Composition of "average" rain water

Concentration	Na^+	K^+	Mg^{2+}	Ca^{2+}	Cl^-	SO_4^{2-}	HCO_3^-	SiO_2	Al*	H^+
ppm	1.98	0.30	0.27	0.09	3.79	0.58	0.12	0.29	0.01	
μmol/l	86	8	12	2	107	6	2	5	0.3	
−log (mol/l)	4.1	5.1	4.9	5.7	4.0	5.2	5.7	5.3	6.5	5.7

SiO_2: Whitehead and Feth (1964), quoted by Garrels and Mackenzie (1971).
Al: the lowest of the stations in Cawse (1974). Other data: best estimate by Garrels and Mackenzie (1971, p.107).
Al* denotes total soluble aluminum.

hydrolysis with formation of cryptocrystalline gibbsite) thus appears to have taken place even before the rain reaches the ground. The aluminum-containing particles in rain were found by Cawse (1974) to have a mass median diameter of about 4 μm: probably small clay aggregates. Table 12.2 shows an estimate of "average" rain only. Concentrations range from about an order of magnitude lower to about two orders of magnitude higher, depending on location and wind direction with respect to the sea and to sources of pollution, for example.

Where leaching is less rapid and where clay minerals such as aluminum chlorites, smectites or illites are present, the silica activity may rise to slightly above the equilibrium value for gibbsite (or cryptocrystalline gibbsite) and kaolinite. Kaolinite is then formed and the gibbsite first formed is dissolved again, together with continued dissolution of the other clay minerals present. Many well drained soils in the humid tropics contain kaolinitic or ferruginous kaolinitic material of pedogenic origin. In soils with a strongly heterogeneous pore system, gibbsite might occur along large pores or on ped faces and kaolinite in the soil mass, owing to differences in local leaching rates.

In all cases of weathering where the silica activity exceeds 10^{-4} mol/l, quartz is stable over the clay minerals present, but normally is not formed at a measurable rate. Silica activities in soil solutions appear to be generally governed by clay minerals or amorphous silica rather than by quartz. Quartz synthesis is not inherently slow, however. Quartz crystals were produced experimentally at earth-surface temperature and pressure in sea water within 3 years (Mackenzie and Gees, 1971) and in an aqueous system with amorphous iron hydroxide and silica within 14 days (Harder and Flehmig, 1970). Secondary quartz has been demonstrated in soils as well, for example by Kovda and Rode (1967). Probably, there are factors inhibiting quartz crystallization in most soil environments. These may include polymerization of the silica in solution and the presence of organic matter.

With a further rise in silica activity owing to continued dissolution of, for example, a montmorillonite under conditions of slow leaching, the equilibrium value between kaolinite and aluminum-interlayered montmorillonite (aluminum chlorite, pedogenic chlorite) at the prevailing magnesium activity may be exceeded. From then on, aluminum-interlayered montmorillonite may be formed at the expense of the original montmorillonite and the kaolinite formed previously. After dissolution of the kaolinite, the activities in the percolating solution may change to near-equilibrium between the aluminum-interlayered and the original montmorillonite, and remain there as long as montmorillonite is present. Pedogenic aluminum chlorites are common in many subtropical and tropical soils.

In many cases, especially in well drained, humid temperate environments and in some tropical soils, a large proportion of organic matter from the vegetation decays within the soil rather than on the surface, causing much higher carbon dioxide pressures (of the order of 10^{-2}—10^{-1} bar) than in the atmosphere and hence more acid soil solutions, with pH between about 5 and 4.5. Under these conditions, gibbsite is not formed even if leaching is rapid. Initial dissolution of most clay minerals proceeds congruently until the kaolinite solubility product is reached. Kaolinite dissolves congruently under these conditions if it is the only clay mineral present. Dissolution of a smectite normally proceeds incongruently with formation of kaolinite or aluminum chlorite (the latter reviewed by Blum, 1976) depending on the leaching rate, as described above. At very high CO_2 pressures, montmorillonite may coexist with amorphous silica rather than with kaolinite. Such a system (also involving Mg and Fe carbonates) in a tropical environment was described by Dirven et al. (1976).

It may be noted that the presence of soluble polynuclear aluminum species lowers the maximum CO_2 pressure at which gibbsite may still be formed from other clay minerals. Without polynuclear Al species the equilibrium CO_2 pressure between gibbsite and kaolinite would be about 0.7 bar: an unrealistically high value. Under the assumption (section 12.3) that $Al_6(OH)_{15}^{3+}$ is present, with log K about −36, the equilibrium pressure would be of the order of 10^{-3} bar. A similar value results from calculations without polynuclear species but with cryptocrystalline rather than crystalline gibbsite.

Different weathering sequences may thus be expected during hydrolysis by water and carbon dioxide depending on the CO_2 pressure and the leaching rate, but also on the minerals initially present. For example, illite may be transformed to vermiculite or a smectite where K is the main ion removed, under weakly acidic conditions (review by Tributh, 1976), or to an aluminum chlorite at a lower pH. Feldspars may be hydrolysed with formation of illite, or of smectite in the presence of Mg ions, e.g. from dolomite (Bronger and Kalk, 1976), or of kaolinite at higher leaching rates as discussed above.

Hydrolysis by water containing carbon dioxide of a sodium-saturated clay appears to proceed more rapidly than the hydrolysis processes mentioned above. (A similar rapid acidification was observed in the laboratory during dialysis of Li-clays.) Such a process may cause loss of cation exchange capacity of the clay fraction as well as clay loss in solods and solodized soils, as described below.

When rain water leaches through a (saline) sodic soil, the Na activity and the pH are lowered. When the Na activity has reached low values comparable with the hydrogen activity in the rain water, hydrogen has displaced an appreciable proportion of the Na from exchange positions. Hydrogen clay is unstable, however (Eeckman and Laudelout, 1961), and is transformed into an Al—Mg-clay by diffusion of structural cations to exchange sites formerly occupied by hydrogen.

If this process alternates with accumulation of sodium salt and rise of pH, e.g. during a dry season, Mg may be exchanged and removed with the passing salt solution and the exchangeable Al may be partially neutralized, vacating some exchange sites, and polymerized in non-exchangeable form in the interlayer spaces of clay minerals, blocking part of the C.E.C. Sodium once again occupies the exchange sites vacated by removal of Mg and neutralization of Al.

During a next cycle of leaching by rain water, hydrogen may again displace exchangeable Na, leading to further attack on the clay mineral structure.

12.5. DECOMPOSITION BY STRONG ACIDS

Dissolution of clay minerals by strong (mineral) acids at extremely low pH values (for example, 2.5—3.5) is initially congruent, as long as the activity of dissolved silica remains below $10^{-2.7}$ mol/l (about 120 mg SiO_2/l). If the soil solution is not renewed by leaching before the silica activity reaches that value, amorphous silica is produced during further dissolution, in contrast to the gibbsite or kaolinite (and iron oxide) commonly produced during clay decomposition by water containing carbon dioxide only. Where leaching is sufficiently slow, dissolution of clay minerals may continue until the rising activities of aluminum and other cations from the clay have brought about (partial) equilibrium with the partly dissolved clay mineral.

Under very acid conditions, a smectite or an illite may be stable over kaolinite at relatively high activities of divalent cations or potassium, but unstable with respect to kaolinite at low activities of these basic cations (Fig. 12.1). There is no single weathering sequence for clay minerals in very acid conditions, therefore.

Decomposition by strong acids takes place in soil solutions of relatively high ionic strength, hence activity coefficients (section 12.3) also need to be taken into account, in contrast to the situation during hydrolysis by water and carbonic acid discussed earlier.

In the presence of sulfate, total Al concentrations exceed Al^{3+} activities by formation of the soluble complexes $AlSO_4^+$ and $Al(SO_4)_2^-$ (dissociation constants $10^{-3.2}$ and $10^{-5.1}$, respectively, Sillén and Martell, 1964). At an

SO_4^{2-} activity of $10^{-2.5}$, for example, total soluble aluminum is about 7 times the Al^{3+} activity, disregarding activity coefficients.

Where sulfuric acid is the agent of decomposition, Al activities in solution appear to be limited by the relation:

$$\log(Al^{3+}) + pH + \log(SO_4^{2-}) = -3.2 \text{ (Van Breemen, 1973b)}$$

This could point to the possible existence of a solid with bulk composition $AlOHSO_4$. No mineralogical evidence for this is available, however. The suggestion that the relation could be explained by the simultaneous presence of basaluminite and Al hydroxide (Adams and Rawajfih, 1977) does not apply to Van Breemen's data, which show a considerable range of undersaturation with respect to these two minerals. At an SO_4^{2-} activity of about $10^{-2.5}$ and pH about 3.5, for example, $\log(Al^{3+}) + 3\,pH$ would be limited to 6.3.

12.6. CHELUVIATION

In solutions of chelating acids (fulvic acids, or low-molecular ones such as salicylic or p-hydroxybenzoic acid) at pH values around 4, virtually all of the trivalent iron and aluminum is present in chelated form, hence their total concentrations may be several orders of magnitude higher than the Al^{3+} and Fe^{3+} activities. Therefore, decomposition of kaolinite (or gibbsite) by a chelating acid at a pH about 4 proceeds in a similar way as decomposition by a non-chelating acid at a pH about 3.

At a low pH, chelation of divalent cations or potassium is negligible, hence the stability of clay minerals such as smectites or illite is less affected by the presence of chelating anions than the stability of kaolinite, owing to their lower aluminum (and higher silica and magnesium or potassium) contents. Small quantities of a smectite or vermiculite are in fact present in the eluvial horizons of many podzols, which may have aluminum chlorite, kaolinite, illite or other clay minerals in deeper horizons not affected by cheluviation (e.g. Brown and Jackson, 1958).

Little quantitative information is available about the mobile chelating acids operating in soils. Extraction and analysis of the acids moving in an eluvial horizon as well as of the acids precipitated in the underlying accumulation horizon (Schnitzer and Desjardins, 1968 and 1962) has indicated the predominance of fulvic acid with molecular weight about 670, and containing about 6 carboxyl and 5 phenolic OH groups per molecule. Low-molecular complexing organic acids have also been found in canopy drip, extracts of fallen leaves and, small quantities, in upper soil horizons. These may participate in cheluviation at low temperatures (near $0^\circ C$), but seem to be susceptible to rapid consumption by microorganisms or polymerization to fulvic acid at higher temperatures (Bruckert and Jacquin, 1969). Apparent complex constants (K_{app}) of several metals with fulvic acid estimated by Schnitzer (1969) indicate that aluminum and ferric iron are complexed most strongly, closely followed by copper. For other divalent metals, $\log K_{app}$ is about 3—6 units lower at pH 3.5 and 5. The $\log K_{app}$ increases by 2 units per unit increase of pH for some metals, such as Cu, Pb, and possibly Al and Fe, but increases by about 1 unit or less for most other metals. This suggests that these complexes involve fulvic acid species of different degrees of dissociation.

The different metal/fulvic acid ratios in complexes with different metals (ranging from about 2 for Mg or Zn to less than 1 for Cu or divalent iron) indicate that the complexes are of different kinds as well.

Al and Fe chelates with a molar metal/fulvic acid ratio of 1 are soluble, but those with higher ratios (3—6) have a very low solubility (Schnitzer and Skinner, 1963, 1964). The latter, which apparently contain the metals in partially hydrolyzed form (e.g. $Al(OH)_2^+$), are presumably formed during passage of the mobile chelates through material containing a further supply of Al or Fe. Therefore, the Al and Fe chelates normally do not move very far from their source in upper soil horizons, and are generally precipitated in an accumulation horizon. There, part of the organic acids is oxidized by microbial action. This results in a locally high value of log (Al^{3+}) + 3 pH (still at a relatively low pH) and stability of, for example, kaolinite, aluminum chlorite or an aluminum-interlayered smectite over other clay minerals. Such aluminous materials and, locally, iron oxides are common in the B horizons of podzols.

12.7. FERROLYSIS

Clay decomposition by ferrolysis requires two alternating sets of circumstances: water-saturated conditions with some leaching (during which the soils are reduced and the pH slowly rises) alternating with oxidizing conditions (at a lower pH). During the latter period, exchangeable ferrous iron is oxidized, producing ferric oxides and exchangeable hydrogen. This attacks the clay minerals as do other strong acids, but the aluminum and other cations from the clay become exchangeable, not soluble, while the corresponding silica may remain in the form of unsupported edges of the clay mineral structure.

During the reducing period, ferric oxides are reduced to ferrous ions, part of which displace exchangeable aluminum and other cations into the soil solution. Part of the silica in the unsupported edges is dissolved as well. Leaching water may remove dissolved silica as well as some ferrous iron, displaced basic cations, and part of the displaced aluminum together with dissolved HCO_3^-.

Non-equilibrium concentrations of aluminum may be removed in solution under conditions of rising pH in the form of low-molecular polynuclear ions with 'half-lives' of several weeks or months (section 12.3). Part of the aluminum is not leached out but is polymerized in the interlayer spaces of swelling clay minerals, forming incomplete octahedral sheets. Thus, the originally swelling minerals are brought to a fixed 1.4 nm basal spacing: i.e., soil chlorites are formed. Unlike aluminum chlorites originating in well drained soils, those formed by ferrolysis contain varying amounts of trapped, originally exchangeable, ferrous ions in their interlayers (Brinkman, 1977a). This might increase their stability with respect to kaolinite.

In kaolinitic soils subject to ferrolysis there is no interlayering in sofar as the necessary swelling clays are lacking, but the kaolinite is dissolved (Brinkman, 1977b).

12.8. CLAY TRANSFORMATION IN ALKALINE CONDITIONS

In an alkaline environment, tetrahedral $Al(OH)_4^-$ is the dominant Al species in solution, in contrast to the octahedral species at low pH values. At low silica activities, tetrahedral Al oxides (e.g. bayerite) are formed instead of gibbsite, with $\log Al(OH)_4^- - pH$ around -13.5 to -14 (Hayden and Rubin, 1974). These are metastable with respect to (well crystallized) gibbsite. In most alkaline soils, the silica activity is relatively high, however, so that aluminosilicates are stable over aluminum oxides (Fig. 12.2).

In strongly alkaline conditions and at high Na activities, the sodium zeolite analcime is stable with respect to other clay minerals. The lines of equilibrium for analcime in Fig. 12.2 were derived from Thompson (1973). Older data (Helgeson, 1969) suggest a somewhat lower stability (log K being 1.3 units higher).

There is field evidence for analcime formation (Baldar and Whittig, 1968) and for degradation of montmorillonite concurrent with analcime formation (Frankart and Herbillon, 1971) in strongly (Na carbonate) saline alluvial soils. The analcime tends to occur in the coarse clay and silt fractions, whereas mainly the fine clay fraction ($< 0.2\,\mu m$) of the montmorillonite is degraded.

In alkaline conditions and at high Mg activities in soils, palygorskite is the expected stable phase. Magnesium chlorite may be stable at moderate Mg activities, or may be formed as a metastable phase, as discussed in section 12.10, below. Pedogenic palygorskite seems to be more common than sepiolite in arid soils. In a Mediterranean climate, for example, palygorskite appears to be formed in arid plains with less than 300 mm annual rainfall, and to be transformed into montmorillonite in wetter conditions (Paquet and Millot, 1972).

The acid dissolution rate of Mg from palygorskite is about 240 times slower than from sepiolite (Abdul-Latif and Weaver, 1969), suggesting that palygorskite may be stable over sepiolite at relatively low values for $\log (Mg^{2+}) + 2\,pH$. Stability lines for sepiolite were calculated from Christ et al. (1973), and for palygorskite from Singer (1977) assuming equilibrium with solid $Fe(OH)_3$ (Fig. 12.2). These indicate that palygorskite would be stable over sepiolite except at very high values for $\log (Mg^{2+}) + 2\,pH$ and very low silica activities.

The list of clay minerals stable in intervals between near-neutral and extremely acid conditions, as illustrated in Fig. 12.1, varies with changes in activities of divalent cations (especially magnesium) and potassium. At low activities of these ions — or rather, low activity products of metal ions and OH — the list from extremely acid to near-neutral conditions (from high to low log (H_4SiO_4) and concurrently from low to high log (Al^{3+}) + 3 pH) may read as follows: amorphous silica; Al-saturated beidellite-type smectite; Al-interlayered smectite or soil chlorite (aluminum chlorite); kaolinite; gibbsite. At higher potassium and calcium activities, this may change to: amorphous silica; Ca-saturated illite; Al-saturated illite; possibly Al-interlayered vermiculite; kaolinite; gibbsite. At high magnesium and calcium activities, the list may be: amorphous silica; Ca-saturated montmorillonite; possibly Al-saturated montmorillonite; Al-interlayered montmorillonite or soil chlorite (aluminum chlorite); kaolinite; gibbsite.

The range of log (Al^{3+}) + 3 pH and log (H_4SiO_4) over which each of these minerals is stable (or dissolves congruently) varies with the activities of basic cations in solution and with the composition of the clay minerals.

Clay decomposition in soils under conditions far from those under which the decomposing mineral is stable appears to proceed generally by a series of local, partial near-equilibrium, incongruent dissolution steps. The mineral dissolves with production of the adjacent mineral in a stability diagram, while the soil solution remains near equilibrium between the two. After completion of this dissolution the composition of the percolating soil solution changes to near-equilibrium between the mineral just formed, which is dissolved again, and the next mineral. Decomposition would thus proceed until the mineral is formed that is stable over all others under the circumstances. This may then dissolve congruently with further leaching (extremely slowly in many cases). For example, a Ca-smectite under extremely acid conditions might give rise to the following decomposition sequence: Ca-smectite → Al-smectite → amorphous silica, and under near-neutral conditions: Ca-smectite → Al-interlayered smectite → kaolinite → gibbsite.

Since a soil generally contains pores of various diameters, local leaching rates may differ over short distances, and a soil sample may well contain minerals from several steps in a decomposition sequence.

The concepts of "weathering sequence" and "weathering stage" (Jackson and Sherman, 1953, pp. 235—240) probably originated from mineral sequences observed in nature mainly under one decomposition process: hydrolysis by water at low activities of carbonic acid. The fact that the decomposition sequence varies with the weathering conditions, e.g. the presence of strong mineral acids, or of chelating acids, makes it desirable to specify details of the weathering system (for example, hydrolysis or acidolysis, Hénin et al., 1968; Pédro et al., 1969) or of the main transformation process postulated to be active, as well as the soil mineralogy.

12.10. RATES OF CLAY TRANSFORMATION

The rates of clay transformation in soils appear to be generally dependent on the rate of leaching (or solution renewal) and on the concentrations in the soil solution in equilibrium with dissolving and newly formed (clay) minerals, rather than on dissolution kinetics. (These may be first-order in many cases, and diffusion-controlled through a layer of newly formed material in some incongruent dissolutions.) Even at leaching rates of some 2000 mm/year, contact times of the soil solution with the solid phases are of the order of a week per 10 cm thickness of soil material, and of one or two months over the thickness of a soil profile: generally sufficient to achieve near-equilibrium between clay minerals and the soil solution at the high solid/solution ratios prevailing in soils. Huang and Keller (1971, 1973), for example, estimated dissolution rates of the order of 10^{-3}—10^{-2}% per day for clay minerals in organic acids, and of the order of 10^{-3}% per day for silica from clay minerals in water.

Hydrolysis with production of residual gibbsite or kaolinite typically results in soil solutions with concentrations of the order of 1—5 mg SiO_2/l — roughly equivalent to a loss of 4 mm soil material per 1000 years under 2000 mm annual percolation — derived from the decomposition of 1—10 mg clay mineral per liter of percolating water. (Concentrations of aluminum and iron in solution are some orders of magnitude smaller than 1 mg/l.) Similar concentrations, resulting from the decomposition of 1—5 mg clay per liter of percolating water, were derived by Pédro (1970) and Pédro et al. (1970) from laboratory experiments at 65°C with strongly accelerated leaching.

Clay decomposition by chelating acids may give rise to soil solutions containing between 5 and more than 100 mg SiO_2/l, and similar total concentrations of aluminum and iron species (Huang and Keller, 1971). This would be roughly equivalent to the loss of 10—200 mm soil material per 1000 years at 2000 mm annual percolation.

Not all of this material is lost from the soil profile. Part of the aluminum is precipitated in the B horizon of podzols (section 12.6). The high silica activities, combined with high Al activity products (log (Al^{3+}) + 3 pH) after partial decomposition of organic acids in the lower B horizon, may give rise to formation of clay minerals. The poorly crystalline material with Si/Al ratios about 0.4, interpreted by Veen and Maaskant (1971) as decomposition remnants, might contain newly formed clay as well as aluminum hydroxide.

In strong mineral acids, silica concentrations are about 100 ppm and total aluminum concentrations generally similar, although these may be up to an order of magnitude higher (Van Breemen, 1973b). Even though silica may be residually enriched during clay decomposition by chelating or strong mineral acids, silica is still removed faster than is normally the case during hydrolysis by water and carbonic acid. The rate of acid supply combined with the leaching rate appear to be important factors determining rates of clay decomposition by chelating or strong mineral acids.

A recent study of weathering in Luxemburg (Verstraten, 1977) gives an example of the combined action of CO_2 and small amounts of strong acid. Locally produced CO_2 and the transition from NH_4^+ in the rain water to NO_3^- in the drainage water are the main, equally important, sources of acid involved in chemical weathering. Direct acidity of the rain water accounts for about 10 per cent. The drainage water contains about 0.12 mmol, 7 mg, silica/l. Total acid consumed during weathering amounts to about 0.5 mmol/l drainage water. This is balanced by basic cations, mainly Mg, Ca and K. Aluminum was not reported but is presumably conserved in the solid phases because the pH of the drainage water is about 7. The composition of the drainage water is in accordance with the possible transformation of illite, Mg-montmorillonite and Mg-chlorite to kaolinite.

An apparently absolute increase of kaolinite in the fine earth of the upper soil horizons was found, together with a strong decrease in sericite and a (probably relative) increase in quartz. The clay fractions show a corresponding strong increase in kaolinite, decrease in illite, chlorite and albite, and (probably relative) increase in quartz and undifferentiated vermiculite/smectite.

In ferrolysis, the leaching rate and either the quantity of iron reduced in the wet season or the cation exchange capacity probably determine the rate of clay decomposition, because the amount of exchangeable hydrogen formed by oxidation is equivalent to the exchangeable ferrous iron minus the bicarbonate not yet leached out. This crude estimate leads to maximum possible rates of clay decomposition of the order of 0.1% per year, roughly equivalent to a loss of 200 mm soil material per 1000 years if the layer subject to ferrolysis remains about 20 cm thick.

Actual weathering rates may well be about an order of magnitude smaller than the indicated maxima for several reasons. Annual leaching rates are more commonly several hundred mm rather than 2000 mm, water normally does not move uniformly through the soil mass, cation exchange does not proceed with 100% efficiency, and the cation exchange capacity decreases as ferrolysis progresses.

Under strongly alkaline conditions, formation of analcime seems to be a relatively rapid process, presumably because tetrahedral Al is the dominant Al species in solution. Analcime seems to persist as a metastable mineral in alkaline soils even after the very high Na activity and pH are lowered by leaching. The extremely low activity of octahedral Al in solution could be a factor inhibiting formation of clay minerals such as kaolinite or montmorillonite in these conditions, even if they might be stable over analcime.

Under alkaline conditions at high Mg activities, Mg chlorite may be produced rapidly (within a day) from smectite (Papenfuss, 1976, confirming results by earlier workers). This is metastable with respect to smectite plus brucite, into which it is transformed within a year (Gupta and Malik, 1969).

Palygorskite presumably is the stable phase at high log (Mg^{2+}) + 2 pH and moderately high log (Al^{3+}) + 3 pH, but may be formed much more slowly because this involves greater structural changes.

Clay minerals expected as end products in a given soil-forming process may in fact be formed at measurable rates. In other cases, a steady state of dissolution of a clay mineral and formation of an intermediate in the weathering sequence may persist for long periods. Kaolinite, for example, may be formed in acid solutions from montmorillonite within 3—4 years (Kittrick, 1970), but very little was apparently formed over a period exceeding 10 000 years in soils containing illite (muscovite) weathering to vermiculite (Henderson et al., 1976). The latter process involves loss of structural K and expansion of the layer structure, but no further structural changes, whereas kaolinite formation would require reorganization of the structure through dissolution and precipitation.

Compositions of soil solutions (or solute compositions in suspensions of soil clay fractions) may fall on either side of the equilibrium point between, for example, montmorillonite and kaolinite, but may show supersaturation with respect to the stable mineral or even to both minerals. This may indicate that their formation is slow relative to that of metastable amorphous phases of similar bulk compositions. Weaver et al. (1976) demonstrated such conditions, as well as the control of solute composition by the amorphous (reactive) fraction.

REFERENCES

Abdul-Latif, N. and Weaver, C.E., 1969. Kinetics of acid-dissolution of palygorskite (attapulgite) and sepiolite. Clays Clay Miner., 17: 169—178.

Adams, F. and Z. Rawajfih, 1977. Basaluminite and Alunite: a possible cause of sulfate retention by acid soils. Soil Sci. Soc. Am. J., 41: 686—692.

Bache, B.W. and Sharp, G.S., 1976. Soluble polymeric hydroxy-aluminium ions in acid soils. J. Soil Sci., 27: 167—174.

Baldar, N.A. and Whittig, L.D., 1968. Occurrence and synthesis of soil zeolites. Soil Sci. Soc. Amer. Proc., 32: 235—238.

Blum, W., 1976. Bildung Sekundärer Al-(Fe-)Chlorite. Z. Pflanzenernaehr. Bodenkd., 1976: 107—125.

Brinkman, R., 1970. Ferrolysis, a hydromorphic soil forming process. Geoderma, 3: 199—206.

Brinkman, R., 1977a. Surface-water gley soils in Bangladesh: Genesis. Geoderma, 17: 111—144.

Brinkman, R., 1977b. Problem hydromorphic soils in North-east Thailand: 2. Physical and chemical aspects, mineralogy and genesis. Neth. J. Agric. Sci., 25: 170—181.

Bronger, A. and Kalk, E., 1976. Zur Feldspatverwitterung und ihrer Bedeutung für die Tonmineralbildung. Z. Pflanzenernaehr. Bodenkd., 1976: 37—55.

Brown, B.E. and Jackson, M.L., 1958. Clay mineral distribution in the Hiawatha soils of northern Wisconsin. Proc. 5th Natl. Conf. Clays and Clay Min., pp. 213—226.

Brown, T.H. and Skinner, B.J., 1974. Theoretical prediction of equilibrium phase assemblages in multi-component systems. Am. J. Sci., 274: 961—986.

Bruckert, S. and Jacquin, F., 1969. Interaction entre la mobilité de plusieurs acides organiques et de divers cations dans un sol à mull et dans un sol à mor. Soil Biol. Biochem., 1: 275—294.

Cawse, P.A., 1974. A Survey of Atmospheric Trace Elements in the U.K. (1972—73). AERE-R7669, U.K. Atomic Energy Authority Harwell. ISBN 070-580334-1, H.M.S.O., London. 95pp.

Christ, C.L., Hostetler, P.B. and Siebert, R.M., 1973. Studies in the system $MgO—SiO_2—CO_2—H_2O$ (III): the activity-product constant of sepiolite. Am. J. Sci., 273: 65—83.

Dirven, J.M.C., Van Schuylenborgh, J. and Van Breemen, N., 1976. Weathering of serpentinite in Matanzas Province, Cuba: mass transfer calculations and irreversible reaction pathways. Soil Sci. Soc. Am. J., 40: 901—907.

Droubi, A., Fritz, B. and Tardy, Y., 1976. Equilibres entre minéraux et solutions. Programmes de calcul appliqués à la prédiction de la salure des sols et des doses optimales d'irrigation. Cah. ORSTOM Série Pédologie, XIV (1): 13—38.

Eeckman, J.P. and Laudelout, H., 1961. Chemical stability of hydrogen-montmorillonite suspensions. Kolloid Z., 178: 99—107.

Frankart, R.P. and Herbillon, A.J., 1971. Aspects de la pédogénèse des sols halomorphes de la Basse Ruzizi (Burundi). Présence et génèse de l'analcime. Annales Musée Royal de l'Afrique Centrale, Tervuren, Belgique. Série-in-8°. Sci Géol. no. 71., 125pp.

Garrels, R.M. and Mackenzie, F.T., 1971. Evolution of Sedimentary rocks. Norton, New York, N.Y., 397pp.

Garrels, R.M. and Thompson, M.E., 1962. A chemical model for sea water at 25°C and one atmosphere total pressure. Am. J. Sci., 260: 57—66.

Gupta, G.C. and Malik, W.U., 1969. Chloritisation of montmorillonite by its coprecipitation with magnesium hydroxide. Clays Clay Miner., 17: 331—338.

Harder, H. and Flehmig, W., 1970. Quartzsynthese bei tiefen Temperaturen. Geochim. Cosmochim. Acta, 34: 295—305.

Hayden, P.L. and Rubin, A.J., 1974. Systematic investigation of the hydrolysis and precipitation of aluminum (III). In: A.J. Rubin (Editor), Aqueous-Environmental Chemistry of Metals. Ann Arbor Science Publ., Ann Arbor, Mich./Wiley, U.K., pp. 317—381.

Helgeson, H.C., 1969. Thermodynamics of hydrothermal systems at elevated temperatures and pressures. Am. J. Sci., 267: 729—804.

Helgeson, H.C., Garrels, R.M. and Mackenzie, F.T., 1969. Evaluation of irreversible reactions in geochemical processes involving minerals and aqueous solutions. II. Applications. Geochim. Cosmochim. Acta, 33: 453—461.

Helgeson, H.C., Brown, T.H., Nigrini, A. and Jones, T.A., 1970. Calculations of mass transfer in geochemical processes involving aqueous solutions. Geochim. Cosmochim. Acta, 34: 569—592.

Henderson, J.H., Doner, H.E., Weaver, R.M., Syers, J.K. and Jackson, M.L., 1976. Cation and silica relationships of mica weathering to vermiculite in calcareous Harps soil. Clays Clay Miner., 24: 93—100.

Hénin, S., Pédro, G. and Robert, M., 1968. Considérations sur les notions de stabilité et instabilité des minéraux en fonction des conditions du milieu; essai de classification des 'systèmes d'aggression'. Trans. Int. Cong. Soil Sci., 9th, III: 79—90.

Huang, W.H. and Keller, W.D., 1971. Dissolution of clay minerals in dilute organic acids at room temperature. Am. Mineral., 56: 1083—1095.

Huang, W.H. and Keller, W.D., 1973. Kinetics and mechanism of dissolution of Fithian illite in two complexing organic acids. Proc. Int. Clay Conf. Madrid, 1972: 321—331.

Jackson, M.L. and Sherman, G.D., 1953. Chemical weathering of minerals in soils. Adv. Agron., 5: 219—318.
Kittrick, J.A., 1966. Free energy of formation of kaolinite from solubility measurements. Am. Mineral., 51: 1457—1466.
Kittrick, J.A., 1970. Precipitation of kaolinite at 25°C and 1 atm. Clays Clay Miner., 18: 261—267.
Kittrick, J.A., 1971. Stability of montmorillonites: I. Belle Fourche and Clay Spur montmorillonites. Soil Sci. Soc. Am. Proc., 35: 140—145
Kovda, V.A. and Rode, O.D., 1967. Secondary quartz in weathering products in the deserts of Egypt. Soviet Soil Sci., 1967: 1408—1409.
Mackenzie, F.T. and Gees, R., 1971. Quartz: Synthesis at earth-surface conditions. Science, 173: 533—535.
Papenfuss, K.H., 1976. Bildung sekundärer Mg-Chlorite. Z. Pflanzenernaehr. Bodenkd., 1976: 3—6.
Paquet, H. and Millot, G., 1972. Geochemical evolution of clay minerals in the weathered products and soils of Mediterranean climates. Proc. Int. Clay Conf. Madrid, 1972, pp. 199—206.
Pédro, G., 1970. Sur l'altération des matériaux calcaires en conditions "laterisantes": étude expérimentale de l'évolution d'un marne illitique. C.R. Acad. Sci., Paris, Série D, 270: 36—38.
Pédro, G., Jamagne, M. and Begon, J.C., 1969. Mineral interactions and transformations in relation to pedogenesis during the Quaternary. Soil Sci., 107: 462—469.
Pédro, G., Berrier, J. and Tessier, D., 1970. Recherches expérimentales sur l'alteration "allitique" des argiles dioctaedriques de type kaolinite et illite. Bull. Groupe Fr. Argiles, 22: 29—50.
Pitzer, K.S., 1973. Thermodynamics of electrolytes. I. Theoretical basis and general equations. J. Phys. Chem., 77: 268—277.
Pitzer, K.S., 1975. Thermodynamics of electrolytes. V. Effects of higher-order electrostatic terms. J. Solution Chem., 4: 249—265.
Pitzer, K.S. and Mayorga, G., 1973. Thermodynamics of electrolytes. II. Activity and osmotic coefficients for strong electrolytes with one or both ions univalent. J. Phys. Chem., 77: 2300—2308.
Pitzer, K.S. and Kim, J.J., 1974. Thermodynamics of electrolytes. IV. Activity and osmotic coefficients for mixed electrolytes. J. Am. Chem. Soc., 96: 5701—5707.
Pitzer, K.S. and Mayorga, G., 1974. Thermodynamics of electrolytes. III. Activity and osmotic coefficients for 2—2 electrolytes. J. Solution Chem., 3: 539—546.
Reynolds, R.C., 1971. Clay mineral formation in an alpine environment. Clays Clay Miner., 19: 361—374.
Richburg, J.S. and Adams, F., 1970. Solubility and hydrolysis of aluminum in soil solutions and saturated-paste extracts. Soil Sci. Soc. Am. Proc., 34: 728—734.
Robinson, R.A. and Stokes, R.M., 1968. Electrolyte Solutions. Butterworths, London, 2nd rev. ed., 571 pp.
Ruppert, H., 1975. Geochemical investigations on atmospheric precipitation in a medium-sized city (Göttingen, F.R.G.). Water, Air Soil Pollut., 4: 447—460.
Schnitzer, M., 1969. Reactions between fulvic acid, a soil humic compound and inorganic soil constituents. Soil Sci. Soc. Am. Proc., 33: 75—80.
Schnitzer, M. and Desjardins, J.G., 1968. Chemical characteristics of a natural soil matter of a Podzol. Soil Sci. Soc. Am. Proc., 26: 362—365.
Schnitzer, M. and Desjardins, J.G., 1968. Chemical characteristics of a natural soil leachate from a humic podzol. Can. J. Soil Sci., 49: 151—158.
Schnitzer, M. and Skinner, S.I.M., 1963. Organo-metallic interactions in soils: 1. Reactions between a number of metal ions and the organic matter of a podzol Bh horizon. Soil Sci., 96: 86—93.

Schnitzer, M. and Skinner, S.I.M., 1964. Organo-metallic interactions in soils: 3. Properties of iron- and aluminum-organic-matter complexes, prepared in the laboratory and extracted from a soil. Soil Sci., 98: 197—203.

Sillén, L.G. and Martell, A.E., 1964. Stability Constants of Metal-Ion Complexes. The Chemical Society, London, 754 pp.

Singer, A., 1977. Dissolution of two Australian palygorskites in dilute acid. Clays Clay Miner., 25: 126—130.

Smith, R.W., 1971. Relations among equilibrium and non-equilibrium aqueous species of aluminum hydroxy complexes. In: Non-Equilibrium Systems in Natural Water Chemistry. Adv. in Chem. Ser., 106. Am. Chem. Soc., Washington D.C., pp. 250—279.

Stumm, W. and Morgan, J.J., 1970. Aquatic Chemistry. Wiley, New York, N.Y., 583 pp.

Swindale, L.D. and Jackson, M.L., 1956. Genetic processes in some residual podzolized soils of New Zealand. Proc. 6th Int. Cong. Soil Sci., Paris, E: 233—239.

Thompson, A.B., 1973. Analcime: free energy from hydrothermal data. Implications for phase equilibria and thermodynamic quantities for phases in $NaAlO_2$—SiO_2—H_2O. Am. Mineral., 58: 277—286.

Tributh, H., 1976. Die Umwandlung der glimmerartigen Schichtsilikate zu aufweitbaren Dreischicht-Tonmineralen. Z. Pflanzenernaehr. Bodenkd., 1976: 7—25.

Turner, R.C., 1968. Conditions in solution during the formation of gibbsite in dilute Al salt solutions. II. Effect of length of time of reaction on the formation of polynuclear hydroxyaluminum cations, the substitution of other anions for OH^- in amorphous $Al(OH)_3$ and the crystallization of gibbsite. Soil Sci., 106: 338—344.

Turner, R.C. and Brydon, J.E., 1967. Removal of interlayer aluminum hydroxide from montmorillonite by seeding the suspension with gibbsite. Soil Sci., 104: 332—335.

Van Breemen, N., 1973a. Calculation of ionic activities in natural waters. Geochim. Cosmochim. Acta, 37: 101—107.

Van Breemen, N., 1973b. Dissolved aluminum in acid sulfate soils and in acid mine waters. Soil Sci. Soc. Am. Proc., 37: 694—697.

Veen, A.W.L. and Maaskant, P., 1971. Electron microprobe analysis of plasma in an impervious horizon of a tropical groundwater podzol. Geoderma, 6: 101—107.

Verstraten, J.M., 1977. Chemical erosion in a forested watershed in the Oesling, Luxemburg. Earth Surface Proc., 2: 175—184.

Weaver, R.M., Jackson, M.L. and Syers, J.K., 1976. Clay mineral stability as related to activities of aluminum, silicon, and magnesium in matrix solutions of montmorillonite-containing soils. Clays Clay Miner., 24: 246—252.

Whitehead, H.C. and Feth, J.H., 1964. Chemical composition of rain, dry fallout, and bulk precipitation at Menlo Park, California, 1957—1959. J. Geophys. Res., 69: 3319—3339 (quoted from Garrels and Mackenzie, 1971).

AUTHOR INDEX

Numbers in italics refer to the pages on which the authors are listed in reference sections of the relevant chapters.

SUBJECT INDEX